Communications in Computer and Information Science 2805

Rationale
The CCIS series is devoted to the publication of proceedings of computer science conferences. Its aim is to efficiently disseminate original research results in informatics in printed and electronic form. While the focus is on publication of peer-reviewed full papers presenting mature work, inclusion of reviewed short papers reporting on work in progress is welcome, too. Besides globally relevant meetings with internationally representative program committees guaranteeing a strict peer-reviewing and paper selection process, conferences run by societies or of high regional or national relevance are also considered for publication.

Topics
The topical scope of CCIS spans the entire spectrum of informatics ranging from foundational topics in the theory of computing to information and communications science and technology and a broad variety of interdisciplinary application fields.

Information for Volume Editors and Authors
Publication in CCIS is free of charge. No royalties are paid, however, we offer registered conference participants temporary free access to the online version of the conference proceedings on SpringerLink (http://link.springer.com) by means of an http referrer from the conference website and/or a number of complimentary printed copies, as specified in the official acceptance email of the event.

CCIS proceedings can be published in time for distribution at conferences or as post-proceedings, and delivered in the form of printed books and/or electronically as USBs and/or e-content licenses for accessing proceedings at SpringerLink. Furthermore, CCIS proceedings are included in the CCIS electronic book series hosted in the SpringerLink digital library at http://link.springer.com/bookseries/7899. Conferences publishing in CCIS are allowed to use our online conference service (Meteor) for managing the whole proceedings lifecycle (from submission and reviewing to preparing for publication) free of charge.

Publication process
The language of publication is exclusively English. Authors publishing in CCIS have to sign the Springer CCIS copyright transfer form, however, they are free to use their material published in CCIS for substantially changed, more elaborate subsequent publications elsewhere. For the preparation of the camera-ready papers/files, authors have to strictly adhere to the Springer CCIS Authors' Instructions and are strongly encouraged to use the CCIS LaTeX style files or templates.

Abstracting/Indexing
CCIS is abstracted/indexed in DBLP, Google Scholar, EI-Compendex, Mathematical Reviews, SCImago, Scopus. CCIS volumes are also submitted for the inclusion in ISI Proceedings.

How to start
To start the evaluation of your proposal for inclusion in the CCIS series, please send an e-mail to ccis@springer.com

Jiguo Yu · Yanli Chen
Editors

Frontiers in Cyber Security

8th International Conference, FCS 2025
Guiyang, China, October 24–26, 2025
Proceedings

Springer

Editors
Jiguo Yu
University of Electronic Science
and Technology of China
Sichuan, China

Yanli Chen
Guizhou University
Guiyang, China

ISSN 1865-0929 ISSN 1865-0937 (electronic)
Communications in Computer and Information Science
ISBN 978-981-95-9513-6 ISBN 978-981-95-9514-3 (eBook)
https://doi.org/10.1007/978-981-95-9514-3

This Springer imprint is published by the registered company Springer Nature Singapore Pte Ltd.
The registered company address is: 152 Beach Road, #21-01/04 Gateway East, Singapore 189721, Singapore

Preface

Cyber security is garnering increasing attention by industry and academia. New cyber security technologies are emerging. Many universities and institutions have set up research centers or schools for cyber security. Both government and enterprises strongly support education and research in cyber security. However, the methodologies for conducting effective cybersecurity research and education remain largely untested. How to run a center for cyber security is also worth exploring since it involves many departments and has a multidisciplinary nature. The FCS conference provides a good platform for researchers and practitioners to exchange the latest research results in cyber security.

The Eighth International Conference on Frontiers in Cyber Security (FCS 2025), held in Guiyang, China, from October 24 to 26, 2025, served as an international platform for researchers and practitioners to exchange the latest advancements in cyber security and related areas of communication systems.

The conference received 75 paper submissions, with 31 ultimately selected for inclusion in the proceedings after undergoing a rigorous double-blind peer review process. Each paper was evaluated by 2-3 anonymous reviewers. For submissions co-authored by Program Committee members, the review process was conducted independently to ensure objectivity.

All accepted papers are included in this single volume of proceedings, which has been organized and edited by the program committee chairs.

On behalf of the FCS 2025 organizing committee, we would like to express our heartfelt gratitude to the committee members, reviewers, session chairs, attendees, and volunteers for their invaluable contributions to this event.

We sincerely appreciate the editors and publishing team for their invaluable contributions to the conference's success.

October 2025

Jiguo Yu
Neal N. Xiong
Yanli Chen

Organization

Organizing Committee

Steering Chairs

Fagen Li	University of Electronic Science and Technology of China, China
Daiyuan Peng	Southwest Jiaotong University, China
Hongxia Wang	Sichuan University, China
Yonghui Zhou	Guizhou Normal University, China

General Chairs

Weihua Ou	Guizhou Normal University, China
Changgen Peng	Guizhou University, China
Ying Tan	Southwest Minzu University, China
Xiaohu Tang	Southwest Jiaotong University, China
Xiaosong Zhang	University of Electronic Science and Technology of China, China

Program Chairs

Yanli Chen	Guizhou Normal University, China
Neal N. Xiong	Southern New Hampshire University, USA
Jiguo Yu	University of Electronic Science and Technology of China, China

Local Chairs

Lunzhi Deng	Guizhou Normal University, China
Jinrong Mo	Guizhou Normal University, China
Anzhi Wang	Guizhou Normal University, China

Publicity Chairs

Shuanggen Liu	Xi'an University of Posts & Telecommunications, China
Xiaoguang Liu	Southwest Minzu University, China
Hanzhou Wu	Shanghai University, China
Ling Xiong	Xihua University, China

Publications Chairs

Donghong Cai	Jinan University, China
Chaoyang Li	Zhengzhou University of Light Industry, China
Yong Ma	Jiangxi Normal University, China
Huan Wang	Guizhou University of Finance and Economics, China

Financial Chairs

Weian Li	Guizhou Normal University, China
Yongrong Wang	Guizhou Normal University, China

Program Committee Members

Yi Chen	Guizhou University, China
Minquan Cheng	Guangxi Normal University, China
Yunhe Cui	Guizhou University, China
Mingze He	Sichuan Police College, China
Chunhua Jin	Huangyin Institute of Technology, China
Yue Li	Huaqiao University, China
Zhenghui Liu	Xinyang Normal University, China
Yong Liu	Hubei University, China
Qing Qian	Guizhou University of Finance and Economics, China
Canghong Shi	Xihua University, China
Neal N. Xiong	Southern New Hampshire University, Manchester, USA
Xiangguang Xiong	Guizhou Normal University, China
Zheng Yang	Fujian Normal University, China

Limengnan Zhou	University of Electronic Science and Technology of China, China
Weixi Zhou	Sichuan Normal University, China
Hui Zhu	Hainan University, China

External Reviewers

Franziska Boenisch	CISPA Helmholtz Center for Information Security, Germany
Julia Hesse	IBM Research Europe, Zurich
Qiming Hu	University of Electronic Science and Technology of China, China
Ioanna Karantaidou	George Mason University, USA
John Kirchenbauer	University of Maryland, College Park, USA
Zhenhua Liu	Xidian University, China
Asad Malik	Monash University, Australia
Antoni Viros Martin	IBM Watson, USA
Omid Mir	Johannes Kepler University Linz, Austria
Jianting Ning	Wuhan University, China
Hendrik Norkowski	Grey Rook Gmbh, Germany
Mengdi Ouyang	University of Electronic Science and Technology of China, China
Zhaosen Shi	University of Electronic Science and Technology of China, China
Li Tang	Yunnan University, China
Yilei Wang	Qufu Normal University, China
Xinyan Wu	University of Electronic Science and Technology of China, China
Fu Xiao	Nanjing University of Posts and Telecommunications, China
Shengmin Xu	Fujian Normal University, China
Jian Xu	Shenyang Fire Research Institute of the Ministry of Public Security, China
Yang Yang	Fuzhou University, China
Xue Yang	Southwest Jiaotong University, China
Hexuan Yu	Virginia Polytechnic Institute and State University, USA
Mengyuan Zhang	Beihang University, China
Mingwu Zhang	Hubei University of Technology, China
Yuyang Zhou	Hainan University, China

Contents

Applied Cryptography

Privacy Protection

AI Security

Applied Cryptography

Threshold Signature-Based Roaming Framework Leveraging Blockchain Systems

Xutang Fan[1,2,3], Chong Zhou[4], Lingfeng Wang[1,2,3](✉), and Fan Li[1,2,3]

[1] College of Artificial Intelligence, CUIT Shuangliu Industrial College, Chengdu University of Information Technology, Chengdu 610225, China
sicwolf@163.com

[2] Advanced Cryptography and System Security Key Laboratory of Sichuan Province, CUIT Shuangliu Industrial College, Chengdu 610225, China

[3] National Intelligent Society Comprehensive Governance Experimental Base, CUIT Shuangliu Industrial College, Chengdu 610225, China

[4] Graduate School of Information Sciences, Tohoku University, Sendai, Japan

Abstract. Roaming authentication remains constrained by centralized trust anchors and bilateral federation contracts, which together introduce two fundamental limitations: a single compromised signing key can enable large-scale credential forgery, and the number of trust agreements scales quadratically with the number of domains, undermining scalability and revocation agility.

To address those issues, this paper presents BTRS, the threshold-signaturebased blockchain roaming scheme that eliminates single points of failure and enables verifiable, decentralized credential management. BTRS combines the two-round FROST threshold-signature protocol with a permissioned Hyperledger Fabric ledger. The Home Operator (HO)'s signing key is distributed across n authentication servers using Pedersen DKG, such that any t-subset can jointly generate a Schnorr signature, tolerating up to $n - t$ faults without recovery.

To address dynamic threshold reconfiguration, BTRS introduces a lightweight threshold update mechanism. Instead of rerunning full DKG or exchanging secret shares, the HO samples new polynomial terms and publishes them on-chain, allowing CAs to locally update shares non-interactively. This avoids synchronization pitfalls and network leakage, while preserving information-theoretic secrecy.

The blockchain serves as a tamper-resistant audit log, storing public-key versions, revocation flags, and signed session records. A prototype shows that BTRS authenticates roaming sessions with only two off-chain message rounds and one asynchronous log commit, incurring sub-millisecond cryptographic overhead.

BTRS is the system to integrate threshold cryptography with blockchain-led roaming governance. It delivers scalable, audit-friendly, and compromise-resilient authentication for next-generation cellular, Wi-Fi, and IoT networks.

J. Yu and Y. Chen (Eds.): FCS 2025, CCIS 2805, pp. 3–24, 2027.
https://doi.org/10.1007/978-981-95-9514-3_1

Keywords: Blockchain · Cryptography · Network Roaming · Threshold Signatures

1 Introduction

Roaming authentication enables a user device to securely attach to a visited network managed by a different administrative domain [1]. This capability underpins cellular services [2], Wi-Fi roaming [3], and large-scale IoT deployments [4]. Current roaming practices typically rely on bilateral trust agreements and centrally managed signing keys, such as home location registers (HLR) [5], AAA servers [6], or conventional PKI [7].

However, these approaches exhibit three major shortcomings. First, a single compromised private key can enable attackers to issue fraudulent credentials at scale [8], as illustrated by recent SIM-swap attacks [9] and WPA-Enterprise breaches [10]. Second, the number of bilateral agreements increases quadratically as more domains participate, making key management and policy coordination increasingly complex and unsustainable [11,12]. Third, traditional revocation and key rotation processes are slow and opaque; compromised credentials may take hours to revoke fully, allowing adversaries substantial time for lateral movement [13,14].

In recent years, blockchain-based roaming authentication frameworks have emerged [15–17], leveraging the ledger's intrinsic properties of decentralised consensus, immutability, tamper-evident auditability [18], and programmable smart-contract logic to reduce dependence on bilateral trust relationships. Hojjati et al. (2020) [19] proposed an Ethereum-based [20] Authentication and Key Agreement (AKA) protocol [21] that routes authentication messages through on-chain transactions, thereby eliminating out-of-band key exchanges but re-introducing single-key vulnerabilities and offering no mechanism for seamless key rotation. Mafakheri et al. (2021) [22] presented a Hyperledger Fabric [23] solution that confines the ledger's remit to billing and settlement, yet still depends on a single centralised signing key and lacks an efficient, fine-grained revocation strategy. Accordingly, existing proposals inherit useful blockchain characteristics yet fail to eliminate the single-key point of failure [24].

Threshold signatures, notably the Flexible Round-Optimized Schnorr Threshold (FROST) scheme proposed by Komlo and Goldberg (2020) [25], offer a compelling approach for distributing private signing keys across multiple nodes. Any subset of at least t out of n nodes can collaboratively produce a standard Schnorr signature, while fewer than t nodes reveal no information about the key [26]. This eliminates single points of failure and remains compatible with existing cryptographic systems. However, existing roaming authentication designs have yet to fully leverage threshold signing in conjunction with blockchain coordination to achieve the combined goals of security, low latency, high avai lability, and on-chain auditability.

Threshold signatures, particularly the two-round FROST scheme [25], provide an elegant method for distributing signing authority among multiple parties, mitigating single points of failure. However, existing threshold signing frameworks face critical challenges in dynamic settings. In particular, adjusting the signing threshold t typically requires rerunning a full distributed key generation (DKG) protocol [27], which involves redistributing shares and updating public commitments. These operations are highly synchronous, error-prone under churn, and risk inconsistency across parties [28]. Moreover, during resharing, secret values may traverse the network, increasing the attack surface and weakening share confidentiality.

Such limitations render existing threshold schemes difficult to adopt in cross-domain roaming scenarios, where trust must be decentralized, devices must authenticate across heterogeneous networks, and operational conditions evolve over time.

To address these issues, this research proposes a complete blockchain-based roaming authentication framework built on FROST threshold signatures. Our design introduces a lightweight, blockchain-coordinated threshold update mechanism that reconfigures signing parameters without rerunning a full distributed key generation (DKG). All updates are published via a tamper-resistant ledger, ensuring global consistency and eliminating share exposure. This approach enables scalable, fault-tolerant, and auditable authentication, while addressing long-standing challenges in threshold signature deployment for dynamic cross-domain environments.

The main contributions of this paper are fourfold:

(a) This research proposes the first threshold-signaturebased blockchain roaming scheme (BTRS), which integrates on-chain public-key management with off-chain FROST signatures to eliminate single-key compromise and simplify threshold reconfiguration.
(b) This research presents a formal security analysis showing that BTRS resists key compromise, replay attacks, and collusion by up to $t-1$ malicious certification authorities, while maintaining correctness and availability.
(c) This research design a blockchain-driven lightweight threshold update mechanism that avoids full DKG reruns, guarantees share secrecy, and prevents network leakage or version inconsistency.
(d) This research implements a working prototype of BTRS using Hyperledger Fabric and demonstrate its feasibility through cryptographic validation and latency measurements.

2 Related Work

2.1 Smart Contracts in the 5G Roaming Architecture: the Fusion of Blockchain with 5G Networks

In 2021, Mafakheri *et al.* [22] proposed a consortium blockchain enhancement for LTE/5G wholesale roaming in which every home and visited 5G Core (5GC) operates as an endorsed peer within a Hyperledger-Fabric consortium. The design supplants the bilateral clearing-house required by local breakout

(LBO) arrangements and, simultaneously, circumvents the long-haul latency characteristic of full home-routed (HR) traffic by settling usage directly between the two mobile-network operators (MNOs). At the data/network layer, detach-transaction proposals (DTPs) that encapsulate cell-ID, volume counters, and tariff class are cryptographically time-stamped and propagated to the channel shared by the transacting MNO pair. A Raft-based Orderer cluster linearises endorsed DTPs into blocks, after which chaincode autonomously invokes the *Discovery*, *IdentityManagement*, and *BillingSettlement* routines. Fabric's private-channel abstraction restricts call-detail records to the two contracting operators, while an escrow deposit denominated in a native token guarantees settlement finality when the session teardown event is recorded.

2.2 FROST Threshold Signature Scheme

FROST, introduced by Komlo and Goldberg in 2020 [25], transforms a traditional multi-round Schnorr [29] threshold protocol into a single on-line round procedure while preserving standard security guarantees. The scheme proceeds in three logically distinct phases. Participants execute the PedersenFeldman [30] distributed-key-generation protocol [31], resulting in individual secret shares s_i and a joint public key $Y = g^x$. No further interaction is required until a signature is requested.

Each signer pre-computes multiple nonce pairs (d_i, e_i) and publicly commits to the corresponding group elements once per pair. Because this step is message-independent, it can be performed in advance and amortised over many future signatures, removing it from the critical on-line path.

To sign a message m, any subset S of at least t participants reveals one cached commitment, derives a binding factor that couples the commitment to (m, S), and produces a local response z_i that mixes the pre-computed nonces with the secret share s_i. Summing the responses yields the aggregate value z, and the final signature is the pair $\sigma = (R, z)$, which is verified by checking whether $g^z = RY^{H(R\|Y\|m)}$ holds.

FROST eliminates interactive rounds from the on-line phase, enabling high-throughput, low-latency signing and unbounded concurrency. Its binding construction thwarts known multi-party forgery attacks, and the protocol simply aborts upon detecting inconsistent behaviour, avoiding heavyweight recovery subroutines. These properties make FROST a practical choice for decentralised identity systems, multi-party wallets, and other settings that require efficient, robust threshold signatures.

3 The Architecture and The Definitions of BTRS

3.1 The Definitions of BTRS

Tables 1 and 2.

Table 1. System roles definitions.

Role	Description
Home Operator (HO)	Holds the master private key x with public key $PK = g^x$; publishes PK on-chain with version identifier `verID`; runs Pedersen-DKG to distribute Shamir shares
Certification Authority (CA_i)	Keeps private share s_i; in each session broadcasts nonce commitment $R_i = g^{k_i}$, computes local scalar z_i, and participates in threshold aggregation; a designated CA records the final log on-chain
Visited-Network Filter (VN-Filter)	Edge AAA/SEAF server that pre-screens the UE before admission to the visited network: performs freshness, blacklist, and key-version checks; discards invalid or replayed requests before forwarding
User Equipment (UE)	Device with unique identifier UID; builds the challenge message m from UID, timestamp T_s, random nonce rand, and `verID`

Table 2. Notation definitions.

Symbol	Description	Symbol	Description
q	Prime order of the elliptic-curve group $\mathbb{F}_q$	g	Generator of group $\mathcal{G}$
t, n	Threshold and total number of shares ($t \leq n$)	x	Master private key of the Home Operator (HO)
$\text{PK} = g^x$	Master public key, stored on-chain	`verID`	Version identifier of PK (UUID or hash)
$f(X)$	Shamir polynomial of degree $t-1$	s_i	Private key share i, held by CA_i
$\mathcal{S}$	Indices of online, non-revoked CAs	λ_i	Lagrange coefficient for $i \in \mathcal{S}$
k_i	One-time nonce of CA_i	$R_i = g^{k_i}$	Commitment of nonce k_i
$R = \prod_{i \in \mathcal{S}} R_i$	Aggregate nonce	UID	Device unique identifier
T_s	UE timestamp	rand	128-bit random nonce generated by UE
m	Challenge hash to be signed	c	Schnorr challenge scalar
z_i	Local signature scalar from CA_i	$z = \sum_{i \in \mathcal{S}} z_i$	Aggregate signature scalar
$\sigma = (R, z)$	Final threshold Schnorr signature	$\rho_{\text{key}}, \rho_{\text{UE}}, \rho_i \in \{0, 1\}$	Revocation flags (1 = revoked)

3.2 Overall Framework of BTRS

The overall framework (Fig. 1) outlines a blockchain-based roaming authentication system that enhances security, scalability, and trust. The Home Operator (HO) holds a master key x and uses Pedersen DKG to distribute Shamir shares s_i to Certification Authorities (CAs). Each CA independently stores its share, publishes nonce commitments $R_i = g^{k_i}$, computes local responses z_i, and jointly generates a FROST threshold signature, with one CA recording results on-chain.

The Visited-Network Filter (VN-Filter) performs initial checks—verifying timestamp freshness, blacklist status, and key version—before forwarding valid User Equipment (UE) requests. Each UE constructs a challenge m from its identifier, timestamp T_s, nonce $rand$, and key version $verID$. The blockchain ensures integrity, non-repudiation, and timely auditability through decentralized, tamper-resistant storage.

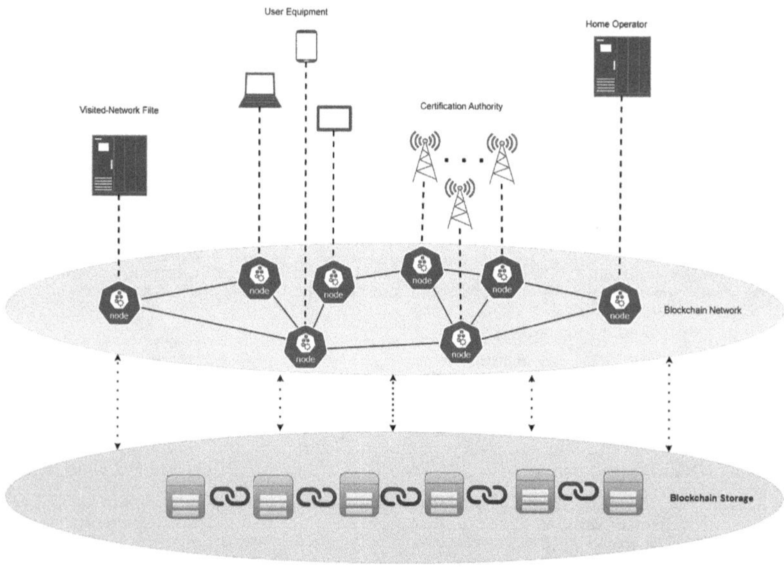

Fig. 1. The Structure of BTRS Framework.

4 The Procedure of BTRS

4.1 UE Constructs Challenge

The challenge message m is derived by concatenating the device's unique identifier (UID), a timestamp T_s, a random nonce rand, and the current public-key version identifier verID into a raw byte string:

$$\text{raw} = \text{UID}\|T_s\|\text{rand}\|\text{verID}. \tag{1}$$

This concatenation uniquely binds the device identity, freshness of the request, and key version, protecting against replay attacks. Hashing the byte string with a cryptographic hash function H (e.g. SHA-256) yields the fixed-length challenge value.

$$m = H(\text{raw}), \tag{2}$$

which is used as input to the threshold signing process.

4.2 Master Key Generation and Secret Sharing

In the Pedersen Distributed Key Generation (DKG) process, the Home Operator (HO) serves as the coordinator responsible for initiating and overseeing the secure generation and distribution of the master private key x. Initially, the HO samples a random polynomial $f(X)$ of degree $t-1$ over $\mathbb{F}_q$:

$$f(X) = x + \sum_{k=1}^{t-1} b_k X^k \pmod{q}, \qquad x, b_k \xleftarrow{\$} \mathbb{F}_q. \tag{3}$$

To ensure verifiability and secrecy, the HO simultaneously selects an auxiliary random polynomial $h(X)$ of the same degree:

$$h(X) = r_0 + \sum_{k=1}^{t-1} r_k X^k \pmod{q}, \qquad r_k \xleftarrow{\$} \mathbb{F}_q, \tag{4}$$

and posts the commitments on-chain:

$$C_k = g^{b_k} h^{r_k}, \qquad k = 0, \ldots, t-1, \tag{5}$$

with generators g, h chosen from a suitable cryptographic group. These commitments are publicly visible, immutable, and timestamped.

Subsequently, the HO securely delivers secret shares to each of the n certification authorities (CAs) via encrypted channels:

$$s_i = f(i), \qquad u_i = h(i), \qquad i = 1, \ldots, n. \tag{6}$$

Upon receiving (s_i, u_i), each CA fetches the on-chain commitments C_k and verifies its shares:

$$g^{s_i} h^{u_i} \stackrel{?}{=} \prod_{k=0}^{t-1} C_k^{i^k}. \tag{7}$$

Equivalence Proof. Expanding the left-hand side gives

$$g^{s_i} h^{u_i} = g^x h^{r_0} \prod_{k=1}^{t-1} \left(g^{b_k} h^{r_k}\right)^{i^k}. \tag{8}$$

Since $x = b_0$ and r_0 are the constant coefficients, the first factor can be rewritten as $\left(g^{b_0} h^{r_0}\right)^{i^0}$. Consequently,

$$g^{s_i} h^{u_i} = \prod_{k=0}^{t-1} \left(g^{b_k} h^{r_k}\right)^{i^k} = \prod_{k=0}^{t-1} C_k^{i^k}, \tag{9}$$

which is exactly the right-hand side; the equality therefore holds.

Finally, the HO derives the master public key and records it on-chain:

$$\mathrm{PK} = g^x, \tag{10}$$

$$\text{KeyRegistry.add}\big(\text{versionID},\ g^x,\ t,\ n\big). \tag{11}$$

The workflow is illustrated in Fig. 2.

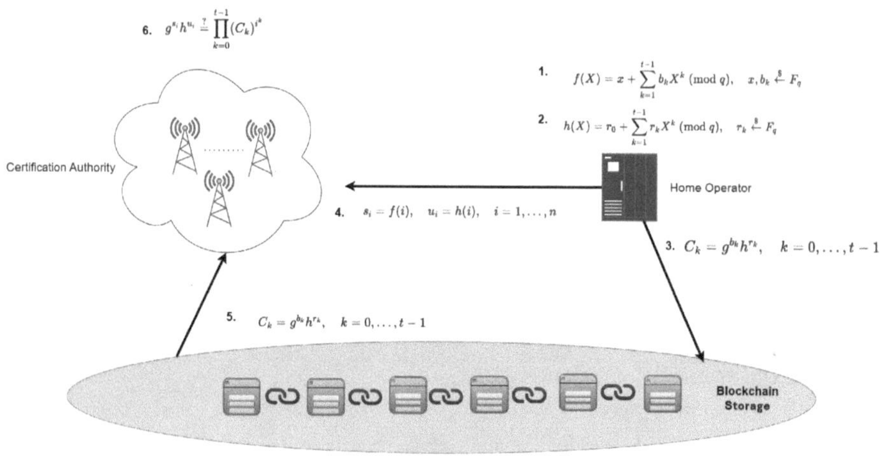

Fig. 2. Master Key Generation and Secret Sharing via On-Chain Commitments.

4.3 Two-Round FROST Threshold Schnorr (Broadcast Aggregation)

For Lagrange basis polynomials at points j:

$$\ell_i(x) = \prod_{\substack{j \in \mathcal{S} \\ j \neq i}} \frac{x - j}{i - j}, \qquad \lambda_i = \ell_i(0) = \frac{\prod_{j \neq i} j}{\prod_{j \neq i} (j - i)} \pmod q. \tag{12}$$

Round-1: Nonce Commitments. Each $\mathrm{CA}_i \in \mathcal{S}$ samples an ephemeral nonce and submits its commitment on-chain:

$$k_i \xleftarrow{\$} \mathbb{F}_q, \qquad R_i = g^{k_i}. \tag{13}$$

After at least t valid on-chain commitments are recorded, every CA reads them and computes

$$R = \prod_{i \in \mathcal{S}} R_i \pmod q, \qquad c = H(R \| \mathrm{PK} \| m) \pmod q. \tag{14}$$

The process of Nonce Commitments is shown in Fig. 3.

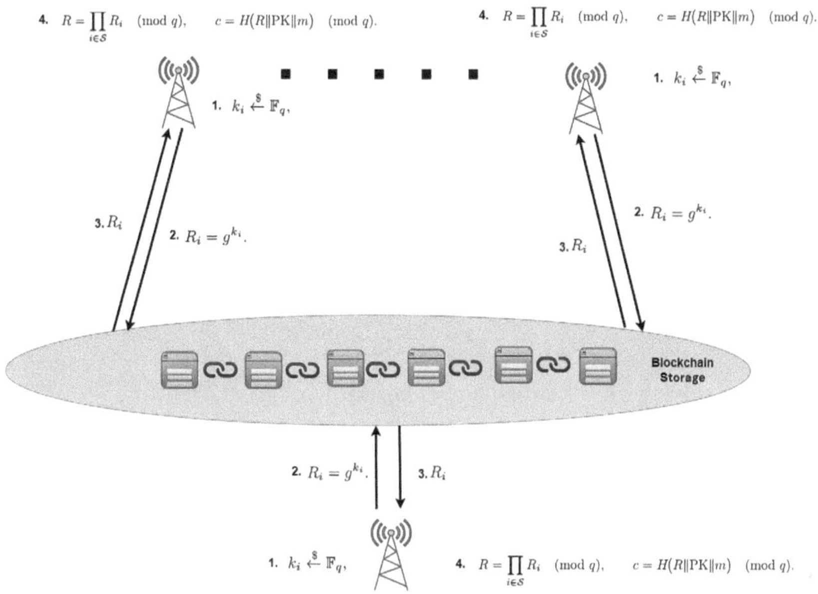

Fig. 3. The Process of Nonce Commitments.

Round-2: Local Signatures. Each CA submits its response on-chain:

$$z_i = k_i + c\lambda_i s_i \pmod q. \tag{15}$$

Signature Aggregation. Once any designated CA has retrieved t valid on-chain pairs $\{(R_i, z_i)\}$, it computes

$$z = \sum_{i \in \mathcal{S}} z_i \pmod q, \qquad \sigma = (R, z), \tag{16}$$

the process of Local Signatures and Signature Aggregation is shown in Fig. 4.

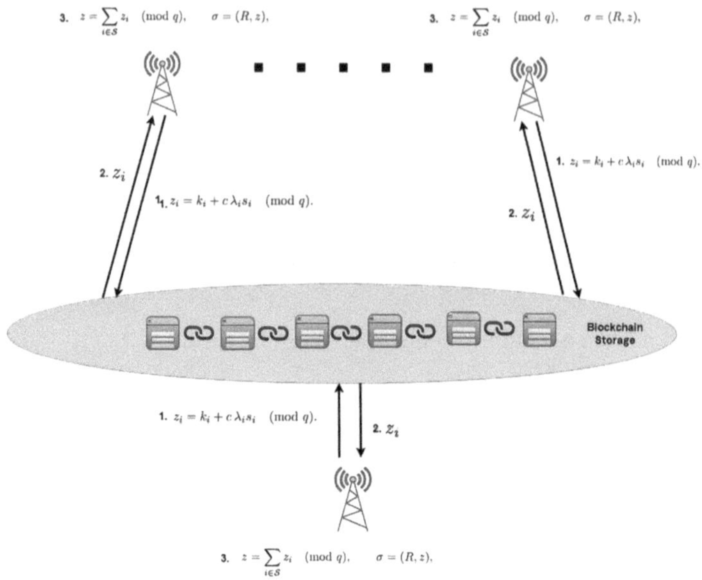

Fig. 4. The Process of Local Signatures and Signature Aggregation.

The verifier checks

$$g^z = R\,\mathrm{PK}^c, \tag{17}$$

and then records the final tuple $\langle \mathrm{ueSn}, \mathrm{verID}, R, z \rangle$ to the Fabric ledger as the threshold signature.

4.4 VN-Filter Pre-Check

In the verification phase, VN-Filter performs a series of pre-checks to ensure the authenticity and freshness of the authentication request. First, it validates that the timestamp T_s included in the challenge message falls within an acceptable freshness window Δ_t to mitigate replay attacks:

$$\left| T_{\text{now}} - T_s \right| < \Delta_t. \tag{18}$$

Next, the VN-Filter verifies that the hash of the device's unique identifier (UID) matches the corresponding entry in the on-chain UE registry—confirming both the legitimacy of the device and that it has not been revoked:

$$H(\mathrm{UID}) = \mathrm{UERegistry}[\mathrm{ueSn}].\mathrm{didHash}. \tag{19}$$

Simultaneously, it ensures that the version identifier verID attached to the request corresponds to the latest public-key version stored in the Key Registry:

$$\mathrm{verID} = \mathrm{KeyRegistry.latestVersionID}. \tag{20}$$

Finally, to guarantee message integrity, the VN-Filter recomputes the challenge hash m locally—hashing the concatenation of the device UID, timestamp, random nonce, and public-key version—and compares it against the value received from the UE:

$$m \stackrel{?}{=} H\big(\mathrm{UID} \| T_s \| \mathrm{rand} \| \mathrm{verID}\big). \tag{21}$$

These sequential checks collectively prevent replay attacks, unauthorized devices, outdated key usage, and message tampering, thereby strengthening the overall security posture of the roaming-authentication process.

4.5 Verification by VN-Filter

The terminalside verification logic is split into three consecutive tests: a revocation filter, a Schnorr equality check, and a succinct correctness argument that binds the distributed computation to the single-key security proof of Schnorr signatures.

Revocation Filter

$$\rho_{\mathrm{key}} = 1 \ \vee \ \rho_{\mathrm{UE}} = 1 \implies \text{Reject} \tag{22}$$

The verifier aborts immediately if either the current public key or the requesting UE is flagged as revoked on-chain.

Schnorr Equation. After receiving the aggregated signature $\sigma = (R, z)$, VN-Filter recomputes the challenge

$$c^* = H\big(R \,\|\, \mathrm{PK} \,\|\, m\big) \pmod q, \tag{23}$$

and validates the Schnorr relation

$$g^{\,z} \stackrel{?}{=} R\,\mathrm{PK}^{\,c^*} \pmod q. \tag{24}$$

A mismatch indicates message tampering in transit.

Correctness Proof. Recall that each share is $z_i = k_i + c\,\lambda_i s_i$ and the aggregator computes $z = \sum_{i \in \mathcal{S}} z_i$. Hence

$$z = \sum_{i \in \mathcal{S}} \big(k_i + c\,\lambda_i s_i\big);\ g^{\,z} = g^{\,K + c\,x} = g^{\,K} g^{\,c\,x} = R\,\mathrm{PK}^{\,c} \pmod q, \tag{25}$$

because $R = g^K$ and $x = \sum_{i \in \mathcal{S}} \lambda_i s_i = f(0)$. Thus, any signature produced by at least t honest CAs is indistinguishable from a standard Schnorr signature under the master key, preserving unforgeability while removing the single-signer point of failure.

4.6 Lightweight Threshold Update

The Home Operator (HO) may raise the Shamir threshold to any desired value $t' \geq \tau_0$ (where τ_0 is the system's initial deployment threshold). Define the increment

$$\Delta = t' - t \quad (\Delta \geq 0). \tag{26}$$

If $\Delta = 0$ no action is required; otherwise proceed as follows.

Sample the additional coefficients.

$$c_k \xleftarrow{\$} \mathbb{F}_q, \quad k = t, \ldots, t' - 1 \tag{27}$$

Chain update. Publish $\langle \mathtt{versionID}^{\text{new}}, t', n, \{c_k\} \rangle$ to the on-chain `KeyRegistry`; set `threshold` $= t'$ and `revoke` the previous version.

Form the update polynomial.

$$g_\Delta(X) = \sum_{k=t}^{t'-1} c_k X^k \pmod q, \qquad g_\Delta(0) = 0. \tag{28}$$

Derive the new sharing polynomial.

$$f'(X) = f(X) + g_\Delta(X) \pmod q, \tag{29}$$

Local share update.

$$s'_i = s_i + g_\Delta(i) \pmod q, \qquad i = 1, \ldots, n. \tag{30}$$

Each CA performs Δ extra modular multiplications and exchanges no additional network traffic. For example, if the HO wants a new threshold of $t + 3$, it samples three coefficients; for a threshold of $t + 2$, it samples two (Fig. 5).

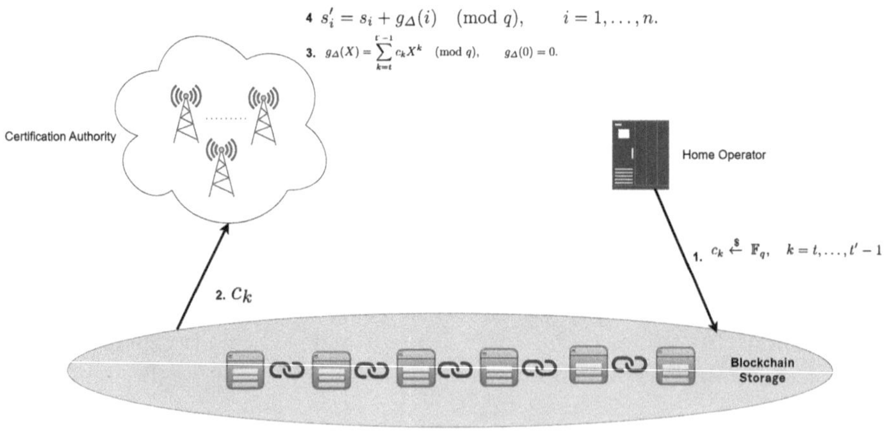

Fig. 5. Lightweight Threshold Update.

4.7 Automatic Resharing and Key Rotation

To compensate for irrevocable share loss or to enact periodic proactive hardening, the **HO** performs a full key rotation that refreshes the master key and every dependent on-chain parameter. The refreshed secrets are redistributed with Pedersen Distributed Key Generation (DKG), ensuring that every certification authority (CA) can publicly verify the correctness of its new share.

Sample a fresh master key.

$$x^{\text{new}} \xleftarrow{\$} \mathbb{F}_q. \tag{31}$$

Construct new ShamirPedersen polynomials.

$$f_{\text{new}}(X) = x^{\text{new}} + \sum_{k=1}^{t'-1} a_k X^k \pmod q, \qquad a_k \xleftarrow{\$} \mathbb{F}_q, \tag{32}$$

$$h_{\text{new}}(X) = r_0 + \sum_{k=1}^{t'-1} r_k X^k \pmod q, \qquad r_k \xleftarrow{\$} \mathbb{F}_q. \tag{33}$$

Publish commitment vectors. For each coefficient index $k \in \{0, \ldots, t'-1\}$ the HO publishes

$$C_k = g^{a_k} h^{r_k}, \tag{34}$$

where g, h are fixed generators of a prime-order group. These commitments allow every CA to verify its received shares.

issue new verifiable shares to every non-revoked CA.

$$s_i^{\text{new}} = f_{\text{new}}(i) \pmod q, \qquad r_i^{\text{new}} = h_{\text{new}}(i) \pmod q, \qquad i \notin \texttt{revoked}. \tag{35}$$

Each CA checks

$$g^{s_i^{\text{new}}} h^{r_i^{\text{new}}} \stackrel{?}{=} \prod_{k=0}^{t'-1} C_k^{i^k}, \tag{36}$$

accepting its share only if the equality holds.

Extend the index space for newly admitted CAs, if any. Authorities joining during this epoch receive fresh shares at indices $j > n$ by evaluating (32) and (35).

Publish the new public key and parameters on-chain.

$$\texttt{KeyRegistry.add}\big(\texttt{versionID}^{\text{new}},\ PK^{\text{new}},\ t',\ n'\big). \tag{37}$$

The transaction implicitly marks the previous key record as `revoked`. Subsequent VN-Filter pre-checks therefore bind every challenge to the latest, uncompromised master key without service interruption.

5 Formal Security Analysis

This section presents poofs of the security about BTRS, covering key generation and share distribution, the two-round FROST signing protocol, and the lightweight threshold update procedure.

5.1 Key Generation And Share Distribution

The sharing protocol stays secure for three clear reasons. First, if an attacker learns no more than $t-1$ shares $\mathcal{S} = \{s_i\}$, those values give at most $t-1$ equations for the t unknowns $\{x, b_1, \dots, b_{t-1}\}$. The secret therefore remains uniformly random,

$$\Pr\big[x = \hat{x} \mid \mathcal{S}\big] = \frac{1}{q}. \tag{38}$$

This means that even after observing up to $t-1$ shares, the attacker's knowledge of x is unchanged: for each possible value $\hat{x} \in \mathbb{F}_q$, the probability that $x = \hat{x}$ remains exactly $1/q$. In other words, with at most $t-1$ linear equations in the t unknowns $\{x, b_1, \dots, b_{t-1}\}$, the system is underdetermined and x is still uniformly distributed over all q field elements.

Second, any full set of t different shares can rebuild the polynomial with the Lagrange formula

$$f(X) = \sum_{j \in T} s_j \prod_{k \in T \setminus \{j\}} \frac{X-k}{j-k} \pmod q, \tag{39}$$

so reading (39) at $X = 0$ returns the master key x. Third, every share is tied to its on-chain commitment:

$$g^{s_i} h^{u_i} = \prod_{k=0}^{t-1} C_k^{i^k}. \tag{40}$$

If a malicious node could swap in a different pair $(s_i^\star, u_i^\star) \neq (s_i, u_i)$ that still passes (40), it would expose $\log_g h$ and break the discrete-log assumption. Together, (38)–(40) show that the real protocol gives the same secrecy, availability, and authenticity as the ideal functionality: fewer than t honest shares reveal nothing about x, and forged shares are rejected unless discrete log is solved.

5.2 Security Analysis of the Two-Round FROST Signing Protocol

The security of the FROST threshold signing protocol is founded on three guarantees: correctness, unforgeability under adaptive chosen-message attacks, and robustness against partial faults. We define a set of t participating certification authorities $\mathcal{S} \subseteq \{1, \dots, n\}$ and denote the final signature as $\sigma = (R, z)$. All operations are performed over a prime-order group $\mathbb{G}$ with generators g, h and scalar field $\mathbb{F}_q$.

Correctness: Each signer $P_i \in \mathcal{S}$ uses its share s_i and Lagrange coefficient λ_i to compute the response:

$$z_i = k_i + c\,\lambda_i\, s_i \pmod q, \tag{41}$$

where $c = H(R \| \mathrm{PK} \| m)$ is the Fiat-Shamir challenge and k_i is the signer's ephemeral nonce. The aggregated response is:

$$z = \sum_{i \in \mathcal{S}} z_i \pmod q. \tag{42}$$

Given that $\sum \lambda_i s_i = x$ (the master secret), the verifier checks:

$$g^z \stackrel{?}{=} R \cdot \mathrm{PK}^c, \tag{43}$$

confirming that valid signatures from honest signers always pass verification.

Unforgeability: Suppose a probabilistic adversary $\mathcal{A}$ produces a forgery $\sigma^* = (R^*, z^*)$ on a new message m^* without t honest CAs.

The Forking Lemma guarantees the extraction of two transcripts with distinct challenges $c_1 \neq c_2$ and the same R^*:

$$z_1 = \sum_i k_i + c_1 x, \quad z_2 = \sum_i k_i + c_2 x. \tag{44}$$

Solving for x yields:

$$x = \frac{z_1 - z_2}{c_1 - c_2} \pmod q, \tag{45}$$

thus breaking the discrete logarithm assumption. Hence, the protocol achieves existential unforgeability under chosen-message attacks (EUF-CMA).

Robustness: Since the aggregation in Eq. (42) only requires responses from any t active CAs, the protocol tolerates up to $n - t$ Byzantine or crashed nodes without service interruption. No single signer is essential to protocol liveness, eliminating single points of failure.

Together, Eqs. (41)–(45) imply that the execution in the real world securely realizes ealizes the ideal signing functionality $\mathcal{F}_{\mathrm{Sign}}$, providing correctness, unforgeability, and fault tolerance under standard cryptographic assumptions.

5.3 Security of Lightweight Threshold Update

In the threshold-update phase, the Home Operator (HO) raises the degree of the Shamir polynomial by appending new high-degree coefficients $\{c_k\}_{k=t}^{t'-1}$ drawn uniformly from $\mathbb{F}_q$, resulting in an updated polynomial:

$$f'(X) = f(X) + \sum_{k=t}^{t'-1} c_k X^k \pmod q. \tag{46}$$

Each certification authority P_i locally computes its updated share as:

$$s_i' = s_i + \sum_{k=t}^{t'-1} c_k i^k \pmod q. \tag{47}$$

This update protocol satisfies the following security properties:

Share Secrecy Preservation. Let $\mathcal{A}$ be an adversary corrupting at most $t-1$ authorities and observing the set of updated shares $\{s_i'\}_{i \in \mathcal{A}}$. Since the additive term $g_\Delta(X) = \sum_{k=t}^{t'-1} c_k X^k$ satisfies $g_\Delta(0) = 0$, the constant term of $f'(X)$ is unchanged:

$$f'(0) = f(0) = x.$$

As the adversary learns at most $t-1$ values of a degree-$(t'-1)$ polynomial, the posterior distribution of the master secret remains uniform:

$$\Pr[x = \hat{x} \mid \{s'_i\}_{i\in\mathcal{A}}] = \frac{1}{q}. \tag{48}$$

Zero Network Leakage. Each updated share s'_i is computed locally by CA P_i as $s'_i = s_i + g_\Delta(i)$, where $g_\Delta(i) = \sum_{k=t}^{t'-1} c_k i^k$ and $\{c_k\}$ are broadcast by the HO via an authenticated blockchain ledger. As a result, the full process of threshold elevation does not require the transmission of any secret s_i or s'_i over the network. Even under the assumption of a global passive eavesdropper capable of recording all public broadcasts, the adversary observes only the public coefficients $\{c_k\}$, while all private shares remain confined to local memory.

Let CHANNEL denote the adversary's view, which includes all on-chain data. The mutual information between any updated share and the adversary's view satisfies:

$$\mathsf{I}(s'_i; \textsc{Channel}) = \mathsf{H}(s'_i) - \mathsf{H}(s'_i \mid \textsc{Channel}) = 0, \tag{49}$$

which formally proves that eavesdroppers learn no additional information about any private share.

Update Consistency. Because the update polynomial $g_\Delta(X) = \sum_{k=t}^{t'-1} c_k X^k$ is publicly committed via the blockchain, all honest CAs use the same coefficients $\{c_k\}$ at the same block height. Therefore, each updated share satisfies:

$$s'_i = s_i + \sum_{k=t}^{t'-1} c_k i^k = f'(i), \tag{50}$$

ensuring that all updated shares lie on the same curve $f'(X)$. Since no coordination or message exchange is needed, this method avoids asynchronous state divergence, unlike traditional DKG-based refreshes where out-of-sync shares may arise due to message delays or clock skews.

Equations (46)–(50) together show that the lightweight threshold update retains perfect secrecy, leaks no private data over the network, and guarantees global consistency among all non-revoked CAs.

6 Experiment

6.1 Experimental Setup

To evaluate the feasibility and robustness of BTRS, we implement a full prototype in Go, integrating the FROST threshold signature protocol with the Hyperledger Fabric v2.5 blockchain platform. The smart contract (chaincode) is developed in Java and is responsible for managing public key versions, revocation flags, and broadcasting threshold update parameters. On the authentication layer, each certification authority (CA) runs a Go-based implementation of the FROST signing protocol, including share management, nonce commitments, and response aggregation, which interacts with the blockchain via Fabric's SDK.

The system is deployed on a local cluster equipped with Intel AMD Ryzen™9 7900 CPU and 32 GB RAM, running Ubuntu 22.04.

In this section, mainly measure two key performance indicators:

Latency: The average end-to-end time from a UE's request to the final threshold signature being committed on-chain.

Throughput: The number of successful authentication requests processed per second.

To evaluate fault tolerance, we inject Byzantine faults into a subset of nodes—such as incorrect signature shares or massive latency—and vary the proportion of faulty nodes from 0% to 30%. This setup allows us to assess how BTRS performs under adversarial conditions and validates its robustness in dynamic, partially trusted environments.

6.2 Latency

Figure 6a reports authentication latency as the number of CAs and the fraction of Byzantine nodes vary. Latency grows with the number of nodes, reflecting the extra aggregation and verification work. More importantly, malicious behavior is emulated by withholding or deliberately delaying partial signatures, thereby capturing the impact of unreliable networks and adversarial traffic manipulation. Even in this worst-case model, BTRS sustains low delay: with 25 CAs and 30% Byzantine nodes the average latency stays below 85 ms.

Performance degradation is progressive and bounded. Moving from a benign configuration to 10% malicious nodes increases latency by only 19.7% on average—the cost of timeout detection, signer eviction, and re-aggregation. When the adversarial share reaches 30%, the delay rises by 70.3% relative to the all-honest baseline yet remains well within the sub-second threshold required by roaming use cases. These results confirm that BTRS degrades gracefully under hostile conditions.

Figure 6b shows a dotted horizontal line for each curve, indicating the global mean delay of that series: 1.86, 2.16, 2.63 and 3.09 ms for 0%, 10%, 20% and 30% Byzantine signers, respectively. Once the number of nodes reaches about twelve CAs every curve flattens and stays close to its own average, so adding more CAs imposes almost no extra per-node cost even under attack. With only three to five CAs the curves nearly overlap, confirming that a single malicious signer cannot noticeably inflate the delay. Overall, each honest CA contributes just a few milliseconds, and the full 25-CA nodes still completes within 0.1 s while degrading gracefully in the presence of Byzantine interference.

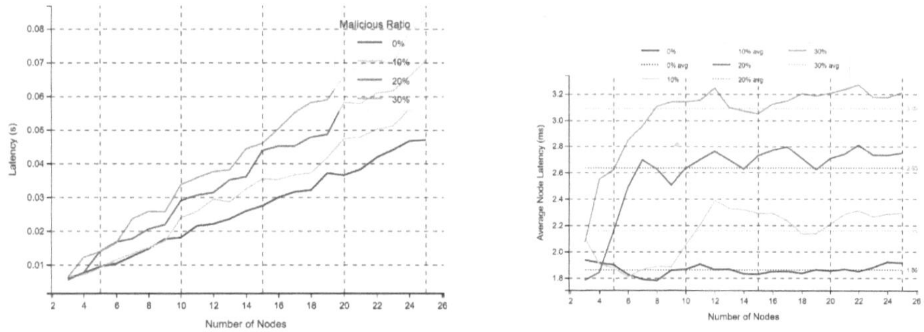

(a) Latency of BTRS (b) Average Node Latency

Fig. 6. Latency comparison of BTRS under different malicious ratio.

6.3 Throughput

Figure 7 shows system throughput (transactions per second, TPS) under the same experimental conditions. Throughput naturally decreases as the nodes grows, yet remains acceptable even when adversaries aggressively refuse to contribute signatures. With no Byzantine nodes present, the prototype sustains over 1600 TPS at 3 CAs and about 200 TPS at 25 CAs. Injecting 10–30% malicious nodes—again emulated via signature withholding and artificial delays—reduces TPS moderately: at 10 CAs the drop is 24.8% for a 10% malicious share and 37.1% for 30%. Even in the worst case (25 CAs, 30% Byzantine) BTRS still exceeds 150 TPS, underlining its robustness to link-layer interference and non-cooperative peers.

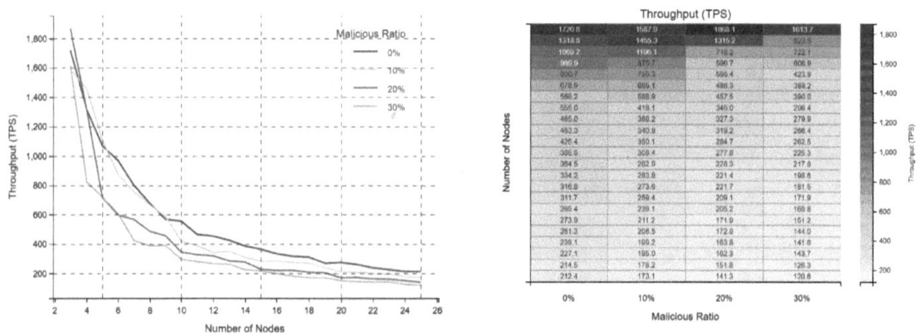

(a) Throughput of BTRS (b) Heat-map of Throughput

Fig. 7. Throughput comparison of BTRS under different malicious ratio.

Collectively, these results demonstrate that BTRS meets the latency and throughput requirements of large-scale roaming deployments while withstanding adversarial delays, packet loss, and signer misbehaviour.

6.4 Threshold Update Latency

Figure 8 shows per-update maintenance latency for full Distributed Key Generation (DKG) and the Lightweight Threshold Update under the same experimental conditions. Latency naturally increases as the number of CAs grows and as the Byzantine fraction rises, yet the Lightweight Threshold Update remains practical even when Byzantine nodes deliberately inject large delays. With no Byzantine nodes present, the prototype completes the Lightweight Threshold Update in $\approx 0.5 - -0.6\,\mathrm{s}$ at 12 CAs, whereas full DKG requires $\approx 1.0 - -1.2\,\mathrm{s}$. Injecting 10–30% malicious nodes—here emulated by deliberate large-delay injection inflates latency moderately: at 25 CAs the worst case (30% Byzantine) yields $\approx 2.8 - -3.0\,\mathrm{s}$ for the Lightweight Threshold Update versus $\approx 9\,\mathrm{s}$ for full DKG. Even in this adverse setting, the Lightweight Threshold Update remains within a few seconds, indicating suitability for frequent key maintenance in medium/large or adversarial deployments, while full DKG is best reserved for infrequent rekeying or smaller groups.

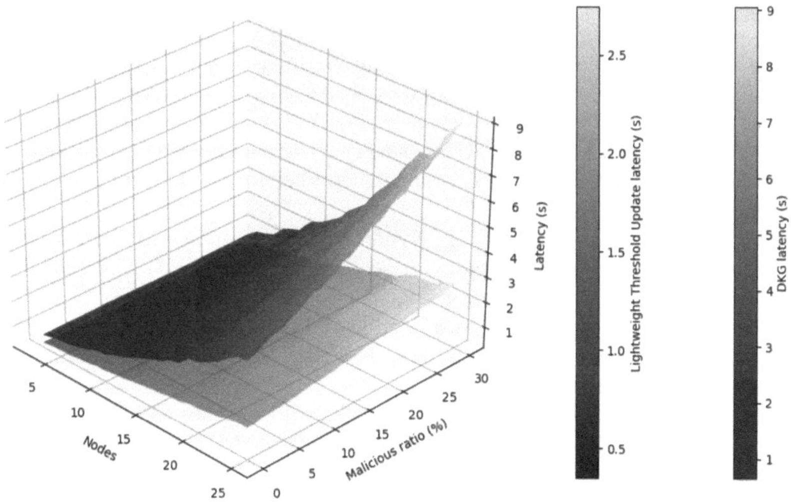

Fig. 8. Lightweight Threshold update contrast with DKG.

7 Conclusion

This paper proposes BTRS, a threshold-signature-based blockchain roaming scheme that combines FROST with Hyperledger Fabric to enable scalable, auditable, and fault-tolerant cross-domain authentication. By distributing the signing key via Pedersen DKG and introducing a blockchain-coordinated lightweight threshold update mechanism, BTRS eliminates single-key vulnerabilities while supporting seamless reconfiguration without the need to rerun full

DKG protocols. All update coefficients are broadcast on-chain, ensuring consistency across nodes and preserving share secrecy with zero network leakage.

This paper formally proves the security of BTRS under standard cryptographic assumptions and demonstrate its practical feasibility through a working prototype implemented in Go. Experimental results show that BTRS sustains low-latency authentication (<80 ms) and acceptable throughput (>150 TPS) even with up to 30% Byzantine nodes among 25 participants, highlighting strong resilience and performance scalability. The system exhibits graceful degradation as the number of nodes increases.

Future work includes two main directions. The first direction is to explore hierarchical threshold structures that enable multilevel authentication across regional or organizational domains. Another is to improve interoperability with existing identity management protocols and to evaluate the performance of BTRS in edge computing environments, where constrained bandwidth and mobility introduce additional challenge.

References

1. Kang, D., Lee, H., Lee, Y., Won, D.: Lightweight user authentication scheme for roaming service in glomonet with privacy preserving. PLoS ONE **16**(2), e0247441 (2021)
2. Scalise, P., Hempel, M., Sharif, H.: A survey of 5g core network user identity protections, concerns, and proposed enhancements for future 6g technologies. Future Internet **17**(4), 142 (2025)
3. Okabe, Y., Nakamura, M., Goto, H.: Dynamic vlan assignment for local users under external idp management in radius-based wi-fi roaming. In: 2024 International Conference on Information Networking (ICOIN), pp. 484–489. IEEE (2024)
4. Jouhari, M., Saeed, N., Alouini, M.S., Amhoud, E.M.: A survey on scalable lorawan for massive iot: recent advances, potentials, and challenges. IEEE Commun. Surv. Tutorials **25**(3), 1841–1876 (2023)
5. Kalejahi, B.K., Eminov, R., Guliyev, A.: Using blockchain technology in mobile network to create decentralized home location registry (hlr). Comput. Syst. Sci. Eng. **39**(2) (2021)
6. Goto, H.: Disruption-tolerant local authentication method for network roaming systems. J. Inf. Process. **32**, 407–416 (2024)
7. Ramadan, M., Du, G., Li, F., Xu, C.: A survey of public key infrastructure-based security for mobile communication systems. Symmetry **8**(9), 85 (2016)
8. Ellison, C., Schneier, B.: Ten risks of pki: what you're not being told about public key infrastructure. Comput. Secur. J. **16**(1), 1–7 (2000)
9. Cheruyot, C.C., Kouame, R.M., Inaba, H.: Securing sim toolkit-based mobile money applications against sim swap attacks using user location data. In: 2024 IEEE 13th Global Conference on Consumer Electronics (GCCE), pp. 100–104. IEEE (2024)
10. Palamà, I., Amici, A., Bellicini, G., Gringoli, F., Pedretti, F., Bianchi, G.: Attacks and vulnerabilities of wi-fi enterprise networks: User security awareness assessment through credential stealing attack experiments. Comput. Commun. **212**, 129–140 (2023)

11. Gegenhuber, G.K., Mayer, W., Weippl, E., Dabrowski, A.: {MobileAtlas}: Geographically decoupled measurements in cellular networks for security and privacy research. In: 32nd USENIX Security Symposium (USENIX Security 23), pp. 3493–3510 (2023)
12. Goto, H.: Inter-federation roaming architecture for large-scale wireless lan roaming systems. J. Inf. Process. **29**, 103–112 (2021)
13. Cerenius, D., Kaller, M., Bruhner, C.M., Arlitt, M., Carlsson, N.: Trust issue (r) s: certificate revocation and replacement practices in the wild, pp. 293–321 (2024)
14. Halim, A., Danielsson, M., Arlitt, M., Carlsson, N.: Temporal analysis of x. 509 revocations and their statuses. In: 2022 IEEE European Symposium on Security and Privacy Workshops (EuroS&PW), pp. 258–265. IEEE (2022)
15. Gong, Y., et al.: Toward next-generation networks: a blockchainbased approach for core network architecture and roaming identity verification. Digit. Commun. Netw. (2024)
16. Chaer, A., Salah, K., Lima, C., Ray, P.P., Sheltami, T.: Blockchain for 5g: opportunities and challenges. In: 2019 IEEE Globecom Workshops (GC Wkshps), pp. 1–6. IEEE (2019)
17. Sharma, V., You, I., Palmieri, F., Jayakody, D.N.K., Li, J.: Secure and energy-efficient handover in fog networks using blockchain-based dmm. IEEE Commun. Mag. **56**(5), 22–31 (2018)
18. Salman, T., Zolanvari, M., Erbad, A., Jain, R., Samaka, M.: Security services using blockchains: a state of the art survey. IEEE Commun. Surv. Tutorials **21**(1), 858–880 (2019). https://doi.org/10.1109/COMST.2018.2863956
19. Hojjati, M., Shafieinejad, A., Yanikomeroglu, H.: A blockchain-based authentication and key agreement (aka) protocol for 5g networks. IEEE Access **8**, 216461–216476 (2020)
20. Buterin, V.: Ethereum white paper. https://github.com/ethereum/wiki/wiki (2015)
21. Lai, C., Li, H., Lu, R., Shen, X.S.: Se-aka: a secure and efficient group authentication and key agreement protocol for lte networks. Comput. Netw. **57**(17), 3492–3510 (2013)
22. Mafakheri, B., Heider-Aviet, A., Riggio, R., Goratti, L.: Smart contracts in the 5g roaming architecture: the fusion of blockchain with 5g networks. IEEE Commun. Mag. **59**(3), 77–83 (2021)
23. Androulaki, E., et al.: Hyperledger fabric: a distributed operating system for permissioned blockchains. In: Proceedings of the thirteenth EuroSys Conference, pp. 1–15 (2018)
24. Xue, K., Luo, X., Ma, Y., Li, J., Liu, J., Wei, D.S.: A distributed authentication scheme based on smart contract for roaming service in mobile vehicular networks. IEEE Trans. Veh. Technol. **71**(5), 5284–5297 (2022)
25. Komlo, C., Goldberg, I.: Frost: Flexible round-optimized schnorr threshold signatures. In: International Conference on Selected Areas in Cryptography, pp. 34–65. Springer (2020)
26. Connolly, D., Komlo, C., Goldberg, I., Wood, C.: Rfc 9591: the flexible round-optimized schnorr threshold (frost) protocol for two-round schnorr signatures (2024)
27. Cimatti, A., et al.: Dynamicfrost: Schnorr threshold signatures with a flexible committee. Cryptology ePrint Archive (2024)
28. Yurek, T., Xiang, Z., Xia, Y., Miller, A.: Long live the honey badger: robust asynchronous {DPSS} and its applications. In: 32nd USENIX Security Symposium (USENIX Security 23), pp. 5413–5430 (2023)

29. Crites, E., Komlo, C., Maller, M.: Fully adaptive schnorr threshold signatures. In: Annual International Cryptology Conference, pp. 678–709. Springer (2023)
30. Feldman, P.: A practical scheme for non-interactive verifiable secret sharing. In: 28th Annual Symposium on Foundations of Computer Science (sfcs 1987), pp. 427–438. IEEE (1987)
31. Pedersen, T.P.: A threshold cryptosystem without a trusted party. In: Davies, D.W. (ed.) EUROCRYPT 1991. LNCS, vol. 547, pp. 522–526. Springer, Heidelberg (1991). https://doi.org/10.1007/3-540-46416-6_47

Strategic Analysis of 51% Attack in Mining Pools via Repeated Games

Guangxu Xie[1], Xiao Liu[2](✉), Haibo Huang[1], and Qingqi Pei[3]

[1] School of Artificial Intelligence, Hubei University of Automotive Technology, Shiyan 442002, China

[2] School of Computer Science, Zhongyuan University of Technology, Zhengzhou 450007, China

cantona@zut.edu.cn

[3] The State Key Laboratory of Integrated Services Networks, Xidian University, Xi'an 710071, China

qqpei@mail.xidian.edu.cn

Abstract. Blockchain systems rely on decentralized consensus mechanisms to ensure security and trust, but are vulnerable to 51% attacks, where a malicious mining pool controls the majority of the network's hash power. Such attacks threaten transaction integrity and system stability. Existing research mainly employs static or symmetric game-theoretic models to analyze miner behavior, which often fails to capture long-term strategic dynamics and real-world asymmetry between mining pools. To address this, we propose an asymmetric dual-mining-pool strategy model (RG5A) based on repeated games. The model captures strategic interactions between large and small mining pools under both unregulated and regulated conditions. In the absence of regulation, we derive a mixed-strategy equilibrium and identify the threshold of attack success probability, showing that higher attack costs discourage malicious actions. With regulation, the model incorporates punishment and reporting incentives. Simulation results demonstrate that regulatory mechanisms significantly raise the attack threshold and reduce the likelihood of malicious behavior. This study provides theoretical insights into mining pool strategies and introduces a quantifiable framework for designing incentive-penalty mechanisms. Future work may extend the model to multi-party, incomplete-information, or dynamically regulated environments to better reflect real blockchain systems.

Keywords: Blockchain security · 51% attacks · Repeated games · Proof of work (PoW)

1 Introduction

As a decentralized distributed ledger technology, blockchain has been widely used in digital currency, Internet of Things, supply chain finance, and other fields due

J. Yu and Y. Chen (Eds.): FCS 2025, CCIS 2805, pp. 25–41, 2027.
https://doi.org/10.1007/978-981-95-9514-3_2

to its trustworthiness, traceability, and data immutability [1]. Public chains represented by Bitcoin use the Proof of Work (PoW) mechanism as a consensus protocol to ensure network security and data consistency by introducing computing power [2]. However, although the PoW mechanism has shown promising results in anti-tampering and anti-sybil attacks, blockchain systems still face a variety of security threats, including block withholding attacks, sybil attacks, and denial-of-service (DoS) attacks. These attacks not only threaten the stability of the network but also shake users' confidence in the fairness and security of the system [3]. Therefore, in-depth research on the potential security risks and their causes in the blockchain system is of great significance to improving the robustness and credibility of the entire ecosystem.

Among many security threats, the 51% attack has garnered widespread attention due to its direct impact on the core mechanism of blockchain technology. This attack refers to the attacker gaining control of more than 50% of the computing power of the entire network, thereby being able to control the block generation process, including rolling back confirmed transactions, implementing double payments, and even preventing other miners from producing blocks normally [4]. Such attacks not only pose a threat to the immutability of transactions but also seriously undermine the basic concept of "decentralization" of the blockchain system. With the increasingly obvious trend of mining pool collaboration and computing power concentration, some large mining pools have the potential to launch such attacks, making the 51% attack no longer a theoretical hypothesis but a real-world problem that needs to be addressed.

In response to the 51% attack problem, academia and industry have conducted research that can be categorized into three areas [5]. The first category focuses on reducing the possibility of attacks at the mechanism level by improving consensus mechanisms such as Proof of Stake (PoS) and DPoS. The second category uses game theory methods to characterize miners' behavior, benefits, and the equilibrium state of the game to suppress attack motivation. The third category focuses on detection and response mechanisms to improve the system's ability to identify and defend against attacks. These studies have improved system security and expanded the theoretical analysis framework to some extent [6]. However, these methods generally have issues such as idealized assumptions, rigid strategies, and high computing resource overhead. They have not fundamentally solved the systemic risks caused by rational miners' profit-seeking behavior. Therefore, analyzing the behavior and revenue game of mining pools from a more dynamic and strategically adaptable perspective remains a major challenge in the current 51% attack defense field and constitutes the so-called "mining dilemma" problem [7].

To more effectively model and solve the strategy selection behavior between two asymmetric mining pools under the threat of a 51% attack, this paper adopts a repeated game framework to capture the long-term strategic interactions in a dynamic and adversarial environment. In this setting, mining pools are subject to asymmetric resources, risk exposure, and incentives, leading to differentiated responses when facing potential attack behaviors. While these pools may seek cooperation to preserve overall network stability, they are simultaneously incentivized to compete for block rewards and dominance in computing power. Traditional static or one-shot game models are limited in capturing the temporal evolution and adaptive behaviors arising from such asymmetric competition. By leveraging repeated game theory, this paper incorporates long-term payoff optimization, strategic retaliation, and cooperation enforcement mechanisms to describe the evolution of mining strategies over time. Based on the constructed asymmetric two-player repeated game model, we systematically analyze subgame perfect equilibria and profit trajectories under various regulatory and attack scenarios. The goal is to explore how mining pools dynamically adjust their strategies to achieve improved payoffs, mitigate the "mining dilemma" caused by 51% threats, and provide a robust theoretical foundation for enhancing blockchain security. In particular, the main contributions of this paper are as follows:

- To the best of our knowledge, this is the first work to employ a repeated game-theoretic approach to analyze mining pools encountering 51% attack threats and their revenue evolution under different entanglement conditions.
- This paper proposes RG5A (Repeated Game for 51% Attack), a non-symmetric dual mining pool decision model based on repeated games. By modeling large and small mining pools as asymmetric players, RG5A captures the long-term strategy evolution and payoff dynamics under 51% attack scenarios. The model provides a new perspective for analyzing the behavioral adaptation and revenue trends of attackers in blockchain systems.
- We introduce a regulatory mechanism and quantitatively analyze how its punitive measures suppress attack strategies. By extending the traditional model to include on-chain regulators, we incorporate both punishment mechanisms and reporting incentives into the game framework. Through the establishment of a mixed-strategy equilibrium, we reveal the dynamic relationship between regulatory intensity and the attacker's willingness to launch 51% attacks.
- The validity of the model is verified through MATLAB numerical simulation experiments, and the quantitative changes in the attack success rate threshold are analyzed. Under both unregulated and regulated scenarios, the system examines the influence of factors such as attack cost and penalty severity on the threshold of attack success rate. It finds that as attack costs or regulatory penalties increase, the willingness to attack decreases significantly. These findings provide a theoretical basis for designing effective incentive-penalty mechanisms in blockchain systems.

The rest of the paper is organized as follows: Sect. 2 reviews the related work. Section 3 provides a detailed introduction to the proposed RG5A methodology. Section 4 presents the experimental results and corresponding analyses. Section 5 concludes the paper and outlines directions for future work.

2 Related Work

To address the security threats posed by 51% attacks, academia, and industry have proposed various defense strategies at different levels. Existing research can be summarized into three categories: consensus mechanism design optimization, game theory modeling analysis, and attack detection and response mechanisms. These categories are committed to suppressing the root causes of attacks, regulating miners' behavior, and improving system robustness, respectively. The following will provide a brief review of these three types of representative research, examining their objectives, advantages, and existing issues.

Some recent works aim to enhance the PoW consensus mechanism itself to resist 51% attacks. Prasad et al. [8] proposed a behavior-aware consensus framework that used miner reputation scores and weighted block acceptance, dynamically adjusting participation and penalties via real-time classification with a TCN-attention model. While effective in reducing attack success rates, it requires substantial system modifications and computational overhead. Babur et al. [9] proposed a modified PoW protocol that disallows consecutive block acceptance from the same miner and introduces a Safe Mode Detection Algorithm to compare UTXO sets for detecting double-spending. This approach strengthens confirmation reliability and prevents private chain dominance, thereby enhancing blockchain resistance to majority attacks. Khatoon et al. [10] introduced a Modified Proof of Stake (MPoS) mechanism to enhance blockchain resistance to double-spending and majority attacks. By restructuring the validation process, this method seeks to improve reliability while reducing vulnerability inherent in traditional PoW-based systems. Yang et al. [11] proposed incorporating miners' historical weights and total computation difficulty into block validation. This modification increases the cost of executing a majority attack by orders of magnitude, thereby improving the system's resistance to double-spending.

Some studies employ game-theoretic models to analyze miner behavior and system dynamics under potential 51% attacks. Zhang et al. [12] modelled the competition for computational resources among edge-based user devices in a blockchain-enabled IoT setting as an evolutionary game, incorporating practical constraints like propagation delay and mining rewards. The study analyzes evolutionary stable strategies (ESS), predicts double-spend attack probabilities, and provides risk indicators based on simulation results. Srivastava et al. [13] proposed bootstrapping and centralization games to evaluate mechanisms and stakeholder behavior. The study introduces the C-NORM metric, which quantifies centralization. The study shows that incorrect initialization of stakes can lead to validator dominance, creating a PoS variant of the 51% attack risk. It also proves that PoW-based bootstrapping (W2SB) better satisfies decentralization and incentive compatibility.

Some works adopt simulation-based approaches to explore the impact of 51% attacks on blockchain systems. Ye et al. [14] implemented a Java-based simulation framework to model blockchain processes and examine the relationship between attack power and system state transitions. This approach provides intuitive insights into security thresholds and can be extended as a middleware tool for attack detection. Novoa et al. [15] analyzed historical hash rate distribution in Bitcoin and Ethereum, establishing miner profiles to identify anomalous behaviors and assess the practical risk of 51% attacks. The study reveals a growing concentration of mining power, suggesting the need for real-time detection and proactive mitigation strategies. Hao et al. [16] analyzed the motivations of rational and irrational attackers, arguing that while financially motivated attackers lack incentives to perform 51% attacks, ideologically driven actors may still pose a threat. To address this, the study proposed a set of user-oriented monitoring indicators, including abnormal block generation frequency, sudden increase in mining pool hash power, and large-scale transactions associated with concentrated pools. Users are advised to adopt adaptive countermeasures such as increasing confirmation times for high-value transactions and tracking mining pool activity in real time. These strategies provide a lightweight, behavior-aware defense mechanism that helps users identify and respond to potential 51% attacks at the application level.

Despite the significant progress made in addressing 51% attacks, existing research still faces several limitations. In the area of consensus mechanism optimization, many proposed schemes, such as miner reputation scoring or hybrid consensus protocols, require substantial modifications to existing blockchain infrastructure or introduce additional computational and operational overhead, limiting their real-world applicability. In game-theoretic modeling, although evolutionary and incentive-based analyses provide valuable theoretical insights, most models assume idealized conditions and symmetric participants, which may not reflect the real-world asymmetry in mining power distribution or long-term strategic behaviors under repeated interactions. As for detection and response mechanisms, while simulation-based tools and behavior monitoring strategies help identify anomalies, these approaches often rely on offline analysis or user intervention and lack a formalized model to predict strategic adaptations over time.

Therefore, there remains a gap in modeling the dynamic and asymmetric strategic evolution between mining pools under 51% attack scenarios, particularly in long-term adversarial environments. To bridge this gap, this paper proposes an asymmetric dual-mining-pool strategy model based on repeated games, aiming to analyze the incentive conditions, revenue evolution, and equilibrium outcomes of attack behaviors over time. By capturing both asymmetry and dynamic adaptation, this model provides a lightweight yet theoretically grounded framework for understanding and mitigating 51% attacks from a behavioral and incentive perspective.

3 The Proposed RG5A Model

This section analyzes whether a mining pool with significant computational power would launch a 51% attack. Initially, we define the term"significant computational power" refers to a mining pool whose computing power exceeds 51% of the total computational power of all mining pools combined. While mining pools with less than 51% of the total power may also have the potential to carry out a 51% attack by exploiting vulnerabilities in the blockchain system, this scenario is not the focus of this paper. This paper considers two scenarios: whether a mining pool would choose to launch an attack in the presence or absence of a regulator. In the absence of a regulator, we consider two types of participants. One is a large mining pool A, and the other consists of one or more small mining pools B (with the number of B pools being at least one). Pool A, which possesses significant computational power, is considered a potential attacker. In contrast, due to their limited computational power, pool B is unable to launch a 51% attack and is therefore assumed to be an honest participant.

For clarity, we first present the list of symbols in Table 1 below.

Table 1. List of Symbols

Symbol	Explanation
A	Mining pool with high hash power, possibly launching a 51% attack; also denotes pool A choosing to attack
B	Mining pool without high hash power, ordinary mining pool
H	Represents honest mining
p	Probability that pool A successfully launches a 51% attack
R	Extra revenue obtained from a successful attack
C	Attack cost lost when the attack fails
r	Stable payoff from honest mining
$E_A(A)$	Expected payoff for pool A when it attacks in a single game
$E_A(H)$	Expected payoff for pool A when it mines honestly in a single game
L	Direct loss of pool B when pool A succeeds in an attack
d_1	Penalty imposed by the regulator when pool A succeeds in an attack
d_2	Penalty imposed by the regulator when pool A fails in an attack
h	Compensation given by the regulator to pool B
a	Reward given by the regulator to pool B for reporting
z	Number of blocks to be caught up

3.1 Without Regulator

In the absence of a regulator, mining pool A can choose to either attack (A) or act honestly (H), while mining pool B, due to its limited computational power, can only act honestly (H). In addition to the above setup, other factors also influence the decision-making of mining pool A. First, the cost of launching an attack, denoted as C, represents the resources spent on the attack. Second, the probability of a successful attack is denoted as p, making the failure probability $1 - p$. Third, if the attack succeeds, A gains additional profit R; if it fails, A loses the attack cost C. Fourth, acting honestly yields a stable mining reward r.

Taking the above factors into account, the expected payoff for mining pool A under the two strategies—attack or act honestly—is analyzed as follows:

- If pool A chooses to attack, it does not receive the reward from honest mining. Its total payoff is: $E_A(A) = p \cdot R - (1 - p) \cdot C$.
- If pool A chooses to mine honestly, its total payoff is: $E_A(H) = r$.

Mining pool B plays a passive role in the game: it cannot significantly influence A's strategy, yet its payoff is directly affected by A's decisions.

- If A launches an attack and it succeeds, B incurs a loss, denoted as L.
- If A's attack fails, or A chooses to mine honestly, B's payoff is r.

Based on the above setup, we derive the mixed-strategy equilibrium. If mining pool A is indifferent between the two strategies, i.e., $E_A(A) = E_A(H)$, then we have:

$$p \cdot R - (1 - p) \cdot C = r. \tag{1}$$

We obtain the following result by solving this equation:

$$p^* = \frac{r + C}{R + C}. \tag{2}$$

The derived value serves as a critical threshold p^* for the success probability of a 51% attack. It characterizes the point at which the expected payoff from launching an attack equals that of mining honestly. If the actual success probability p falls below this threshold, the potential reward from attacking is insufficient to offset the associated risks and penalties, making honest mining the rational and payoff-maximizing choice for mining pool A. Conversely, when $p \geq p^*$, the expected gain from the attack surpasses that of honest behavior, thus providing mining pool A with an economic incentive to initiate the attack. This threshold therefore acts as a decision boundary, delineating under which conditions rational attackers will choose to deviate from honest participation. It also highlights the importance of controlling system parameters—such as attack cost, detection risk, and regulatory pressure—to ensure that p^* remains sufficiently high to discourage malicious behavior.

3.2 With Regulator

In the presence of a regulator, this paper considers a zero-tolerance approach in which the regulator that imposes on-chain or off-chain penalties on the attacker and provides rewards to mining pool nodes that submit evidence of the attack. In addition to the two types of mining pools mentioned earlier, the participants now also include the regulator. As before, mining pool A is assumed to have large computational power, while mining pool B has relatively small computational power.

Under regulation, mining pool A can choose to either attack (A) or act honestly (H), while mining pool B, due to its limited computational power, can only act honestly (H). Considering the same additional factors as previously described, the expected payoffs for mining pool A under the attack and honest strategies are analyzed as follows:

- If mining pool A chooses to attack, it does not receive the honest mining reward r. Its total payoff is $E_A(A) = p \cdot (R - d_1) - (1 - p) \cdot (C + d_2)$. Specifically, d_1 and d_2 represent the punishments imposed by the regulator on A in the cases of a successful and failed attack, respectively.
- If mining pool A chooses to mine honestly, its total payoff is $E_A(H) = r$.

Although mining pool B has limited influence on pool A's decision-making, its own outcome is significantly affected by A's strategic choice:

- If A chooses to attack and the attack succeeds, and B is not the reporter, B's loss is denoted as L, and the compensation provided by the regulator is denoted as h. Therefore, B's net loss is $L - h$.
- If A chooses to attack and the attack succeeds, and B is the reporter, B's loss is denoted as L, the compensation provided by the regulator is denoted as h, and the reporting reward is denoted as a, Therefore, B's net loss is $L - h - a$.
- If A's attack fails, or if A chooses to mine honestly, B's payoff is r.

Based on the above setup, we find the mixed strategy equilibrium. Let mining pool A be indifferent between the two strategies, i.e., $E_A(A) = E_A(H)$, then we have:

$$p \cdot (R - d_1) - (1 - p) \cdot (C + d_2) = r. \tag{3}$$

We obtain the following result by solving this equation:

$$p^* = \frac{r + C + d_2}{R + C - d_1 + d_2}. \tag{4}$$

This represents the threshold for the attack success probability. If the actual success probability is below p^*, A will choose to mine honestly; otherwise, it will choose to attack.

According to [17], if the mining pool A's hash rate exceeds 50%, the probability of a successful 51% attack initiated by pool A approaches 1. In the absence of regulation, the reward for a successful attack is not limited, and if pool A's hash rate is large enough, A is more likely to launch a 51% attack. Unlike the unregulated scenario, in a regulated environment, pool A will consider the following two situations before launching an attack:

- The success probability of the attack is 100%, and if pool A chooses to attack, there is no honest mining reward, r. The total revenue is $E_A(A) = R - d_1$, where d_1 is the penalty imposed by the regulator if pool A successfully attacks.
- The success probability of the attack is 100%, and if pool A chooses to mine honestly, the total revenue is $E_A(H) = r$.

Considering the goal of preventing A's attack, it is necessary to ensure that A's total revenue from honest mining is greater than the revenue from attacking, i.e., $E_A(A) < E_A(H)$, which leads to:

$$R - d_1 < r. \tag{5}$$

In general, assuming R remains unchanged, effective ways to prevent a large-hash-rate A from attacking may include increasing the reward for honest mining and raising the penalty for dishonest mining.

4 Experiments

The experimental platform is a computer equipped with an Intel(R) Core(TM) Ultra 9 285K 3.70 GHz processor and 128 GB of RAM, running MATLAB R2025b. This chapter conducts an experimental analysis based on the proposed scheme.

4.1 Without Regulator

In the absence of regulation, we set $r = 1$ or 2, and $R = 8, 9$ or 10, to examine how the threshold varies with the attack cost C , which is assumed to continuously change from 0 to 10. If the attack cost exceeds R, mining pool A has no incentive to attack. Here, the threshold refers to the probability threshold for a successful attack. If the success probability exceeds this threshold, A is more likely to attack; otherwise, A is more likely to mine honestly. Let p^* represent the success probability threshold, and the threshold's variation with the attack cost is shown in Fig. 1.

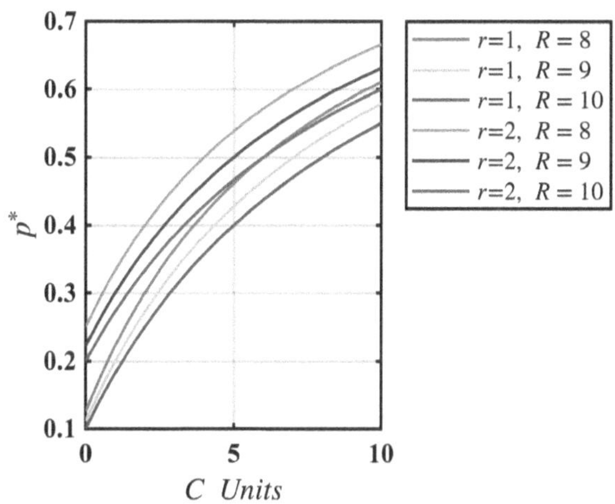

Fig. 1. In the absence of regulation, the variation of the attack success probability threshold with the attack cost under different parameter settings.

4.2 With Regulator

As can be seen from Fig. 1, as the attack cost increases, the attack success probability threshold also continues to rise. For mining pool A, the determination required to launch an attack becomes greater, which aligns with common reasoning. Additionally, for the same attack cost, the higher the honest mining reward, the lower the reward for a successful attack, and p^* also becomes larger. In this case, mining pool A needs even greater determination to launch an attack. In Fig. 1, the unit of C is $Unit$, representing a unit of cost, which could be in terms of "ten thousand yuan," "hundred thousand yuan," or "million yuan," etc. The unit can be adjusted according to actual needs.

In the presence of regulation, we set $d_1 = 5$, meaning that when the attack is successful, the regulator imposes a penalty of 5 units on A. We examine how the threshold varies with the attack cost C, d_2, r, and R. Here, we assume that C changes continuously from 0 to 10, d_2 takes values from the set $\{0, 5\}$, r takes values from the set $\{1, 2\}$, and R takes values from the set $\{9, 10\}$. The variation of the threshold p^* with C, d_2, r, and R is shown in Fig. 2.

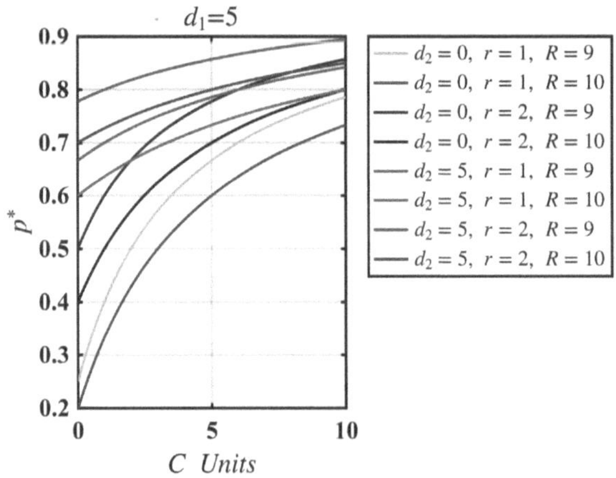

Fig. 2. Under regulation with an attack penalty, where a fixed penalty of 5 units is imposed by the regulator when A's attack is successful, the variation of the attack success probability threshold with C, d_2, r, and R.

As can be seen from Fig. 2, when d_1 remains unchanged, as the attack cost increases, the attack success probability threshold also continues to rise. For mining pool A, the determination required to launch an attack becomes greater, which aligns with common reasoning. When C remains constant, if d_2 is larger, r is larger, and R is smaller, the attack success probability threshold also increases. In this case, mining pool A requires even greater determination to launch an attack, which is also consistent with common reasoning.

In the presence of regulation, we further set d_2, meaning that when the attack fails, the regulator imposes a penalty of 5 units on A. We examine how the threshold varies with the attack cost C, d_1, r, and R. Here, we assume that C changes continuously from 0 to 10, d_1 takes values from the set $\{0, 5\}$, rrr takes values from the set $\{1, 2\}$, and R takes values from the set $\{9, 10\}$. The variation of the threshold p^* with C, d_1, r, and R is shown in Fig. 3.

From Fig. 3, it can be seen that when d_2 remains unchanged, as the attack cost increases, the threshold for the attack success rate also increases. For mining pool A, this means that the decision to launch an attack requires a greater level

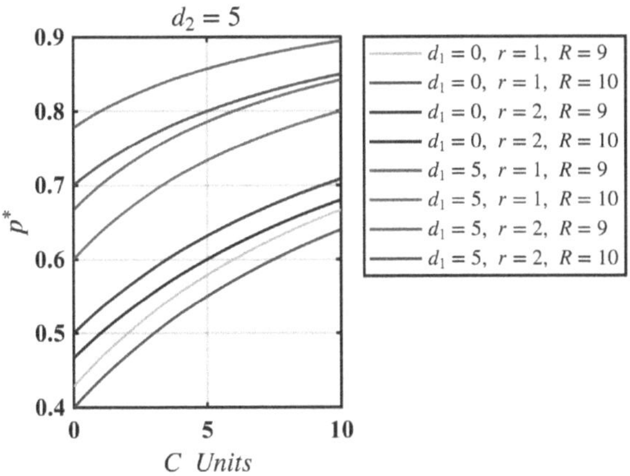

Fig. 3. Under regulation with an attack penalty, where a fixed penalty of 5 units is imposed by the regulator when A's attack fails, the variation of the attack success probability threshold with C, d_2, r, and R.

of determination. When C remains unchanged, as d_1 increases, r increases, and R decreases, the threshold for the attack success rate also continues to increase. For mining pool A, this also means that the decision to launch an attack requires a greater level of determination.

In the presence of a regulator, this paper sets $R = 10$ and $r = 2$. The threshold is examined as it changes with attack cost C, d_1, and d_2, where it is assumed that C varies continuously from 0 to 10. Here, d_1 takes values from the set $\{0, 5\}$, and d_2 takes values from the set $\{0, 5\}$. The threshold as a function of C and d_2 is shown in Fig. 4.

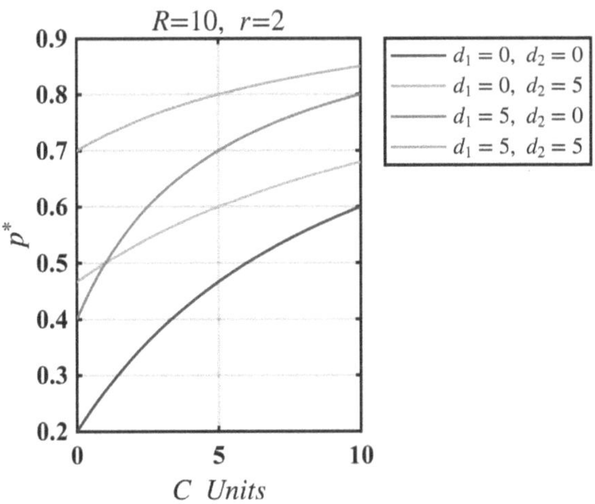

Fig. 4. Under regulation and attack penalties, with the additional reward R for a successful attack and the honest mining reward r, the threshold for the attack success rate changes with C, d_1, and d_2.

From Fig. 4, it can be seen that when R and r remain unchanged, as the attack cost increases, the threshold for the attack success rate also increases. For mining pool A, this means that the decision to launch an attack requires a greater level of determination. When C remains unchanged, as d_1 and d_2 increase, the threshold for the attack success rate also continues to increase. For mining pool A, this also means that the decision to launch an attack requires a greater level of determination.

In addition to the above discussion, the threshold for the success rate as a function of R was also examined. Under this setting, two scenarios were considered: one where $d_1 = d_2 = 5$, and another where $d_1 = d_2 = 0.2R$. The threshold for the success rate as a function of R is shown in Fig. 5.

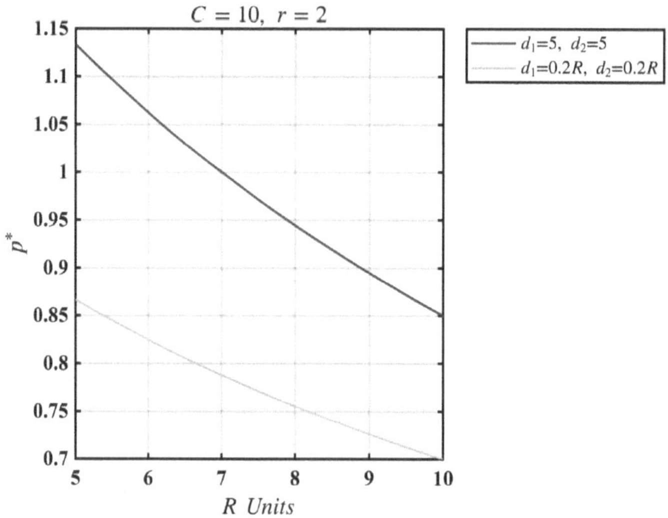

Fig. 5. Under regulation and attack penalties, with C and r fixed, the threshold for the attack success rate as a function of R: one where $d_1 = d_2 = 5$, and the other where $d_1 = d_2 = 0.2R$.

As shown in Fig. 5, as R increases, the threshold for the attack success rate continuously decreases, meaning that mining pool A requires less determination to launch an attack. Since $d_1 = d_2 = 0.2R \leq 2 < 5$, it can be concluded that, under the same values of C, r, and R, the smaller the values of d_1 and d_2, the smaller the threshold for the attack success rate, and the less determination mining pool A needs to launch an attack.

According to [17], the probability of mining success for mining pool A is related to A's hash rate and the number of blocks caught up with, denoted as z. The relationship between the mining success probability of pool A, its hash rate, and the number of blocks caught up with is shown in Fig. 6 below.

As shown in the Fig. 6, when the hash rate of mining pool A exceeds 50% of the total hash rate, the probability of a successful attack by pool A approaches

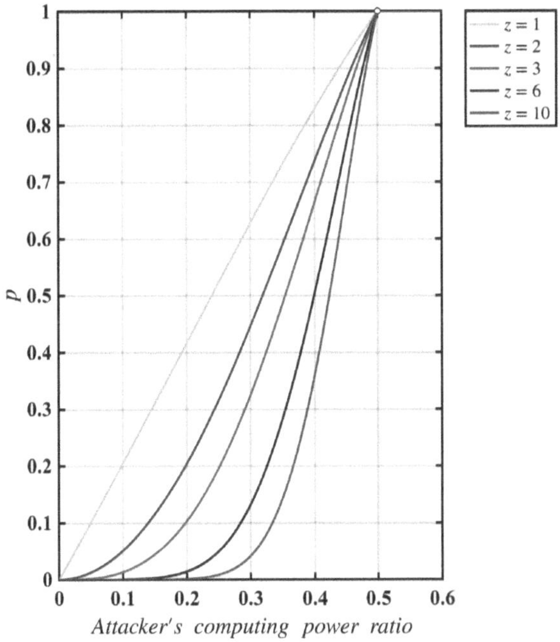

Fig. 6. The relationship between the attack success probability of mining pool A, its hash rate, and the number of blocks caught up with z, as obtained from [17].

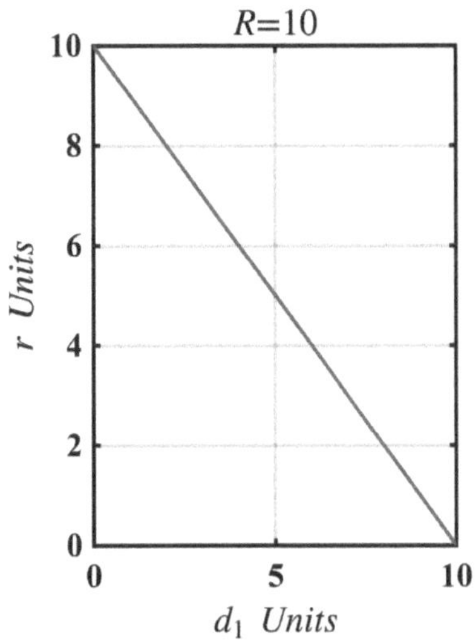

Fig. 7. In the high hash rate scenario, with $R = 10$, the variation of r with d_1.

1. The more blocks that need to be caught up with, the slower the increase in the attack success probability. When the hash rate of pool A exceeds 50%, under

regulation, pool A's decision to launch an attack will take R, r, and d_1 into account. If R is fixed, the decision curve forms a line segment. When $R = 10$, the decision line segment is illustrated in Fig. 7.

In the high hash rate scenario, the probability of a successful attack by mining pool A can be considered 100%; therefore, the possibility of pool A's attack failing, as well as the corresponding parameters C and d_1, are not considered. As shown in Fig. 7, when the values of r and d_1 lie on the line segment, mining pool A receives the same payoff whether it chooses to attack or mine honestly, resulting in a 50% probability of either action. When the values of r and d_1 lie below the line segment, the feasible parameter region forms a triangle, and honest mining yields a higher payoff, so the pool tends to mine honestly. When the values of r and d_1 lie above the line segment, the feasible parameter region is the infinite area above the line segment, and attacking yields a higher payoff, so the pool tends to choose an attack.

4.3 Comparison with and Without Regulation

Under the same parameter settings, the comparison of the success rate threshold as a function of C between the regulated and unregulated scenarios is shown in Fig. 8 As can be seen from the figure, in the regulated scenario, the attacker requires greater determination to launch an attack, whereas under the same settings in the unregulated scenario, the attacker is more inclined to attack. This highlights the important role of regulation.

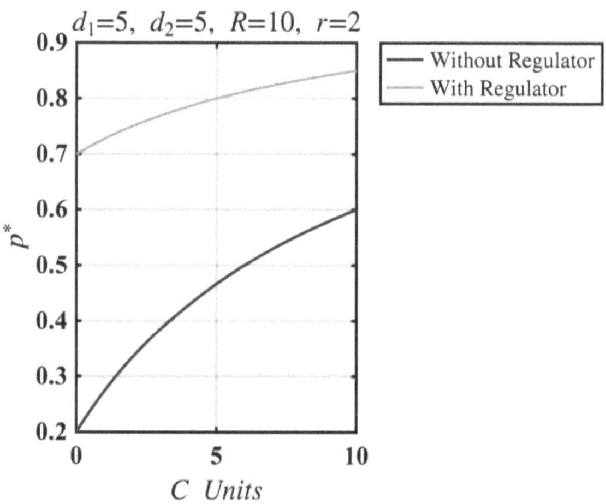

Fig. 8. Comparison of the success rate threshold as a function of C under regulated and unregulated scenarios with the same parameter settings.

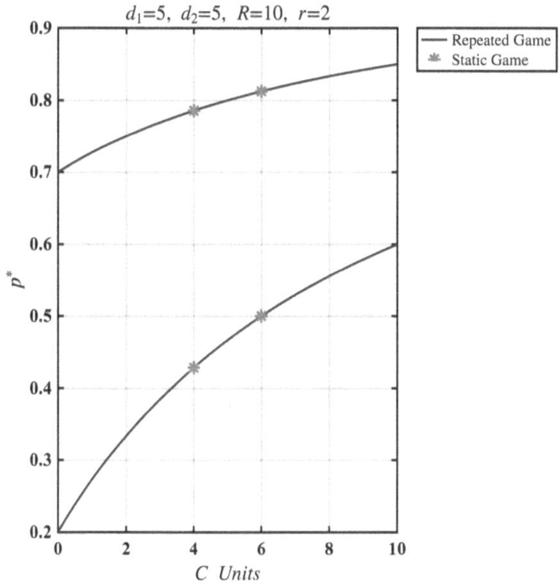

Fig. 9. Comparison between the traditional static game and the repeated game. The traditional static game cannot capture the complete variation of the success rate threshold with parameters; it can only provide values at specific points. For clarity in the figure, four points are shown for comparison.

4.4 Comparison Between Static and Repeated Games

As shown in Fig. 9, the static game can determine the success rate threshold of mining pool A under fixed parameter settings; however, as a static game, it cannot show how the success rate threshold varies with multiple parameter changes. This is something that the repeated game model can accomplish, which is why this paper chooses to use the repeated game model.

In summary, compared to the no-penalty scenario (equivalent to having no regulator), in the presence of penalties, a high-hash-rate attacker requires greater determination to launch an attack. Therefore, regulators play a significant role in preventing 51% attacks on blockchains. Consequently, in systems without regulators, such as the Bitcoin system, 51% attacks could result in more severe consequences.

5 Conclusion and Future Works

This paper focuses on the potential 51% attack behavior in blockchain systems and constructs a non-symmetric dual mining pool strategy model based on repeated games for 51% attacks (RG5A). It systematically analyzes the attack motives and cost-benefit trade-offs of large-scale mining pools under different game scenarios. In the absence of a regulatory authority, the threshold for the probability of a successful attack is derived through mixed strategy equilibrium analysis, revealing that higher attack costs make mining pools less inclined to launch attacks. When a regulatory authority is introduced, the model further incorporates attack penalties and reporting rewards. Simulation results show that regulatory mechanisms significantly raise the decision threshold for attack success rates, thereby effectively suppressing attack intentions. This study not only provides theoretical support for understanding mining pool behavior in blockchain systems but also offers a quantifiable analytical framework for designing incentive and punishment mechanisms in system design. Future research could further explore more complex game models involving multiple participants, incomplete information, or dynamic regulatory mechanisms to better align with real-world network environments.

References

1. Liu, X., et al.: When blockchain meets auction: a comprehensive survey. IEEE Trans. Comput. Soc. Syst. **11**, 4242–4254 (2024)
2. Rebello, G.A.F., et al.: A survey on blockchain scalability: from hardware to layer-two protocols. IEEE Commun. Surv. Tutorials **26**, 2411–2458 (2024)
3. Zheng, Z., Xie, S., Dai, H.N., et al.: Blockchain challenges and opportunities: a survey. Int. J. Web Grid Serv. **14**(4), 352–375 (2018)
4. Raju, R.S., Gurung, S., Rai, P.: An overview of 51% attack over bitcoin network. In: Sarma, H.K.D., Balas, V.E., Bhuyan, B., Dutta, N. (eds.) Contemporary Issues in Communication, Cloud and Big Data Analytics. LNNS, vol. 281, pp. 39–55. Springer, Singapore (2022). https://doi.org/10.1007/978-981-16-4244-9_4
5. Baruwa, Z., Bhattacherjee, S., Chandnani, S.R., et al.: Social media perceptions of 51% attacks on proof-of-work cryptocurrencies: a natural language processing approach. arXiv preprint arXiv:2310.14307 (2023)
6. Zhang, Y., Ma, Z., Meng, J.: Auditing in the blockchain: a literature review. Front. Blockchain **8**, 1549729 (2025)
7. Moosavi, N., Taherdoost, H., Mohamed, N., et al.: Blockchain technology, structure, and applications: a survey. Procedia Comput. Sci. **237**, 645–658 (2024)
8. Prasad, V.S.R., Harshitha, G., Sujitha, A.V., et al.: Strengthening blockchain security: countering 51% attacks using dynamic miner reputation and weighted block acceptance (DRW-BA). Synth. Multidisc. Res. J. **3**(1s), 1–13 (2025)
9. Babur, S.M., Khan, S.U.R., Yang, J., et al.: Preventing 51% attack by using consecutive block limits in Bitcoin. IEEE Access **12**, 77852–77869 (2024)
10. Khatoon, N., Rishu, R., Verma, S., et al.: Proposing a modified proof of stake system to counter 51% attack in blockchain. In: 2nd International Conference on Device Intelligence, Computing and Communication Technologies (DICCT), pp. 01–04. IEEE (2024)
11. Yang, X., Chen, Y., Chen, X.: Effective scheme against 51% attack on proof-of-work blockchain with history weighted information. In: IEEE International Conference on Blockchain (Blockchain), pp. 261–265. IEEE, Atlanta (2019)
12. Zhang, H., Wang, J., Zhao, Z., et al.: Security resource allocation in blockchain-based IoT. Peer-to-Peer Netw. Appl. **18**(1), 45 (2025)
13. Srivastava, V., Damle, S., Gujar, S.: Centralization in proof-of-stake blockchains: a game-theoretic analysis of bootstrapping protocols. arXiv preprint arXiv:2404.09627 (2024)

14. Ye, C., Li, G., Cai, H., et al.: Analysis of security in blockchain: case study in 51%-attack detecting. In: 5th International Conference on Dependable Systems and Their Applications (DSA), pp. 15–24. IEEE, Guangzhou (2018)
15. Aponte-Novoa, F.A., Orozco, A.L.S., Villanueva-Polanco, R., et al.: The 51% attack on blockchains: a mining behavior study. IEEE Access **9**, 140549–140564 (2021)
16. Hao, Y.: Research of the 51% attack based on blockchain. In: 3rd International Conference on Computer Vision. Image and Deep Learning (CVIDL) & International Conference on Computer Engineering and Applications (ICCEA), pp. 278–283. IEEE, China (2022)
17. Nakamoto S. Bitcoin: a peer-to-peer electronic cash system. SSRN 3440802 (2008)

A Quantum-Resistant Authentication Protocol for Power IoT Networks

Jinsong Hou[1], Jun Liu[1], Yu Wang[2], Rui Cheng[1], and Hongyan Wang[2](✉)

[1] Information and Communication Branch, State Grid Anhui Electric Power Co., Ltd., Hefei, China
{houjs0431,liuj641X,chengr0011}@ah.sgcc.com.cn
[2] Anhui Jiyuan Software Co., Ltd., Hefei, China
yuwang0551@sina.com, 1260468391@qq.com

Abstract. We present a novel authentication protocol that integrates Identity Public Key (IPK) technology with Quantum Key Distribution (QKD) to enhance the quantum resistance of SM2-based systems in power IoT networks. By employing a dual-protection mechanism—where IPK enables local public key computation to prevent interception, and QKD provides secure symetric encryption for signatures—the protocol eliminates the preconditions for quantum attacks without modifying the SM2-based authentication protcol itself. Security analysis confirms attackers cannot simultaneously obtain complete public keys (hidden by IPK matrix operations) and valid signature pairs (protected by QKD), while performance evaluations show 2× faster verification than SM9. Our solution offers a practical transition path for critical infrastructure like smart grids, maintaining backward compatibility with existing SM2 deployments.

Keywords: Quantum Resistance · SM2 Algorithm · Identity Public Key (IPK) · Quantum Key Distribution (QKD) · Authentication

1 Introduction

1.1 Research Background and Motivation

Information security is the cornerstone of the digital era, and public key cryptosystems are the central pillar of the modern information security framework. However, the dawn of quantum computing is casting a long shadow over this edifice. Quantum algorithms, represented by Shor's algorithm [1], have been theoretically proven to be capable of solving mathematical problems like large integer factorization and discrete logarithms in polynomial time, which form the security foundation of mainstream public key algorithms such as RSA and Elliptic Curve Cryptography (ECC). The SM2 algorithm [2], as China's independently controlled commercial cryptography standard, is built upon ECC theory and is equally vulnerable to the potential threat of quantum computing.

In response to this challenge, the global cryptographic community has initiated a research boom in Post-Quantum Cryptography (PQC). The U.S. National

J. Yu and Y. Chen (Eds.): FCS 2025, CCIS 2805, pp. 42–55, 2027.
https://doi.org/10.1007/978-981-95-9514-3_3

Institute of Standards and Technology (NIST) began soliciting and reviewing PQC algorithms in 2016 and selected the first batch of standardized algorithms in 2022 [3]. However, the development of PQC technology [4–7] has not been without its setbacks. For instance, lattice-based cryptography, one of the mainstream technical routes, was challenged by a new quantum cracking algorithm in 2024 [8], prompting a re-evaluation of its security. This clearly indicates that post-quantum technology is still in a phase of continuous development and refinement, and its theoretical and practical maturity requires more time.

1.2 Problem Statement

In critical national information infrastructures such as power grids, finance, and government services, the SM2 algorithm has been widely deployed and applied, forming a vast ecosystem. Before PQC standards are fully mature and widely adopted, how to protect these massive existing SM2 application systems to withstand potential quantum attacks is an urgent and practical issue. Adopting a "rip-and-replace" model to completely substitute the SM2 algorithm with new PQC algorithms would entail enormous engineering costs, long transformation cycles, and unknown compatibility risks. Therefore, researching a solution that is compatible with the existing SM2 system and can achieve quantum resistance through enhanced protection measures holds significant theoretical and practical value.

1.3 Contributions

To solve the aforementioned problems, this paper proposes an innovative, quantum-resistant authentication protocol based on Identity Public Key (IPK) and SM2. The main contributions of this paper are:

1. **A Non-Replacement Quantum Security Enhancement Scheme:** The core of this scheme is not to design a new signature algorithm, but to protect the application process of the existing SM2 algorithm through a systematic design, making it immune to quantum attacks.
2. **A Dual Security Protection Mechanism:** The protocol cleverly combines IPK technology [9] and Quantum Key Distribution (QKD) technology [10]. The IPK mechanism ensures that the user's complete public key is not transmitted over the channel but is computed locally by the verifier; QKD provides physical-layer security encryption for the authentication message. This design makes it impossible for an attacker to simultaneously obtain the two key elements required for a quantum attack: the public key and the corresponding signature data.
3. **A Balance of Security and Practicality:** The protocol provides a smooth, low-cost quantum-resistant upgrade path for existing SM2 systems, which is conducive to ensuring the security of critical information systems before PQC technology is fully mature.

1.4 Paper Structure

The structure of this paper is as follows: Sect. 2 will introduce the foundational technologies required for the research. Section 3 will focus on the proposed quantum-resistant authentication protocol. Section 4 will conduct an in-depth security analysis of the protocol. Section 5 provides a performance comparison with other schemes. Finally, Sect. 6 will summarize the entire paper and look forward to future research directions.

2 Preliminaries

2.1 SM2 Public Key Algorithm

The SM2 algorithm is a commercial cryptography standard issued by the China State Cryptography Administration [2]. Its core is a suite of public key cryptographic algorithms based on Elliptic Curve Cryptography (ECC). It includes digital signatures, key exchange, and public key encryption. Its security relies on the computational difficulty of the Elliptic Curve Discrete Logarithm Problem (ECDLP). In the classical computing model, breaking ECDLP requires exponential time complexity, hence it is considered secure. However, the emergence of Shor's quantum algorithm [1] reduces the problem of solving ECDLP on a quantum computer to polynomial time complexity, thus posing a fundamental threat to the security of the SM2 algorithm.

A standard SM2 digital signature process requires the signer's private key d and the message to be signed M. The signer computes a signature value (r, s). The verifier then needs to use the signer's public key P (where $P = [d]G$, and G is the base point) and the message M to verify the validity of the signature (r, s). In a traditional PKI system, the public key P needs to be bound to the signer's identity through a digital certificate, which is publicly published or transmitted during verification.

2.2 Identity Public Key (IPK) Technology

Identity Public Key (IPK) technology [9] is an innovative, identity-based lightweight key management scheme. It is also based on ECC theory but solves the problems of complex certificate management in traditional PKI and key escrow in Identity-Based Cryptography (IBC) by introducing matrix operations and combinatorial ideas. Its core mechanism is detailed in Ref. [9].

IPK Technical Principle. The core principle of IPK technology is based on a fundamental property of ECC: if (d_1, P_1) and (d_2, P_2) are two valid key pairs, then their sum, $((d_1 + d_2) \pmod{n}, P_1 + P_2)$, is also a valid key pair (where n is the order of the base point G). IPK utilizes this property by decomposing a user's final key into a combination of multiple partial keys, which are jointly generated by the user and a key center. The final public key is then locally synthesized by the verifier based on the user's public information.

Key Generation Process. The IPK key system involves three key processes, as shown in Fig. 1 and Fig. 2:

1. **Matrix Generation:** A trusted Key Center (KC) generates an $m \times h$ private key matrix $\mathbf{SK} = \{sk_{i,j}\}$, where each element $sk_{i,j}$ is a 256-bit random number. Concurrently, the corresponding public key matrix $\mathbf{PK} = \{PK_{i,j}\}$ is generated through elliptic curve point multiplication, where $PK_{i,j} = [sk_{i,j}]G$. The private key matrix **SK** is kept secret by the KC, while the public key matrix **PK** is distributed as a public system parameter to all participants.
2. **Private Key Generation:** When a user, Alice, requests a private key, the process is as follows (see Fig. 1):
   - Alice randomly generates a temporary key pair (r_1, R_1) and sends her identity ID_A and R_1 to the KC.
   - The KC also randomly generates a temporary key pair (r_2, R_2) and computes a "declared public key" $R_A = R_1 + R_2$.
   - The KC uses ID_A and R_A as input to a Hash-based Message Authentication Code function (e.g., HMAC-SM3 [11]) to compute a hash value. This hash value is used to select h elements from the private key matrix **SK**.
   - The KC sums these h private key elements to obtain the identity private key $isk_m = \sum_{k=0}^{h-1} sk_{I_k,k} \pmod{n}$.
   - The KC computes the first-order composite private key $isk' = (isk_m + r_2) \pmod{n}$ and sends it securely to Alice.
   - Upon receiving isk', Alice combines it with her own secret r_1 to generate the final, complete second-order composite private key $isk_A = (isk' + r_1) \pmod{n}$.
3. **Public Key Computation:** When a verifier, Bob, needs to verify Alice's signature, he only needs to know her identity ID_A and her declared public key R_A. The process is as follows (see Fig. 2):
   - Bob performs the same hash computation as the KC, using ID_A and R_A to derive the index values.
   - Bob selects the corresponding h public key elements from his locally stored public key matrix **PK** based on the indices.
   - Bob sums these h public key elements to obtain the identity public key $IPK_m = \sum_{k=0}^{h-1} PK_{I_k,k}$.
   - Finally, Bob computes Alice's complete public key $IPK_A = R_A + IPK_m$. This IPK_A and Alice's private key isk_A form a valid SM2 key pair.

IPK Security Features. A significant advantage of the IPK technical system is that the user's complete public key, IPK_A, is dynamically computed locally by the verifier and is never directly transmitted over any network channel. Only the identity ID_A and the declared public key R_A are transmitted. Since R_A is only a part of the final public key and is strongly bound to ID_A through a hash function, an attacker, even if they intercept ID_A and R_A, cannot deduce the complete public key IPK_A, nor can they substitute it. This feature provides a natural barrier against quantum attacks.

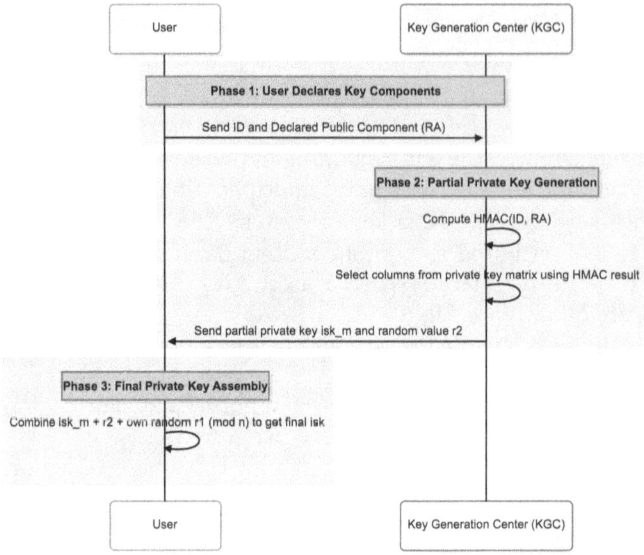

Fig. 1. IPK Private Key Generation Process

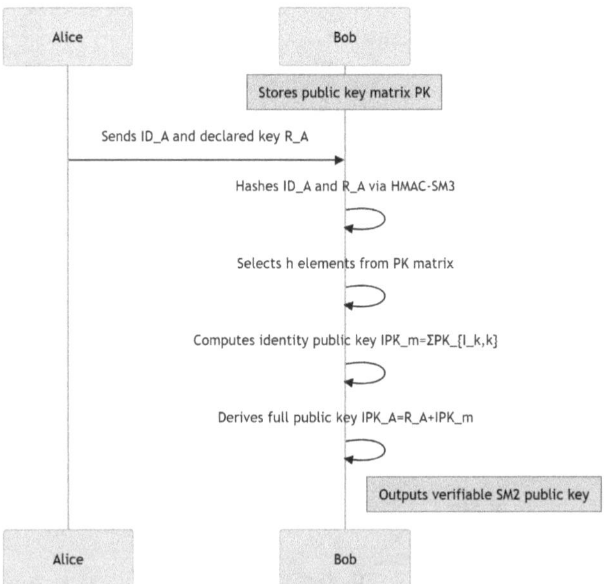

Fig. 2. IPK Public Key Computation Process

3 Proposed Quantum-Resistant Authentication Protocol

3.1 Protocol Design Philosophy and Objectives

The design objective of this protocol is to construct an authentication framework capable of withstanding attacks from quantum computers, withoutaltering the core of the SM2 signature algorithm. Its core design philosophy is to **eliminate the attack preconditions**, meaning that through systematic design, it prevents an attacker from obtaining all the information necessary to launch a quantum attack.

A quantum algorithm (like Shor's) for breaking an ECC system typically requires two key inputs:

1. The user's complete public key P.
2. A valid signature (r, s) for a certain message M, generated by the corresponding private key d.

This protocol aims to simultaneously sever the attacker's access to these two inputs through a dual-protection mechanism:

- **Protection 1 (Hiding the Public Key):** Using IPK technology, the user's complete public key IPK_A is no longer transmitted over the channel but is computed locally by the verifier based on the publicly transmitted identity ID_A and declared public key $APKA$. An attacker cannot directly intercept IPK_A on the channel.
- **Protection 2 (Encrypting the Payload):** Using a symmetric key provided by Quantum Key Distribution (QKD) technology, which offers information-theoretic security, the authentication payload containing the declared public key $APKA$ and the SM2 signature is encrypted. An attacker cannot obtain the signature value and the signed data because they cannot break the quantum key.

With these two layers of protection, an attacker is left in a "can't make bricks without straw" situation, thereby endowing the SM2-based authentication process with quantum resistance.

3.2 Protocol Participants and System Architecture

In the context of power IoT networks [12,13], Alice represents an edge-side device (e.g., a smart meter or local controller) deployed at the distribution or substation level, while Bob represents a cloud-based master station device responsible for centralized control and data aggregation. This distinction clarifies the deployment topology and communication paths for both IPK and QKD operations in the protocol.

The system architecture of this protocol mainly consists of three logical entities, as shown in Fig. 3:

- **Authentication Initiator (Alice):** For example, an edge computing device in a power system. It holds its own IPK identity private key isk_A and can communicate with the quantum key management system.

- **Authentication Verifier (Bob):** For example, a cloud-based master station device. It stores the IPK public key matrix and can communicate with the quantum key management system.
- **Quantum Key Management System (QSS-ME/QKD):** Responsible for generating quantum keys and distributing a symmetric key K_Q to Alice and Bob through a secure quantum channel or a post-quantum key agreement protocol over a classical channel.

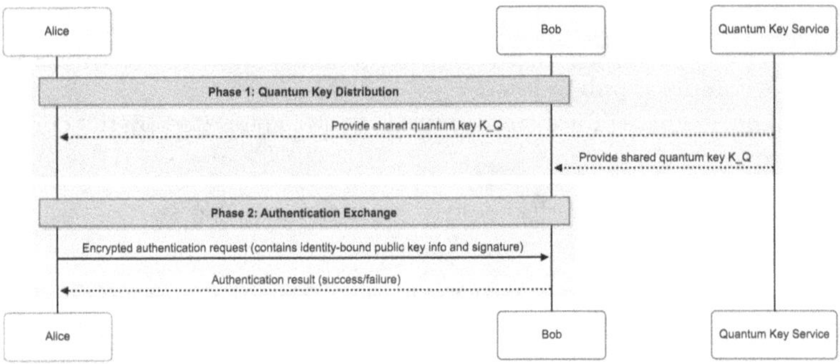

Fig. 3. System Architecture

3.3 Detailed Protocol Flow

The entire interaction process of the protocol is displayed through the sequence diagram in Fig. 4.

The detailed interaction flow of the protocol is as follows, assuming Alice needs to initiate an authentication to Bob:

1. **Initialization:** Alice and Bob obtain a shared symmetric encryption key K_Q through negotiation with the quantum key management system. Alice holds her IPK private key isk_A and declared public key $APKA$. Bob holds the IPK public key matrix **PK**.
2. **Initiator (Alice) Operations:**
   - Alice generates a one-time random number r_1 (Nonce) and obtains the current trusted timestamp t.
   - Alice concatenates r_1 and t as the message to be signed and uses her IPK private key isk_A to perform an SM2 signature computation, generating the signature value Sig.

   $$\text{Sig} = \text{SM2.Sign}(isk_A, H(t \parallel r_1))$$

   where $H(\cdot)$ is a cryptographic hash function, such as SM3 [11].

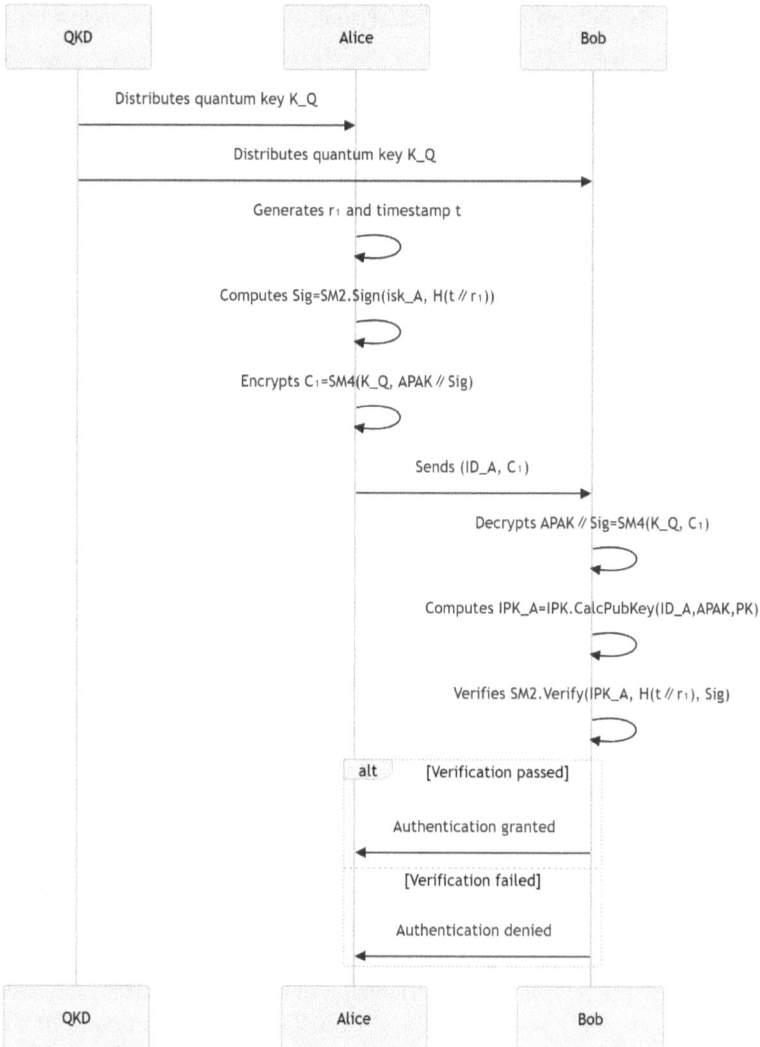

Fig. 4. Proposed Quantum-Resistant Authentication Protocol

- Alice concatenates her declared public key $APKA$ and the signature value Sig, and encrypts it using the quantum key K_Q with a symmetric cipher (e.g., SM4 algorithm [14]) to generate the encrypted payload C_1.

$$C_1 = \text{Encrypt.SM4}(K_Q, APKA \parallel Sig)$$

- Alice sends her identity ID_A and the encrypted payload C_1 to Bob.

3. **Verifier (Bob) Operations:**
   - Bob receives ID_A and C_1 from Alice.

- Bob uses the previously negotiated quantum key K_Q to decrypt C_1, recovering Alice's declared public key $APKA$ and her signature value Sig.

$$APKA \parallel Sig = \text{Decrypt.SM4}(K_Q, C_1)$$

- Bob computes Alice's complete public key IPK_A from his local public key matrix **PK**, according to the IPK public key computation procedure, using the received ID_A and the decrypted $APKA$.

$$IPK_A = \text{IPK.CalcPubKey}(ID_A, APKA, \mathbf{PK})$$

- Bob uses the computed IPK_A and the agreed-upon message construction method (i.e., hashing the timestamp t and random number r_1 from Alice's message) to verify the validity of the signature Sig.

$$\text{Result} = \text{SM2.Verify}(IPK_A, H(t \parallel r_1), Sig)$$

- If the verification is successful, Bob confirms Alice's identity is legitimate and returns a successful authentication result to Alice. Otherwise, the authentication fails.

4 Security Analysis

The security of this protocol is mainly reflected in its ability to resist quantum computing attacks, while also inheriting the original security features of IPK and SM2.

4.1 Resistance to Quantum Computing Attacks

As previously stated, a quantum computer's attack on an ECC algorithm depends on simultaneously obtaining the public key and the corresponding signature data. The design of this protocol is intended to fundamentally prevent an attacker from achieving this goal.

Assume an attacker, Eve, equipped with a quantum computer, can eavesdrop on all communications between Alice and Bob. The data packet content that Eve can intercept is (ID_A, C_1).

1. **Signature and Message Unobtainable:** The first obstacle Eve faces is the encrypted payload C_1. C_1 is encrypted with the quantum key K_Q. According to the security principles of QKD (based on the fundamental laws of quantum mechanics, such as the uncertainty principle and the no-cloning theorem) [10], any eavesdropping on the quantum channel will be detected, thus ensuring the secure distribution of K_Q. Therefore, the attacker Eve cannot obtain the key K_Q. In both classical and quantum computing models, breaking a symmetric encryption algorithm with a sufficiently random key (like SM4) is extremely difficult. Consequently, Eve cannot decrypt the declared public key $APKA$ and the signature value Sig from C_1. Since the signature value itself cannot be obtained, the second prerequisite for launching a quantum attack is not met.

2. **Full Public Key Unobtainable:** To take it a step further, even if we assume that in some extreme scenario, the attacker obtains the signature value Sig and the signed message digest $H(t \parallel r_1)$ through other means, she still faces a second obstacle: obtaining the complete public key IPK_A used to verify the signature. In the protocol, only the declared public key $APKA$ is transmitted over the channel. According to the principles of IPK technology, $IPK_A = APKA + IPK_m$. Here, the identity public key IPK_m is computed locally by the verifier Bob based on ID_A and $APKA$. Its computation process depends on the initial setup of the secret private key matrix **SK** and the public key matrix **PK**. Even if the attacker Eve possesses the public key matrix **PK**, she cannot correctly map and compute from $(ID_A, APKA)$ to IPK_A because she does not know the private key matrix **SK**. Therefore, the attacker cannot reconstruct the unique and correct complete public key IPK_A used for signature verification.

In summary, under no circumstances can an attacker simultaneously satisfy the two necessary conditions for launching a quantum attack. This protocol establishes an effective quantum attack defense system through the dual mechanisms of "payload encryption" and "hiding the public key."

4.2 Defense-in-Depth for Key Materials

To achieve a higher level of security, this scheme also supports a defense-in-depth strategy. A key material in the IPK system is the public key matrix **PK**. Although it is a public parameter, it is the foundation for computing all public keys. The **PK** itself can also be encrypted with a quantum key or a pre-set master key and stored in the secure chip of the terminal device. When public key computation is needed, it is decrypted and processed inside the chip. This further increases the difficulty for an attacker to obtain critical system parameters, effectively protecting key materials even if the terminal device is physically captured.

4.3 Security Against Classical Adversaries

Beyond quantum resistance, the proposed protocol also ensures robustness against classical adversaries. In scenarios where quantum capabilities are absent, the security relies on the SM2 signature scheme combined with timestamp-based freshness and partner identity binding. The inclusion of a trusted timestamp (t) and a nonce (r_1) prevents replay attacks. Moreover, the explicit binding of Bob's identity within the signature scope ensures that Alice confirms her partner's identity before proceeding, mitigating the risk of man-in-the-middle impersonation. Under the standard hardness assumption of the Elliptic Curve Discrete Logarithm Problem (ECDLP), the signature remains unforgeable and the authentication process is secure even without QKD.

- **Anti-Replay Attack:** The protocol incorporates a timestamp t and a one-time random number r_1. The verifier Bob checks the freshness of the timestamp t during verification and can record used r_1 values to prevent message replay.
- **Resistance to Collusion and Substitution Attacks:** The protocol inherits the intrinsic security features of IPK technology [9]. Since the user's final private key isk_A includes a random factor r_1 generated by the user, the key center cannot escrow or forge the user's private key. At the same time, the declared public key $APKA$ is strongly bound to the identity ID_A through a hash function, effectively preventing an attacker from performing a public key substitution attack.
- **Non-repudiation:** Since the signature is generated with Alice's private key isk_A, and only she holds this private key, the signing act is non-repudiable.

5 Performance and Comparison

To evaluate the practicality of the proposed scheme, which is built upon the IPK framework, it is useful to compare IPK with other major identity-based cryptosystems. The comparison, adapted from [9], is based on a security strength of 128 bits. The performance analysis primarily focuses on the computational cost, where the time for one 256-bit elliptic curve point multiplication is defined as a single unit of time (UT).

Table 1 presents a comparison between IPK, IBC (SM9), TF-CPK, and SM2 with implicit certificates.

Table 1. Comparison of Identity-Based Cryptosystems (Adapted from [9])

Identity-based cryptosystem	IBC(SM9)	TF-CPK	SM2 implicit cert.	IPK
Length of signed private key	64 B	32 B	32 B	32 B
Length of signed public key	Arbitrary	64 B	64 B	64 B
Length of private key (encryption)	128 B	32 B	32 B	32 B
Length of public key (encryption)	Arbitrary	64 B	64 B	64 B
Key encapsulation	96 B	128 B	128 B	128 B
Signature length	96 B	128 B	128 B	128 B
Signature time	13 UT	1 UT	1 UT	1 UT
Verification time	12 UT	4 UT	3 UT	2 UT
Storage space (client)	128+64 B	64 KB	64 B	64 KB

As shown in Table 1, the computational cost of the SM9 cryptographic algorithm is significantly higher than that of the other schemes based on the SM2 algorithm. The IPK scheme demonstrates the lowest computational cost for signature verification (2 UT). While the signature generation time is consistent across SM2-based schemes (1 UT), IPK's verification process is more efficient than TF-CPK (4 UT) and SM2 with implicit certificates (3 UT) because the public key computation involves only simple additions, with negligible time consumption.

To further illustrate the end-to-end impact of adopting our proposed protocol, we compare the total authentication process (including QKD operations and payload encryption) with a traditional SM2 authentication process. The results are summarized in Table 2, showing that although the proposed protocol adds a negligible symmetric encryption overhead and requires pre-established quantum keys, it reduces verification time from 3 UT to 2 UT thanks to IPK's efficient public key computation, while simultaneously achieving quantum resistance.

Table 2. Performance Comparison between Traditional SM2 Authentication and Proposed Protocol

Metric	Traditional SM2 Authentication	Proposed Protocol (IPK + QKD)
Authentication Delay	Signature (1 UT) + Verification (3 UT) $\approx$ 4 UT	Signature (1 UT) + Verification (2 UT) + SM4 Enc/Dec ($\approx$ 0.01 UT) $\approx$ 3.01 UT (+ QKD pre-setup)
Computation Cost	ECC scalar multiplications for signature & verification	ECC scalar multiplications + simple point additions (IPK) + symmetric enc/dec
Communication Overhead	Transmission of full public key + signature	Transmission of declared public key + encrypted signature (slightly larger payload due to encryption)
Quantum Resistance	No	Yes (dual-protection: public key hiding + QKD encryption)

However, it is worth noting that the client-side storage cost for IPK and TF-CPK is larger (64KB) due to the need to store the public key matrix. This makes them less suitable for embedded terminal devices with extremely limited storage space, although this is becoming less of a constraint with modern hardware. The proposed quantum-resistant protocol inherits these performance characteristics, with the main additional overhead coming from the QKD process and a single symmetric encryption/decryption operation, which are typically very fast.

6 Conclusion

In response to the severe challenges posed by quantum computing to the existing SM2 cryptosystem, this paper has proposed a feasible, quantum-resistant authentication protocol based on IPK and quantum keys. The greatest innovation of this protocol is that it does not attempt to replace or modify the SM2 signature algorithm itself. Instead, by re-architecting the authentication process and introducing advanced protective technologies, it creatively eliminates the preconditions required for a quantum attacker to launch an attack.

This scheme achieves localized computation of the verification public key through IPK technology, avoiding the direct exposure of the complete public key in the communication channel. Simultaneously, it uses QKD technology to encrypt the critical authentication payload, ensuring the confidentiality of the signature data during transmission. This dual-protection mechanism of "hiding the public key" and "encrypting the payload" makes it impossible for an attacker to obtain the necessary information to build a quantum attack model, thus providing a solid quantum-safe barrier for the application of SM2.

This solution is not only highly secure in theory but, more importantly, it offers a smooth, low-cost path for upgrading the quantum resistance of the vast number of currently deployed SM2-based critical information systems. This holds significant practical importance for ensuring the security of national critical information infrastructure before post-quantum cryptography technology fully matures and becomes standardized.

Future research can focus on the following points: first, a detailed performance evaluation of the protocol to analyze the specific impact on authentication latency and system overhead after introducing QKD and IPK computations; second, research on the lightweight implementation and optimization of this protocol on resource-constrained Internet of Things (IoT) terminals.

Acknowledgments. This work is supported by the Science and Technology Project of State Grid Anhui Electric Power Co., Ltd. (521207250006).

References

1. Shor, P. W.: Algorithms for quantum computation: discrete logarithms and factoring. In: Proceedings 35th Annual Symposium on Foundations of Computer Science, pp. 124–134. IEEE (1994)
2. State Cryptography Administration of China: GM/T 0003–2012 SM2 Elliptic Curve Public Key Cryptography Algorithm (2012)
3. National Institute of Standards and Technology: Selected Algorithms for the Post-Quantum Cryptography Standardization Process. NISTIR 8413 (2022)
4. Bahache, A.N., Chikouche, N., Akleylek, S.: Securing cloud-based healthcare applications with a quantum-resistant authentication and key agreement framework. Internet Things **26**, 101200 (2024)
5. Xia, T., et al.: A quantum-resistant identity authentication and key agreement scheme for UAV networks based on kyber algorithm. Drones **8**(8), 359 (2024)
6. Wang, W., Tan, S. F.: Quantum resistant authentication and key agreement protocol (AKA) for autonomous vehicle. In: 2025 5th International Conference on Neural Networks, Information and Communication Engineering (NNICE). IEEE (2025)
7. Tawfeeq, N.H., et al.: Lightweight and quantum-resistant authentication for the internet of drones (iod) using dilithium signatures. Int. J. Innov. Res. Sci. Stud. **8**(2), 2842–2853 (2025)
8. Chen, Y.: Quantum algorithms for lattice problems. arXiv preprint arXiv:2404.06255 (2024)

9. Sun, X., Han, J., Lv, B., Sun, C., Zeng, C.: A novel method for generating public keys involving matrix operations. PLoS ONE **19**(10), e0312690 (2024)
10. Bennett, C.H., Brassard, G.: Quantum cryptography: public key distribution and coin tossing. Theor. Comput. Sci. **560**, 7–11 (2014)
11. State Cryptography Administration of China: GM/T 0004–2012 SM3 Cryptographic Hash Algorithm (2012)
12. Mishra, R., Mishra, A.: Current research on internet of things (IoT) security protocols: a survey. Comput. Secur. 104310 (2025)
13. Singamaneni, K. K., Budati, A. K., Bikku, T.: An efficient q-kpabe framework to enhance cloud-based IoT security and privacy. Wirel. Pers. Commun. 1-29 (2024)
14. State Cryptography Administration of China: GM/T 0002–2012 SM4 Block Cipher Algorithm (2012)

Verifiable Multi-level Delegatable Attribute-Based Proxy Re-encryption Scheme for Cloud Computing

Duanyang Ji(✉) and Zhenhua Liu(✉)

School of Mathematics and Statistics, Xidian University, Xi'an 710071, China
1725395402@qq.com, zh_liu@mail.xidian.edu.cn

Abstract. The rapid development of cloud computing has facilitated data outsourcing and sharing, while also posing challenges to data security and access control, especially in collaborative scenarios such as healthcare. Attribute-based proxy re-encryption enables fine-grained access control for data sharing, yet existing schemes lack efficient verification for multi-level delegation and overlook the verifiability of re-encrypted ciphertexts. This paper proposes a verifiable multi-level delegatable attribute-based proxy re-encryption scheme. The proposed scheme supports complex access structures based on linear secret sharing. To achieve verifiability, a commitment mechanism is employed to ensure the correctness of re-encrypted ciphertexts. A delegation token mechanism based on BLS aggregate signatures enables efficient multi-level delegation. The proposed scheme is proved to be indistinguishable under chosen plaintext attacks based on the q-parallel BDHE assumption.

Keywords: Cloud computing · Attribute-based encryption · Proxy re-encryption · Multi-level delegation · Verifiability

1 Introduction

Cloud computing provides sufficient storage and computing power, allowing users to outsource their data to the cloud instead of managing it locally, eliminating concerns about data management [6]. Data security and privacy, particularly access control for shared data, are critical concerns due to the reliance on potentially untrusted third-party cloud providers [22,23].

Attribute-based encryption (ABE) is a cryptographic primitive that enables fine-grained access control by allowing data to be decrypted only when the attributes of a user satisfy the access policy specified by the data owner. In collaborative environments, such as electronic health record (EHR) systems, the capability of ABE to facilitate secure and flexible data sharing among multiple parties is of paramount importance. Attribute-based proxy re-encryption (ABPRE) allows ciphertext to be transformed into another ciphertext without revealing plaintext, facilitating secure data sharing. Traditional methods support

J. Yu and Y. Chen (Eds.): FCS 2025, CCIS 2805, pp. 56–76, 2027.
https://doi.org/10.1007/978-981-95-9514-3_4

single transformations, but real-world scenarios often require multiple transformations with controlled delegation rights. Moreover, it is essential to verify the correctness of transformation to ensure the validity of delegation operations.

Consider the following healthcare scenario in a cloud computing environment: A patient's data from Department C in Hospital A is shared with Department D, which subsequently intends to share the data with Department E in Hospital B. During this multi-step data sharing process, several critical questions must be addressed: whether Department D has the legitimate authority to further grant access to Department E in Hospital B, whether Department E is permitted to redelegate access rights, and whether each data transfer is executed accurately and in compliance with established access control policies. The correct handling of these issues is essential not only for safeguarding patient privacy, but also for ensuring that medical treatment remains both timely and effective.

Although some existing schemes supported multi-level delegated access control, they often lack verifiability mechanism to ensure the correctness of re-encrypted ciphertexts. For instance, Pussewalage et al.'s scheme [19] in 2023 enabled multi-level delegation for secure sharing outsourced electronic health records, but it did not verify whether each re-encryption in the delegation chain is performed correctly. This limitation raises trust concerns in practical cloud-based environments where re-encryption operations are delegated to untrusted servers. Furthermore, existing solutions typically required separate verification for each level of delegation, resulting in significant computational overhead in multi-hop settings. They also failed to support delegation rights aggregation, which is crucial for efficient enforcement of access control across multiple levels.

Inspired by Ge et al.'s scheme [10] and Pussewalage et al.'s scheme [19], this paper aims to develop a verifiable multi-level delegated access control scheme based on attribute-based proxy re-encryption. The proposed scheme will address complex access control structures, enabling flexible and efficient access control. The main contributions of this paper are summarized as below:

1. A verifiable multi-level delegatable attribute-based proxy re-encryption (VMD-ABPRE) scheme is proposed.
2. The proposed scheme is based on linear secret sharing matrices, making it suitable for complex access control structures.
3. We introduce a novel delegation token construction based on BLS aggregate signatures, which significantly improves the verification efficiency by reducing the verification cost from linear to constant time.
4. The proposed scheme is proven to be indistinguishability under chosen plaintext attacks based on the q-parallel BDHE assumption.

The rest of this paper is organized as follows. In Sect. 2, we introduce the related works. The preliminaries of the proposed scheme is presented in Sect. 3. And the system model and security model are defined in Sect. 4. In Sect. 5, we describe the proposed scheme. In Sect. 6, security proof are presented. In Sect. 7, the conclusions of this study are presented. Finally, performance evaluation and the correctness analysis of the proposed scheme are provided in the appendix.

2 Related Works

In recent years, with the widespread adoption of cloud computing, data security and access control have become critical research focuses. ABE and proxy re-encryption (PRE) have received extensive attention from academia and industry as core techniques for solving the challenges of cloud data sharing and access control.

ABE is a public key encryption technique based on identity and attributes, which allows data owners to define flexible access control policies based on a set of user attributes [20]. Goyal et al. [12] first proposed key-policy attribute-based encryption (KP-ABE) model, enabling decryption permissions by matching encrypted data attributes with access policies embedded in users' keys. Later, Bethencourt et al. [2] introduced ciphertext-policy attribute-based encryption (CP-ABE) model, reversing the assignment of policies, where the encryptor defines the access policy, and user keys correspond to the set of attributes. While ABE enables fine-grained access control, ABE lacks flexibility in implementing data sharing in complex scenarios. Traditional ABE schemes primarily support single-level access control, making them unsuitable for cross-organizational or multi-level data sharing scenarios.

PRE is a key technique for enabling data sharing, which allows a proxy node to convert data from one user's encrypted form to another's using a re-encryption key, without decrypting the ciphertext [16]. Blaze et al. [3] first proposed the concept of PRE and demonstrated its potential applications in data sharing and decentralized storage. Combining ABE with PRE addresses the challenges of complex data sharing and access control in cloud environments. In 2009, Liang et al. [15] integrated ABE with PRE to support fine-grained access control and dynamic data sharing, achieving attribute-based multi-user sharing and access delegation. Subsequently, some attribute-based proxy re-encryption schemes [10,11,13,14] have also been proposed. However, since traditional PRE schemes rely on proxy nodes to perform re-encryption operations, they cannot guarantee correctness, making verification essential [17,21,24]. Ge et al. [10] proposed a verifiable and fair attribute-based proxy re-encryption scheme for cloud data sharing in 2022, which allows the sharing users to verify the correctness of re-encrypted ciphertext returned by the server using commitment technique [5], and ensures that the cloud server is not maliciously accused when performing re-encryption honestly using message locking encryption technique [1]. However, most existing schemes [10,13,14] are tailored to single-level or fixed delegation models and do not support multi-level delegation scenarios.

Multi-level delegation is an essential requirement for data sharing in cloud environments, especially in distributed systems and cross-organizational collaborations, where access rights must be flexibly transferred across different levels [7,9,18,19]. Ding et al. [7] proposed an ABE scheme that allows users to delegate access to encrypted data through attribute delegation but did not address subsequent access control. In 2023, Pussewalage et al. [19] proposed a controlled multi-level delegation ABE scheme for user data sharing, combining proxy re-encryption. However, Pussewalage et al.'s scheme lacks flexibility and efficiency and does not consider verifiability. In 2024, Duan et al. [8] introduced a multi-authority ABE scheme supporting access delegation for cross-blockchain data sharing. In summary, although these works supported delegation in various multi-authority or cross-domain settings, they largely overlooked the issue of delegation correctness verification and impose significant verification overhead in multi-hop scenarios. To address the verification overhead, this paper introduces a delegation token mechanism based on BLS aggregate signatures [4]. Unlike existing approaches that require verifying each delegation individually, our method allows all delegation signatures to be compressed and verified via a single pairing operation, significantly improving the verification efficiency.

In summary, ABE and PRE play critical roles in cloud data sharing and access control. Based on the above analysis, this paper proposes a verifiable multi-level delegation attribute-based proxy re-encryption scheme, aiming to combine ABE and PRE to achieve a fine-grained, multi-level, and verifiable flexible access control scheme.

3 Preliminaries

3.1 Bilinear Maps

Let $\mathbb{G}$ and $\mathbb{G}_T$ be two multiplicative cyclic groups of the same prime order p, where g is a generator of $\mathbb{G}$. A bilinear map [25] e: $\mathbb{G} \times \mathbb{G} \to \mathbb{G}_T$ satisfies

1. Bilinearity: For $a_1, a_2 \in \mathbb{Z}_p$ and $g_1, g_2 \in \mathbb{G}$, $e(g_1^{a_1}, g_2^{a_2}) = e(g_1, g_2)^{a_1 \cdot a_2}$.
2. Non-degeneracy: For any $g_1, g_2 \in \mathbb{G}$, $e(g_1, g_2) \neq 1$, implying that the map does not send all elements in $\mathbb{G} \times \mathbb{G}$ to the identity in $\mathbb{G}_T$.
3. Computability: There exists an efficient algorithm to compute $e(g_1, g_2)$ for any $g_1, g_2 \in \mathbb{G}$.

3.2 Linear Secret Sharing Scheme (LSSS)

Each participant's sharing can be represented as a vector over $\mathbb{Z}_p$. A share-generating matrix $\mathbb{A}$ is used to compute the sharings [25]:

1. $\mathbb{A}$ is an $l \times n$ matrix, where l is the number of sharings, and n is the number of parameters for sharing generation.
2. Each row $i \in \{1, \ldots, l\}$ corresponds to a participant, indexed by $\rho(i)$.
3. The secret $s \in \mathbb{Z}_p$ is represented as a vector $\mathbf{v} = (s, v_2, \ldots, v_n)^T \in \mathbb{Z}_p^n$. The sharings are then generated as $\mathbb{A}\mathbf{v}$, where each entry in the resulting vector corresponds to a participant's sharing.

For an LSSS associated with an access structure $(\mathbb{A}, \rho)$, any authorized set S satisfies $(\mathbb{A}, \rho)$ must satisfy the following condition: Let $I = \{i : \rho(i) \in S\}$ denote the indices of participants in the authorized set S. There exists a set of constants $\{\omega_i\}_{i \in I}$ that can be computed in polynomial time, such that for any valid set of shares $\{\lambda_i\}_{i \in I}$, the equation $\sum_{i \in I} \omega_i \lambda_i = s$ holds true. This ensures that any authorized set can reconstruct the secret s from their sharings.

3.3 Hardness Assumptions

Definition 1. *Let $\mathbb{G}$ and $\mathbb{G}_T$ be two multiplicative cyclic groups of the same prime order p, with a bilinear map $e : \mathbb{G} \times \mathbb{G} \to \mathbb{G}_T$. $a, r \in \mathbb{Z}_p$, $g \in \mathbb{G}$ are random elements. Let $b_1, b_2, \ldots, b_q \in \mathbb{Z}_p^*$ be randomly chosen elements. Given:*

$$g, g^r, g^a, g^{a^2}, \ldots, g^{a^q}, g^{a^{q+2}}, \ldots, g^{a^{2q+1}},$$

$$\forall j \in [1, q] \quad g^{r \cdot b_j}, g^{a/b_j}, g^{a^2/b_j}, \ldots, g^{a^q/b_j}, g^{a^{q+2}/b_j}, \ldots, g^{a^{2q+1}/b_j},$$

$$\forall j, t \in [1, q],\ t \neq j, \quad g^{a \cdot r \cdot b_t/b_j}, \ldots, g^{a^q \cdot r \cdot b_t/b_j}$$

the challenge is to distinguish $e(g, g)^{\alpha^{q+1}}$ from a random element in $\mathbb{G}_T$.

The q-parallel BDHE assumption [10] states that no probabilistic polynomial-time algorithm can solve this problem with non-negligible advantage.

Definition 2. *Let $\mathbb{G}$ be a multiplicative cyclic group of prime order p, and g be a generator of $\mathbb{G}$. Given $g \in \mathbb{G}$ and $y = g^x$, where $x \in \mathbb{Z}_p$ is unknown, it is computationally infeasible to determine x in polynomial time.*

The discrete logarithm assumption states that no probabilistic polynomial-time algorithm can solve this problem with non-negligible advantage.

4 System Architecture

4.1 System Model

The system model of the proposed scheme is illustrated in Fig. 1 and its delegation model is illustrated in Fig. 2.

- Central Authority (CA). Central authority is considered fully trustworthy. Central authority can publish public parameters PP, keep master secret key MSK, and generate a secret key SK based on user attributes.
- Data Owner (DO). Data owner can share data with data user. Data owner encrypts a plaintext message M to generate a ciphertext CT, and uploads the generated ciphertext CT to cloud server.
- Cloud Server (CS). Cloud server provides a platform for storing and sharing encrypted data, verifies a signed delegation token ST, and performs the ciphertext CT re-encryption operations to obtain re-encrypted ciphertext CT'.
- Data User (DU). Data user decrypts the ciphertext CT, generates a re-encryption key RK for cloud server. As a delegator, DU creates a delegation token DT and obtain a signed delegation token ST for a delegatee. As a delegatee, DU submits ST to cloud server for verification, obtains a re-encrypted ciphertext CT', and decrypts CT'. Furthermore, the delegatee can act as a delegator to generate a signed delegation token ST for the next delegatee to perform the same operations.

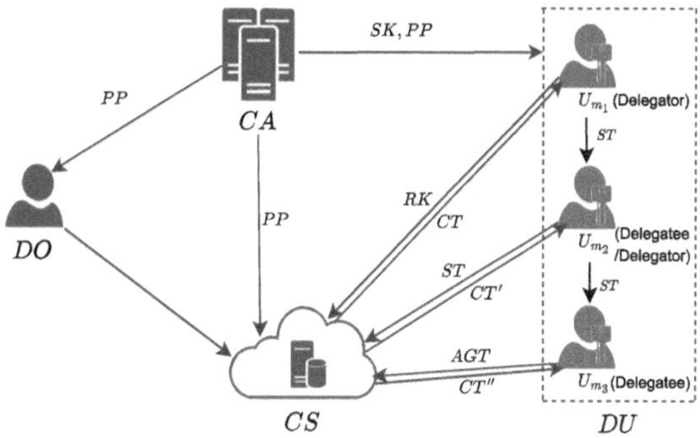

Fig. 1. The system model of VMD-ABPRE scheme

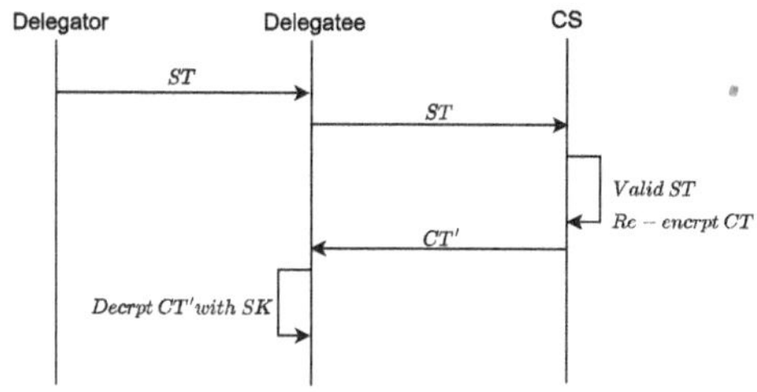

Fig. 2. The delegation model of VMD-ABPRE scheme

4.2 Definition of VMD-ABPRE Scheme

A VMD-ABPRE scheme consists of nine algorithms.

- **Setup**$(1^\lambda) \to \{MSK, PP\}$. The algorithm is executed by CA. CA inputs the security parameter 1^λ, and outputs public parameters PP and the master secret key MSK.
- **KeyGen**$(S, PP, MSK) \to SK$. The key generation algorithm is executed by CA. Assume the attribute set associated with DU is S. CA takes S of DU, PP and MSK as inputs, and outputs a secret key SK to DU.
- **Encrypt**$(PP, M, (\mathbb{A}, \rho)) \to CT$. The encryption algorithm is executed by DO. DO takes public parameters PP, plaintext message M and access control structure $(\mathbb{A}, \rho)$ as inputs, and outputs a ciphertext CT.
- **Decrypt**$(CT, PP, SK) \to M$. The decryption algorithm is executed by DU. DU uses SK to decrypt CT downloaded from CS to obtain M.
- **ReKeyGen**$(SK, PP, (\mathbb{A}', \rho')) \to RK$. The re-encryption key generation algorithm is executed by DU. DU generates a re-encryption key RK using SK, PP, and new access control structure $(\mathbb{A}', \rho')$, and sends RK to CS.
- **Signed Delegation Token Generation**. This algorithm is executed by a delegator. The delegator constructs a delegation token DT containing access information and signs DT using secret key to generate a signed delegation token ST, which is then sent to the delegatee. The delegatee then sends ST to CS for verification.

- **Verification of Signed Delegation Token.** This algorithm is executed by CS. Upon receiving a multi-level signed delegation token ST, CS aggregates the signatures and verifies the correctness of the aggregated signature. If the verification succeeds, CS uses the embedded re-encryption key to perform ciphertext re-encryption.
- **ReEnc**$(CT, PP, RK) \rightarrow CT'$. The re-encryption algorithm is executed by CS. CS uses PP, CT and RK to generate a re-encrypted ciphertext CT'.
- **ReDec**$(CT', PP, SK) \rightarrow M$. The algorithm is executed by DU. DU downloads the re-encrypted ciphertext CT' from CS and decrypts CT' using secret key SK and public parameters to obtain M.

4.3 Security Model

In this section, we consider two types of security properties: confidentiality and verifiability [10,25]. A selective security model is adopted, where the adversary must declare the access control structure to be challenged before the game initiates. When considering confidentiality, there are original ciphertext and re-encrypted ciphertext in VMD-ABPRE. Then, we define Game-Or game and Game-Re game. Each game can be described as an IND-CPA game, where an adversary $\mathcal{A}$ interacts with a simulator (challenger) $\mathcal{B}$.

Game-Or.

- **Init:** $\mathcal{A}$ selects a target access structure $(\mathbb{A}^*, \rho^*)$ and sends it to $\mathcal{B}$.
- **Setup:** $\mathcal{B}$ runs the setup algorithm to generate the public parameters PP and the master secret key MSK. The public parameters PP are sent to $\mathcal{A}$.
- **Query Phase I:** $\mathcal{A}$ queries attribute set S. If S does not satisfy access structure $(\mathbb{A}^*, \rho^*)$, $\mathcal{B}$ generates a secret key SK_S for the attribute set S using key generation algorithm and sends it to $\mathcal{A}$.
- **Challenge:** $\mathcal{A}$ submits two equal-length messages m_0 and m_1 to $\mathcal{B}$. $\mathcal{B}$ randomly selects a bit $t \in \{0, 1\}$ and encrypts m_t using access structure $(\mathbb{A}^*, \rho^*)$. The resulting ciphertext CT^* is sent to $\mathcal{A}$.
- **Query Phase II:** $\mathcal{A}$ queries like in Query phase I.
- **Guess:** $\mathcal{A}$ outputs a guess $t' \in \{0, 1\}$. $\mathcal{A}$ wins Game-Or game if $t' = t$. The advantage of $\mathcal{A}$ in Game-Or game is defined as:

$$\mathrm{Adv}_{\mathcal{A}-Or}^{IND-CPA} = \left| \Pr[t' = t] - \frac{1}{2} \right|.$$

Game-Re.

- **Init:** $\mathcal{A}$ selects a target access structure $(\mathbb{A}^*, \rho^*)$, and sends it to $\mathcal{B}$.
- **Setup:** $\mathcal{B}$ runs the setup algorithm to generate public parameters PP and master secret key MSK. And the public parameters PP are sent to $\mathcal{A}$.
- **Query Phase I:** $\mathcal{A}$ queries attribute set S that does not satisfy access structure $(\mathbb{A}^*, \rho^*)$. $\mathcal{B}$ responds as follows:
  1. $O_{SK}(S)$: $\mathcal{A}$ queries on attribute set S. $\mathcal{B}$ executes $SK = KeyGen(MSK, S, PP)$, and sends SK to $\mathcal{A}$.

2. $O_{RK}(S, (\mathbb{A}', \rho'))$: $\mathcal{A}$ makes a re-encryption key query on $(S, (\mathbb{A}', \rho'))$, where S does not satisfy access structure $(\mathbb{A}', \rho')$. $\mathcal{B}$ executes $SK = KeyGen(MSK, S, PP)$, $RK = ReKeyGen(SK, PP, (\mathbb{A}', \rho'))$, and sends RK to $\mathcal{A}$.

- **Challenge:** $\mathcal{A}$ submits two equal-length messages m_0 and m_1 to $\mathcal{B}$. $\mathcal{B}$ randomly selects a bit $t \in \{0, 1\}$ and performs the following:
  1. $\mathcal{B}$ encrypts m_t under the access structure $(\mathbb{A}^*, \rho^*)$ to produce the original ciphertext CT^*.
  2. $\mathcal{B}$ re-encrypts CT^* using re-encryption key to produce a re-encrypted ciphertext $CT^{*'}$.

  And the re-encrypted ciphertext $CT^{*'}$ is sent to $\mathcal{A}$.
- **Query Phase II:** $\mathcal{A}$ queries like in Query phase I.
- **Guess:** $\mathcal{A}$ outputs a guess $t' \in \{0, 1\}$. $\mathcal{A}$ wins Game-Re game if $t' = t$.

The advantage of $\mathcal{A}$ in Game-Re game is defined as:

$$\text{Adv}_{\mathcal{A}-Re}^{IND-CPA} = \left| \Pr[t' = t] - \frac{1}{2} \right|.$$

Definition 3. *A VMD-ABPRE scheme is IND-CPA secure if the advantage of an adversary $\mathcal{A}$ in Game-Or game is negligible and the advantage of an adversary $\mathcal{A}$ in Game-Re game is negligible.*

Verifiability. The verifiability of a VMD-ABPRE scheme is captured by the following game.

- **Setup.** In this step, a challenger $\mathcal{B}$ runs the Setup algorithm and gets public parameter and master secret key. Then, $\mathcal{B}$ passes public parameters to $\mathcal{A}$.
- **Query phase I.** $\mathcal{A}$ queries:
  1. $O_{SK}(S)$: $\mathcal{A}$ queries on attribute set S. The challenger $\mathcal{B}$ executes $SK = KeyGen\ (MSK, S, PP)$, and sends SK to $\mathcal{A}$.
  2. $O_{RK}(S, (\mathbb{A}', \rho'))$: $\mathcal{A}$ makes a re-encryption key query on $(S, (\mathbb{A}', \rho'))$, where S does not satisfy $(\mathbb{A}', \rho')$. $\mathcal{B}$ executes $SK = KeyGen(MSK, S, PP)$ and $RK = ReKeyGen\ (SK, (\mathbb{A}', \rho'))$, sends RK to $\mathcal{A}$.
- **Challenge.** $\mathcal{A}$ chooses a message m^* and access policy $(\mathbb{A}^*, \rho^*)$ and sends them to $\mathcal{B}$. $\mathcal{B}$ computes $CT^* = Enc(m^*, (\mathbb{A}^*, \rho^*))$ and sends CT^* to $\mathcal{A}$.
- **Query phase II.** $\mathcal{A}$ queries like in Query phase I.
- **Guess.** $\mathcal{A}$ outputs an attribute set S^*, where S^* satisfies $(\mathbb{A}^*, \rho^*)$, and a re-encrypted ciphertext CT'^*. The adversary $\mathcal{A}$ wins if the decryption of the re-encrypted ciphertext fails to recover the plaintext m^*.

The adversary $\mathcal{A}$'s advantage to win the verifiability security game is defined as $\text{Adv}_{\mathcal{A}}^{Ver}(\lambda) = |\Pr[\mathcal{A}\, wins]|$.

Definition 4. *A VMD-ABPRE scheme is verifiability if for any probabilistic polynomial-time adversary $\mathcal{A}$, its advantage $Adv_{\mathcal{A}}^{Ver}$ is negligible.*

5 Construction of Our VMD-ABPRE

In this section, we describe our VMD-ABPRE scheme in detail. The proposed scheme is divided into the following nine algorithms.

- **Setup** $(1^\lambda) \rightarrow \{MSK, PP\}$. To initialize the system, a set of global public parameters is first generated. The system selects two multiplicative cyclic groups: a group $\mathbb{G}$ with a generator g and a group $\mathbb{G}_T$ with the same prime order p. A bilinear pairing $e : \mathbb{G} \times \mathbb{G} \rightarrow \mathbb{G}_T$ is defined. $H_1 : \mathbb{G}_T \rightarrow \{0,1\}^{2k}$ (k denotes the bit-length of the message M), $H_2 : \{0,1\}^* \rightarrow \mathbb{Z}_p^*$, $H_3 : \mathbb{G}_T \rightarrow \mathbb{Z}_p^*$ and $H_4 : \{0,1\}^* \rightarrow \mathbb{G}$ are four hash functions, where H_4 is used in the BLS signature scheme for signing delegation tokens. The attribute set managed by CA is denoted as S, and then CA is initialized as described below.
  1. CA randomly chooses $\alpha, \beta, a \in \mathbb{Z}_p^*, f_1, \cdots, f_U, X, Y \in \mathbb{G}$, where U denotes the size of the attribute universe. CA computes g^β and $e(g,g)^\alpha$.
  2. CA will keep $\{\alpha, \beta\}$ as the master secret denoted by MSK and publish $g^a, g^\beta, g^{a\cdot\beta}, f_1^\beta, \cdots, f_U^\beta, e(g,g)^\alpha$. Then public parameters are published as

$$PP = \left\{ \begin{array}{c} e, \mathbb{G}, \mathbb{G}_T, g, p, H_1, H_2, H_3, H_4, g^a, f_1, \cdots, \\ f_U, X, Y, f_1^\beta, \cdots, f_U^\beta, g^\beta, g^{a\cdot\beta}, e(g,g)^\alpha \end{array} \right\}.$$

  For each participant involved in signing delegation tokens, a signing key pair is generated $sk_i \in \mathbb{Z}_p^*$ and $pk_i = g^{sk_i} \in \mathbb{G}$. These keys are used exclusively for BLS signature generation and verification over the delegation tokens.
- **KeyGen**$(S, PP, MSK) \rightarrow SK$. Assume the attribute set related to CA is denoted as S. CA generate the corresponding key for DU. The key generation process is as follows:
  1. U_m is associated with ID_m, which is mapped to a unique identifier $r_m = H_2(ID_m)$ using hash function H_2.
  2. CA generates the secret key $SK = \{S, sk_1, sk_2, \{sk_{3,x}\}_{x=1,2,\dots,|S|}\}$, such that $sk_1 = g^{\alpha/\beta} g^{a\cdot r_m/\beta}, sk_2 = g^{r_m/\beta}, sk_{3,x} = f_x^{r_m/\beta}, \forall x \in S$.
  3. Finally, the secret keys SK generated by AA are sent to DU.
- **Encrypt** $(PP, M, (\mathbb{A}, \rho)) \rightarrow CT$. Taking public parameters PP, message $M \in \{0,1\}^k$ and LSSS access structure $(\mathbb{A}, \rho)$ as input, where $\mathbb{A}$ is an $l \times n$ matrix and ρ is a function that maps each row of $\mathbb{A}$ to attributes.
  1. DO randomly selects $d \in \{0,1\}^k$ as auxiliary input for verifiability, and computes $r = H_2(M||d)$. Then DO generates a random vector $\boldsymbol{v} = (r, v_2, \dots, v_n)^T$, where $v_2, \dots, v_n \in \mathbb{Z}_p$ are chosen randomly. Next, DO calculates the sharings of r as $(\lambda_1, \dots, \lambda_l)^T = \mathbb{A} \cdot \boldsymbol{v}$. For $i \in [1, l]$, DO randomly selects $r_i \in \mathbb{Z}_p$.
  2. DO computes C, C_1, C' as

$$C = (M||d) \oplus H_1(e(g,g)^{\alpha\cdot r}),$$
$$C' = X^{H_2(M)} \cdot Y^{H_2(d)},$$
$$C_1 = (g^\beta)^r.$$

In the encryption phase, DO computes the commitment $C' = X^{H_2(M)} \cdot Y^{H_2(d)}$, implementing a cryptographic commitment to the plaintext M. During decryption, the integrity of C' is verified to ensure the data has not been tampered with during re-encryption.
3. For each $i \in [1, l]$, DO computes $C_{2,i} = g^{a \cdot \lambda_i} \cdot f_{\rho(i)}^{-r_i}$, $C_{3,i} = g^{r_i}$.
4. Finally, DO outputs the following ciphertext

$$CT = \left\{(\mathbb{A}, \rho), C, C', C_1, \left\{C_{2,i}, C_{3,i}\right\}_{i \in [1,l]}\right\}.$$

- **Decrypt**$(CT, PP, SK) \rightarrow M$. Taking SK, CT and PP as input. DU executes the following operations.
  1. If S matches the access structure $(\mathbb{A}, \rho)$, DU defines $I = \{i : \rho(i)\}$, calculates the constants $\{w_i\}_{i \in I}$ to $\sum_{i \in I} w_i \cdot \lambda_i = r$, and computes

$$\begin{aligned} F &= \frac{e(sk_1, C_1) \cdot e(g, g^\beta)^{-1}}{\prod_{i \in I} \left(e(sk_2, C_{2,i}) \cdot e(sk_{3,x}, C_{3,i})\right)^{w_i}} \\ &= \frac{e(g^{\alpha/\beta} \cdot g^{a \cdot r_m/\beta}, (g^\beta)^r) \cdot e(g, g^\beta)^{-1}}{\prod_{i \in I} \left(e(g^{r_m/\beta}, g^{a \cdot \lambda_i} \cdot f_{\rho(j)}^{-r_i}) \cdot e(f_x^{r_m/\beta}, g^{r_i})\right)^{w_i}} \\ &= e(g, g)^{\alpha \cdot r}. \end{aligned}$$

  2. Then, DU calculates $M||d = C \oplus H_1(F)$.
  3. Finally, DU outputs M if $X^{H_2(M)} \cdot Y^{H_2(d)} = C'$. Otherwise, output $\perp$.
- **ReKeyGen**$(SK, PP, (\mathbb{A}', \rho')) \rightarrow RK$. Taking public parameters PP, SK and a new LSSS access structure $(\mathbb{A}', \rho')$ as input, where $\mathbb{A}'$ is an $l' \times n'$ matrix the re-encryption key generation process is as follows:
  1. DO randomly selects $R \in \mathbb{G}_T$ and computes

$$\begin{aligned} rk_1 &= (sk_1)^{H_3(R)} = (g^{\alpha/\beta} g^{a \cdot r_m/\beta})^{H_3(R)}, \\ rk_2 &= (sk_2)^{H_3(R)} = (g^{r_m/\beta})^{H_3(R)}, \\ rk_{3,x} &= (sk_{3,x})^{H_3(R)} = (f_x^{r_m/\beta})^{H_3(R)}, \forall x \in S. \end{aligned}$$

  2. Then DO randomly selects r' and generates a random vector $\boldsymbol{v'} = (r', v_2', \ldots, v_n')^T$, where $v_2', \ldots, v_n' \in \mathbb{Z}_p$ are chosen randomly. Next, DO calculates the sharings of r' as $(\lambda_1', \ldots, \lambda_{l'}')^T = \mathbb{A}' \cdot \boldsymbol{v'}$. For $j \in [1, l']$, DO randomly selects $r_j' \in \mathbb{Z}_p$.
  3. Next, DO computes

$$\begin{aligned} rk_4 &= R \cdot e(g, g)^{\alpha \cdot r'}, \\ rk_5 &= (g^\beta)^{r'}, \\ rk_{6,j} &= g^{a \cdot \lambda_j'} \cdot f_{\rho(j)}^{-r_j'}, \ rk_{7,j} = g^{r_j'}, j \in [1, l']. \end{aligned}$$

  4. Finally, DO outputs re-encryption key as

$$RK = \left\{rk_1, rk_2, \left\{rk_{3,x}\right\}_{\forall x \in S'}, rk_4, rk_5, \left\{rk_{6,j}, rk_{7,j}\right\}_{j \in [1,l']}, (\mathbb{A}', \rho')\right\}.$$

- **Signed Delegation Token Generation**. Let DT_i denote the delegation token issued at level i, which encapsulates: the delegator and delegatee access structures $(\mathbb{A}, \rho)$ and $(\mathbb{A}', \rho')$, the re-encryption key RK, the delegation permission flag $DP \in \{0, 1\}$, and the token expiration timestamp EXP. The delegator randomly generates a signing key pair $(sk_i, pk_i = g^{sk_i})$, and signs the token as:
$$\sigma_i = H_4(DT_i)^{sk_i} \in \mathbb{G}.$$
The signed delegation token is then represented as $ST_i = (DT_i, \sigma_i,)$ and sent to the delegatee.
For multi-level delegation $(DT_1, \sigma_1), \ldots, (DT_n, \sigma_n)$, the delegatee computes the aggregate signature $\sigma_{agg} = \prod_{i=1}^{n} \sigma_i$, and then generate an aggregated delegation token $AGT = \{DT_1, \ldots, DT_n, \sigma_{agg}\}$ to CS.
- **Verification of Aggregated Delegation Token**. Upon receiving the AGT, CS first checks the validity of expiration time in each DT_i, that $DP_i = 1$ for all $i < n$ (i.e., further delegation is permitted), consistency of the delegation chain: the delegatee in DT_i must match the delegator in DT_{i+1}. Then, CS verifies the validity of the aggregate signature $\sigma_{agg} = \prod_{i=1}^{n} \sigma_i$. Verification is done by checking:
$$e(\sigma_{agg}, g) \stackrel{?}{=} \prod_{i=1}^{n} e(H_4(DT_i), pk_i).$$
If the equation holds, the aggregated token is accepted as valid.
- **ReEnc** $(CT, PP, RK) \rightarrow CT'$. Taking PP, CT and RK as input. If S matches the access structure $(\mathbb{A}, \rho)$, CS defines $J = \{j : \rho(j)\}$, calculates the constants $\{w_j\}_{j \in J}$ to $\sum_{j \in J} w_j \cdot \lambda_j = r$, and computes
$$C_1' = \frac{e(rk_1, C_1) \cdot e(g, g^{\beta})^{-1}}{\prod_{j \in J} \left(e(rk_2, C_{2,j}) \cdot e(rk_{3,x}, C_{3,j})\right)^{w_j}} = e(g, g)^{\alpha \cdot r \cdot H_3(R)}.$$
Then, $C'' = C', \overline{C} = C, C_1', C_2' = rk_4, C_3' = rk_5, C_{4,j}' = rk_{6,j}, C_{5,j}' = rk_{7,j}, \forall j \in J$ are set. Finally, CT' is output
$$CT' = \left\{C'', \overline{C}, C_1', C_2', C_3', \{C_{4,j}', C_{5,j}'\}_{j \in [1,l']}, (\mathbb{A}', \rho')\right\}.$$
- **ReDec** $(CT', SK, PP) \rightarrow M$. Taking CT', SK and PP as input. Delegatee executes the following operations.
  1. If S' matches the access structure $(\mathbb{A}', \rho')$, DU defines $J = \{J : \rho(j)\}$, calculates constants $\{w_j'\}_{j \in J}$ to make $\sum_{j \in J} w_j' \cdot \lambda_j' = r'$, and calculates
$$Z = \frac{e(sk_1, C_3') \cdot e(g, g^{\beta})^{-1}}{\prod_{j \in J} \left(e(sk_2, C_{4,j}') \cdot e(sk_{3,x}, C_{5,j}')\right)^{w_j'}} = e(g, g)^{\alpha \cdot r'}.$$
  2. Then, DU calculates $R = \frac{C_2'}{Z}$ and $M||d = \overline{C} \oplus H_1({C_1'}^{1/H_3(R)})$.
  3. Finally, DU outputs M if $X^{H_2(M)} \cdot Y^{H_2(d)} = C''$. Otherwise, output $\perp$.

6 Security Proof

This section presents the security and verifiability proof of the proposed scheme.

In the proposed scheme, the cloud server is considered curious. It is assumed that an unauthorized adversary$\mathcal{A}$, may collude with the cloud server to attempt to obtain the original data. For both the original ciphertext and the re-encrypted ciphertext as targets, two chosen access structure CPA security games under standard model are considered. We give Theorem 1 and Theorem 2 ,and prove it as follows.

Theorem 1. *The proposed scheme is IND-CPA secure with respect to original ciphertext, assuming q-parallel BDHE assumption holds.*

Proof. Assuming that adversary $\mathcal{A}$ breaks the proposed scheme with a non-negligible advantage, the simulator $\mathcal{B}$ can be constructed using a non-negligible advantage ϵ of $\mathcal{A}$, enabling $\mathcal{B}$ to solve q-parallel-BDHE assumption with a non-negligible advantage.

- **Init:** $\mathcal{A}$ selects $(\mathbb{A}^*, \rho^*)$ to challenge and sends $(\mathbb{A}^*, \rho^*)$ to $\mathcal{B}$.
- **Setup:** $\mathcal{B}$ selects a random number $\alpha', \beta' \in \mathbb{Z}_p^*$, then sets $g^{\beta} = g^{\beta'}$ and $e(g,g)^{\alpha} = e(g,g)^{\alpha'} \cdot e(g^a, g^{a^q})$, which implies $\alpha = a^{q+1} + \alpha'$ and $\beta = \beta'$. By implicitly setting $\alpha = a^{q+1} + \alpha'$, the unknown term a^{q+1} of q-parallel-BDHE problem is embedded into the master secret key. $\mathcal{B}$ randomly selects $t_x \in \mathbb{Z}_p^*$. Assuming I is the set of i and $\rho(i) = x$, $\mathcal{B}$ computes $f_x = g^{t_x} \prod_{i \in I} g^{a\mathbb{A}_{i,1}^*/b_i} \cdot g^{a^2\mathbb{A}_{i,2}^*/b_i} \cdots g^{a^{n^*\mathbb{A}_{i,n^*}^*/b_i}}$.
  If I is an empty set, $f_x = g^{t_x}$. $\mathcal{B}$ selects two multiplicative cyclic groups: a group $\mathbb{G}$ with a generator g and a group $\mathbb{G}_T$ with the same prime order p. $e: \mathbb{G} \times \mathbb{G} \to \mathbb{G}_T$ is a bilinear mapping. $H_1 : \mathbb{G}_T \to \{0,1\}^{2k}$ $H_2 : \{0,1\}^* \to \mathbb{Z}_p^*$, $H_3 : \mathbb{G}_T \to \mathbb{Z}_p^*$ and $H_4 : \{0,1\}^* \to \mathbb{G}$ are four hash functions. Randomly choose $X, Y \in \mathbb{G}$. Then, public parameters are published as

$$PP = \{e, \mathbb{G}, \mathbb{G}_T, g, p, H_1, H_2, H_3, H_4, g^a, f_1, \cdots, f_U,$$
$$X, Y, f_1^{\beta}, \cdots, f_U^{\beta}, g^{\beta}, g^{a \cdot \beta}, e(g,g)^{\alpha}\}.$$

  Then public parameters PP is provided to $\mathcal{A}$.
- **Query phase I:** $\mathcal{A}$ queries $\mathcal{B}$ for secret key corresponding to a attribute set S, where S isn't satisfy access control structure $(\mathbb{A}^*, \rho^*)$. Then, $\mathcal{B}$ chooses $\boldsymbol{v} = (v_1, v_2, \ldots, v_{n^*})^T$, where $v_1 = -1$ and $\boldsymbol{v} \cdot A_i^* = 0$. Since S does not satisfy $(\mathbb{A}^*, \rho^*)$, there exists a vector $\boldsymbol{v} = (v_1, v_2, \ldots, v_{n^*})^\top$ such that $v_1 = -1$ and $\boldsymbol{v} \cdot \mathbb{A}_1^* = 0$ based on LSSS. For all i, where $\rho(i) \in S$, $\mathcal{B}$ randomly selects $r' \in \mathbb{Z}_p$, and computes $sk_2 = g^{r'/\beta'} \cdot \prod_{j=1}^{n^*} (g^{a^{q+1-i}})^{\theta_i/\beta'} = g^{r_m/\beta}$. This implicitly sets $r_m = r' + \sum_{i=1}^{n^*} \theta_i \cdot a^{q+1-i}$. Then, we have

$$sk_1 = g^{\alpha/\beta} g^{a \cdot r_m/\beta} = g^{a^{q+1} + \alpha'/\beta'} g^{a \cdot r' + \sum_{i=1}^{n^*} \theta_i \cdot a^{q+1-i}/\beta'},$$

$$sk_{3,x} = f_x^{r_m/\beta} = g^{\frac{r_m}{\beta} \cdot t_x} \prod_{i=I} \prod_{j=1}^{n^*} (g^{\frac{a^j \cdot r}{b_i \cdot \beta}} \prod_{\substack{k=1 \\ k \neq j}}^{n^*} (g^{a^{q+1-j-k/b_i}})^{\frac{v_k}{\beta}})^{A_{i,j}^*}, \forall x \in S.$$

$\mathcal{B}$ returns SK to $\mathcal{A}$.

- **Challenge:** $\mathcal{A}$ selects two messages M_0 and M_1 of equal length and sends to $\mathcal{B}$. $\mathcal{B}$ randomly selects M_t, where $t \in \{0, 1\}$ and then encrypt message M_t using access control structure $(\mathbb{A}^*, \rho^*)$.
  $\mathcal{B}$ then returns CT^* which encrypts M_t to $\mathcal{A}$. Firstly, $\mathcal{B}$ computes $C = M_t \cdot T \cdot e(g, g)^{\alpha \cdot r}$, $C_1 = (g^{\beta})^r$. Then, $\mathcal{B}$ randomly selects $y'_2, \ldots, y'_{n^*} \in \mathbb{Z}_p$ and $t'_1, \ldots, t'_{l^*} \in \mathbb{Z}_p$, where $-t_j = t'_j + R \cdot b_j$. For $j \in [1, l^*]$, define F_j, where $k \neq j$ and $\rho^*(k) = \rho^*(j)$. $\mathcal{B}$ implicitly sets $\boldsymbol{w} = (r, ra + y'_2, , \ldots, ra^{n^*-1} + y'_{n^*})$ and $\lambda_j = \boldsymbol{w} \cdot A^*_j, j \in [1, l^*]$. $\mathcal{B}$ computes

$$
\begin{aligned}
C_{2,j} &= g^{a \cdot \lambda_j} \cdot f_{\rho(j)}^{-r_j} = g^{a \cdot \{\boldsymbol{w} \cdot A^*_j\}} \cdot f_{\rho(j)}^{-r_j} \\
&= g^{a \cdot \{\boldsymbol{w} \cdot A^*_j\}} \cdot g^{t_j} \prod_{k \in F_j} g^{a \mathbb{A}^*_{k,1}/b_k} \cdot g^{a^2 \mathbb{A}^*_{k,2}/b_k} \ldots g^{a^{n^*} \mathbb{A}^*_{k,n^*}/b_k}{}^{-r_j} \\
&= f_{p^*(j)}^{t_j} \cdot (g^{r_j \cdot b_j})^{t_j} \cdot (\prod_{k \in F_j} \prod_{i=1}^{n^*} (g^{a_j \cdot r_j \cdot b_j / b_k})^{A^*_{k,i}}) (\prod_{i=2}^{n^*} (g^{a \cdot y'_j \cdot A^*_{j,i}}), \\
C_{3,j} &= g^{r_j}.
\end{aligned}
$$

  Finally, $\mathcal{B}$ outputs ciphertext CT^* to $\mathcal{A}$.
- **Query phase II:** $\mathcal{A}$ performs a query-response process similar to that of Query phase I.
- **Guess:** $\mathcal{A}$ provides guess v' for v. If $v' = v$, $T = e(g, g)^{r \cdot a^{q+1}}$, $\mathcal{B}$ outputs 1. Otherwise, T is randomly selected from $\mathbb{G}_T$, $\mathcal{B}$ outputs 0.
- **Analysis.** If $T = e(g, g)^{r \cdot a^{q+1}}$, the correct form of CT^*, then $\Pr[v' = v] = \frac{1}{2} + \epsilon$. Otherwise, T is randomly selected from $\mathbb{G}_T$, then $\Pr[B(\boldsymbol{y}, T) = 0] = 0$. Thus, $|\Pr[B(\boldsymbol{y}, T) = 1] - \Pr[B(\boldsymbol{y}, T) = 0| = \epsilon$, which implies that $\mathcal{B}$ can solve the q-parallel BDHE assumption with non-negligible advantage ϵ.

Theorem 2. *The proposed scheme is IND-CPA secure with respect to re-encryption ciphertext, assuming q-parallel BDHE assumption holds,*

Proof. Assuming that $\mathcal{A}$ breaks the proposed scheme with a non-negligible advantage, $\mathcal{B}$ can be constructed using a non-negligible advantage ϵ of $\mathcal{A}$, enabling $\mathcal{B}$ to solve the q-parallel-BDHE assumption with a non-negligible advantage. Init and Setup phases are similar to proof of Theorem 1.

- **Query phase I:** $\mathcal{A}$ queries $\mathcal{B}$, and $\mathcal{B}$ responds. Apart from secret key queries, which are the same as in Theorem 1, re-encryption key queries and re-encryption queries also need to be performed.
  - Re-encryption key queries: $\mathcal{A}$ queries $\mathcal{B}$ for re-encryption key RK. If S satisfies access control structure $(\mathbb{A}^*, \rho^*)$ or $(\mathbb{A}', \rho')$, the process terminates. Termination occurs because allowing such queries would trivially break the security notion. Otherwise, $\mathcal{B}$ obtains secret key SK through a secret key extraction query and then calculates RK using re-encryption key generation algorithm ReKeyGen.

- Re-encryption queries: $\mathcal{A}$ queries $\mathcal{B}$ for re-encryption. If S satisfies access control structure $(\mathbb{A}^*, \rho^*)$ and S satisfies access control structure $(\mathbb{A}', \rho')$, the process terminates. Otherwise, $\mathcal{B}$ performs a re-encryption key extraction query to obtain RK, and then re-encrypts CT using re-encryption algorithm ReEnc.

- **Challenge:** $\mathcal{A}$ selects two messages M_0 and M_1 of equal length and sends to $\mathcal{B}$. $\mathcal{B}$ randomly selects M_t, where $t \in \{0, 1\}$ and then encrypt M_t. $\mathcal{B}$ selects attribute set S that does not satisfy $(\mathbb{A}^*, \rho^*)$ to generate SK. $\mathcal{B}$ uses ReKeyGen to generate RK. $\mathcal{B}$ generates CT, similar to challenge phase in Theorem 1, using $(\mathbb{A}, \rho)$. Finally, $\mathcal{B}$ re-encrypts CT using RK to obtain CT'^*, and returns CT'^* to $\mathcal{A}$.

Finally, Query phase II, Guess, and Analysis phases are similar to those in Theorem 1. Thus, the data confidentiality can be proved.

Theorem 3. *The proposed scheme is verifiable under the Discrete Logarithm Assumption*

Proof. Assume that there exists an adversary $\mathcal{A}$ that breaks the verifiability of the proposed scheme with non-negligible advantage ϵ. We can construct a simulator $\mathcal{B}$ such that $\mathcal{B}$ solves discrete logarithm assumption with non-negligible advantage, where $\mathcal{B}$ is given $(e, \mathbb{G}, \mathbb{G}_T, p, g, g^\omega)$ and its goal is to output ω.

- **Init:** $\mathcal{A}$ selects $(\mathbb{A}^*, \rho^*)$ to challenge and sends $(\mathbb{A}^*, \rho^*)$ to $\mathcal{B}$.
- **Setup:** $\mathcal{B}$ sets a bilinear mapping $e : \mathbb{G} \times \mathbb{G} \to \mathbb{G}_T$, three hash functions H_1, H_2, H_3, a group $\mathbb{G}$ with a generator g and a group $\mathbb{G}_T$ with the same prime order p. $\mathcal{B}$ randomly chooses $a, \alpha, \beta, \delta \in \mathbb{Z}_p^*$, $f_1, \cdots, f_U, X, Y \in \mathbb{G}$. $\mathcal{B}$ sets $X = g^\omega$ and $Y = g^\delta$, where g^ω is from the discrete logarithm challenge. Then, $\mathcal{B}$ outputs public parameters

$$PP = \{e, \mathbb{G}, \mathbb{G}_T, g, p, H_1, H_2, H_3, H_4, g^a, f_1, \cdots, \\ f_U, X, Y, f_1^\beta, \cdots, f_U^\beta, g^\beta, g^{a \cdot \beta}, e(g, g)^\alpha\}$$

  to $\mathcal{A}$.
- **Query phase I:** When $\mathcal{A}$ performs a secret key query, since $\mathcal{B}$ knows master secret key, $\mathcal{B}$ can generate secret key SK using key generation algorithm. Additionally, $\mathcal{B}$ can generate the re-encryption key RK using re-encryption key generation algorithm. Then, $\mathcal{B}$ provides both SK and RK to $\mathcal{A}$.
- **Challenge:** $\mathcal{A}$ sends M^* to $\mathcal{B}$. $\mathcal{B}$ uses encryption algorithm to compute challenge ciphertext $CT^* = Enc(M^*, (\mathbb{A}^*, \rho^*))$ under access control structure $(\mathbb{A}^*, \rho^*)$, and then sends CT^* to $\mathcal{A}$.
- **Query phase II:** $\mathcal{A}$ performs a query-response process similar to that of Query phase I.
- **Output:** $\mathcal{A}$ outputs challenge re-encrypted ciphertext $CT^{*'}$. $\mathcal{B}$ selects an attribute set $S^{*'}$ that satisfies access control structure $(\mathbb{A}^{*'}, \rho^{*'})$ and then generates the corresponding secret key $SK^{*'}$. $\mathcal{B}$ decrypts $CT^{*'}$ to obtain

$M^{*'}$ and $d^{*'}$. If $\mathcal{A}$ wins this verifiable game, then $M^{*'} \neq \{M^*, \perp\}$. Therefore, $C^{*'} = C^*$. Thus, $\mathcal{B}$ can ensure

$$\begin{aligned} C^{*'} &= C^* \\ X^{H_2(M^{*'})} \cdot Y^{H_2(d^{*'})} &= X^{H_2(M^*)} \cdot Y^{H_2(d^*)} \\ g^{\omega \cdot H_2(M^{*'}) + \delta \cdot H_2(d^{*'})} &= g^{\omega \cdot H_2(M^*) + \delta \cdot H_2(d^*)} \\ \omega \cdot (H_2(M^{*'}) - H_2(M^*)) &= \delta \cdot (H_2(d^*) - H_2(d^{*'})). \end{aligned}$$

Since $M^{*'} \neq M^*$ and H_2 is a collision-resistant hash function, the probability that $H_2(M^{*'}) - H_2(M^*) = 0$ is negligible. Then, $\mathcal{B}$ can output $\omega = \dfrac{\delta \cdot (H_2(d^*) - H_2(d^{*'}))}{H_2(M^{*'}) - H_2(M^*)}$ as the answer.

– **Analysis.** If reduction is perfect, then the advantage of $\mathcal{B}$ solving discrete logarithm assumption is same as the advantage of $\mathcal{A}$ winning the verifiable game described above attacking the scheme.

7 Conclusion

This paper has proposed a verifiable multi-level delegatable attribute-based proxy re-encryption scheme for secure and flexible access control in cloud-based healthcare systems. A commitment mechanism ensures the correctness of re-encrypted ciphertexts, and a BLS-based aggregated delegation token enables single-pairing verification. The proposed scheme is proved to be IND-CPA security based on the q-parallel BDHE assumption. The performance evaluation reveals that the proposed scheme offers extensive functionality, though additional features may need to be incorporated. Future work will consider efficient attribute and user revocation, and traceable delegation to enhance accountability against malicious re-encryption.

Acknowledgments. This work is supported by the National Cryptologic Science Fund of China(2025NCSF02032), Natural Science Foundation of Shaanxi Province (2022JZ-38), National Natural Science Foundation of China (61807026), Secure Communication Key Laboratory Stabilization Program Support Project (2023).

Disclosure of Interests. The authors have no competing interests to declare that are relevant to the content of this article.

A Performance Analysis

This section evaluates the performance of the proposed scheme. And we define several parameters, as shown in Table 1.

Table 1. Parameter definition

Parameters	Definition
N	The number of attributes
l, l'	The number of attribute authorities that manage delegator's attributes and delegate's attributes
$\|p\|$	The size of element in $\mathbb{G}$ and $\mathbb{G}_T$
$\|A_1\|$	The number of attributes in delegator's substructure
$\|A_2\|$	The number of attributes in delegate's substructure
T_e	The computational overhead of exponentiation operation
T_p	The computational overhead of pairing operation on group $\mathbb{G}$

A.1 Functional Comparisons

This section compares multiple schemes in Table 2, highlighting our scheme's functional advantages in verifiability, flexible access structure, delegated control and multiple delegation. In contrast, Ge et al.'s scheme [10] supports verifiability and LSSS but lacks delegated control and multiple delegation. Pussewalage et al.'s scheme [19] provides delegated control and multiple delegation but lacks verifiability and employs a less flexible Boolean statement for access control. Liang et al.'s scheme [15] is the most limited, supporting only a basic "And" gate access structure without verifiability, delegated control, or multiple delegation. Through comparative analysis, the proposed scheme demonstrates significantly enhanced versatility and comprehensiveness over existing approaches.

Table 2. Functional comparisons of schemes

Schemes	Verifiable	Access structure	Delegated control	Multiple delegation
Ours	✓	LSSS	✓	✓
[10]	✓	LSSS	×	×
[19]	×	Boolean statement	✓	✓
[15]	×	"And" gate	×	×

A.2 Theoretical Analysis

Table 3 provides a systematic comparison of computational overhead among schemes for cryptographic operations. Under the single-attribute-authority configuration, the proposed scheme achieves superior efficiency compared to Liang et al.'s scheme [15] and Pussewalage et al.'s scheme [19] by streamlining exponential operations in KeyGen and Encrypt phases and optimizing operation allocation for ReEnc and Decrypt phases. Although the ReKeyGen phase incurs higher costs, its support for dynamic attribute expansion and stronger security design compensates for this limitation.

Table 3. Computation overhead

Schemes	[15]	[19]	Ours
KeyGen	$(2\|A_1\|+1)T_e$	$(\|A_1\|+1)T_e$	$(\|A_1\|+3)T_e$
Encrypt	$(\|A_1\|+3)T_e+T_p$	$(2\|A_1\|+2)T_e+T_p$	$(\|A_1\|+2)T_e+T_p$
ReKeyGen	$(2\|A_1\|+\|A_2\|+4)T_e$	$(\|A_1\|+3)T_e$	$(\|A_1\|+3\|A_2\|+4)T_e+T_p$
ReEnc	$(2\|A_1\|+\|A_2\|+1)T_p$	$(\|A_1\|+1)T_e+(\|A_1\|+1)T_p$	$(2\|A_2\|+2)T_p$
Decrypt	$(\|A_1\|+1)T_p$	$(\|A_1\|+1)T_p$	$(2\|A_1\|+2)T_p$
ReDec	$T_e+(\|A_2\|+4)T_p$	$(\|A_1\|+\|A_2\|+2)T_p$	$3T_e+(2\|A_2\|+2)T_p$

A.3 Experimental Analysis

The performance of Liang et al.'s scheme [15], Pussewalage et al.'s scheme scheme [19] , and the proposed scheme under varying numbers of attributes are illustrated in Fig. 3. The results show that the proposed scheme achieves better performance in key generation and encryption, with significantly improved encryption efficiency. Notably, in the verification phase, the proposed scheme maintains constant-time complexity, compared to the linear growth in the other two schemes.

B Correctness Analysis

Next, we describe the correctness of the decryption algorithm based on the properties of bilinear pairings.

For decryption of original ciphertext CT, correctness analysis is as follows:

$$\begin{aligned} F &= \frac{e(sk_1, C_1)\cdot e(g, g^{\beta})^{-1}}{\prod_{i\in I}\left(e(sk_2, C_{2,i})\cdot e(sk_{3,x}, C_{3,i})\right)^{w_i}} \\ &= \frac{e(g^{\alpha/\beta}\cdot g^{a\cdot r_m/\beta}, (g^{\beta})^{r})\cdot e(g, g^{\beta})^{-1}}{\prod_{i\in I}\left(e(g^{r_m/\beta}, g^{a\cdot\lambda_i}\cdot f_{\rho(j)}^{-r_i})\cdot e(f_x^{r_m/\beta}, g^{r_i})\right)^{w_i}} \\ &= \frac{e(g^{\alpha/\beta}, (g^{\beta})^{r})\cdot e(g^{a\cdot r_m/\beta}, (g^{\beta})^{r})\cdot e(g, g^{\beta})^{-1}}{\prod_{i\in I}\left(e(g^{r_m/\beta}, g^{a\cdot\lambda_i})\cdot e(g^{r_m/\beta}, f_{\rho(j)}^{-r_i})\cdot e(f_x^{r_m/\beta}, g^{r_i})\right)^{w_i}} \\ &= \frac{e(g,g)^{\alpha\cdot r}\cdot e(g,g)^{a\cdot r_m\cdot r}}{e(g,g)^{a\cdot r_m\cdot(\sum_{i\in I}\lambda_i\cdot w_i)}} \\ &= e(g,g)^{\alpha\cdot r}. \end{aligned}$$

$$M||d = C\oplus H_1(F) = (M||d)\oplus H_1(e(g,g)^{\alpha}\cdot r).$$

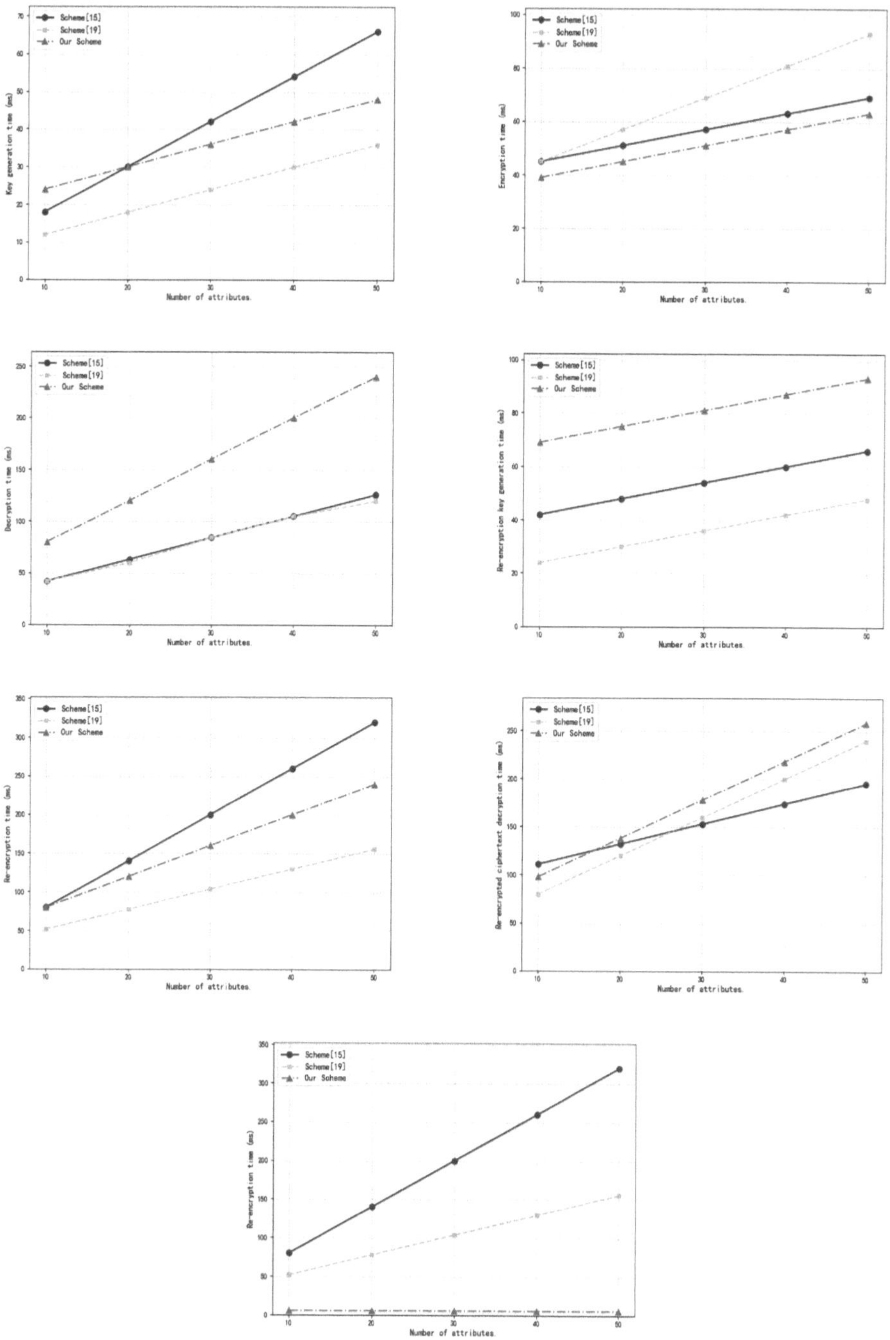

Fig. 3. Computational cost of different phases

For decryption of CT', correctness analysis is as follows:

$$
\begin{aligned}
Z &= \frac{e(sk_1, C'_3)\cdot e(g, g^{\beta})^{-1}}{\prod_{j\in J}\left(e(sk_2, C'_{4,j})\cdot e(sk_{3,x}, C'_{5,j})\right)^{w'_j}} \\
&= \frac{e(g^{\alpha/\beta}g^{a\cdot r_m/\beta}, (g^{\beta})^{r'})\cdot e(g, g^{\beta})^{-1}}{\prod_{j\in J}\left(e(g^{r_m/\beta}, g^{a\cdot\lambda'_j}\cdot f_{\rho(j)}^{-r'_j})\cdot e(f_x^{r_m/\beta}, g^{r'_j})\right)^{w'_j}} \\
&= \frac{e(g^{\alpha}, g^{r'})\cdot e(g^{a\cdot r_m}, g^{r'})\cdot e(g, g^{\beta})^{-1}}{\prod_{j\in J}\left(e(g^{r_m/\beta}, g^{a\cdot\lambda'_j})\cdot e(g^{r_m/\beta}, f_{\rho(j)}^{-r'_j})\cdot e(f_x^{r_m/\beta}, g^{r'_j})\right)^{w'_j}} \\
&= \frac{e(g,g)^{\alpha\cdot r'}\cdot e(g,g)^{a\cdot r_m\cdot r'}}{e(g,g)^{a\cdot r_m\cdot(\sum_{j\in J}\lambda'_J\cdot w'_j)}} \\
&= e(g,g)^{\alpha\cdot r'}.
\end{aligned}
$$

$$
\begin{aligned}
R &= \frac{C'_2}{Z} = \frac{R\cdot e(g,g)^{\alpha\cdot r'}}{e(g,g)^{\alpha\cdot r'}}, \\
M||d &= \overline{C}\oplus H_1({C'_1}^{1/H_3(R)}) \\
&= (M||d)\oplus H_1(e(g,g)^{\alpha\cdot r})\oplus H_1({e(g,g)^{\alpha\cdot r\cdot H_3(R)}}^{1/H_3(R)}).
\end{aligned}
$$

References

1. Bellare, M., Keelveedhi, S., Ristenpart, T.: Message-locked encryption and secure deduplication. In: Johansson, T., Nguyen, P.Q. (eds.) EUROCRYPT 2013. LNCS, vol. 7881, pp. 296–312. Springer, Heidelberg (2013). https://doi.org/10.1007/978-3-642-38348-9_18
2. Bethencourt, J., Sahai, A., Waters, B.: Ciphertext-policy attribute-based encryption. In: 2007 IEEE Symposium on Security and Privacy, pp. 321–334 (2007)
3. Blaze, M., Bleumer, G., Strauss, M.: Divertible protocols and atomic proxy cryptography. In: Nyberg, K. (ed.) EUROCRYPT 1998. LNCS, vol. 1403, pp. 127–144. Springer, Heidelberg (1998). https://doi.org/10.1007/BFb0054122
4. Boneh, D., Gentry, C., Lynn, B., Shacham, H.: Aggregate and verifiably encrypted signatures from bilinear maps. In: Biham, E. (ed.) EUROCRYPT 2003. LNCS, vol. 2656, pp. 416–432. Springer, Heidelberg (2003). https://doi.org/10.1007/3-540-39200-9_26
5. Brassard, G., Chaum, D., Crépeau, C.: Minimum disclosure proofs of knowledge. J. Comput. Syst. Sci. **37**(2), 156–189 (1988)
6. Chen, X., Huang, X., Li, J., Ma, J., Lou, W., Wong, D.S.: New algorithms for secure outsourcing of large-scale systems of linear equations. IEEE Trans. Inf. Forensics Secur. **10**(1), 69–78 (2015)
7. Ding, S., Zhao, Y., Zhu, H.: Extending fuzzy identity-based encryption with delegating capabilities. In: 2011 6th IEEE Joint International Information Technology and Artificial Intelligence Conference (2011)

8. Duan, P., Ma, Z., Gao, H., Tian, T., Zhang, Y.: Multi-authority attribute-based encryption scheme with access delegation for cross blockchain data sharing. IEEE Trans. Inf. Forensics Secur. (2024)
9. Gai, K., She, Y., Zhu, L., Choo, K.K.R., Wan, Z.: A blockchain-based access control scheme for zero trust cross-organizational data sharing. ACM Trans. Internet Technol. **23**(3), 25–25 (2023)
10. Ge, C., Susilo, W., Baek, J., Liu, Z., Xia, J., Fang, L.: A verifiable and fair attribute-based proxy re-encryption scheme for data sharing in clouds. IEEE Trans. Dependable Secure Comput. **19**(5), 2907–2919 (2022)
11. Ge, C., Susilo, W., Liu, Z., Baek, J., Luo, X., Fang, L.: Attribute-based proxy re-encryption with direct revocation mechanism for data sharing in clouds. In: Proceedings of the ACM Turing Award Celebration Conference - China 2023, pp. 164–165. Association for Computing Machinery, New York (2023)
12. Goyal, V., Pandey, O., Sahai, A., Waters, B.: Attribute-based encryption for fine-grained access control of encrypted data. In: Proceedings of the 13th ACM Conference on Computer and Communications Security – CCS '06, pp. 89–98. Association for Computing Machinery, New York (2006)
13. Liang, K., Au, M.H., Liu, J.K., Susilo, W., Wong, D.S., Yang, G., Yu, Y., Yang, A.: A secure and efficient ciphertext-policy attribute-based proxy re-encryption for cloud data sharing. Futur. Gener. Comput. Syst. **52**, 95–108 (2015)
14. Liang, K., Susilo, W.: Searchable attribute-based mechanism with efficient data sharing for secure cloud storage. IEEE Trans. Inf. Forensics Secur. **10**, 1981–1992 (2015)
15. Liang, X., Cao, Z., Lin, H., Shao, J.: Attribute based proxy re-encryption with delegating capabilities, pp. 276–286. Association for Computing Machinery, New York (2009)
16. Liu, X., Song, H., Li, Z., Yin, Y., Han, G., Liu, S.: Identity-based threshold broadcast proxy re-encryption plus scheme. In: Chen, B., Fu, X., Huang, M. (eds.) Frontiers in Cyber Security, pp. 396–412. Springer, Singapore (2024). https://doi.org/10.1007/978-981-96-0151-6_26
17. Ohata, S., Kawai, Y., Matsuda, T., Hanaoka, G., Matsuura, K.: Re-encryption verifiability: how to detect malicious activities of a proxy in proxy re-encryption. In: Nyberg, K. (ed.) CT-RSA 2015. LNCS, vol. 9048, pp. 410–428. Springer, Cham (2015). https://doi.org/10.1007/978-3-319-16715-2_22
18. Pal, S., Hill, A., Rabehaja, T., Hitchens, M.: Controlling resource allocation using blockchain-based delegation. In: 2021 IEEE Globecom Workshops, pp. 1–7 (2021)
19. Pussewalage, H.S.G., Oleshchuk, V.: A delegatable attribute based encryption scheme for a collaborative e-health cloud. IEEE Trans. Serv. Comput. **16**(2), 787–801 (2023)
20. Sahai, A., Waters, B.: Fuzzy identity-based encryption. In: Cramer, R. (ed.) EUROCRYPT 2005. LNCS, vol. 3494, pp. 457–473. Springer, Heidelberg (2005). https://doi.org/10.1007/11426639_27
21. Tang, Y., Jin, M., Meng, H., Yang, L., Zheng, C.: Attribute-based verifiable conditional proxy re-encryption scheme. Entropy **25**(5), 822 (2023)
22. Wang, C., Wang, Q., Ren, K., Lou, W.: Privacy-preserving public auditing for data storage security in cloud computing. In: 2010 Proceedings IEEE INFOCOM, pp. 1–9 (2010)
23. Xie, M., Hu, J., Hong, H.: Blockchain-based CP-ABE with publicly verifiable outsourced decryption in IoT. In: Xu, G., Liang, K., Su, C. (eds.) FCS 2020. CCIS, vol. 1286, pp. 3–18. Springer, Singapore (2020). https://doi.org/10.1007/978-981-15-9739-8_1

24. Yang, C.C., Tso, R., Liu, Z.Y., Hsu, J.C., Tseng, Y.F.: Improved proxy re-encryption scheme with equality test. In: 2021 16th Asia Joint Conference on Information Security, pp. 37–44 (2021)
25. Zhang, Q., Fu, Y., Cui, J., He, D., Zhong, H.: Efficient fine-grained data sharing based on proxy re-encryption in IIoT. IEEE Trans. Dependable Secure Comput. **21**(6), 5797–5809 (2024)

A Lattice-Based Designated Verifier Signcryption Scheme

Bingfan Gao, Chaoyang Li(✉), and Xiangjun Xin

College of Software Engineering, Zhengzhou University of Light Industry, Zhengzhou 450001, China
lichaoyang2013@163.com

Abstract. With the rapid development of quantum computing and quantum computers, most current cryptographic algorithms based on traditional mathematical hard problems are seriously threatened. In a designated verifier signature cryptography, the only person who can verify the validity of the signcrypted message is the designated verifier, and this designated verifier can also generate an indistinguishable simulated cryptography message. In this paper, a Lattice-based Designated Verifier Signcryption (L-DVSC) scheme is constructed using learning with errors (LWE) and small integer solution (SIS) problems. L-DVSC performs reject sampling in a bimodal Gaussian distribution, which can reduce the standard deviation of the resulting signature with an asymptotic square root factor in the safety parameters. The L-DVSC scheme, under the random oracle model, enables adaptive selection of the indistinguishability under cryptography message attack and adaptive choice message attack unforgeability. Comparative efficiency analysis shows that the L-DVSC scheme has a smaller key size and less communication consumption.

Keywords: Lattice cryptography · Designed verifier signcryption · Bimodal Gaussian · LWE problem · SIS problem

1 Introduction

As information technology and networks advance, the need for signcryption technology in contemporary information networks has become increasingly clear. For example, in the process of information transmission, it is necessary to ensure the confidentiality and integrity of network data; in scenarios such as online transactions and online payments, it is necessary to ensure the authenticity and reliability of the identities of both parties involved in the transaction. Cryptographic technology plays an important role in security protection in current Internet and information systems. In 1976, Diffie and Hellman [1] proposed the concept of public-key cryptography. Ron Rivest, Adi Shamir, and Leonard Adleman [2] (RSA) proposed the first practically usable public-key algorithm in 1977 and implemented the RSA public-key cryptosystem. In 1985, N. Koblitz and V. Miller

J. Yu and Y. Chen (Eds.): FCS 2025, CCIS 2805, pp. 77–90, 2027.
https://doi.org/10.1007/978-981-95-9514-3_5

applied the theory of Elliptic Curves Cryptography (ECC) to public-key cryptosystems and proposed the ECC algorithm. The RSA and ECC cryptographic systems are two very important basic cryptographic theoretical algorithms.

However, with the continuous development of quantum computing technology, especially the increase in the number of quantum bits and the stability of quantum computers, the threat posed by quantum algorithms to most current classical cryptographic algorithms is growing day by day. Cryptographic algorithms like RSA and ECC, which rely on challenging problems in number theory (such as integer factorization and discrete logarithm), are vulnerable to attacks from quantum algorithms. Shor's algorithm [3], introduced in 1994, is a quantum computing-based method for integer factorization. Grover's algorithm [4] can be used to accelerate the search for unstructured data. It achieves a faster search speed than classical algorithms through the characteristics of quantum computing. Based on quantum computers, both Shor's and Grover's algorithms can solve these hard problems in polynomial time, thereby easily obtaining encrypted information.

Post-quantum cryptographic algorithms have received extensive attention in recent years. In 2022, the NIST officially released relevant algorithm standards, including hash-based, code-based, multivariate-based, and lattice-based cryptographic algorithms, among others. Among various encryption technologies, lattice-based encryption stands out as the most effective and secure. It has smaller-sized public and private keys, and features higher computational efficiency and more reliable security. The Rijndael algorithm, proposed by cryptographers Vincent Rijmen and Joan Daemen [5] in 1997, is the origin of lattice cryptosystems. Around 2001, the famous cryptographer Serge Vaudenay proposed a new scheme called lattice cryptography. In lattice cryptography, the main purpose of digital signature technology is to verify the integrity of data and the identity of the sender; the main purpose of digital encryption technology is to protect the confidentiality of data; lattice-based digital signcryption technology combines signature and encryption technologies, ensuring both the authenticity and integrity of messages and the confidentiality of messages. In designated verifier signcryption, the signer encrypts the message with a private key to generate an encrypted ciphertext. And only the designated verifier can verify the correctness of the ciphertext through the private key, thereby determining the source and correctness of the message. Literatures [6–25] show that although quantum computing has potential in some fields, there are currently no known quantum algorithms that can effectively solve the computational problems in lattices, and compared with classical algorithms, the advantages of quantum algorithms are still not obvious. Among them, in 2002, Steinfeld et al. [6] laid the foundation for a formal security model for signcryption technology. Tsai et al. [7] designed a signcryption scheme using user identity information. The signature scheme by Singh et al. [8] allows signers to sign some hidden information such as additional conditions. The scheme by Ye et al. [9] proposed an image encryption scheme based on blind signature and ImproLS. The signcryption scheme proposed by Ming et al. [12] integrates generalized designated verification technology, thereby eliminating the dependence on secure channels between both signing parties in the process of information transmission. The signcryption scheme by Lu et al. [13] does not use core algorithms such as trapdoor generation algorithms, and is the first signcryption scheme that does not rely on core algorithms.

Yu et al. [14] developed a lattice-based proxy signcryption scheme that employs a certificateless mechanism to address the key escrow issue. Hundera et al. [15] separated the signcryption process into two phases: offline and online. In the offline phase, most computationally intensive tasks are completed in advance, while in the online phase, only lightweight computations are required, which reduces the real-time computational burden on vehicle nodes. Ashok et al. [16] utilized signcryption technology to reduce computational and communication overheads, addressing the security and efficiency issues of resource-constrained devices in the Internet of Medical Things (IoMT) environment. Zhang et al. [17] constructed a lattice-based signcryption scheme that satisfies security of post-quantum. The signcryption scheme proposed by Yu et al. [18] realizes both certificateless signature and encryption functions.

Based on the LWE and SIS problems on lattice, a Lattice-based Designated Verifier Signcryption (L-DVSC) scheme is proposed. The L-DVSC scheme can achieve confidentiality, unforgeability, and non-transferability under the random oracle model.

2 The L-DVSC Scheme

The L-DVSC scheme is shown in Fig. 1. It is divided into four steps in total, namely Setup (Sect. 2.1), KeyGen (Sect. 2.2), Signcryption (Sect. 2.3), and Unsigncryption (Sect. 2.4). The key generation center (KGC) distributes the keys to the signcryption user and the designated verifier. After the signcryption user completes the signcryption, the ciphertext is uploaded to the cloud server. The designated verifier downloads the ciphertext and uses the key to decrypt it to obtain the plaintext.

2.1 Setup

The Key Generation Center (KGC) selects security parameters that include 1^n, a prime number $q \geq 3$, complexity $L = O(\sqrt{n\log_2 q})$, and positive integers $m(m \geq 3n\log q), k(k \geq 5n\log q), d$, along with a real number $\delta = 12d\sqrt{k}$ and a Gaussian parameter $s = L \cdot \omega\left(\sqrt{\log_2 n}\right)$. The Gaussian distribution D_σ has a mean of 0 and a standard deviation of σ. The pair $(E_\kappa(\cdot), D_\kappa(\cdot))$ represents secure encryption and decryption algorithms, with κ being a symmetric key from the key space Σ. Then select a hash function that satisfies the conditions ($\prod$ is the multicoin-tossing space):

$$H_1 : \{0,1\}^* \to \{c : c \in \{-1,0,1\}^n\}$$

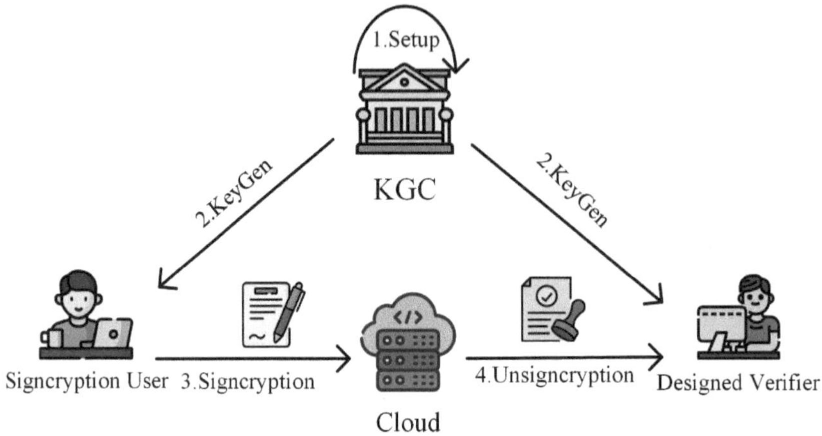

Fig. 1. The L-DVSC scheme

$$H_2 : \{0,1\}^n \to \sum$$

$$H_3 : \{0,1\}^* \times \{0,1\}^* \to \prod$$

Finally, the KGC announces the global parameter $\widetilde{w} = \{H_1, H_2, H_3, q, m, d, \delta\}$ of the system.

2.2 KeyGen

In this L-DVSC scheme, there are two users, namely user Alice and designated user Bob. Based on the system parameter $\widetilde{w}$ generated in the above steps, the following outlines the steps for generating public and private keys for users.

After inputting (n, m, q) into $TrapGen(n, m, q)$ [27], the trapdoor generation algorithm is used to obtain a matrix A following a uniform distribution and a related secret trapdoor S_A as user Alice's private key, where $AS_A = qI_n \mod 2q$ and I_n is the $n \times n$ identity matrix. In the same manner, (B, S_B) is generated as the public-private key pair of the designated verifier Bob, which also satisfies $BS_B = qI_n \mod 2q$.

2.3 Signcryption

Alice performs the following steps to signcrypt the message μ.

- Selects a vector y from the Gaussian distribution D_σ^m.
- Calculates $c = H_1 (Ay \mod 2q, \mu)$.

- Randomly selects $b \in \{0,1\}$.
- Calculates $e_1 = y + (-1)^b S_A c$ and $e_2 = (-1)^b B^T c$.
- Outputs (e_1, e_2, c) with a probability of $\min\left\{\frac{D_\sigma^m(e_1)D_\sigma^m(e_2)}{MD_{cS_A}^m(e_1)D_{cS_A}^m(e_2)}\right\}$, where M is some fixed positive real number.
- Randomly selects $\lambda \in \{0,1\}^n$ and calculates $\theta = E_{\kappa=H_2(\lambda)(\mu,e_1,e_2,c)}$.
- Calculates $\eta = H_3(\lambda,\theta)$, ${k_1}^T = -{z_1}^T B$, and ${k_2}^T = {z_1}^T + \lambda\lfloor q/2\rfloor$. Noise vector $z_1 \leftarrow D_\sigma^n$.
- Outputs the ciphertext $\gamma = (\theta, k_1, k_2)$.

2.4 Unsigncryption

The operations of the designated verifier Bob in the signcryption decryption algorithm are as follows.

- Calculates $\hat{\lambda} = k_2^T + k_1^T S_B$, lets $\hat{\lambda} = (\lambda_1', \lambda_2', \cdots \lambda_n')$. For $i = 1, 2\cdots, n$, if $|\lambda_i' - \lfloor q/2\rfloor| = \lfloor q/2\rfloor$, $\lambda_i = 0$, otherwise $\lambda_i = 1$.
- Calculates $D_{\kappa \leftarrow H_{3(\hat{\lambda})}(\theta)=(\mu,e_1,e_2,c)}$.
- If both e_1 and e_2 satisfy $\|e_1\| \le 2\sigma\sqrt{n}$ and $\|e_2\| \le 2\sigma\sqrt{n}$, then perform step (4); otherwise, stop the unsigncryption algorithm.
- Calculates $c' \leftarrow H_1\left(Ae_1 + S_B^T e_2 + 2qc \bmod 2q, \mu\right)$.
- If $c' = c$, then accepts the message μ; otherwise reject.

In conclusion, the proposed L-DVSC scheme performs signcryption on the plaintext, and only the designated verifier can decrypt it using the private key.

3 Security Analysis

3.1 Correctness

From $\hat{\lambda} = k_2^T + k_1^T S_B$, the following Eq. (1) can be derived.

$$\begin{aligned}\hat{\lambda} &= k_2^T + k_1^T S_B \\ &= z_1^T + \lambda\lfloor q/2\rfloor - z_1^T B S_B \\ &= z_1^T + \lambda\lfloor q/2\rfloor - z_1^T \\ &= \lambda\lfloor q/2\rfloor\end{aligned} \tag{1}$$

If $\left|\hat{\lambda} - \lfloor q/2\rfloor\right| = \lfloor q/2\rfloor$, then $\lambda_i = 0$; otherwise $\lambda_i = 1$. Based on this judgment condition, λ can be restored. The following Eq. (2), (3) can verify that the algorithm in unsigncryption is valid.

$$\begin{aligned}Ae_1 + S_B^T e_2 + 2qc \bmod 2q &= A\left(y + (-1)^b S_A c\right) + S_B^T(-1)^b B^T c + 2qc \\ &= Ay + (-1)^b\left(AS_A + (BS_B)^T\right)c + 2qc \\ &= Ay \bmod 2q\end{aligned} \tag{2}$$

$$H_1\left(Ae_1 + S_B^T e_2 + 2qc \bmod 2q, \mu\right) = H_1\left(Ay \bmod 2q, \mu\right) \tag{3}$$

3.2 Confidentiality

Theorem 1. Under the IND-CCA2-I attack, suppose that the adversary Eve can break the L-DVSC scheme with an advantage probability of $\frac{\varepsilon'}{\varepsilon q_{H_1} e(q_J+q_G)}$. Then there exists a challenger Challenger that can solve the decisional LWE problem with a non-negligible probability, where q_{H_1}, q_G, and q_G are the number of queries made by the adversary Eve to H_1, public key replacement, and the unsigncryption algorithm within polynomial time, respectively.

Proof. In the scenario of the decisional LWE problem, the challenger is required to determine whether a given random instance belongs to one of the following two cases: (1) It follows a uniform distribution; (2) $p = \langle o, s \rangle + e$, where s follows the Gaussian distribution D_σ^n and e follows the Gaussian distribution D_σ. To accomplish this task, the Challenger treats an adversary Eve as a subroutine and interacts with it. During this process, the Challenger maintains an initially empty list to record all interactive queries with Eve. The Challenger will select a number of queries $t \in \{1, 2, \cdots, q_{H_1}\}$ as needed and judge whether the instance belongs to case (1) or case (2) based on the results of these queries.

In the initial stage of the game, the challenger first executes the set-up algorithm to generate a set of system parameters $\widetilde{w}$, and passes it to the adversary Eve. Subsequently, in the first phase, Eve can submit a polynomial number of query requests to the Challenger in an adaptive manner according to its own strategy.

***H*₁ Query.** When making H_1 query, the adversary Eve requests the hash value of $(A, y, m, -, -)$. If the hash value already exists in the initialmpty list $List_1$ established by the Challenger, the Challenger will directly return the corresponding hash value (e_i, c) to Eve; IfIfhe hash value is not recorded in $List_1$, the Challenger will sample e_i, y from the Gaussian distribution D_σ^m, select a parameter c, then return (A, y, m, e_i, c) to EvEve and addthe list $List_1$ for subsequent query use.

H_2 Query. When the Challenger receives a query about H_2, it first checks whether the corresponding $(\lambda, H_2(\lambda))$ already exists in the initially empty list $List_2$ it has established. If $(\lambda, H_2(\lambda))$ exists in $List_2$, the Challenger will directly return the corresponding κ to Eve; if κ does not exist in $List_2$, the Challenger will randomly select $H_2(\lambda) \leftarrow \sum$, calculate $\kappa \leftarrow H_2(\lambda)$ based on this value and return it to Eve, while adding $(\lambda, H_2(\lambda))$ to $List_2$.

H_3 Query. When the Challenger receives a query about H_3, it first checks whether the corresponding $(\lambda, \theta, \eta \leftarrow H_3(\lambda, \theta))$ has already been recorded in the initinitialty list $List_3$ it established. If $(\lambda, \theta, \eta \leftarrow H_3(\lambda, \theta))$ exists in $List_3$, the Challenger will directly return this value η to Eve; if $(\lambda, \theta, \eta \leftarrow H_3(\lambda, \theta))$ is not in $List_3$, the Challenger will randomly select $H_3(\lambda, \theta) \leftarrow \prod$, calculate the corresponding $\eta \leftarrow H_3(\lambda, \theta)$ based on $H_3(\lambda, \theta) \leftarrow \prod$ and return it to Eve, while adding $(\lambda, \theta, \eta \leftarrow H_3(\lambda, \theta))$ to $List_3$.

Public Key Query. When Eve requests the public key of a designated identity. The Challenger first checks whether the public key B already exists in the initially empty list $List_4$ it created. If the $List_4$ contains the public key B, the Challenger will directly send the public key B to Eve; if the $List_4$ does not contain the public key B, the Challenger randomly selects a matrix B from $\mathbf{Z}_{2q}^{n\times m}$, sends it to Eve, and adds $(B, -)$ to $List_4$.

Private Key Query. When Eve requests the private key of a designated user, the Challenger will randomly select a value S_B and ensure that it satisfies the equation requirement $B \times S_B = qI_n \mod 2q$. Subsequently, the Challenger will return S_B to Eve and update the initially empty list $List_4$ it created using (B, S_B).

Signcryption Query. When Eve requests the ciphertext of message μ. The Challenger will execute the signcryption algorithm to generate the ciphertext and return the generated ciphertext to Eve. If the Challenger has processed the key query in previous inquiries and the list $List_4$ already stores the public-private key pair (B, S_B) of the designated verifier, then the Challenger will directly retrieve (B, S_B) from the list $List_4$, with the specific steps as follows:

- Selects a vector y from the Gaussian distribution D_σ^m.
- Calculates $c = H_1\left(Aymod2q, \mu\right)$.
- Randomly selects $b \in \{0, 1\}$.
- Calculates $e_1 = y + (-1)^b S_A c$ and $e_2 = (-1)^b B^T c$.
- Outputs (e_1, e_2, c) with a probability of $\min\left\{\frac{D_\sigma^m(e_1)D_\sigma^m(e_2)}{MD_{cS_A}^m(e_1)D_{cS_A}^m(e_2)}\right\}$, where M is some fixed positive real number.
- Randomly selects $\lambda \in \{0, 1\}^n$ and calculates $\theta = E_{\kappa=H_2(\lambda)(\mu,e_1,e_2,c)}$.
- Calculates $\eta = H_3(\lambda, \theta)$, ${k_1}^T = -{z_1}^T B$, and ${k_2}^T = {z_1}^T + \lambda\lfloor q/2\rfloor$. Noise vector $z_1 \leftarrow D_\sigma^n$.
- Returns the ciphertext $\gamma = (\theta, k_1, k_2)$ to Eve.

Unsigncryption Query. Eve inquires about the result of the unsigncryption of γ. The Challenger responds as follows:

- Calculates $\hat{\lambda} = k_2^T + k_1^T S_B$, lets $\hat{\lambda} = (\lambda_1', \lambda_2', \cdots \lambda_n')$. For $i = 1, 2 \cdots, n$, if $|\lambda_i' - \lfloor q/2\rfloor| = \lfloor q/2\rfloor$, $\lambda_i = 0$, otherwise $\lambda_i = 1$.
- Calculates $D_{\kappa \leftarrow H_{3(\hat{\lambda})}(\theta)=(\mu,e_1,e_2,c)}$.
- If both e_1 and e_2 satisfy $\|e_1\| \le 2\sigma\sqrt{n}$ and $\|e_2\| \le 2\sigma\sqrt{n}$, then performs step (4); otherwise, stops the unsigncryption algorithm.
- Calculates $c' \leftarrow H_1\left(Ae_1 + S_B^T e_2 + 2qc \bmod 2q, \mu\right)$.
- If $c' = c$, then accepts the message μ; otherwise reject.

After that, Eve outputs μ', (A', B') to the Challenger, where $d \in \{0,1\}$. Before the challenge, Eve is not allowed to query the private key of the designated verifier, and the public key of the signcryptor cannot be replaced. The Challenger responds as follows:

- Selects a vector y' from the Gaussian distribution D_σ^m.
- Calculates $c = H_1\left(Ay' mod 2q, \mu\right)$.
- Randomly selects $b \in \{0,1\}$.
- Calculates $e_1' = y + (-1)^b S_A c$ and $e_2' = (-1)^b B^T c$.
- Randomly selects $\lambda' \in \{0,1\}^n$ and calculates $\theta' = E_{\kappa = H_2(\lambda)(\mu, e_1, e_2, c)}$.
- Calculates $\eta' = H_3\left(\lambda', \theta'\right)$, ${k_1}'^T = -{z_1}^T B$, and ${k_2}'^T = {z_1}^T + \lambda' \lfloor q/2 \rfloor$. Noise vector $z_1 \leftarrow D_\sigma^n$.
- Returns the ciphertext $\gamma' = \left(\theta, {k_1}', {k_2}'\right)$ to Eve.

Eve still makes a polynomially bounded number of queries. However, Eve is subject to the following restrictions: (1) Before the challenge phase, Eve cannot obtain the private key of the designated verifier, and the public key of the signcryptor is not allowed to be replaced; (2) After the challenge phase, Eve cannot make queries to the unsigncryption oracle for the ciphertext γ'.

Finally, Eve outputs its guess d' for d to the Challenger. If $d' = d$, the Challenger learns that $A' \in \mathbf{Z}_{2q}^{n \times m}$ and $B' \in \mathbf{Z}_{2q}^{n \times m}$ exist, where all components of A' and B' do not exceed 7σ. Otherwise, the Challenger learns that A' and B' follow a uniform distribution. This indicates that the Challenger has solved an instance of the decisional LWE problem; however, under current computational conditions, this problem cannot actually be broken, which contradicts the facts. Therefore, in the game, the probability $\frac{\varepsilon'}{\varepsilon q_{H_1} e (q_J + q_G)}$ that the Challenger solves the decisional LWE problem is negligible. Hence, the adversary's assumption is invalid, the problem cannot be broken, and further, it shows that the designed L-DVSC scheme is secure.

3.3 Unforgeability

Theorem 2. If the adversary Eve breaks the UF-CMA-I security of L-DVSC with a significant advantage ε, then there must exist a challenger Challenger that can utilize Eve's attack capability to solve the SIS problem with a probability $\varepsilon' \geq \frac{\varepsilon}{\left(1 - 2^{-\omega(\log_2 n)}\right)}$.

Proof. Given a random instance of the SIS problem, the Challenger can find a non-zero short vector v', v'' such that $Av' + v'' = 0 \mod q$.

At the start of the game, the Challenger runs the setup algorithm to generate the global parameters $\widetilde{w}$ of the system and returns them to Eve. Then Eve submits the exact same queries as in Phase 1 of Theorem 1.

Finally, Eve outputs a forged ciphertext γ^* for message μ^*. Based on the ciphertext, it can forge a signature (e_1^*, e_2^*, c^*) for any message μ^*. Eve forges another ciphertext γ^{**} and calculates the corresponding signature $(e_1^{**}, e_2^{**}, c^{**})$ such that $(c^* \neq c^{**})$. The following equation Eq. (4) can be obtained.

$$Ae_1^* + S_B^T e_2^* + 2qc^* = Ae_1^{**} + S_B^T e_2^{**} + 2qc^{**} \mod 2q \tag{4}$$

The above formula can be simplified to equation $A(e_1^* - e_1^{**}) + S_B^T(e_2^* - e_2^{**}) = 0 \mod q$.

The SIS problem is to find a non-zero short vector v', v'' such that $Av' + v'' = 0 \mod q$. From the above formula, then can get $v' = (e_1^* - e_1^{**})$ and $v'' = S_B^T(e_2^* - e_2^{**})$. This indicates that the Challenger has found a solution to the SIS problem. However, since this problem cannot be solved under existing computational conditions, the assumption is invalid. Because S_B is not unique, Eve cannot determine S_B. Therefore, the advantage $\varepsilon' \geq \frac{\varepsilon}{(1-2^{-\omega(\log_2 n)})}$ of the Challenger in solving the SIS problem is negligible.

4 Performance Evaluation

This section analyzes the computational efficiency of L-DVSC and existing cryptographic algorithms. An in-depth analysis is conducted from both spatial and temporal perspectives, focusing on the performance differences of various cryptographic algorithms in terms of storage space occupation and computational efficiency. Specifically, the trapdoor-based matrix preimage sampling algorithm (T_t) takes 35.42 ms per operation, the Gaussian sampling operation (T_D) takes 23.03 ms per operation, and the matrix-vector multiplication operation (T_v) takes 5.32 ms per operation. These time data are average values obtained based on the operation of the open-source cryptographic library PBC in the VC 6.0 environment.

In the simulation experiments, this scheme selects three different parameters. For Parameter I, the corresponding values of n, m, and q are 128, 8832, and 2^{23} respectively; for Parameter II, the corresponding values of n, m, and q are 256, 19200, and 2^{25} respectively; for Parameter III, the corresponding values of n, m, and q are 256, 20736, and 2^{27} respectively. and three simulation experiments are conducted. Different setting numbers correspond to different parameter configurations, which in turn affect the security level of the cryptographic algorithm.

In the comparison of computational efficiency, the operations of vector addition and hash functions are ignored. Table 1 shows the comparison of public key size, private key sizes and ciphertext lengths among various cryptographic algorithms. It can be seen from Fig. 2 and Fig. 3. The ciphertext length of L-DVSC is reduced under different settings in Fig. 4. Table 2 compares the number of signcryption operations, the number of unsign-cryption operations, and the total number of operations. Scheme [17] combines lattice cryptography, chameleon signcryption, and CRT key management for the first time, and the proposed scheme achieves both quantum-safe privacy protection and resistance to signature abuse. Scheme [18] is based on the hardness of the LWE and SIS problems in lattice cryptography. It is proven to be resistant to quantum computing attacks, with specific parameter settings provided, thus meeting the conditions for comparison.

Table 1. Comparison of L-DVSC in terms of spatial resources

Items	Public Key	Private Key	Cipertext
Scheme [17]	$4n^3 \log q$	$2n^3 \log q$	$12n^2 \log q$
Scheme [18]	$2n^3 \log q$	$n^3 \log q$	$2mn \log q$
This scheme	$nm \log q$	$mn \log q$	$nm \log q$

Table 2. Performance comparison of execution time

Items	Signcryption	Unsigncryption	Total
Scheme [17]	$T_t + 5T_D + 6T_v$	$T_t + 5T_v$	$2T_t + 6T_D + 13T_v$
Scheme [18]	$T_t + 6T_D + 6T_v$	$T_t + 4T_v$	$2T_t + 6T_D + 10T_v$
This scheme	$T_t + 4T_D + 4T_v$	$T_t + 3T_v$	$2T_t + 4T_D + 7T_v$

This section analyzes the computational time cost through three simulation experiments. The experiments are conducted based on the MATLAB R2022a software platform, and comparative simulation graphs of the running time of the signcryption, unsigncryption, and the overall scheme of the cryptographic algorithm are generated. It can be seen in Fig. 5 that the L-DVSC scheme shows significant optimization effects in terms of signcryption time, unsigncryption time, and total time.

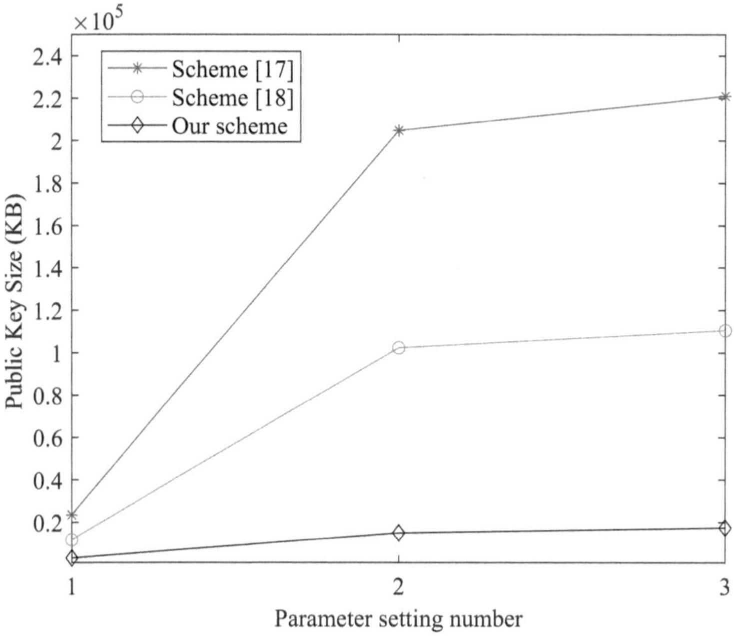

Fig. 2. Public key size comparison

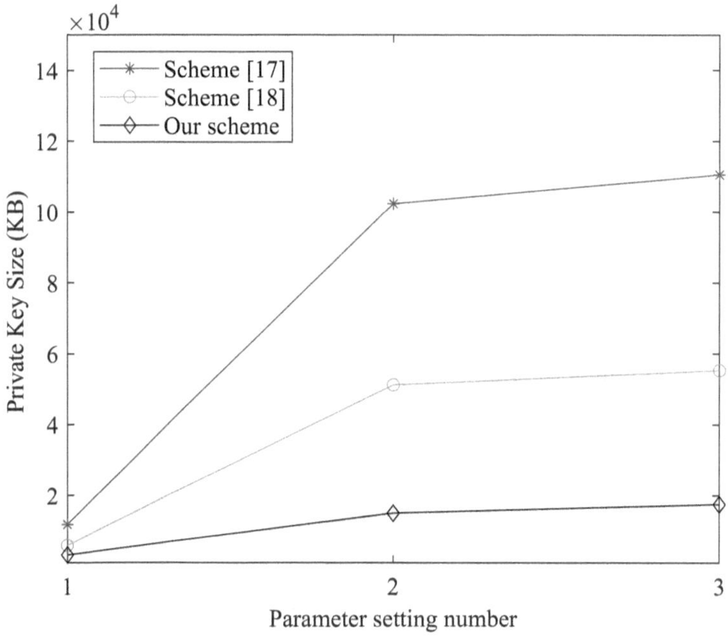

Fig. 3. Private key size comparison

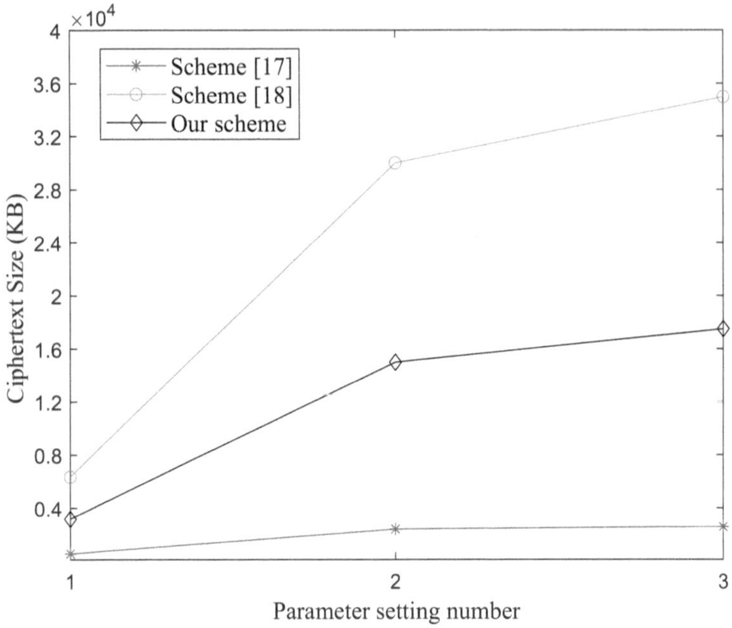

Fig. 4. Ciphertext size comparison

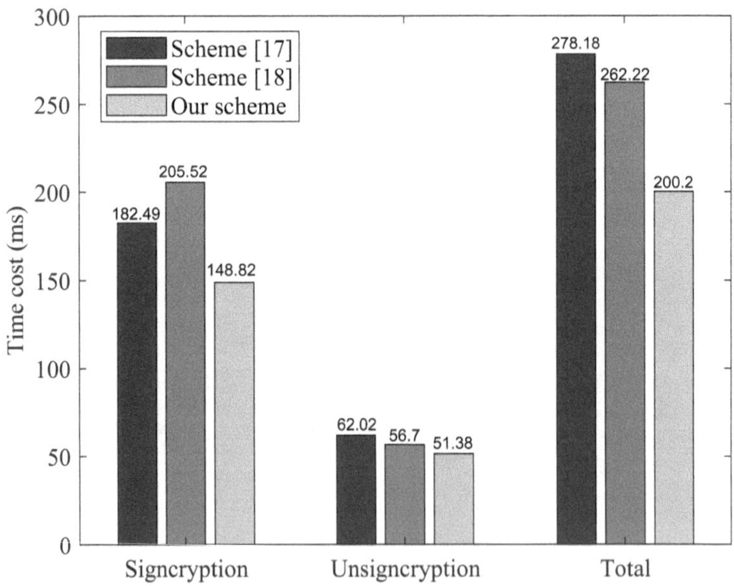

Fig. 5. Time cost comparison in different schemes

5 Conclusion

The proposed L-DVSC scheme utilizes a rejection sampling algorithm based on the bimodal Gaussian distribution to efficiently generate private keys with extremely small norms. This not only lays the foundation for its resistance to quantum computing attacks (relying on the hardness of lattice problems) but also significantly reduces the computational and communication overheads of the scheme, thereby improving efficiency. In terms of security, the scheme strictly meets the two major standards of IND-CCA2 and UF-CMA, which means it can not only provide the strongest encryption confidentiality under chosen-ciphertext attacks but also ensure the unforgeability of digital signatures under chosen-message attacks. Therefore, the L-DVSC scheme, with its quantum security, high efficiency, and robust comprehensive security guarantees, demonstrates important application value in the field of post-quantum cryptography and provides a candidate scheme that combines security and efficiency for building future secure infrastructure.

Acknowledgments. This work was supported by the National Natural Science Foundation of China under Grant Numbers 72293583, 72293580, the Foundation and Cutting-Edge Technologies Research Program of Henan Province (CN) under Grant Numbers 262102210202, 252102110182, 252102110178, and the Doctor Scientific Research Fund of Zhengzhou University of Light Industry under Grant 2021BSJJ033.

References

1. Whitfield, D., Martine, E.H.: New directions in cryptography. IEEE Trans. Inf. Theory **22**(6), 644–654 (1976)
2. Rivest, R.L., Shamir, A., Adleman, L.: A method for obtaining digital signatures and public-key cryptosyste. Commun. ACM **21**(2), 120–126 (1978)
3. Shor, P.W.: Algorithms for quantum computation: discrete logarithms and factoring. In: Proceedings 35th Annual Symposium on Foundations of Computer Science, USA, pp. 124–134 (1994)
4. Grover, L.K.: A fast quantum mechanical algorithm for database search. In: Symposium on the Theory of Computing (1996). https://doi.org/10.1145/237814.237866
5. Bedoui, M., Mestiri, H., Bouallegue, B.: An improvement of both security and reliability for AES implementation. J. King Saud Univ.-Comput. Inf. Sci. **34**(10), 9844–9851 (2022)
6. Zhan, Q., Luo, M., Qiu, M.: An efficient multi-mode certificateless ring signcryption scheme in vanets. IEEE Internet Things J. **11**(20), 33508–33524 (2024)
7. Tsai T., Lin H., Huang W.: Anomaly detection through outsourced revocable identity-based signcryption with equality test for sensitive data in consumer IoT environments. IEEE Trans. Consum. Electron. **71**(2) 6327–6339 (2025)
8. Singh, T., Ali, R., Tyagi, V.: An efficient identity based multi-proxy multi-signcryption scheme for electronic commerce using bilinear pairing. Procedia Comput. Sci. **259**, 1592–1601 (2025)

9. Ye, G., Wu, H., Liu, M.: Image encryption scheme based on blind signature and an improved Lorenz system. Expert Syst. Appl. **205**(C) 0957–4174 (2022)
10. Wei, J., Zhu, Q.: IDTRSC: ID-based traceable ring signcryption framework for data sharing without key escrow. IEEE Trans. Veh. Technol. **74**(7), 11207–11220 (2025)
11. Ali, M., Liu, X.: A novel approach to cloud security: publicly verifiable remote signcryption framework. IEEE Internet Things J. **12**(10), 14828–14844 (2025)
12. Ming, Y., Zhang, L., Han, J.: Universal designated verifier signcryption scheme in standard model. J. Comput. Appl. **34**(2), 464–468 (2014)
13. Lu, X., Wen, Q., Wang, L.C.: A lattice-based signcryption scheme without trapdoors. J. Electron. Inf. Technol. **38**(9), 2287–2293 (2016)
14. Yu, H., Wang, N.: Certificateless proxy signcryption scheme from lattice. J. Electron. Inf. Technol. **44**(7), 2584–2591 (2022)
15. Hundera, N., Aftab, M., Mesfin, D.: An efficient heterogeneous online/offline anonymous certificateless signcryption with proxy re-encryption for Internet of Vehicles. Veh. Commun. **49**, 100811 (2024)
16. Ashok, K., Gopikrishnan, S.: A hybrid secure signcryption algorithm for data security in an internet of medical things environment. J. Inf. Secur. Appl. **85**, 103836 (2024)
17. Zhang J., Cui X.: LBCSC: lattice-based chameleon signcryption scheme for secure and privacy-preserving vehicular communications. Trans. Emerg. Telecommun. Technol. **35**(10) (2024)
18. Yu, H., Bai, L., Hao, M.: Certificateless signcryption scheme from lattice. IEEE Syst. J. **15**(2), 2687–2695 (2021)
19. Zhang, P., Chi, H., Li, J.B.: Lattice based strong designated verifier signature scheme. J. Beijing Univ. Aeron. Astronaut. **49**(6), 1294–1300 (2023)
20. Yu, H., Lv, Z.: Lattice-based ring signcryption for consortium blockchain. J. King Saud Univ. Comput. Inf. Sci. **35** (2023)
21. Li, C., Chen, Y., Dong, M., Li, J., Huang, M., Xiang, X.: Quantum-safe identity-based designated verifier signature for BIoMT. J. Syst. Architect. **160**, 103362 (2025)
22. Lindner R. Peiker C.: Better key sizes (and attacks) for LWE-based encryption. The Cryptographers' Track at the RSA Conference on Topics in Cryptology, USA, pp. 319–339 (2011)
23. Zhu, H., Wang, Y., Wang, C., Cheng, X.: An efficient identity-based proxy signcryption using lattice. Futur. Gener. Comput. Syst. **117**, 321–327 (2021)
24. Li, C., Jiang, B., Dong, M.: Efficient designated verifier signature for secure cross-chain health data sharing in BIoMT. IEEE Internet Things J. **11**, 19838–19851 (2024)
25. Li, C., Jiang, B., Dong, M., Chen, Y., Huang, M., Xiang, X.: Cross-chain privacy preserving for BIoMT with designated verifier proxy signature. IEEE Internet Things J. **12**(12), 20042–20055 (2025)
26. Zang, W., Zhang, Y., Zheng, Y.: Evaluation of a semi-analytical bimodal Gaussian model for the estimation of the lateral velocity distribution behind a tidal stream turbine. Ocean Eng. **337**, 121888 (2025)
27. Albrecht M., Lai R., Lapiha O.: Partial Lattice Trapdoors: How to Split Lattice Trapdoors, Literally. Cryptology ePrint Archive (2025). https://ia.cr/2025/367

A Lattice Based IBE Scheme for Automatic Key Update

Zixuan Yan, Shanqiang Feng, and Fengyin Li(✉)

School of Computer Science, Qufu Normal University, Rizhao 276826, China
lfyin318@qfnu.edu.cn

Abstract. The Internet of Things (IoT) connects various "things" in the physical world to the Internet, collecting data and communicate with each other. However, during the data collection process, security and privacy issues cannot be guaranteed due to the high exposure of the links. In response, numerous cryptographic tools have been proposed to establish secure communication between IoT devices. In this work, we investigate Identity Based Encryption (IBE) for the IoT. In particular, considering security and privacy issues, key updates are required. In Identity-Based Encryption, the Private Key Generator (PKG) generates and distributes private Key, which causes a workload for the PKG. For this reason, we propose a scheme that enables autonomous key updating without the involvement of the PKG. In the Chosen Plaintext Attack (CPA) model, we proved the security of the proposed scheme under the hard assumption of Learning With Errors (LWE) problems.

Keywords: public key encryption · Identity based encryption · key update · lattice

1 Introduction

The Internet of Things (IoT) is considered a bridge between the real world and the Internet [3,16]. It connects to the Internet or local networks via embedded sensors, actuators, and communication modules (e.g., Wi-Fi, Bluetooth, Zigbee, LoRaWAN, NB-IoT, cellular networks). By embedding these devices into real world objects, the interconnection between the real world and the internet is achieved. When deploying IoT devices, security and Privacy is considered during data transmission. In recent years, some security algorithms have been proposed to achieve security in communication between IoT devices. It is well known that a public key cryptosystem can be used to establish a secure channel between IoT devices. Although public key encryption tools serve as a secure and efficient means of securing IoT communications, they usually involve expensive certificate management, which can burden IoT devices with limited bandwidth resources. To eliminate certificate management from public key encryption, Identity Based Encryption (IBE) [6,18] encryption tools are proposed. In this application scenario, the public key of an IoT device serves as its own identity, and the private

J. Yu and Y. Chen (Eds.): FCS 2025, CCIS 2805, pp. 91–105, 2027.
https://doi.org/10.1007/978-981-95-9514-3_6

key of each device is generated by the PKG (Private Key Generator). The tedious process of certificate issuance, validation, and storage is avoided [12,20]. Unfortunately, even though IBE can effectively secure IoT communications, existing IBE schemes still suffer from the following security issues:

- **Key Exposure.** Cryptosystems can only guarantee the security and privacyof communication data when the key is not exposed. Unfortunately, IoT devices have highly exposed links, which can serve as entry points for attackers to gain access and steal keys. In particular, side-channel attacks can be launched against IoT devices to monitor power consumption and radio waves during the device communication process, allowing the recovery of the key [13,23].
- **Quantum Attacks.** With the development of quantum computers, some encryption schemes built on traditional hardness assumptions can be broken in polynomial time [19].

In this work, we briefly discuss some possible solutions. To prevent key leakage in the communication process, key update is required. However, the key update is the responsibility of the PKG, which imposes computational overhead on PKG. To solve this problem, the IBE scheme is proposed which can autonomously update the key, allowing the user key to be updated independently without the involvement of the PKG. Overall, in the scheme, the PKG only needs to distribute decryption keys to each IoT device during the initialization phase. They enable updates of keys on a regular interval. To resist quantum attacks, quantum resistant cryptographic methods can be employed in the design of IBE schemes. Moreover, cryptographic schemes based on the lattice exhibit high efficiency and low storage costs [15].

The contributions of this paper are as follows:

- By introducing binary trees into the construction of IBE, we propose an IBE scheme based on the lattice LWE problems. In the proposed scheme, the autonomous key update was achieved. The key update process does not require PKG's participation, which reduces the computational overhead of PKG.
- Under our proposed selective ID, chosen plaintext attack(sID-CPA) model, we give a formal security reduction. It proved that the scheme can achieve CPA security.

Other Related Work. There are some currently known PKE schemes. The first non-interactive PKE scheme with forward security was proposed by Canetti et al. In addition, based on the work of Gentry and Silverberg [14], this scheme implements forward-secure CPA security under the Standard Model. In [8], the authors propose an fs-PKE scheme based on the idea of Hierarchical IBE (HIBE). Compared with [9], this scheme reduces the size of the main parameter and thus achieves higher efficiency. In [21], the authors define the fs-HIBE model for the first time, propose the first fs-HIBE, and show some applications of fs-HIBE. In [24], the authors first proposed fs-IBE scheme, and show some applications of fs-IBE.

2 Preliminaries

In this section, we introduce the syntax and security model. Then, we briefly introduce the basic definitions and related algorithms for lattices, Discrete Gaussian, and Learning with errors.

2.1 Notations

- Denote the ring of integers by $\mathbb{Z}_q$ and the ring $\mathbb{Z}/q\mathbb{Z}$ by $\mathbb{Z}_q$, where q is a positive integer.
- Let $\mathcal{C}$ be a probabilistic polytemporal (PPT) algorithm. By $y \leftarrow \mathcal{C}(x)$, we mean the random variables defined as the output of $\mathcal{C}$, the input x and a coin chosen at random.
- Write $s \leftarrow D$, if s samples uniformly from the distribution D.
- Let $y \leftarrow \mathbb{Z}_q^n$, denotes the random vector uniformly sampled from $\mathbb{Z}_q^n$.
- $\|\cdot\|$: For a set of vectors $\mathcal{S} := \{s_0, s_1, \cdots, s_{k-1}\}$, we define $\|s_i\|$ as the s_i about ℓ_2 paradigm and write $\|\mathcal{S}\| := \max_{i\in\{0,1,\cdots,k-1\}} \|s_i\|$ which is the paradigm of $\mathcal{S}$.
- Suppose that $\widetilde{\mathcal{S}} := \{\widetilde{s}_1, \cdots, \widetilde{s}_k\}$ is the Gram-Schmidt quadrature of $s_1, \cdots, s_k$ with respect to the same order. Denote by $\|\widetilde{\mathcal{S}}\|$ the Gram-Schmidt paradigm of $\widetilde{\mathcal{S}}$.

2.2 Integer Lattice

Suppose that $\mathbf{L} = [\mathbf{l}_1, \mathbf{l}_2, \cdots, \mathbf{l}_m] \in \mathbb{R}^{m\times m}$ is a matrix of $m \times m$ which columns are linearly independent. The m-dimensional lattice generated by the matrix $\mathbf{L}$ is defined as:

$$\mathcal{L} := \mathbf{L}\cdot\mathbf{c}\left\{\sum_{i\in\{1,\cdots,m\}} c_i\mathbf{l}_i : \mathbf{c} = (c_1, \cdots, c_m) \in \mathbb{Z}^m\right\}$$

Note that only integer lattices are used in this paper, with $\mathbf{c}$ integer components. For a prime q, a random matrix $\mathbf{A} \in \mathbb{Z}_q^{n\times m}$, a vector $\mathbf{u} \in \mathbb{Z}_q^n$ set:

$$\Lambda_q(\mathbf{A}) := \{\mathbf{e} \in \mathbb{Z}^m \text{ s.t. } \exists \mathbf{s} \in \mathbb{Z}_q^n \text{ where } \mathbf{A}^T\mathbf{s} = \mathbf{e} \pmod{q}\},$$

$$\Lambda_q^{\perp}(\mathbf{A}) := \{\mathbf{e} \in \mathbb{Z}^m \text{ s.t. } \mathbf{A}\mathbf{e} = 0 \pmod{q}\},$$
$$\Lambda_q^{\mathbf{u}}(\mathbf{A}) := \{\mathbf{e} \in \mathbb{Z}^m \text{ s.t. } \mathbf{A}\mathbf{e} = \mathbf{u} \pmod{q}\}.$$

It is clear that both $\Lambda_q^{\perp}(\mathbf{A})$ and $\Lambda_q^{\mathbf{u}}(\mathbf{A})$ are q-ary lattices, and $\Lambda_q^{\mathbf{u}}(\mathbf{A})$ can be seen as a coset on $\Lambda_q^{\perp}(\mathbf{A})$.

2.3 Discrete Gaussian and Learning with Errors

For a parameter $r > 0$ and a vector $\mathbf{x}$ randomly chosen from $\mathbb{R}^n$, denoted on $\mathbb{R}^n$ by the following function:

$$\rho_{r,\mathbf{c}}(\mathbf{x}) := \exp\left(-\pi\frac{\|\mathbf{x}-\mathbf{c}\|^2}{r^2}\right),$$

where $\mathbf{c}$ is the centre. We note that r and $\mathbf{c}$, when the values of these two parameters are equal to 1 and $\mathbf{0}$, respectively, can be omitted. For an n-dimensional lattice Λ and a randomly selected vector $\mathbf{x}$ on the lattice. The discrete Gaussian distribution on the lattice Λ is:

$$\mathcal{D}_{\Lambda,r,c}(\mathbf{x}) := \left(\frac{\rho_{r,\mathbf{c}}(\mathbf{x})}{\rho_{r,\mathbf{c}}(\Lambda)}\right).$$

Learning with Error. Regev [17] introduced Learning with Error (LWE), which is an intermediate problem between the worst case problem of lattice structure and cryptographic structure. The LWE problem not only facilitates the construction of public-key cryptosystems, but also reduces to the lattice-difficulty problem in the worst case. For $q \geq 2$ some suitable parameters and some distributions (discrete Gaussian distribution), the decision LWE is expressed as follows:

1. Given a pair $(\mathbf{A}, \mathbf{A}^T\mathbf{s}+\mathbf{e})$, where a uniform distribution $\mathbf{A} \in \mathbb{Z}_q^{n\times m}$, select a uniform random vector $\mathbf{s} \in \mathbb{Z}_q^n$ and a discrete Gaussian distribution are chosen to sample $\mathbf{e} \in \chi^m$.
2. Given another pair $(\mathbf{A}, \mathbf{b})$, where $\mathbf{A}$ is chosen uniformly at random from $\mathbb{Z}_q^{n\times m}$ and $\mathbf{b} \in \mathbb{Z}_q^m$ is a uniformly random vector.

Theorem 1 ([17]). *Suppose $\alpha \in (0,1)$ and q are a prime number satisfying $\alpha\cdot q > 2\sqrt{n}$. If an efficient algorithm exists for solving the LWE, then an efficient quantum algorithm also exists for approximating the SIVP and the GapSVP on any n-dimensional lattice, in the worst case within a factor of $\widetilde{O}(\frac{n}{\alpha})$.*

In the above theorem, Regev proved that solving the average-case LWE problem with quantum algorithms is not easier than some worst-case lattice problems.

LWE Based on the Dual Regev Scheme. In [15], the authors proposed a Dual Regev PKE scheme, which differs from the Regev cryptosystem [24]. This scheme is more suitable for constructing lattice IBE, which is described below. First, we need a shared matrix $\mathbf{A} \leftarrow \mathbb{Z}_q^{n\times m}$ and a function $\mathbf{f_A} := \mathbf{Ax} \pmod q$, where $\mathbf{f_A}$ is a one-way function with a trapdoor ([15], Theorem 4.9).

- **DualKeyGen:** Sample an error vector $\mathbf{e} \leftarrow \mathcal{D}_{\mathbb{Z}^m,r}$ (choose the input from $\mathbf{f_A}$) and set $\mathbf{e}$ to be the secret key. The public key is represented by $\mathbf{u} = \mathbf{f_A}(\mathbf{e})$.

- **DualEnc:** The algorithm takes $\mathbf{u}$ and a random bit b as input. It chooses a uniform vector $\mathbf{s} \in \mathbb{Z}_q^n$, outputs a ciphertext $(\mathbf{p}, C)$, here $\mathbf{p} := \mathbf{A}^T\mathbf{s}+\mathbf{x} \in \mathbb{Z}_q^m$, $\mathbf{x} \leftarrow \chi^m$, and $C = \mathbf{u}^T + \mathbf{x} + b \cdot \lfloor \frac{q}{2} \rceil \in \mathbb{Z}_q^m \times \mathbb{Z}_q$, $\mathbf{x} \leftarrow \chi$.
- **DualDec:** Compute $b' = C - \mathbf{e}^T\mathbf{p} \in \mathbb{Z}_q$. If b' is closer to $\lfloor \frac{q}{2} \rceil \pmod q$ than 0 output 1, otherwise output 0.

Theorem 2 ([18]). *Suppose $q \geq 5r(m+1)$, $\alpha \leq \frac{1}{r\sqrt{m+1}\cdot\omega(\sqrt{\log n})}$ and set $m \geq 2n\log q$. Assume that $LWE_{q,\chi}$ is difficult and Dual cryptosystem is IND-CCA secure.*

2.4 Delegation Algorithm

Theorem 3 ([1,2]). *Suppose $m := \lceil 6n\log q \rceil$. PPT algorithm runs TrapGen algorithm with inputs n, q and outputs the pair $(\mathbf{A} \in \mathbb{Z}_q^{n\times m}, \mathbf{T_A} \in \mathbb{Z}_q^{m\times m})$. $\mathbf{A}$ satisfy close to a randomly chosen matrix from $\mathbb{Z}_q^{n\times m}$, $\mathbf{T_A}$ with non-negligible probability $\|\widetilde{\mathbf{T}}_\mathbf{A}\| \leq O(\sqrt{n\log q})$.*

To safely extend the lattice base to higher-dimensional extended lattices, the following algorithm is introduced.

Lemma 1 ([11]). *There exists an ExtBasis algorithm that takes a matrix $\mathbf{A}$ over $\mathbb{Z}_q^{n\times m}$ whose columns are generated by $\mathbb{Z}_q^n$, of a basis $\mathbf{S} \in \mathbb{Z}^{m\times m}$ of $\Lambda_q^\perp(\mathbf{A})$, and a random matrix $\widetilde{\mathbf{A}} \in \mathbb{Z}_q^{n\times m}$. The algorithm ExtBasis$(\mathbf{S}, \mathbf{A}' = \mathbf{A}\|\widetilde{\mathbf{A}})$, $\mathbf{S}' \in \mathbb{Z}^{(m+\tilde{m})\times(m+\tilde{m})}$ of $\Lambda_q^\perp(\mathbf{A}')$ that satisfies $\|\widetilde{\mathbf{S}}'\| = \|\widetilde{\mathbf{S}}\|$.*

In order to randomise the lattice basis and enhance security and reduce the number of paradigms, the following algorithm is introduced.

RandBasis. In [11], a technique for randomising lattice bases is proposed:

- Input: a basis on $\Lambda_q^\perp(\mathbf{A})$ $\mathbf{T}$, and a Gaussian parameter $\sigma \geq \|\widetilde{\mathbf{T}}\| \cdot \omega(\sqrt{\log n})$.
- Output: a basis $\mathbf{T}'$ on $\Lambda_q^\perp(\mathbf{A})$ satisfying: with overwhelming probability $\|\widetilde{\mathbf{T}}'\| \leq \sigma\sqrt{m}$ (m is the dimension of $\Lambda_q^\perp(\mathbf{A})$), the distributions of $\mathbf{T}$ and $\mathbf{T}'$ are independent of each other.

Lemma 2 ([10]). *The PPT algorithm GenSamplePre, takes a matrix $\mathbf{A} \in \mathbb{Z}_q^{n\times m}$, a set $\mathcal{S} \subseteq [k]$, and a random short basis about $\Lambda_q^\perp(\mathbf{A}_\mathcal{S})$ $\mathbf{B}_\mathcal{S}$, a vector $\mathbf{y} \in \mathbb{Z}_q^n$, and a positive integer $r \geq \|\mathbf{B}_\mathcal{S}\| \cdot \omega(\sqrt{\log km})$, output $\mathbf{e} \leftarrow$ GenSamplePre$(\mathbf{A}, \mathbf{B}_\mathcal{S}, \mathcal{S}, \mathbf{y}, r)$ satisfies the distribution on $\mathcal{D}_{\Lambda_q^\mathbf{y}(\mathbf{A}),r}$.*

2.5 System Model

The system model is adapted from [24].

The U-IBE (key update identity based encryption) scheme consists of the five algorithms along with a message space and a ciphertext space.

- **Setup:** Taking security parameters λ and N (total period) as input parameters to PKG. The algorithm outputs a pair (mpk, msk) and a public parameter PP, where mpk is the master public key and msk is the master private key, and msk is kept secret by the PKG.
- **KeyGen:** This algorithm inputs msk and an identity id, outputs the initial secret key of the first interval of the system corresponding to this id, denoted $\mathrm{sk}_{\mathrm{id},0}$.
- **Update:** This algorithm inputs id, $\mathrm{sk}_{\mathrm{id},t}$ and t, outputs a secret key that will be used in interval $t+1$ $\mathrm{sk}_{\mathrm{id},t+1}$.
- **Encrypt:** This algorithm takes id, t and a message $M \in \mathcal{M}$, outputs a ciphertext $C \in \mathcal{C}$.
- **Decrypt:** This algorithm takes id, a ciphertext $C \in \mathcal{C}$ and $\mathrm{sk}_{\mathrm{id},t}$, returns a decrypted message $M \in \mathcal{M}$ or a symbol for rejection $\perp$.

2.6 Security Model

Selective ID, Chosen Plaintext Attack (sID-CPA): If there does not exist a PPT adversary that can win the following attack game with non- negligible advantages in polynomial time. The U-IBE system satisfies sID-CPA security.

- **Init:** The attacker submits a id*. The challenger runs Setup and sends the obtained parameters to the adversary.
- **Query Phase 1:** The adversary will make q_i queries to $i = 1, \cdots, m$ and the challenger responds as follows.
  - **Query(id, j):** Run KeyGen and Update algorithm to generate a secret key $\mathrm{sk}_{\mathrm{id},j}$ which corresponds to the pair (id, j) and forward it to the adversary.
- **Challenge:** After the adversary finishs the first stage of the query, he submits $M_0, M_1 \in \mathcal{M}$ of the same length, along with id* and the challenge interval i^*. The challenger randomly chooses $b \leftarrow \{0,1\}$ runs the Decrypt to encrypt it M_b and receives a ciphertext C^* which is then sent to the adversary.
- **Query Phase 2:** The adversary will make q_i queries to $i = m+1, \cdots, n$ and the challenger will respond as follows.
  - **Query(id, j):** Keep the same restrictions as in the challenge step for the pair (id, j) and keep the same response as in Phase 1 for the secret key query.
- **Guess:** The adversary terminates and outputs a random bit $b' \in \{0,1\}$. If $b = b'$, we say that the adversary wins the attack game.

The adversary wins the attack game by the advantage:

$$\mathrm{Adv}[\mathcal{A}, \mathcal{E}] = \left|\Pr[b = b'] - \frac{1}{2}\right|,$$

where $\mathcal{A}$ is the adversary and $\mathcal{E}$ is the scheme. Here is the scheme U-IBE security model we defined.

3 U-IBE Scheme

The U-IBE system relies on the LWE hard problem assumption. Before presenting the construction of the system, we introduce the binary tree structure, the definition of a binary tree, and the rules for representing nodes in the tree on which the system in this paper is based.

3.1 Tree Definition

Binary Tree. As shown in Fig. 1 below, we define a binary tree with depth l.

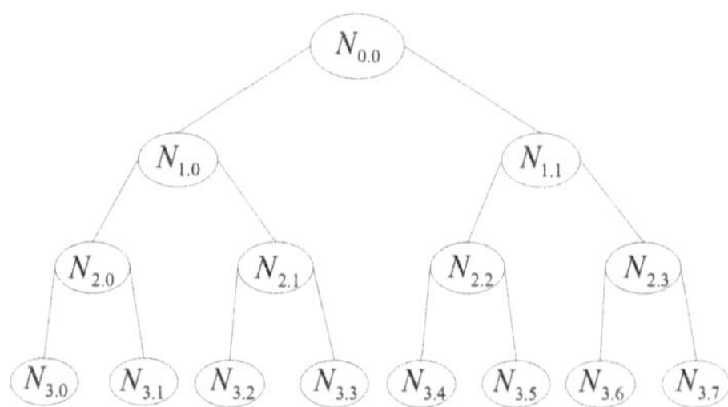

Fig. 1. depth of binary tree l = 4, total epoch N = 8.

Definitions

- In general, a finite string of 0 and 1 represents an id. For each branch of a binary tree, we assume that the branch on the left is denoted as 0 and the branch on the right is denoted as 1. For example, the string associated with $N_{2,1}$ is 001.
- We define three hash functions: $H : \{0,1\}^* \rightarrow \mathbb{Z}_q^{n\times m}$, $H_1 : \{0,1\}^* \rightarrow \mathbb{Z}_q^{n\times m}$ and $G : \{0,1\}^* \rightarrow \mathbb{Z}_q^m$.
- Assume that the depth of the system binary tree is l, $l \geq 3$. Intuitively, for an identity id and a fixed length l, r (defined in Lemma 2), is used to generate a secret key for that user id.

- We define the second leaf of the l layer of the binary tree to define the first time interval of the system. To simplify the representation, we define:

$$N^{(l-1,t)} := t - 1(1 \leq t \leq 2^l - 1).$$

The first t represents the t-node at the l-th level of the binary tree, and the second t represents the period of the U-IBE system.

Leaf Nodes Structure

The leaf nodes at each level of the binary tree are defined:

$$N^{(i,j)} = (N_{0,0}, N_{1,j_1}, \cdots, N_{i,j_i}) \quad (0 \leq i \leq l-1, 0 \leq j_i \leq 2^i - 1).$$

The path vector shows the unique path from the root $N_{0,0}$ to the node N_{i,j_i}. For each layer of leaf nodes, the nodes in this layer are labeled in order from left to right. For any node N_{i,j_i}, we create the corresponding matrix:

$$\mathbf{A}_{N^{(i,j)}} = \left[\mathbf{A} \parallel \mathbf{A}_{N_{1,j_1}} \parallel \cdots \parallel \mathbf{A}_{N_{i,j_i}}\right] \in \mathbb{Z}_q^{n\times(i+1)m},$$

here the matrix $\mathbf{A} \in \mathbb{Z}_q^{n\times m}$ is associated with the $\mathbf{T_A} \in \mathbb{Z}_q^{m\times m}$ generated by the TrapGen algorithm. The matrix $\mathbf{A}_{N_{i,j_i}} \in \mathbb{Z}_q^{n\times m}$ is a random matrix ($0 \leq i \leq l-1$, $0 \leq j_i \leq 2^i - 1$). Each node N_{i,j_i} is associated with its corresponding trapdoor $\mathbf{T}_{N^{(i,j)}}$, and the trapdoor can be computed as follows:

$$\mathbf{T}_{N^{(i,j)}} \leftarrow \text{ExtBasis}(\mathbf{T_A}, \mathbf{A}_{N^{(i,j)}}).$$

In addition, if one knows its ancestor node N_{k,j_k} and the trapdoor associated with it, one can also compute:

$$\mathbf{T}_{N^{(i,j)}} \leftarrow \text{ExtBasis}(\mathbf{T}_{N^{(i,j)}}, \mathbf{A}_{N^{(i,j)}}),$$

where $\mathbf{A}_{N^{(i,j)}} = \left[\mathbf{A} \parallel \mathbf{A}_{N_{1,j_1}} \parallel \cdots \parallel \mathbf{A}_{N_{k,j_k}} \parallel \cdots \parallel \mathbf{A}_{N_{i,j_i}}\right]$.

3.2 Scheme Construction

We call this IBE scheme the U-IBE scheme and its algorithm consists of the following parts.

- **Setup:** Input the security parameter λ, and the total epoch N. Run Trap -Gen(q, n) algorithm to output $(\mathbf{A}, \mathbf{T_A})$, $(\mathbf{A} \in \mathbb{Z}_q^{n\times m}, \mathbf{T_A} \in \mathbb{Z}_q^{m\times m})$. The system defines $(\mathcal{M}, \mathcal{C})$, $\mathcal{M}$ as the plaintext space $\mathcal{M} := \{0,1\}$ and $\mathcal{C}$ as the ciphertext space $\mathcal{C} := \{\mathbb{Z}_q^{ml} \times \mathbb{Z}_q\}$. Output parameters PP $:= \{\lambda, n, q, m, l, r, \mathbf{A}, H, H_1, G\}$ and master private key mpk $= \mathbf{A}$, msk $:= \mathbf{T_A}$.
- **KeyGen:** Input id, msk. The algorithm outputs the user's initial private key $\text{sk}_{\text{id},0}$. At the beginning, for a user identity id, we define the lattice $\mathbf{A}_{\text{id},0}$, and the vector $\mathbf{y}_{\text{id},0}$. The private key consists of two parts: the trapdoor and the decryption key.

$$\mathbf{A}_{\text{id},0} = \left[\mathbf{A} \parallel \mathbf{A}_{\text{id},N_{1,j_1}} \parallel \cdots \parallel \mathbf{A}_{\text{id},N_{i,j_i}} \parallel \mathbf{A}_{\text{id},N_{l-1,0}}\right],$$

$$\mathbf{A}_{\text{id},N_{i,j_i}} = H(\text{id}, N_{i,j_i}),$$
$$\mathbf{A}_{\text{id},N_{l-1,0}} = H(\text{id}, N_{l-1,0}),$$
$$\mathbf{y}_{\text{id},0} = G(\text{id}, t),$$
$$\mathbf{T}_{\text{id},0} \leftarrow \text{ExtBasis}(\mathbf{T}_{\mathbf{A}}, \mathbf{A}_{\text{id},0}).$$

These trapdoors are randomized using the RandBasis algorithm to generate the new trapdoor set.Then the newly trapdoor set is used to calculate the decryption key.

$$\mathbf{e}_{\text{id},0} \leftarrow \text{GenSamplePre}(\mathbf{A}_{\text{id},0}, \mathbf{T}_{\text{id},0}, \mathcal{S}, \mathbf{y}_{\text{id},0}, r),$$

$$\text{sk}_{\text{id},0} = (\mathbf{T}_{\text{id},0}, \mathbf{e}_{\text{id},0}).$$

- **Update:** Input id, current time period t and previous time period $\text{sk}_{\text{id},t}$, output the private key $\text{sk}_{\text{id},t+1}$ for current identity id.

$$\mathbf{A}_{\text{id},t+1} = \left[\mathbf{A} \parallel \mathbf{A}_{\text{id},N_{1,j_1}} \parallel \cdots \parallel \mathbf{A}_{\text{id},N_{i,j_i}} \parallel \mathbf{A}_{\text{id},N_{l-1,t+1}}\right]$$
$$\mathbf{A}_{\text{id},N_{i,j_i}} = H(\text{id}, N_{i,j_i}),$$
$$\mathbf{A}_{\text{id},N_{l-1,t+1}} = H(\text{id}, N_{l-1,t+1}),$$
$$\mathbf{y}_{\text{id},t+1} = G(\text{id}, t+1),$$
$$\mathbf{T}_{\text{id},t+1} \leftarrow \text{ExtBasis}(\mathbf{T}_{\mathbf{A}}, \mathbf{A}_{\text{id},t+1}).$$

These trapdoors are randomized using the RandBasis algorithm to generate the new trapdoor set. Then the newly trapdoor set is used to calculate the decryption key.

$$\mathbf{e}_{\text{id},t+1} \leftarrow \text{GenSamplePre}(\mathbf{A}_{\text{id},t+1}, \mathbf{T}_{\text{id},t+1}, \mathcal{S}, \mathbf{y}_{\text{id},t+1}, r),$$

$$\text{sk}_{\text{id},t+1} = (\mathbf{T}_{\text{id},t+1}, \mathbf{e}_{\text{id},t+1}).$$

- **Encrypt:** Input id, t, and a random bit $b \in \{0,1\}$, randomly select $\mathbf{s} \leftarrow \mathbb{Z}_q^n$, compute $\mathbf{p} = \mathbf{A}_{\text{id},t}^T\mathbf{s} + \mathbf{x} \in \mathbb{Z}_q^{ml}$, $\mathbf{x} \leftarrow \chi^{ml}$, outputs a ciphertext, $(\mathbf{p}, C = \mathbf{y}_{\text{id},t}^T\mathbf{s} + x + b \cdot \lfloor \frac{q}{2} \rceil) \in \mathbb{Z}_q^{ml} \times \mathbb{Z}_q$, $x \leftarrow \chi$.
- **Decrypt:** Input a ciphertext C, and $\text{sk}_{\text{id},t}$ with the user's identity id. Calculate $b^* = C - \mathbf{e}_{\text{id},t}^T\mathbf{p} \in \mathbb{Z}_q$, if b^* is closer to $\lfloor \frac{q}{2} \rceil$ than 0 (mod q) output 1, otherwise output 0.

Correctness: All the algorithms of the system work correctly because the parameters are chosen to satisfy the appropriate conditions. The correctness of the IBE system, which is based on the Regev cryptosystem, can be deduced from [15] Theorem 7.1.

4 Security Analysis

Assume that Theorem 2 holds. The Dual Regev cryptosystem is IND-CPA secure. For a shared matrix, if there are three hash functions, then it is CPA secure in the oracle model. Further,

$$\mathrm{Adv}[\mathcal{A}_{\mathrm{BTE}}, \mathrm{U\text{-}IBE}] \leq \frac{1}{N} \cdot \mathrm{Adv}[\mathcal{B}_{\mathrm{dual}}, \mathrm{PKE}] + \mathrm{negl}(n),$$

N is the total epoch of U-IBE.

Proof. Suppose there exists an adversary $\mathcal{A}$ attacking the system under the CPA secure model. Before attacking the system, $\mathcal{A}$ first submits a target to be attacked id*, and the adversary $\mathcal{A}$ will attack in the challenge phase. We construct a PPT algorithm $\mathcal{B}$ as a challenger to simulate the attack against the adversary under the Regev cryptosystem. Firstly,$\mathcal{B}$ selects $i^* \in \{0, \cdots, N-1\}$ as a guess of the time period in the challenge phase. Subsequently, $\mathcal{B}$ is given as a matrix $\mathbf{A}' \in \mathbb{Z}_q^{n\times ml}$ and a vector $\mathbf{u}^* \in \mathbb{Z}_q^n$, which are the main parameters of the Dual cryptosystem

$$\mathbf{A}' = [\mathbf{A} \parallel \mathbf{A}'_{1,j_1} \parallel \mathbf{A}'_{1,j_2} \parallel \cdots \parallel \mathbf{A}'_{i,j_l}],$$

here, $\mathbf{A} \in \mathbb{Z}_q^{n\times m}$ and $\mathbf{A}'_{i,j_i} \in \mathbb{Z}_q^{n\times m}$, $i = 1, 2, \cdots t$.
In order to simulate the confrontation between the challenger and the adversary. At the start of the simulation, $\mathcal{B}$ sends $\mathbf{A} \in \mathbb{Z}_q^{n\times m}$ as a parameter to the adversary $\mathcal{A}$, who $\mathcal{A}$ can make hash queries. We assume that:

- The adversary $\mathcal{A}$ can make Q_H different $H(\cdot)$ hash queries in polynomial time.
- The adversary $\mathcal{A}$ can make Q_{H_1} different $H_1(\cdot)$ hash queries in polynomial time; and
- The adversary $\mathcal{A}$ can make Q_G different $G(\cdot)$ hash queries in polynomial time; When the adversary $\mathcal{A}$ is asking for a private key or initiating a challenge, the random predictor query was completed.

Next, the simulator $\mathcal{B}$ needs to simulate any type of query answer from the adversary $\mathcal{A}$. First $\mathcal{B}$ create empty lists called H, H_1, G and record the simulated values in subsequent queries.

- **Query $H(\cdot)$:** The adversary $\mathcal{A}$ randomly asks $H(\mathrm{id}, N_{i,j_i})$, and $\mathcal{B}$ responds as follows:
  - If id = id* and N_{i,j_i} is a path vector from $N_{0,0}$ to N_{l-1,i^*+1}, $\mathcal{B}$ returns a matrix $\mathbf{A}'_{i,j_i}$ to the adversary $\mathcal{A}$.
  - If id = id*, N_{i,j_i} is not a path vector from $N_{0,0}$ to N_{l-1,i^*+1}, $\mathcal{B}$ runs TrapGen algorithm to generate $\mathbf{A}_{\mathrm{id},N_{i,j_i}} \in \mathbb{Z}_q^{n\times m}$, corresponding trapdoor $\mathbf{T}_{N_{i,j_i}} \in \mathbb{Z}_q^{n\times m}$, $\mathcal{B}$ returns $\mathbf{A}_{\mathrm{id},N_{i,j_i}}$ to the adversary, stores $((\mathrm{id}, N_{i,j_i}), \mathbf{A}_{\mathrm{id},N_{i,j_i}}, \mathbf{T}_{\mathrm{id},N_{i,j_i}})$ in the H-list.

- If $\text{id} \neq \text{id}^*$ $\mathcal{B}$ responds in the same way as in step two.

- **Query $H_1(\cdot)$:** the adversary $\mathcal{A}$ randomly asks $H_1(\text{id}, N_{i-1,k})$, and $\mathcal{B}$ responds as follows:
  - If $\text{id} = \text{id}^*$ and $k = i^*$, $\mathcal{B}$ responds that the adversary failed or terminated.
  - If $\text{id} \neq \text{id}^*$ or $\text{id} = \text{id}^*$, $k \neq i^*$, $\mathcal{B}$ runs the TrapGen algorithm generate $\mathbf{A}_{\text{id},N_{i-1,k}} \in \mathbb{Z}_q^{n\times m}$, $\mathbf{T}_{\text{id},N_{i-1,k}} \in \mathbb{Z}_q^{n\times m}$, $\mathcal{B}$ returns $\mathbf{A}_{\text{id},N_{i-1,k}}$ to the adversary and stores $((\text{id}, N_{i-1,k}), \mathbf{A}_{\text{id},N_{i-1,k}}, \mathbf{T}_{\text{id},N_{i-1,k}})$ in the H_1 list.
  - If $\text{id} \neq \text{id}^*$ $\mathcal{B}$ responds in the same way as in step two.
- **Query $G(\cdot)$:** the adversary $\mathcal{A}$ randomly asks $G(\text{id}, k)$, and $\mathcal{B}$ responds as follows:
  - If $\text{id} = \text{id}^*$ and $k = i^*$, $\mathcal{B}$ responds that the adversary failed or terminated.
  - If $\text{id} \neq \text{id}^*$ or $\text{id} = \text{id}^*$, $k \neq i^*$, set $\mathbf{y}_{\text{id},k} = G(\text{id}, k)$. Then return the vector $\mathbf{y}_{\text{id},k}$ to the adversary and put $((\text{id}, k), \mathbf{y}_{\text{id},k})$ into the list G.
- **Query Phase 1:** When the adversary $\mathcal{A}$ makes a private key query for (id, k), we assume that we have defined two vectors $\mathbf{A}_{\text{id},k}$, $\mathbf{y}_{\text{id},k}$.
  - If $\text{id} = \text{id}^*$ and $k \neq i^*$, $\mathcal{B}$ first builds the corresponding matrix $\mathbf{A}_{\text{id}^*,j}$. In fact we only need to make sure that $\mathcal{B}$ can correctly simulate the trapdoor $\mathbf{T}_{\text{id}^*,k}$, and then based on the known trapdoor, the rest of the private key is computed by the GenSamplePre algorithm. According to the definition of a binary tree, $\mathbf{A}_{\text{id}^*,k}$ and $\mathbf{A}_{\text{id}^*,j}$ have at least one $n \times m$ with different matrices. Therefore, when answering a query from an adversary, the trapdoors of different $n \times m$ matrices can be simulated with the TrapGen algorithm. For $\mathbf{T}_{\text{id}^*,k}$ other trapdoors, $\mathcal{B}$ are already recorded in the H-list during the hash query phase and they can be obtained using the simple ExtBasis algorithm. Thus $\mathcal{B}$ can generate $\mathbf{T}_{\text{id}^*,k}$, then use the RandBasis algorithm to compute the vector $\mathbf{e}_{\text{id}^*,k}$, and finally $\mathcal{B}$ returns $\text{sk}_{\text{id}^*,k} = (\mathbf{T}_{\text{id}^*,k}, \mathbf{e}_{\text{id}^*,k})$ to the adversary $\mathcal{A}$.
  - If $\text{id} \neq \text{id}^*$, $\mathcal{B}$ get the corresponding trapdoor by ExtBasis algorithm and RandBasis algorithm. Then $\mathbf{e}_{\text{id},k}$ is computed by GenSamplePre algorithm using the trapdoor $\mathbf{T}_{\text{id},k}$ in the G hash list and the vector $\mathbf{y}_{\text{id},k}$. Finally, finally $\mathcal{B}$ returns $\text{sk}_{\text{id},k} = (\mathbf{T}_{\text{id},k}, \mathbf{e}_{\text{id},k})$ to the adversary $\mathcal{A}$.
- **Challenge:** Suppose the adversary's Query1 phase is over and the adversary $\mathcal{A}$ submits a challenge to $(\text{id}^*, j, M_0, M_1)$ and and has not been accessed before (id^*, j). If $j = i^*$, $\mathcal{B}$ sends the message M_0, M_1 to its own challenger and obtains a challenge ciphertext C^* from the challenger and sends C^* to the adversary $\mathcal{A}$. If $j \neq i^*$, $\mathcal{B}$ replies fail or terminate to the adversary $\mathcal{A}$.
- **Query Phase 2:** When the adversary $\mathcal{A}$ performs (id, k) private key query with the same constraints as in the challenge phase, the response of $\mathcal{B}$ is the same as in the query 1 phase.
- **Guess:** When the adversary takes a random bit as output, $\mathcal{B}$ terminates and outputs the same. From the simulation, we can see that the probability of the game aborting is exactly close to $1 - \frac{1}{N}$, from the perspective of the adversary $\mathcal{A}$. The condition for the simulation not to abort is that the scenario is close to a CPA game with real selective ID from the adversary's point of view. Moreover, for queries from the adversary's random preconditioner, the simulator's answers are indistinguishable from those chosen in a uniform

distribution (for almost all variables of the matrix). In particular, the answers to the simulator's extracted queries are indistinguishable from the answers to the KeyGen and Update outputs. Finally, we claim that if the simulation is not aborted, then the advantage of the adversary $\mathcal{A}$ winning the game is equivalent to the advantage of the simulator $\mathcal{B}$ succeeding in the simulation, which proves the theorem.

5 Applications and Analyses of U-IBE

In this section, we present specific application scenarios and provide detailed parameter analyses.

5.1 Applications of U-IBE in IoT

- **Application 1: Data Communication Security.** With the common application of IoT, it is especially important to ensure the security of data during mutual communication. By applying the scheme, we can protect the data during the process of data interaction between communicating parties, prevent sensitive information leaked, and ensure the security and privacy of the data. For example, the transmission of the user's data in the wearable device system or the data circulation in the banking system. The security of data communication is guaranteed, ensuring that private data remains inaccessible to unauthorized parties. It can prevent third parties from intercepting or tampering with data, thus protecting privacy and reliability.
- **Application 2: Smart home remote control.** Devices connected to a smart home in the IoT can be remotely controlled and operated to fulfill some necessary functions. A security mechanism is provided for such remote operations to ensure that only legally authorised users can perform the relevant operations. The management of user authorization time is necessary to ensure that the user's system is not illegally compromised. If implemented, it will prevent unauthorized individuals from taking operational control of the system, providing a safe and secure environment for stable system operation.
- **Application 3: Secure Firmware Updates.** Regular updates of commonly used firmware are particularly important for access to IoT devices, as they can not only improve the reliability of the firmware but also introduce new features to enhance the user experience. Ensure security during firmware updates to prevent unauthorised intrusion of malicious firmware. This protects users from potential security breaches, especially in smart vehicles and smart cities. Ensure that only authenticated firmware is allowed to function properly, thus securing the firmware from third-party firmware coercion.

5.2 Analysing

Parameter Selection. For the parameter selection of quantum-resistant encryption schemes, we follow the standards published by. In particular, we refer to the parameter choices in [7] and the levels of security they achieve. While the hard problem differs from the underlying hard problem in this work, the design documentation treats it as if it were analyzing security. Therefore, we consider the parameter set to be a valuable reference for us.

Size of the Main Parameters. In Table 1, we list the parameter sizes of. The parameters in the table denote the security parameter, the number of rows in the matrix, the number of columns in the matrix, and the total number of cycles, respectively.

Table 1. Parameter selection for U-IBE.

Parameter	mpk	msk	sk	cipherter size
U-IBE	$n \cdot m$	$m \cdot m$	$\leq (m \cdot m) + \log N + m$	$m \cdot \log N + 1$

In our proposed scheme, not all of our parameters are of "constant size". Nevertheless, we still achieve the same "logarithmic size" of the secret key and ciphertext as in the previous work [9,21]. The Assumption on which we have based this work is the hardship Assumption, and most previous work has been constructed based on this traditional Assumption. Compared with the existing scheme, the parameter sizes of the scheme in this paper are not theoretically increased, at least to the same level. Compared to the known realisations of the lattice [4,5,22,24]. Compared to this, our parameter choices are somewhat similar, but our key and ciphertext sizes are larger than those of some known grids, which will affect the efficiency of the scheme. Essentially, this is a trade-off between efficiency and functionality. Furthermore, applying their optimisation techniques may further enhance efficiency.

6 Conclusion

In this paper, we propose a lattice based Identity Based Encryption scheme with selective ID, CPA security. Due to this, the scheme is considered to be resistant to quantum attacks based on the lattice difficulty assumption. In addition, we combine the binary tree structure and the lattice structure to construct the scheme presented in this paper. We present the results of analyzing the scheme parameters and explore and analyze application examples in the Internet of Things. In our future work, we plan to investigate a new secure update method to achieve autonomous key update without relying on binary trees. It will achieve the forward security in the IBE scheme. In addition to this, we will also consider how to reduce the size of the parameters more efficiently in the lattice-based scheme and design new lattice-based cryptographic algorithms to achieve higher security.

References

1. Ajtai, M.: Generating hard instances of lattice problems. In: 28th Annual ACM Symposium on Theory of Computing, pp. 99–108 (1996)
2. Alwen, J., Peikert, C.: Generating shorter bases for hard random lattices. Theory Comput. Syst. **48**(3), 535–553 (2011)
3. Li, S., Xu, L.D., Zhao, S.: The internet of things: a survey. Inf. Syst. Front. **17**(2), 243–259 (2015)
4. Bert, P., Eberhart, G., Prabel, L., Roux-Langlois, A., Sabt, M.: Implementation of lattice trapdoors on modules and applications. In: Cheon, J.H., Tillich, J.-P. (eds.) PQCrypto 2021 2021. LNCS, vol. 12841, pp. 195–214. Springer, Cham (2021). https://doi.org/10.1007/978-3-030-81293-5_11
5. Bert, P., Fouque, P.-A., Roux-Langlois, A., Sabt, M.: Practical implementation of ring-SIS/LWE based signature and IBE. In: Lange, T., Steinwandt, R. (eds.) PQCrypto 2018. LNCS, vol. 10786, pp. 271–291. Springer, Cham (2018). https://doi.org/10.1007/978-3-319-79063-3_13
6. Boneh, D., Franklin, M.: Identity-based encryption from the weil pairing. In: Kilian, J. (ed.) CRYPTO 2001. LNCS, vol. 2139, pp. 213–229. Springer, Heidelberg (2001). https://doi.org/10.1007/3-540-44647-8_13
7. Bos, J., et al.: CRYSTALS - Kyber: a CCA-secure module-lattice-based KEM. In: 2018 IEEE European Symposium on Security and Privacy (EuroS&P), pp. 353–367 (2018)
8. Boneh, D., Boyen, X., Goh, E.-J.: Hierarchical identity based encryption with constant size ciphertext. In: Cramer, R. (ed.) EUROCRYPT 2005. LNCS, vol. 3494, pp. 440–456. Springer, Heidelberg (2005). https://doi.org/10.1007/11426639_26
9. Canetti, R., Halevi, S., Katz, J.: A forward-secure public-key encryption scheme. In: Biham, E. (ed.) EUROCRYPT 2003. LNCS, vol. 2656, pp. 255–271. Springer, Heidelberg (2003). https://doi.org/10.1007/3-540-39200-9_16
10. Cash, D., Hofheinz, D., Kiltz, E.: How to delegate a lattice basis. Cryptology ePrint Archive (2009)
11. Cash, D., Hofheinz, D., Kiltz, E., et al.: Bonsai trees, or how to delegate a lattice basis. J. Cryptol. **25**(4), 601–639 (2012)
12. Wang, C.: An IBE-based security scheme on Internet of Things. In: 2012 IEEE 2nd International Conference on Cloud Computing and Intelligence Systems, Hangzhou, pp. 1046–1049 (2012)
13. Zankl, A., Seuschek, H., Irazoqui, G., et al.: Side-channel attacks in the internet of things: threats and challenges. In: Solutions for Cyber-Physical Systems Ubiquity, pp. 325–357. IGI Global (2018)
14. Gentry, C., Silverberg, A.: Hierarchical ID-based cryptography. In: Zheng, Y. (ed.) ASIACRYPT 2002. LNCS, vol. 2501, pp. 548–566. Springer, Heidelberg (2002). https://doi.org/10.1007/3-540-36178-2_34
15. Gentry, C., Peikert, C., Vaikuntanathan, V.: Trapdoors for hard lattices and new cryptographic constructions. In: 40th Annual ACM Symposium on Theory of Computing (STOC 2008), pp. 197–206 (2008)
16. Daniele, M., Sabrina, S., Francesco, D.P., Imrich, C.: Internet of things: vision, applications and research challenges. Ad Hoc Netw. **10**(7), 1497–1516 (2012)
17. Regev, O.: On lattices, learning with errors, random linear codes, and cryptography. J. ACM **56**(6), 1–40 (2009)
18. Shamir, A.: Identity-based cryptosystems and signature schemes. In: Blakley, G.R., Chaum, D. (eds.) CRYPTO 1984. LNCS, vol. 196, pp. 47–53. Springer, Heidelberg (1985). https://doi.org/10.1007/3-540-39568-7_5

19. Ambainis, A., Rosmanis, A., Unruh, D.: Quantum attacks on classical proof systems: the hardness of quantum rewinding. In: 2014 IEEE 55th Annual Symposium on Foundations of Computer Science. IEEE (2014)
20. Shao, S., et al.: IBE-BCIOT: an IBE based cross-chain communication mechanism of blockchain in IoT. World Wide Web **24**(5), 1665–1690 (2021). https://doi.org/10.1007/s11280-021-00864-9
21. Yao, D., Fazio, N., Dodis, Y., Lysyanskaya, A.: ID-based encryption for complex hierarchies with applications to forward security and broadcast encryption. In: 11th ACM Conference on Computer and Communications Security (CCS 2004), pp. 354–363 (2004)
22. Zhao, R.K., McCarthy, S., Steinfeld, R., Sakzad, A., O'Neill, M.: Quantum-safe HIBE: does it cost a Latte? IEEE Trans. Inf. Forensics Secur. **19**, 2680–2695 (2023)
23. Deogirikar, J., Vidhate, A.: Security attacks in IoT: a survey. In: 2017 International Conference on I-SMAC (IoT in Social, Mobile, Analytics and Cloud) (I-SMAC), pp. 32–37. IEEE (2017)
24. R., Qu, L., Chen, R., et al.: A lattice-based forward secure IBE scheme for internet of things. Inf. Sci. **660**, 120083 (2024)

A Quantum-Resistant Strong Designated Verifier Proxy Signature Scheme

Jincao Meng, Chaoyang Li(✉), and Xiangjun Xin

College of Software Engineering, Zhengzhou University of Light Industry, Zhengzhou 450001, China
lichaoyang2013@163.com

Abstract. Standard digital signatures are publicly verifiable, but certain applications require special privacy-preserving measures. Therefore, there is a need to prevent signatures from being maliciously disseminated by unspecified users. This paper proposes a strong designated verifier proxy signature scheme based on lattice (L-SDVPS), which prevents the theft of privacy in signatures by non-designated verifiers. The L-SDVPS scheme is based on the small integer solution (SIS) lattice problem, which is resistant to quantum computation attacks. The proposed scheme is anonymous and unforgeable under the random oracle model (ROM). Additionally, the scheme has a shorter signature size and a lower storage consumption advantage compared to related schemes.

Keywords: Lattice · Proxy Signature · Strong Designated Verifier · SIS problem

1 Introduction

Many current cryptographic algorithms risk being compromised by quantum attacks due to the accelerated development of quantum computing [1]. Digital signature is a core technology for ensuring the security of network information and can effectively prevent data leakage, forgery, and tampering [2].

RSA is the first cryptographic algorithm that can be used for both encrypting data and creating digital signatures [3].Although anyone can verify the validity of an RSA signature, there is a risk of privacy leakage. Jakobsson first proposed the DVS scheme in 1996 [4]. In this scheme, the verification of the signature is only possible for the designated verifier, which provides better privacy protection. However, if the signature is stolen during transmission, the adversary will know it comes from the signer, posing certain security issues. In the SDVS scheme, the verification process includes the use of the designated verifier's private key, thereby preventing any third party from verifying the signature. In the identity-based cryptosystem [5], users select their identity information as the public key, and a trusted key generation center (KGC) generates the corresponding private key, simplifying the key management process.

J. Yu and Y. Chen (Eds.): FCS 2025, CCIS 2805, pp. 106–117, 2027.
https://doi.org/10.1007/978-981-95-9514-3_7

Quantum computers can break digital signature algorithms based on the difficulty of large integer factorization and discrete logarithm problems with significant advantages. Therefore, post-quantum cryptography algorithms have received extensive attention from cryptographers both domestically and internationally. The lattice-based cryptosystem has advantages such as resistance to quantum computing attacks and simple and efficient operations [7–25]. Among them, Wang et al. [13] utilized the SIS problem and the learning with errors (LWE) problem to construct the first lattice-based SDVS scheme. Yu et al. [14] proposed a certificateless proxy signature scheme from lattice that can resist quantum attacks. Zhang et al. [16] constructed an SDVS scheme based on an improved version of the SIS problem on lattices. Cai et al. [17] utilized the R-SIS problem and employed the filtering technique of uniform sampling to design an SDVS scheme with a shorter signature, and it can also resist side-channel attacks. Zhang et al. [18] utilized untrapdoorable lattice signatures to construct an SDVS scheme with non-delegatability. Wang et al. [19] constructed a lattice-based proxy signature scheme with fewer parameters under the standard model, thereby addressing the problem of excessive parameters in existing schemes. Wu et al. [20] based on the NTRU lattice, proposed an identity-based proxy signature scheme with a message recovery function, providing a new idea for the design of lattice signature schemes in resource-constrained environments. Wang et al. [21] constructed a proxy signature scheme based on lattice and combined it with IoT, which is more efficient in post-quantum IoT compared to related schemes. Zhang et al. [22] combined lattice theory with the medical industry and proposed the first lattice-based designated verifier auditing scheme, providing better privacy protection for medical data.

This paper proposes an L-SDVPS scheme based on the SIS problem and provides a detailed description of the signature process. The signature generated by the signer can only be verified by the pre-designated verifier (not by an arbitrary third party). This ensures authentication of the origin and integrity of the message and prevents the signature from being misused for validation by unauthorised third parties. Proxy signatures enable the original signer to securely delegate its signing authority to a proxy, which can improve flexibility and efficiency and prevent the original signer from becoming a bottleneck or single point of failure. The anonymity and unforgeability of the scheme are then demonstrated through the use of ROM. Compared with the related schemes, this scheme has a shorter signature length and a smaller public-private key size.

2 Preliminaries

This section gives specific definitions of lattice theory.

2.1 Lattice Theories

Definition 1: ***Lattice*** [26]: $v_1, v_2, \cdots, v_n \in \mathbb{R}^m$ is a set of a linearly independent vectors, the lattice Λ is as shown in Eq. (1):

$$\Lambda = \{a_1 v_1 + a_2 v_2 + \cdots + a_n v_n | a_i \in \mathbb{Z}\} \tag{1}$$

Definition 2: *q-ary Lattice* [26]**:** For a matrix A, a positive integer q, and a vector $y \in \mathbb{Z}_q^n$, the q-ary lattice is as shown in Eq. (2):

$$\begin{aligned} \Lambda^{\perp}(A) &= \{e \in \mathbb{Z}^m | Ae = 0 \; modq\} \\ \Lambda_y^{\perp}(A) &= \{e \in \mathbb{Z}^m | Ae = y \; modq\} \end{aligned} \tag{2}$$

Definition 3: *Gaussian distribution* [27]**:** For any vector $x \in \mathbb{R}^n$, the Gaussian distribution centered on the vector c with $s > 0$ is:

$$\rho_{s,c}(x) = exp\frac{-\pi\|x-c\|^2}{s^2} \tag{3}$$

And the discrete Gaussian distribution is defined as:

$$D_{\Lambda,s,c}(x) = \frac{\rho_{s,c}(x)}{\rho_{s,c}(\Lambda)} \tag{4}$$

Definition 4: $\mathbb{Z} - SIS_{q,n,m,\beta}^k$ *problem* [28]**:** For a positive integer q, matrix $A \in \mathbb{Z}_q^{n \times m}$ and real number β, this problem is to find a solution for $\mathbb{Z}-SIS_{q,n,m,\beta}^k$ istance $Av = 0 \; modq$, where v is a non-zero vector and $\|v\| \leq \beta$.

3 L-SDVPS

The detailed construction of the L-SDVPS scheme under the $\mathbb{Z} - SIS_{q,n,m,\beta}^k$ problem is given below, and the specific process is shown in Fig. 1.

Suppose the original signer is O, the proxy signer is P, and the designated verifier is V. Initially, KGC distributes the key to the user, then O grants the signing authority to the legitimate P and generates a corresponding proxy key. After signing, P uploads the signature to the cloud server. Finally, V verifies the legitimacy of the signature.

(1) Setup: Given a security parameter n, a prime number $q \geq 3$, a complexity function $L = O\sqrt{nlog_2 q}$, the Gaussian parameter $\sigma = L \cdot \omega(\sqrt{log_2 n})$, randomly select a uniform matrix $A \in \mathbb{Z}_q^{n \times m}$ $(m \geq 5nlog_2 q)$, KGC obtains a basis $T \in \mathbb{Z}_q^{m \times m}$ from $\Lambda_q^{\perp}(A)$ as $\|T\| \leq L$ by applying $TrapGen(n, m, q)$, then chooses hash functions:

$$H_1 : \{0,1\}^* \rightarrow \mathbb{Z}_q^{n \times k}$$

$$H_2 : \{0,1\}^* \rightarrow \mathbb{Z}_q^{k}$$

$$H_3 : \{0,1\}^* \rightarrow \mathbb{Z}_q^{k \times k}$$

Output $\{A, H_1, H_2, H_3, \sigma\}$ as the system public parameter pp and retain T as the system master private key.

(2) KeyGen: Given pp, users' identity ID_i, KGC computes the private key S_{ID_i} and the public key Q_{ID_i}.

- Computes the public key $Q_{ID_i} = H_1(ID_i) \in \mathbb{Z}_q^{n \times k}$;

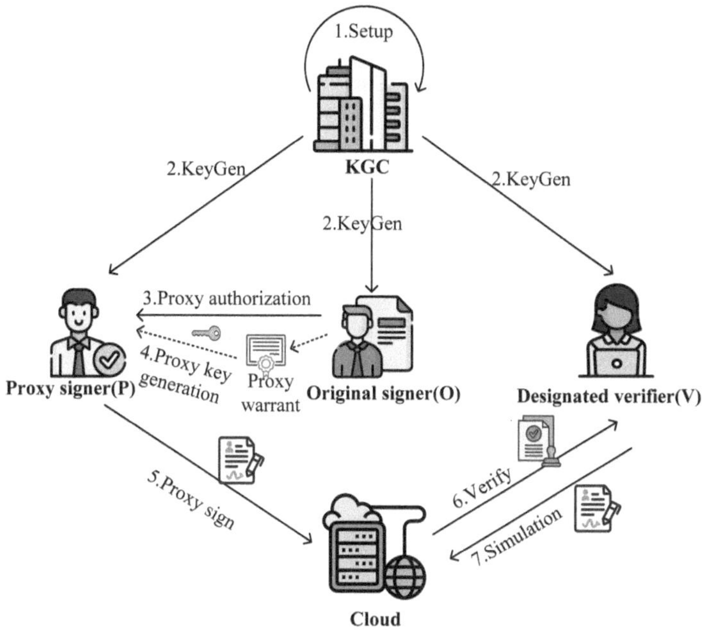

Fig. 1. The L-SDVPS scheme model.

- Runs the $SampleMat(A, T, Q_{ID_i}, \sigma)$ algorithm, and generates a matrix $S_{ID_i} \in \mathbb{Z}_q^{m \times k}$ as the users' private key which satisfies $AS_{ID_i} = Q_{ID_i} modq$, where $\|ID_i\|_\infty \leq \sigma\sqrt{m}$.
- Outputs the key pair (Q_{ID_i}, S_{ID_i}).

(3) Proxy authorization: Given O's S_{ID_O}, proxy warrant W, and *pp*, O authorizes P to sign as a proxy signer.

- O randomly selects $x_0 \leftarrow D_\sigma^m$;
- Computes $e_0 = H_2(Ax_0 modq, W)$;
- Computes $y_0 = x_0 + S_{ID_O} e_0$;
- Sends the proxy warrant $W_{O \to P}(W, e_0, y_0)$ to P with probability $min(\frac{D_\sigma^m(y_0)}{M_0 D_{e_0, \sigma_0(y_0)}^m}, 1)$.

(4) Proxy keys generation: Given the system public parameter *pp*, P's S_{ID_P}, O's Q_{ID_O}, and proxy warrant $W_{O \to P}(W, e_0, y_0)$, P calculates the proxy key pair.

- P checks and rejects $W_{O \to P}$ if $\|y_0\| > T$ or $\|y_0\|_\infty > q/4$;
- Verifies whether $e_0 = H_2(Ay_0 - Q_{ID_O} e_0 modq, W)$ holds or not. If not, rejects. Otherwise, continues.
- Calculates $M = H_3(W_{O \to P})$;
- Calculates proxy public key $G_P = M^{-1}$;
- Calculates proxy private key $S_{G_P} = MS_{ID_P}$.

(5) Proxy sign: Given the message μ, P uses *pp*, proxy key pair (G_P, S_{G_P}), and V's Q_{ID_V} for signing.

- P randomly selects a vector $x_1 \leftarrow D_\sigma^m$;
- Calculates $c = H_2(AG_P x_1 mod q, \mu)$;
- Calculates $v_1 = x_1 + S_{G_P} c$;
- Calculates $v_2 = Q_{ID_V} c$;
- Outputs signature (v_1, v_2, c) with probability $min(\frac{D_{\sigma_1}^m(v_2)}{M_1 D_{v_1,\sigma_1(v_2)}^m}, 1)$.

(6) **Verify**: Given the signature (v_1, v_2, c), pp, G_P, Q_{ID_P}, and V's S_{ID_V}.

- V Checks and rejects (v_1, v_2, c) if $\|v_i\| > T$ or $\|v_i\| > q/4$;
- Verifies whether $c = H_2\{A(G_P v_1 - S_{ID_V} c) + v_2 - Q_{ID_P} c mod q, \mu\}$ holds or not. If this condition is valid, V accepts it; Otherwise, rejects.

(7) **Simulation**: Given the same message μ, pp, Q_{ID_P}, G_V, and S_{G_V}.

- V randomly selects a vector $x_2 \leftarrow D_\sigma^m$;
- Calculates $c' = H_2(AG_V x_2 mod q, \mu)$;
- Calculates $v_1' = x_2 + S_{G_V} c'$;
- Calculates $v_2' = Q_{ID_P} c'$;
- Outputs an indisguishable signature (v_1', v_2', c') with probability $min(\frac{D_{\sigma_1}^m(v_2')}{M_1 D_{v_1',\sigma_1(v_2')}^m}, 1)$.

4 Security Analysis

This subsection analyses the correctness, anonymity, and unforgeability of the proposed L-SDVPS scheme.

4.1 Correctness

Correctness contains two components: the proxy authorization and the proxy signature.

Proxy Authorization. For the correctness of the proxy authorization, first, the proxy warrant is $W_{O \to P}(W, e_0, y_0)$ trusted because it is easy to verify that the proxy warrant satisfies the two conditions $\|y_0\| \leq T$ and $\|y_0\|_\infty \leq q/4$. Meanwhile, equation $H_2(Ay_0 - Q_{ID_O} e_0 mod q, W) = H_2(Ax_0 mod q, W)$ should also be satisfied. This means that $Ay_0 - Q_{ID_O} e_0 = Ax_0 mod q$, the detailed process is shown in Eq. (5).

$$\begin{aligned}
&Ay_0 - Q_{ID_O} e_0 \\
=&A(x_0 + S_{ID_O} e_0) - Q_{ID_O} e_0 \\
=&Ax_0 + AS_{ID_O} e_0 - Q_{ID_O} e_0 \\
=&Ax_0 + Q_{ID_O} e_0 - Q_{ID_O} e_0 \\
=&Ax_0 \mod q
\end{aligned} \tag{5}$$

Proxy Signature. The signature (v_1, v_2, c) satisfies $\|v_i\| \leq T$ and $\|v_i\|_\infty \leq q/4$. Furthermore, $c = H_2\{A(G_P v_1 - S_{ID_V} c) + v_2 - Q_{ID_P} c mod q, \mu\}$ should be fulfilled, which is equivalent to $A(G_P e_1 - S_{ID_V} c) + v_2 - Q_{ID_P} c = AG_P x_1$, the detailed process is shown in Eq. (6).

$$\begin{aligned}
&A(G_P v_1 - S_{ID_V} c) + v_2 - Q_{ID_P} c \\
=&A\{G_P(x_1 + S_{G_P} c) - S_{ID_V} c\} + Q_{ID_V} c - Q_{ID_P} c \\
=&A(G_P x_1 + G_P S_{G_P} c - S_{ID_V} c) + Q_{ID_V} c - Q_{ID_P} c \\
=&AG_P x_1 + AG_P S_{G_P} c - AS_{ID_V} c + Q_{ID_V} c - Q_{ID_P} c \\
=&AG_P x_1 + AS_{ID_P} c - Q_{ID_V} c + Q_{ID_V} c - Q_{ID_P} c \\
=&AG_P x_1 + Q_{ID_P} c - Q_{ID_P} c \\
=&AG_P x_1 \mod q
\end{aligned} \tag{6}$$

At the same time, the simulated signature generated by the designated verifier can also be verified as $\|v_i\| \leq T$, $\|v_i\|_\infty \leq q/4$, and the following equation Eq. (7).

$$\begin{aligned}
&A(G_V v_1' - S_{ID_V} c') + v_2' - Q_{ID_P} c' \\
=&A\{G_V(x_2 + S_{G_V} c') - S_{ID_V} c'\} + Q_{ID_P} c' - Q_{ID_P} c' \\
=&A(G_V x_2 + G_V S_{G_V} c' - S_{ID_V} c') \\
=&AG_V x_2 + AG_V S_{G_V} c' - AS_{ID_V} c' \\
=&AG_V x_2 + AS_{ID_V} c' - Q_{ID_V} c' \\
=&AG_V x_2 + Q_{ID_V} c' - Q_{ID_V} c' \\
=&AG_V x_2 \mod q
\end{aligned} \tag{7}$$

4.2 Anonymity

If all adversaries cannot distinguish the real signer correctly with non-negligible probability, the proposed scheme has anonymity. The anonymity of L-DVPS can be proved by a query-answer game between challenger $\mathcal{D}$ and adversary Γ as follows:

$\mathcal{D}$ and Γ execute two games $Game_0$ and $Game_1$, which simulate the signature process of user $U_0 \rightarrow (E_{k_0}, S_{E_{k_0}})$, $U_1 \rightarrow (E_{k_1}, S_{E_{k_1}})$ respectively. If these two signatures cannot be distinguished by Γ, the proposed scheme is anonymous.

- *$Game_0$*:
  - *Initialize*: $\mathcal{D}$ initializes the parameters, and sends them to Γ.
  - *Constructing signature*: Γ inputs the message $\mu \in \{0,1\}^*$, sends two users U_0 and U_1 to $\mathcal{D}$. $\mathcal{D}$ selects $b = 0 (b \in \{0,1\})$, chooses the user key $(E_{k_b}, S_{E_{k_b}})$, and sends the signature (v_1, v_2, c_0) generated by the **Proxy sign** algorithm to Γ.
  - *Guess*: Γ outputs his guess for the value of b.

- *Game*$_1$: $\mathcal{D}$ selects $b = 1$, and the rest of the steps are the same as *Game*$_0$, obtains signature (v_1^*, v_2^*, c_1) and sends it to Γ. Γ guesses the value of b.
- *Analyze*: From the signature process, as the statistical distances c_0 and c_1 are indistinguishable, the parameter x_1 is randomly selected from the Gaussian distribution D_σ^m, which is uniformly, and the signatures are generated by $v_1 = x_1 + S_{E_{k_i}} c_i$ and $v_2 = Q_{ID_V} c_i$, the statistical distances of these two signatures are also indistinguishable. Therefore, Γ cannot distinguish who generates the signature, the L-SDVPS scheme has anonymity.

4.3 Unforgeability

If all adversaries cannot generate a valid signature with non-negligible probability, then the proposed scheme is unforgeable. The unforgeability of L-SDVPS can be proved by building a similar query-answer game as follows:

Adversary Γ is assumed to be able to successfully forge the proxy signature using q_1 times H_1 *query*, q_2 times *Key query*, q_3 times H_2 *query*, q_4 times *Proxy key query*, q_5 times H_3 *query* and q_6 times *Proxy sign query*, the querying process generates six query lists $\{L_i, i = 1, 2, 3, 4, 5, 6\}$, which are used to store the results of accessing the random oracle model.

- *Initialize*: $\mathcal{D}$ initializes the parameters, and sends them to Γ.
- *Query*: In polynomial time, Γ queried the hash, key, and signature process of the non-target message several times.
  - H_1 *query*: Γ queries $H_1(ID_i)$, $\mathcal{D}$ sends the corresponding value to Γ if it can be found in L_1. Otherwise, $\mathcal{D}$ computes $Q_{ID_i} = H_1(ID_i)$, sends it to Γ, and stores it in L_1.
  - *Key query*: $\mathcal{A}$ queries for key of ID_i, $\mathcal{D}$ sends S_{ID_i} to Γ if it is present in L_2. Otherwise, $\mathcal{D}$ checks if the ID_i has made a H_1 query. If not, returns to H_1 query; if yes, $\mathcal{D}$ computes $S_{ID_i} \leftarrow SampleMat(A, T, Q_{ID_i}, \sigma)$, sends the result to Γ, and stores it in L_2.
  - H_2 *query*: Γ queries the hash value of W, $\mathcal{D}$ checks L_3 and returns the value directly if there is a record in L_3. Otherwise $\mathcal{D}$ randomly selects $x_0 \leftarrow D_\sigma^m$, computes $H_2(Ax_0 mod q, W)$, sends it to Γ, then stores it in L_3.
  - H_3 *query*: Γ queries $W_{O\to P}$ for H_3 value, $\mathcal{D}$ check if the corresponding value exists in L_4, if so, returns to Γ, If not, $\mathcal{D}$ computes $H_3(W_{O\to P})$, returns it to Γ and stores it in L_4.
  - *Proxy key query*: Γ queries the proxy key, $\mathcal{D}$ sends (G_P, S_{G_P}) to Γ if it can be found in L_5. Otherwise, $\mathcal{D}$ executes the **Proxy key generation** algorithm to get (G_P, S_{G_P}), sends it to Γ, and stores it in L_5.
  - *Proxy sign query*: Γ queries for the signature of non-target message μ_i, $\mathcal{D}$ and sends (v_{1_i}, v_{2_i}, c_i) to Γ if it can find the corresponding signature in L_6, otherwise $\mathcal{D}$ performs **Proxy sign** algorithm to computes (v_{1_i}, v_{2_i}, c_i), and stores the result in L_6.
- *Forgery*: Γ forges a legitimate signatures (v_1^*, v_2^*, c^*) on target message μ^*, send it to $\mathcal{D}$.

- *Challenge*: With the query response data from this series of queries and the forged signature (v_1^*, v_2^*, c^*) generated by Γ, $\mathcal{D}$ can generate another valid signature $(v_1^{**}, v_2^{**}, c^{**})$ for the same message.
- *Analyze*: $\mathcal{D}$ uses these two signatures (v_1^*, v_2^*, c^*), and $(v_1^{**}, v_2^{**}, c^{**})$ to find a solution to the SIS problem. The following Eq. (8) holds.

$$\begin{cases} c^* = H_2\{A(G_P v_1^* - S_{ID_V} c^*) + v_2^* - Q_{ID_P} c^* \ modq, \mu^*\} \\ \quad = H_2(AG_P x_1^* modq, \mu^*) \\ c^{**} = H_2\{A(G_P v_1^{**} - S_{ID_V} c^{**}) + v_2^{**} - Q_{ID_P} c^{**} \ modq, \mu^*\} \\ \quad = H_2(AG_P x_1^{**} modq, \mu^*) \end{cases} \tag{8}$$

After simplifying Eq. (8) can obtain Eq. (9).

$$\begin{cases} AG_P(v_1^* - x_1^*) - (Q_{ID_V} + Q_{ID_P})c^* + v_2^* = 0 \\ AG_P(v_1^{**} - x_1^{**}) - (Q_{ID_V} + Q_{ID_P})c^{**} + v_2^{**} = 0 \end{cases} \tag{9}$$

Then it has

$$\begin{cases} AG_P(v_1^* - x_1^*) = Q_{ID_P} c^* modq \\ AG_P(v_1^{**} - x_1^{**}) = Q_{ID_P} c^{**} modq \end{cases} \tag{10}$$

From Eq. (8) can derive $AG_P(v_1^* - v_1^{**} - x_1^* + x_1^{**}) = Q_{ID_P}(c^* - c^{**}) mod2q$, where $c^* \neq c^{**}$, so it can derive $AG_P(v_1^* - v_1^{**} - x_1^* + x_1^{**}) = 0 modq$. Let $v = (v_1^* - v_1^{**} - x_1^* + x_1^{**})$, then v satisfies $AG_P v = 0 \ modq$. Therefore, v is a solution of the SIS problem.

Currently, it is well known that this lattice-based issue cannot be resolved, therefore, the assumption that was previously made is no longer valid. Meanwhile, the probability that Γ generates a valid signature is

$$\varepsilon = Pr[(v_1^*, v_2^*, c^*) = (v_1^{**}, v_2^{**}, c^{**})] = 2^{-\omega logn} \tag{11}$$

As the number of queries increases, the probability that Γ succeeds in forging a valid signature of the target message is negligible, therefore, the L-SDVPS scheme is unforgeable.

5 Performance Evaluation

This section analyses the space efficiency of L-SDVPS and existing lattice-based signature algorithms.

Table 1 gives the comparison of public and private key sizes and signature length of different signature algorithms in spatial dimension. The parameters used for comparison are shown in Table 2. This paper selects two different parameters according to the literature [28], the parameters corresponding to different setup numbers are different, and the security of the corresponding signature algorithms is also different. From Table 1, it can be seen that at the theoretical level, the public key, private key size, and signature length of L-SDVPS are somewhat reduced.

Table 1. Comparison between schemes in the space dimension.

Scheme	Public key	Private key	Signature size
Cai et al. [17]	$m^2 logq$	$m^2 logq$	$3mlogq$
Zhang et al. [18]	$mnlogq$	$m^2 logq$	$3(m + 1 + n)logq$
Wang et al. [21]	$mnlogq$	$m^2 logq$	$2mlogq$
L-SDVPS	$nklogq$	$mklogq$	$(m + n + k)logq$

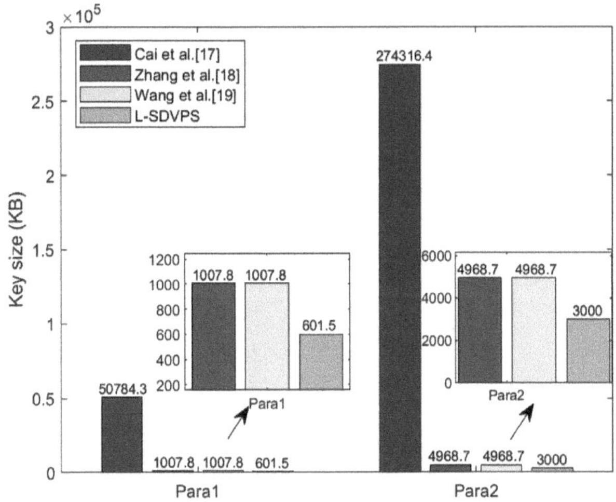

Fig. 2. Public key size comparison.

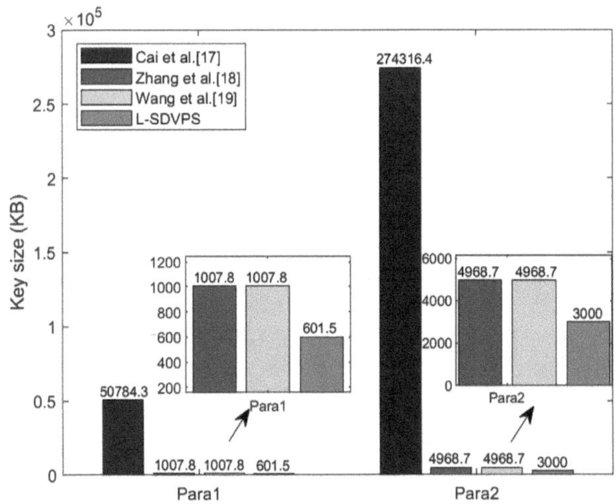

Fig. 3. Private key size comparison.

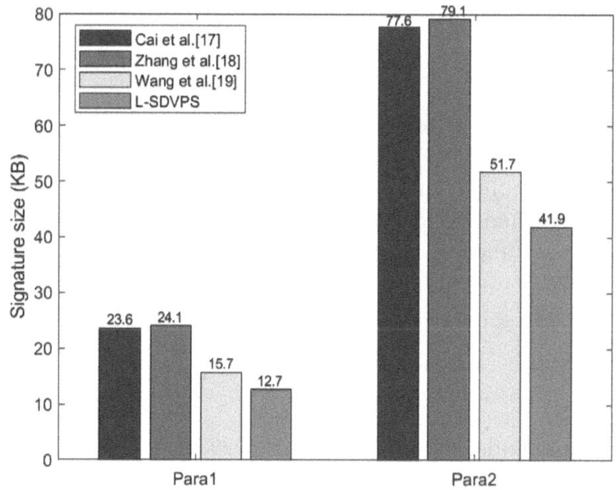

Fig. 4. Signature size comparison.

Table 2. Parameters define

Parameter	1	2
n	128	192
m	6450	10600
q	1024	2048
k	3850	6400

To more easily visualise the gap between several signature algorithms in the spatial dimension, this subsection plots the comparison of public key size, private key size, and signature length between schemes under two different parameters using Matlab 2022 software based on the experimental parameters in Table 2. The results in Fig. 2 and Fig. 3 show that the public and private keys of L-SDVPS are reduced to a great extent under the two parameters, and the gap increases progressively as the security parameters increase. The results in Fig. 4 show that the L-SDVPS signature length is much smaller, and the larger the security parameter, the more obvious the advantage.

From this, it can be shown that the L-SDVPS scheme can take up less storage space with the lowest space consumption.

6 Conclusion

In the context of the continuous evolution in the post-quantum era, the significance of research on lattice-based digital signatures is increasingly prominent. This paper constructs an L-SDVPS scheme based on lattice. This scheme integrates the core features of designated verification signatures and proxy signatures, which not only ensures the privacy of designated verification but also possesses the flexibility of proxy signatures. The correctness of the scheme is analysed. It is proven that the scheme satisfies the anonymity and is non-forgeable under the ROM model. Moreover, by comparing with the existing schemes, it is found that this scheme achieves a significant reduction in storage cost. This advantage enables the scheme to provide users with a relatively more efficient solution in practical application scenarios. At the same time, the privacy protection features of the scheme can fully protect the user's private information and ensure the user's information security in the process of use. The ROM model assumes that the hash function is a perfect and publicly available random function. In the future, the scheme can be further extended to achieve security under the standard model.

Acknowledgments. This work was supported by the National Natural Science Foundation of China under Grant Numbers 72293583, 72293580, the Foundation and Cutting-Edge Technologies Research Program of Henan Province (CN) under Grant Numbers 262102210202, 252102110182, 252102110178, and the Doctor Scientific Research Fund of Zhengzhou University of Light Industry under Grant 2021BSJJ033.

References

1. Yang, Z., Zolanvari, M., Jain, R.: A survey of important issues in quantum computing and communications. IEEE Commun. Surv. Tutor. **25**(2), 1059–1094 (2023)
2. Merkle, R.C.: A digital signature based on a conventional encryption function. In: Pomerance, C. (ed.) CRYPTO 1987. LNCS, vol. 293, pp. 369–378. Springer, Heidelberg (1988). https://doi.org/10.1007/3-540-48184-2_32
3. Wiener, M.J.: Cryptanalysis of short RSA secret exponents. IEEE Trans. Inf. Theory **36**(3), 553–558 (1990)
4. Jakobsson, M., Sako, K., Impagliazzo, R.: Designated verifier proofs and their applications. In: Maurer, U. (ed.) EUROCRYPT 1996. LNCS, vol. 1070, pp. 143–154. Springer, Heidelberg (1996). https://doi.org/10.1007/3-540-68339-9_13
5. Yi, P., et al.: An efficient identity-based signature scheme with provable security. Inf. Sci. **576**, 790–799 (2021)
6. Joseph, D., Misoczki, R., Manzano, M.: Transitioning organizations to post-quantum cryptography. Nature **605**(7909), 237–243 (2022)
7. Li, Q., Shen, J., Lin, C., Wang, Z., He, D.: Two-round identity-based proxy blind signature scheme on lattices. IEEE Internet Things J. **12**(16), 33621–33634 (2025)
8. Yu, H., Wang, N.: Certificateless network coding proxy signatures from lattice. Front. Comput. Sci. **17**(5), 175810 (2023)
9. Lyubashevsky V.: Lattice-based digital signatures. Natl. Sci. Rev. **8**(9), nwab077 (2021)
10. Li, C., et al.: Cross-Chain Privacy-preserving for BIoMT with designated verifier proxy signature. IEEE Internet Things J. **12**(12), 20042–20055 (2025)
11. Zhu, H., Wang, Y., Wang, C., Cheng, X.: An efficient identity-based proxy signcryption using lattice. Futur. Gener. Comput. Syst. **117**, 321–327 (2021)
12. Ji, W., Fan, S., Li, Z.: Identity-based proxy signature on lattices. Comput. Syst. Appl. **32**(10), 301–307 (2023)
13. Wang, F., Hu, Y., Wang, B.: Lattice-based strong designate verifier signature and its applications. Malays. J. Comput. Sci. **25**(1), 11–12 (2012)

14. Yu, H., Wang, N.: Certificateless proxy signcryption scheme from lattice. J. Electron. Inf. Technol. **44**(7), 2584–2591 (2022)
15. Li, C., et al.: Quantum-safe identity-based designated verifier signature for BIoMT. J. Syst. Architect. **160**, 103362 (2025)
16. Zhang, P., Chi, H., Li, J., Shang, Y.: Lattice based strong designated verifier signature scheme. J. Beijing Univ. Aeron. Astron. **49**(6), 1294–1300 (2023)
17. Cai, J., et al.: ID-based strong Designated verifier signature over R-SIS assumption. Secur. Commun. Netw. **2019**(1), 9678095 (2019)
18. Zhang, Y., Susilo, W., Guo, F.: Lattice-based strong designated verifier signature with non-delegatability. Comput. Stand. Interfaces **92**, 103904 (2025)
19. Wang, Z., Bao, H., Zhao, T., Du, F., Chen, X.: New lattice-based short proxy signature scheme with fewer system parameters in the standard model. IEEE Access **10**, 57180–57191 (2022)
20. Wu, F., Zhou, B., Zhang, X.: Identity-based proxy signature with message recovery over NTRU lattice. Entropy **25**(3), 454 (2023)
21. Wang, L., Huang, C., Cheng, H.: Novel proxy signature from lattice for the post-quantum Internet of Things. J. Ambient. Intell. Humaniz. Comput. **14**(8), 9939–9946 (2023)
22. Zhang, X., Huang, C., Zhang, Y., Zhang, J., Gong, J.: LDVAS: lattice-based designated verifier auditing scheme for electronic medical data in cloud-assisted WBANs. IEEE Access **8**, 54402–54414 (2020)
23. Wang, F., Hu, Y., Wang, C.: Identity-based strong designate verifier signature over lattices. J. China Univ. Posts Telecommun. **21**(6), 52–60 (2014)
24. Wang, Q., Wang, K., Chen, H., Xin, C., Yao, Y., Wang, Y.: Identity-based content extraction signature scheme on idea lattices. J. Inf. Secur. Res. **11**(01), 57–65 (2025)
25. Li, C., et al.: Efficient designated verifier signature for secure cross-chain health data sharing in BIoMT. IEEE Internet Things J. **11**(11), 19838–19851 (2024)
26. Micciancio, D.: Lattice-based cryptography. In: Encyclopedia of Cryptography and Security, pp. 713–715. Springer, Boston (2011)
27. Ducas, L., Durmus, A., Lepoint, T., Lyubashevsky, V.: Lattice signatures and bimodal Gaussians. In: Canetti, R., Garay, J.A. (eds.) CRYPTO 2013. LNCS, vol. 8042, pp. 40–56. Springer, Heidelberg (2013). https://doi.org/10.1007/978-3-642-40041-4_3
28. Lyubashevsky, V.: Lattice signatures without trapdoors. In: Pointcheval, D., Johansson, T. (eds.) EUROCRYPT 2012. LNCS, vol. 7237, pp. 738–755. Springer, Heidelberg (2012). https://doi.org/10.1007/978-3-642-29011-4_43

Lattice-Based Boolean Keyword Searchable Encryption for Cloud Server

Shanqiang Feng, Zixuan Yan, and Fengyin Li(✉)

School of Computer Science, Qufu Normal University, Rizhao 276826, China
lfyin318@qfnu.edu.cn

Abstract. With the wide application of cloud storage technology, the contradiction between data security requirements and efficient retrieval needs is becoming more and more prominent. Most of the existing Boolean keyword searchable encryption schemes rely on traditional cryptographic hard problems, which are susceptible to the threat of quantum computation and suffer from the problems of key field leakage and ciphertext length growing with the number of keywords. In this paper, we propose a Boolean keyword search encryption scheme, which is a post-quantum secure identity-based scheme based on lattices. Specifically, the scheme realizes set relation verification by transforming the entire set of signed keywords into vectors and combining them with an inner product encryption structure, which does not require keyword fields. We prove that our scheme satisfies the adaptive chosen-identity chosen-plaintext attack (IND-ID-CPA) under the random oracle model. Theoretical analysis and experimental comparison show that the scheme has advantages in security and efficiency, and can be well suited for the application scenario of multi-keyword Boolean search.

Keywords: Boolean search · Identity-based encryption with keyword search · Lattice

1 Introduction

Today, more and more businesses and individuals chooses to store their data on cloud servers. Compared with traditional local storage, cloud servers are more flexible and efficient. Specifically, users do not need to buy and maintain hardware equipment, storage space can be expanded at any time, and you can access data over the network anytime, anywhere. However, the problem of data leakage is becoming increasingly serious. In 2023, Twitter (now the X platform) had a major data leak, more than 200 million users' information was sold publicly, once again sounding the alarm on cloud data security. How to enjoy the convenience of cloud storage while ensuring data security and reliability has become the focus of more and more people's attention.

In cloud data security, a highly promising technology path is to encrypt data before uploading it to cloud servers. However, the encryption operation inevitably affects data search. Although traditional encryption ensures data confidentiality, it severely restricts the retrieval capability of the cloud, and this contradiction has catalyzed the rapid development of Public-key Encryption with Keyword Search (PEKS). Early searchable

J. Yu and Y. Chen (Eds.): FCS 2025, CCIS 2805, pp. 118–131, 2027.
https://doi.org/10.1007/978-981-95-9514-3_8

encryption schemes only support single-keyword search, which has obvious limitations in applications. For this reason, researchers have continuously explored innovative solutions and proposed diversified search methods such as ranked search, Boolean search, range search, and fuzzy search. Ranked search orders results based on the relevance between keywords and content, enabling users to quickly retrieve the most pertinent information from vast result sets. Boolean search combines differentkeywords using logical operators like "AND," "OR," and "NOT" to precisely filter results that match complex criteria. Fuzzy search uses specialized algorithms to match approximate content, allowing it to accommodate spelling errors or character variations in keywords. This resolves the issue where users fail to find the needed information due to input mistakes. Among many technical solutions, the searchable encryption based on Boolean expressions is more popular in practice.

However, existing Boolean search encryption schemes still face many serious challenges in terms of security and efficiency, which directly affect their practical applications in privacy-sensitive scenarios. Keyword fields are data labels (e.g., "From", "Subject") used to categorize and limit the search scope, and many schemes use keyword fields to achieve accurate and effective Boolean search. The system not only needs to protect the content of the keywords themselves, but also must guard against information leakage through the fields, which brings great difficulties to the design of the scheme. Furthermore, existing PEKS-based Boolean schemes cause the query token size to increase linearly with the number of queried keywords, incurring extra communication and storage overhead. More serious is the post-quantum security threat; most of the current schemes rely on traditional hard problems such as bilinear pairing, which may be efficiently cracked in the quantum computing environment, leading to the failure of the scheme's security.

1.1 Related Works

In 2000, Song et al. [1] pioneered the concept of searchable encryption (SE). In 2004, Boneh et al. [2] proposed the PEKS scheme. Keywords are encrypted by the sender using the receiver's public key. The receiver then generates a trapdoor using the private key and submits it to the cloud server, which returns the matching ciphertexts. Since then, many schemes have emerged in this area. For example, Xu et al. [3]designed a PEKS scheme with a hidden structure, which effectively improves the efficiency of keyword search; Chen et al. [4]proposed a server-assisted PEKS scheme, which is able to resist offline keyword guessing attacks; Lu et al. [5]developed a certificate-less keyword search encryption scheme that protects against both external and internal keyword guessing attacks. However, the Public Key Infrastructure (PKI) architecture makes these schemes less efficient. Shamir et al. [6] first proposed Identity-Based Encryption (IBE), is an intuitive solution to the above problems. In this cryptosystem, the user has a unique identity, which reduces the overhead of key management. Based on IBE, a number of Identity-based encryption with keyword search (IBEKS) schemes have been constructed in [7–10]. The system model of IBEKS is shown in Fig. 1.

Among them, Lin et al. [10] proposed three lattice-based IBEKS schemes that support conjunctive, disjunctive, and range keyword search, respectively. These schemes achieve post-quantum security based on the Learning with Errors(LWE) assumption, do not require keyword fields, and are suitable for multi-keyword scenarios. The schemes are pioneering in converting the relational judgments between sets of keywords to operations between vectors through polynomials with roots.

Cash et al. [11] first proposed a SE scheme to support Boolean search, although the scheme was only applicable to single-user scenarios. Since then, many schemes [12, 13] have extended the work of [11] to support more generalized Boolean queries in multi-user environments to improve search efficiency. Moataz et al. [14] proposed a more generalized Boolean search encryption scheme, which uses inner-product arithmetic to determine whether a document meets the search conditions. Recently, several advanced schemes have been proposed. Li et al. [15]proposed an attribute-based Boolean search encryption scheme, which utilizes user attributes to restrict the permissions of data recipients, but the scheme has the risk of leaking information about keyword fields. Chen et al. [16]proposed a Boolean search encryption scheme that is also based on attributes and does not require keyword fields. Li et al. [17] also proposed a Boolean search encryption scheme that does not require keyword fields; however, it lacks support for multi-user scenarios. These schemes remain limited in practical deployment. In addition to the shortcomings in multi-user adaptation and key field processing, they generally lack the ability to resist quantum attacks and are difficult to cope with the security challenges posed by the emerging quantum computing threats.

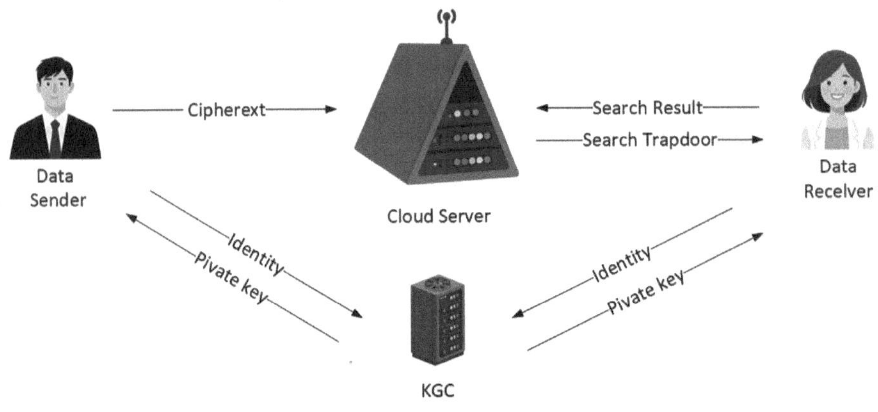

Fig. 1. System model of IBEKS.

1.2 Our Contributions

In this paper, we are committed to solving the problems explored in the previous section by proposing an identity-based Boolean keyword searchable encryption scheme that is based on lattice to achieve post-quantum security. Specifically, our main contributions are as follows:

- We design a lattice-based identity-based Boolean keyword searchable encryption scheme. The scheme enables efficient Boolean search, which eliminates the need for keyword fields, keeps the storage overhead stable, and is resistant to quantum attacks compared to Boolean search schemes in recent years. The core approach is to efficiently verify whether one set is contained in another set by transforming two signed keyword sets into two vectors, and then to support flexible multi-keyword public key searchable encryption schemes by integrating with an inner product encryption structure, which can ultimately realize Boolean search excellently.

- Leveraging the Learning With Errors (LWE) assumption, which ensures post-quantum security, we formally prove that our scheme achieves ciphertext indistinguishability under IND-ID-CPA within the random oracle model.

2 Preliminaries

2.1 Basic Definitions of Lattice

Definition 1 (Lattices): A lattice Λ is formed by the matrix $B = [b_1, b_2, \cdots, b_m] \in \mathbb{R}^{n\times m}$ with linearly independent columns. It can be expressed as $\Lambda = \{Bv = \sum_{i=1}^{m} b_i v_i \mid v_i \in \mathbb{Z}, \forall i = 1, 2, \cdots, n\}$. Given a matrix $A \in \mathbb{Z}_q^{n\times m}$ and a vector $u \in \mathbb{Z}_q^n$, the q-ary lattices are defined as:

$$\Lambda_q^{\perp}(A) = \{e \in \mathbb{Z}^m : Ae \equiv 0 \pmod{q}\},$$

$$\Lambda_q^{u}(A) = \{e \in \mathbb{Z}^m : Ae \equiv u \pmod{q}\}.$$

Definition 2 (Gaussian Distribution [18]): Given a positive parameter, a positive integer m, a vector $c \in \mathbb{R}^m$, and a lattice $\Lambda \subseteq \mathbb{Z}^m$, the Gaussian function is defined as $\rho_{\sigma,c}(x) = \exp\left(-\frac{\pi\|x-c\|^2}{\sigma^2}\right)$, and $\rho_{\sigma,c}(\Lambda) = \sum_{x\in\Lambda} \rho_{\sigma,c}(x)$. The distribution over Λ centered at c with parameter is given by $D_{\Lambda,\sigma,c}(y) = \frac{\rho_{\sigma,c}(y)}{\rho_{\sigma,c}(\Lambda)}$ for all $y \in \Lambda$.

Definition 3 3 (LWE [19]): The problem involves retrieving the pair (s, e) from $(A, A^T s + e)$, where A is randomly selected from $\mathbb{Z}_q^{n\times m}$, s is chosen randomly from $\mathbb{Z}_q^n$, and e is picked randomly from χ^m. The decisional-LWE problem refers to telling apart $(A, A^T s + e)$ and (A, b), with b being randomly selected from $\mathbb{Z}_q^m$.

Definition 4 (ISIS [20]): The problem is to find a vector $e \in \mathbb{Z}_q^m$ such that $Ae = u$ and $0 < \|e\|_\infty \leq \beta$, where β serves as a bounding parameter.

2.2 Basic Algorithms on Lattice

Lemma 1 (TrapGen [21]): *For integers $n \geq 2$, $q \geq 2$ and $m \geq 6n\lceil\log q\rceil$, the algorithm outputs a matrix $A \in \mathbb{Z}_q^{n\times m}$ and its short basis $T_A \in \mathbb{Z}_q^{m\times m}$. Here, $A \in \mathbb{Z}_q^{n\times m}$ has an approximately uniform distribution, and it satisfies $A \cdot T_A \equiv 0 \pmod{q}$ with $\|\tilde{T}_A\| \leq O(\sqrt{n\log q})$.*

Lemma 2 (SamplePre [20]): *For integers $n \geq 2$, $q \geq 2$ and $m \geq 2n\lceil \log q \rceil$, when provided with a matrix $A \in \mathbb{Z}_q^{n\times m}$, its short basis $T_A \in \mathbb{Z}_q^{m\times m}$, a parameter $\sigma \geq \|\tilde{T_A}\| \cdot \omega(\sqrt{\log m})$, and a vector $u \in \mathbb{Z}_q^n$, the algorithm yields a vector $e \in \mathbb{Z}_q^m$ that satisfies* $Ae \equiv u \pmod q$.

Lemma 3 (SampleLeft [22]): *For integers $n \geq 2$, $q \geq 2$ and $m \geq 2n\lceil \log q \rceil$, given matrices $A_1 \in \mathbb{Z}_q^{n\times m}$, $A_2 \in \mathbb{Z}_q^{n\times m'}$, a short basis T_{A_1} of $\Lambda_q^{\perp}(A_1)$, a parameter $\sigma \geq \|\tilde{T_{A_1}}\| \cdot \omega(\sqrt{\log(m+m')})$, and a vector $u \in \mathbb{Z}_q^n$, the algorithm returns a vector $e \in \mathbb{Z}_q^{m+m'}$ such that* $[A_1 \mid A_2]e \equiv u \pmod q$.

Lemma 4 (NewBasisDel [23]): *For integers $n \geq 2$, $q \geq 2$ and $m \geq 2n\lceil \log q \rceil$, suppose there is an invertible matrix $R \in \mathbb{Z}_q^{m\times m}$, a matrix $A \in \mathbb{Z}_q^{n\times m}$ along with its short basis $T_A \in \mathbb{Z}_q^{m\times m}$, and a parameter $\sigma \geq \|\tilde{T_A}\| \cdot \delta \sqrt{m} \cdot \omega(\log^{\frac{3}{2}} m)$ (where $\delta = \sqrt{n \log q} \cdot \omega(\log m)$). The algorithm then returns a short basis T_B for $B = AR^{-1} \in \mathbb{Z}_q^{n\times m}$, with $B \cdot T_B \equiv 0$* $\pmod q$.

Lemma 5 (SampleR [23]): *For integers $n \geq 2$, $q \geq 2$ and $m \geq 2n\lceil \log q \rceil$, with $R \in D_{m\times m}$ and a parameter σ satisfying $\sigma > \|\tilde{T_A}\|\sigma_R \sqrt{m}\omega(\log^{\frac{3}{2}} m)$ (where $\sigma_R = \sqrt{n \log q}\,\omega(\sqrt{\log m})$), the algorithm outputs a $\mathbb{Z}_q$-invertible matrix with a low norm.*

Lemma 6 (SampleRwithBasis [23]): *For integers $n \geq 2$, $q \geq 2$ and $m \geq 2n\lceil \log q \rceil$, when a matrix $A \in \mathbb{Z}_q^{n\times m}$ is given, the algorithm outputs a $\mathbb{Z}_q$-invertible matrix $R \in D_{m\times m}$ with a low norm, along with a short basis T_B where $B = AR^{-1} \in \mathbb{Z}_q^{n\times m}$ satisfying* $B \cdot T_B \equiv 0 \pmod q$.

2.3 Lin's Method

The basic transformation method designed by Lin aims to determine whether a particular plaintext belongs to a certain set of plaintexts. The core idea is to achieve verification through polynomial equations and vector transformations: First, given a single plaintext m and a collection of plaintexts M', the hash function is used to map them into elements of $\mathbb{Z}_q$, and then a polynomial is constructed based on these elements. On this basis, the polynomial takes the value 0 when m belongs to M'. Next, the polynomial coefficients and the power terms of the explicit hash value are converted to vectors b and y respectively, at which time the inner product of the two is 0. Finally, after the introduction of the random vector v, the inclusion relation can be verified by $vb^T y = 0$. The specific steps are as follows:

1) Given the maximum number of keywords N, the hash function $H_2 : \{0,1\}^* \to \mathbb{Z}_q$, a single plaintext m, the set of plaintexts $M' = \{m'_1, m'_2, ..., m'_l\}$ $(1 \leq l \leq N)$.
2) Compute the hash values $H_2(m) = x_m$, $H_2(m'_i) = x_{m'_i}$ $(1 \leq i \leq l)$; if $l < N$, randomize the complement $x_{m'_j}$ $(l < j \leq N)$.
3) Construct polynomial $f(x) = (x - x_{m'_1})(x - x_{m'_2}) \cdots (x - x_{m'_N}) = a_N x^N + ... + a_0$. When $m \in M'$, $f(x_m) = 0$.
4) Let $b = \{a_0, a_1, ..., a_N\}$, $y = \{1, x_m, x_m^2, ..., x_m^N\}$, then $b^T y = 0$.
5) Introduce a random vector $v \in \mathbb{Z}_q^{N+1}$ and verify $m \in M'$ by $vb^T y = 0$.

3 Definition and Security Model of IBEKS

3.1 Definition

The IBEKS scheme for Boolean search comprises a set of algorithms.

- **Setup**$(1^\lambda, N) \rightarrow (pp, msk)$: Given the security parameter λ and the maximum number of keywords$N$, output the public parameterpp and the master key$msk = T_A$.
- **KeyGen**$(msk, id) \rightarrow (pk_{id}, sk_{id})$: Given a master key$msk$ and an identifierid, output the key pair$(pk_{id}, sk_{id}) = (A_{id}, T_{id})$.
- **Encrypt**$(id, W) \rightarrow CT$: Given an identifierid and a set of keywords$W$, output the corresponding ciphertextCT.
- **Trapdoor**$(B, sk_{id}, id) \rightarrow t_{W'}$: Given the keywords Boolean expression$B$, private key$sk_{id} = T_{id}$, and identifierid, output the corresponding trapdoor$t_{W'}$.
- **Test**$(CT, t_{W'}) \rightarrow 1/0$: Given a ciphertext CT, trapdoor $t_{W'}$, if the keyword encapsulated in CT matches the keyword Boolean expression associated with trapdoor $t_{W'}$, then return 1; otherwise, return 0.

3.2 Security Model

Suppose that C is a challenger and $\mathcal{A}$ is a probabilistic polynomial time (PPT) adversary.The IND-ID-CPA game is defined as follows.

- **Setup:** After receiving the security parameter λ and the maximum number of keywords N, the challenger C generates the public parameter pp and the master key msk, and then provides pp to the adversary $\mathcal{A}$.
- **Query Phase 1:** The adversary$\mathcal{A}$ makes multiple polynomial-level queries to the following predictor.
  1) $Q_{KeyGen}(id)$: given an identity id and a Boolean expression $id \neq id^*$, the challenger C generates and provides the secret key T_{id} if $id \neq id^*$; otherwise, return 1.
  2) $Q_{Trapdoor}(id, B')$: given an identity id and a Boolean expression B, the challenger C generates and provides the trapdoor $t_{W'}$.
- **Challenge:** Adversary$\mathcal{A}$ submits an identityid^* as a challenge identity, along with two keyword setsW_0^* and W_1^*. Need to make sure that id^* has not been queried in Query Phase 1. W_0^* and W_1^* either both do not match any Boolean expression queried in Query Phase 1, or both match the same Boolean expression previously queried in Query Phase 1. The challenger C chooses b at random from 0, 1, computes $CT_b \leftarrow Encrypt(pp, id^*, W_b^*)$, and sends CT_b to the adversary $\mathcal{A}$.
- **Query Phase 2:** The adversary $\mathcal{A}$ is queried in the same way as in Query Stage 1, but with one restriction: it is forbidden to submit Boolean expressions matching W_0^* and W_1^* to the prediction machine.
- **Guess:** The adversary $\mathcal{A}$ outputs $b' \in \{0, 1\}$ as a guess at b.
  The adversary $\mathcal{A}$ wins the above game if and only if $b' = b$. We define the advantage of $\mathcal{A}$ winning that game as

$$Adv_A^{IND-ID-CPA} = \left|\Pr[b' = b] - \frac{1}{2}\right|$$

Definition 5 (IND-ID-CPA): An IBEKS scheme is IND-ID-CPA secure if for any PPT adversary $\mathcal{A}$, its advantage in the above game is negligible.

4 Proposed Scheme

- **Setup:** Given the security parameter λ and the maximum number of keywords N, the system initialization performs the following actions:
  1) Choose two hash functions $H_1 : 0, 1^* \to \mathbb{Z}_q^{m\times m}$ and $H_2 : 0, 1^* \to \mathbb{Z}_q$. Among them, the output of H_1 is reversible and follows the $D_{m\times m}$ distribution.
  2) For positive integers q and $n > 3^N$, perform TrapGen (q, n) to get $A \in \mathbb{Z}_q^{n\times m}$ and $T_A \in \mathbb{Z}_q^{m\times m}$.
  3) Let $N' = 3^N - 1$, the matrix $D \in \mathbb{Z}_q^{n\times m}$, and the vectors $u \in \mathbb{Z}_q^n$, b_0, and $y_0 \in \mathbb{Z}_q^{n-N'-1}$ be randomly chosen and $b_0^T y_0 = 0$ needs to be satisfied.
  4) Output public parameters $pp = (A, B, b_0, y_0, u)$, master key $msk = T_A$.
- **KeyGen:** The Key Generation Center (KGC) performs the following processing when given the master key msk and identity information id:
  1) Compute $R_{id} = H_1(id)$, and $A_{id} = A(R_{id})^{-1}$.
  2) Perform NewBasisDel(A, R_{id}, T_A, c) to generate the lattice $\Lambda_q^{\perp}(A_{id})$ corresponding to $T_{id} \in \mathbb{Z}_q^{m\times m}$.
  3) Output key pair $(pk_{id}, sk_{id}) = (A_{id}, T_{id})$
- **Encrypt:** Given the identity id and the set of keywords $W = \{\diamond w_1, \diamond w_2, \ldots, \diamond w_l\}$ ($1 \le l \le N$, "$\diamond$" is the unary operator, $\diamond \in \{\emptyset, \neg\}$, and all keywords in the set W are sorted in ascending binary order), the data sender performs the following operations:
  1) Compute $R_{id} = H_1(id)$, and $A_{id} = A(R_{id})^{-1}$
  2) For $i = 1, 2, \ldots, l$, compute $CW = (H_2(S) \mid S \subseteq \{\diamond w_1, \diamond w_2, \ldots, \diamond w_l\},\ S \neq \emptyset,\ \text{each } w_i \text{ occurs at most once}) = (\hat{x}_1, \hat{x}_2, \ldots, \hat{x}_{3^l-1})$
  3) If $3^l - 1 < N'$ (i.e., the total number of non-empty combinations of the set W is less than N'), then for $i' = 3^l, 3^l + 1, \ldots, N'$, a positive integer is randomly selected from $\mathbb{Z}_q$ $\hat{x}_{i'} \in \mathbb{Z}_q$
  4) Compute the polynomial expansion $(x - \hat{x}_1)(x - \hat{x}_2)\cdots(x - \hat{x}_{N'}) = a_{N'}x^{N'} + a_{N'-1}x^{N'-1} + \cdots + a_0$, form a vector of coefficients of this polynomial $b' = [a_0, a_1, \ cdots, a_{N'}] \in \mathbb{Z}_q^{N'+1}$, and then splice b' with b_0 to form $b = [b'|b_0] \in \mathbb{Z}_q^n$.
  5) Randomly select v,$s \in \mathbb{Z}_q^n$,$R \in \{-1, 1\}^{m\times m}$,and $z_u \leftarrow \chi$,$z_{id} \leftarrow \chi^m$,and let $z_W = Rz_{id} \in \mathbb{Z}_q^m$.
  6) Let $c_u = u^T s + z_u \in \mathbb{Z}_q$, $c_{id} = A_{id}^T s + z_{id} \in \mathbb{Z}_q^m$, and $c_W = (B + vb^T A_{id})^T s + z_W \in \mathbb{Z}_q^m$.
  7) Send the ciphertext $CT = (c_u, c_{id}, c_W)$ to the cloud server (CS).
- **Trapdoor:** Given a Boolean expression $B = \diamond w_1' * \diamond w_2' * \cdots * \diamond w_{l'}'$ "$*$" is the binary operator, $* \in \{\vee, \wedge\}$ and all keywords in W' are sorted in ascending binary order, the secret key $sk_{id} = T_{id}$, and an identity id, the data receiver performs the following processing:
  1) Compute $R_{id} = H_1(id)$, and $A_{id} = A(R_{id})^{-1}$.
  2) Converts a Boolean expression into a parsing paradigm: $B = \bigvee_{i=1}^{k} Q_i = \bigvee_{i=1}^{k} \bigwedge_{j\in I_i} \diamond\omega_j$, where $\bigcup_{i=1}^{k} I_i = \{1, \cdots, l'\}$, such that $W_i' = \{\diamond w_j | j \in I_i\}$
  3) Compute $x_{w_i'} = H_2\left(\|_{j\in I_i} \diamond w_j\right)$,and let $y_i' = \{1, x_{w_i'}, \cdots, x_{w_i'}^{N'}\} \in \mathbb{Z}_q^{N'+1}$,$y_i = (y_i'|y_0) \in \mathbb{Z}_q^n$.
  4) Perform SamplePre$(A_{id}, T_{id}, y, \tau)$ to generate the vector $e_{0i} \in \mathbb{Z}_q^m$, where e_{0i} satisfies $A_{id}e_{0i} = y$.

5) Perform SampleLeft $(A_{id}, Be_{0i}, T_{id}, u, \tau)$ to generate the vector $e_{1i} \in \mathbb{Z}_q^{m+1}$, where e_{1i} satisfies $[A_{id}|Be_{0i}]e_{1i} = u$.
6) Send the trapdoor $t_{W'} = (e_{01}, e_{11}, e_{02}, e_{12}, \ldots, e_{0k}, e_{1k})$ to the CS.

- **Test:** Given the ciphertext CT,trapdoor $t_{W'}$, the CS performs the following processing:
  1) Compute $\mu_i = c_u - e_{1i}^T \begin{pmatrix} c_{id} \\ e_{0i}^T c_W \end{pmatrix}$.
  2) If the inequality $\mu_i \leq \left\lfloor \frac{q}{4} \right\rfloor$ is satisfied, then let $\mu_i = 0$; otherwise, let $\mu_i = 1$.
  3) Compute$\mu = \prod_{i=1}^{k} \mu_i$,return μ.

4.1 Correctness

For ciphertext $CT = (c_u, c_{id}, c_w)$ and trapdoor $t_{W'} = (e_{01}, e_{11}, e_{02}, e_{12}, \ldots, e_{0k}, e_{1k})$, we need to discuss the results of μ_i first in two cases.

1) If $W_i' \subseteq W$, then $vy_i^T b = 0$, we have:

$$\mu_i = u^T s + z_u - e_{1i}^T \begin{pmatrix} A_{id}^T s + z_{id} \\ e_{0i}^T([B + vb^T A_{id}]^T s + z_W) \end{pmatrix}$$

$$= u^T s + z_u - e_{1i}^T \begin{pmatrix} A_{id}^T s + z_{id} \\ (Be_{0i} + vb^T y_i)^T s + e_{0i}^T z_W \end{pmatrix}$$

$$= u^T s + z_u - e_{1i}^T \begin{pmatrix} A_{id}^T s + z_{id} \\ (Be_{0i})^T s + e_{0i}^T z_W \end{pmatrix}$$

$$= z_u - e_{1i}^T \begin{pmatrix} z_{id} \\ e_{0i}^T z_W \end{pmatrix}$$

Where $z_u - e_{1i}^T \begin{pmatrix} z_{id} \\ e_0^T z_w \end{pmatrix}$ is the error term.

2) If $W_i' \nsubseteq W$,then $vy_i^T b \neq 0$,we have:

$$\mu_i = u^T s + z_u - e_{1i}^T \begin{pmatrix} A_{id}^T s + z_{id} \\ e_{0i}^T([B + vb^T A_{id}]^T s + z_W) \end{pmatrix}$$

$$= u^T s + z_u - e_{1i}^T \begin{pmatrix} A_{id}^T s + z_{id} \\ (Be_{0i} + vb^T y_i)^T s + e_{0i}^T z_W \end{pmatrix}$$

$$= z_u - e_{1i}^T \left[0|vb^T y_i\right]^T s - e_{1i}^T \begin{pmatrix} z_{id} \\ e_{0i}^T z_w \end{pmatrix}$$

The probability that $z_u - e_{1i}^T \left[0|vb^T y_i\right]^T s - e_{1i}^T \begin{pmatrix} z_{id} \\ e_{0i}^T z_w \end{pmatrix}$ is within the range allowed by the noise is negligible.

Obviously, if and only if neither μ_i is not 0, μ is not 0; when one or more of μ_i is 0, μ is not 0.

4.2 Security Analysis

Theorem 1 (IND-ID-CPA): If the dLWE assumption holds, then our scheme is IND-ID-CPA secure.

Security Proof. The assumption here is that the adversary $\mathcal{A}$ will choose an identity id^* during the challenge phase to destroy the indistinguishability of the ciphertext. For each identity id, the challenger C maintains two tuples:

$$L_1 = (id, A_{id}, R_{id}, T_{id}),\ L_2 = (W, r_W)$$

Let V be the number of queries to H_2 and K be the maximum number of visits allowed by the adversary $\mathcal{A}$ to H_2. The challenger C randomly guesses a number J^* which represents the index that $\mathcal{A}$ will use to query w_b^* to the H_2 prophecy machine by the first J^* query. If $\mathcal{A}$ does not perform this query, its ability to break the indistinguishability of the ciphertext is negligible. Therefore, we assume that $\mathcal{A}$ will query a series of challenges in its interaction with C. The original IND-CPA game is the initial game in our proof. In the game d, GW_d means that $\mathcal{A}$ accurately predicts the events of the challenge bits. We use $|Pr[GW_d] - \frac{1}{2}|$ to measure thc dominance of $\mathcal{A}$ in the game d.

- **Game 0:** It is indistinguishable from the IND-ID-CPA game defined in Sect. 3.2.
- **Game 1:** It is almost identical to Game 0, the only difference being the selection of the matrix A. First, the challenger C must obtain the LWE samples $u' \in \mathbb{Z}_q^n, U \in \mathbb{Z}_q^{n\times m}$ from the LWE preconditioner, and set $u = u'$. Here, v_u is generated by u, i.e., $v_u = (u')^T s + z_u \in \mathbb{Z}_q$, where $s \xleftarrow{\$} \mathbb{Z}_q^n$, $z_u \xleftarrow{\$} \chi$; and v_U is generated by U, i.e., $v_U = U^T s + z_{id} \in \mathbb{Z}_q^m$, where $z_{id} \xleftarrow{\&} \chi^m$. The challenger C sets $A = UR^* \in \mathbb{Z}_q^{n\times m}$ instead of generating A via Trapen algorithm, where R^* is generated by sampling via the SampleR(m) algorithm with the distribution $D_{m\times m}$. Since R^* obeys the distribution of $D_{m\times m}$ and U is uniformly distributed in $\mathbb{Z}_q^{n\times m}$, we can determine that A is uniformly distributed in $\mathbb{Z}_q^{n\times m}$. Since A is statistically indistinguishable from A in Game 0, we can infer that $Pr[GW_1] = Pr[GW_0]$.
- **Game 2:** It is almost identical to Game 1, the only difference being the selection of the matrix D. First, the challenger C must obtain the LWE sample $U' \in \mathbb{Z}_q^{n\times m}$ from the LWE prediction machine. Here, $v_{U'}$ is generated from U, i.e., $v_{U'} = (U')^T s + z_w \in \mathbb{Z}_q^m$, where $z_w \xleftarrow{\$} \chi^m$. Then,the challenger Cselects a set$W^* = \{x_1, x_2, \cdots, x_{l'}\}$,a random vector $v^* \in \mathbb{Z}_q^n$,and computes $CW = (H_2(S) \mid S \subseteq \{\diamond w_1, \diamond w_2, \ldots, \diamond w_l\}, S \neq \emptyset$, each w_i occurs at most once$) = (\hat{x}_1, \hat{x}_2, \ldots, \hat{x}_{3^l-1})$. If $3^l - 1 < N'$,then for $i' = 3^l, 3^l + 1, \ldots, N'$,a positive integer $\hat{x}_{i'} \in \mathbb{Z}_q$ is randomly selected from $\mathbb{Z}_q$.Compute the polynomial expansion $(x - \hat{x}_1)(x - \hat{x}_2) \cdots (x - \hat{x}_{N'}) = a_{N'}x^{N'} + a_{N'-1}x^{N'-1} + \cdots + a_0$,form the coefficients of this polynomial into the vector $b' = [a_0, a_1, \cdots, a_{N'}] \in \mathbb{Z}_q^{N'+1}$, and then splice b' with b_0 to form $b^* = [b'|b_0] \in \mathbb{Z}_q^n$. Next, the challenger C sets $D = U' - v^*(b^*)^T U \in \mathbb{Z}_q^{n\times m}$. Finally, the challenger C sends pp to the adversary $\mathcal{A}$. Since D is statistically indistinguishable from D in Game 1, we can deduce that $Pr[GW_2] = Pr[GW_1]$.
- **Game 3:** It is almost identical to Game 2, the only difference is that the challenge cipher $CT^* = (c_u^*, c_{id^*}, c_{W_b^*})$ is chosen at random. $c_u^*, c_{id^*}, c_{W_b^*}$ come from $\mathbb{Z}_q, \mathbb{Z}_q^m, \mathbb{Z}_q^m$ respectively. Therefore, we can derive $Pr[GW_3] = Pr[GW_2]$.
- **Reduction from dLWE:** We can construct the dLWE problem solver $\mathcal{A}$ by $\mathcal{A}$ based on the assumption that "the adversary $\mathcal{A}$ can differentiate between the last two games by a non-negligible margin."

1) dLWE instance: The goal of the solver $\mathcal{B}$ is to determine, by obtaining samples of LWE, whether the tuples being processed are randomly selected (v_u, v_U, v_U) or are generated by $(u^T s + z_u, U^T s + z_{id}, U'^T s + z_w)$.
2) Setup: The solver $\mathcal{B}$ generates the public parameters pp to send to the adversary $\mathcal{A}$ in the following way.
   a) Compute $R^* \leftarrow \text{SampleR}(1^m)$ and set $A = UR^* \in \mathbb{Z}_q^{n\times m}$.
   b) The vectors $u, v^* \in \mathbb{Z}_q^n$, $b_0, y_0 \in \mathbb{Z}_q^{n-N'-1}$ are chosen randomly and need to satisfy $b_0^T y_0 = 0$.
   c) Let $W^* = \{x_1, x_2, \cdots, x_{l'}\}$,and compute $CW = (H_2(S) \mid S \subseteq \{\diamond w_1, \diamond w_2, \ldots, \diamond w_l\}, S \neq \emptyset$, each w_i occurs at most once$) = (\hat{x}_1, \hat{x}_2, \ldots, \hat{x}_{3^l-1})$. If $3^l - 1 < N'$, then for $i' = 3^l, 3^l + 1, \cdots, N'$, randomly select the positive integer $\hat{x}_{i'} \in \mathbb{Z}_q$ from $\mathbb{Z}_q$. Then compute the polynomial expansion $(x - \hat{x}_1)(x - \hat{x}_2)\cdots(x - \hat{x}_{N'}) = a_{N'}x^{N'} + a_{N'-1}x^{N'-1} + \cdots + a_0$ by forming the coefficients of the polynomial into a vector $b' = [a_0, a_1, \cdots, a_{N'}] \in \mathbb{Z}_q^{N'+1}$.
   d) Splice b' with b_0 as $b^* = [b'|b_0] \in \mathbb{Z}_q^n$.
   e) Send the public parameter $pp = (A, D, b_0, y_0, u)$ to $\mathcal{A}$.
3) Query Phase 1: Based on the given parameters, the adaptive query solver $\mathcal{B}$ for the adversary $\mathcal{A}$ will be processed as follows.
   (a) $Q_{H_1(id)}$ If $id \neq id^*$, check the list L_1, if there exists the corresponding tuple $(id, A_{id}, R_{id}, T_{id})$, then return R_{id} directly, otherwise generate (T_{id}, R_{id}) by executing SampleRwithBasis (A). Store into L_1 and return R_{id}; if $id = id^*$, also check L_1 first, if there is a corresponding tuple $(id^*, A_{id^*}, R_{id^*}, T_{id^*})$, then return R_{id^*}, else set $R_{id^*} = R^*$,$A_{id} = AR_{id^*}^{-1}$, add $(id^*, A_{id^*}, R_{id^*}, -)$ to L_1 then return R_{id}.
   (b) $Q_{H_2(W)}$ If $V = J^*$, set $H_2(W^*)$ to r^* and send it to $\mathcal{A}$; otherwise, check the list L_2 first, and if the tuple (W, r_W) exists, return r_W directly to $\mathcal{A}$; if it doesn't the tuple exists, pick a random vector $r_W \in \mathbb{Z}_q$ and finally add (W, r_W) to L_2 and send r_W to $\mathcal{A}$.
   (c) $Q_{KeyGen}(id)$ If there is a tuple in the list L_1, the tuple is retrieved from L_1; then, if $id \neq id^*$, return T_{id} to $\mathcal{A}$, otherwise, it will be processed in other ways.
   (d) $Q_{\text{Trapdoor}}(id, B)$ïijŽSelect k random integers $r_1 = x_{w'_1}, r_2 = x_{w'_2}, \cdots, r_k = x_{w'_k} \in \mathbb{Z}_q$, set $y'_i = \{1, x_{w'_i}, \cdots, x_{w'_i}^{N'}\} \in \mathbb{Z}_q^{N'+1}$, $y_i = (y'_i|y_0) \in \mathbb{Z}_q^n$. And check the list L_1 to recover the tuple $(id, A_{id}, R_{id}, T_{id})$. Then, perform SamplePre$(A_{id}, T_{id}, y, \tau)$ to generate the vector $e_{0i} \in \mathbb{Z}_q^m$, perform SampleLeft$(A_{id}, Be_{0i}, T_{id}, u, \tau)$ to generate the vector $e_{1i} \in \mathbb{Z}_q^{m+1}$. Finally, the trapdoor $t_{W'} = (e_{01}, e_{11}, e_{02}, e_{12}, \ldots, e_{0k}, e_{1k})$ is sent to the CS.
4) Challenge: Adversary $\mathcal{A}$ sends an identity id^* as the challenge identity, and a keyword pair (W_0^*, W_1^*). It is necessary to ensure that id^* has not been queried in Query Phase 1 and either (W_0^*, W_1^*) does not match any of the Boolean expressions in Query Phase 1, or both match the same Boolean expression in Query Phase 1. The solver$\mathcal{B}$ randomly selects a bit $b \in \{0, 1\}$ and computes the challenge ciphertext $CT_{id^*,W_b^*} = (c_u^*, c_{id^*}, c_{W_b^*})$ according to the following process.
   a) If $(v_u, v_U, v_{U'})$ is generated by a dLWE instance, i.e., $v_u = (u')^T s + z_u$, $v_U = U^T s + z_{id}$ and $v_{U'} = (U')^T s + z_w$, given that $u = u'$, $A = UR^*$, $A_{id^*} =$

$AR_{id}^{-1}, R_{id} = R^*$ and $D = U' - v^*(b^*)^T U$, then the challenge ciphertext is generated as follows:

$$c_u^* = u^T s + z_u = (u')^T s + z_u$$

$$\begin{aligned} c_{id^*} &= A_{id^*}^T s + z_{id} = \left(AR_{id}^{-1}\right)^T s + z_{id} \\ &= \left(UR^*(R^*)^{-1}\right)^T s + z_{id} \\ &= U^T s + z_{id}, \end{aligned}$$

$$\begin{aligned} c_{W_b^*} &= \left(D + v^*(b^*)^T A_{id^*}\right)^T s + z_w \\ &= \left(D + v^*(b^*)^T U\right)^T s + z_w \\ &= (U')^T s + z_w. \end{aligned}$$

b) If (v_u, v_U, v_U) is randomly selected from $\mathbb{Z}_q \times \mathbb{Z}_q^m \times \mathbb{Z}_q^m$, it can be seen that the challenge ciphertext $CT^* = (c_u^*, c_{id^*}, c_{W_b^*})$ has a distribution is uniform, which is consistent with the case in Game 3.

5) Query Phase 2: This phase is essentially indistinguishable from Query Phase 1, with the only restriction that the adversary $\mathcal{A}$ may not initiate a KeyGen query for the identity id^* and a Trapdoor query for the tuple (id^*, B'), where the Boolean expression B' matches W_0^* or W_1^*.
6) Guess: $\mathcal{B}$ will refer to the output results provided by adversary $\mathcal{A}$ for the dLWE challenge, while adversary $\mathcal{A}$ will use guesswork to determine whether the interaction corresponds to a challenger from game 2 or game 3.

In summary, if $\mathcal{B}$ can guess w_j^* in W_b^* with probability $\frac{1}{K}$, then $\mathcal{B}$ solves the dLWE problem, i.e., $\mathcal{A}$ distinguishes the challengers in the first few rounds. When $\mathcal{A}$ acts as a subroutine, $\mathcal{B}$'s dLWE advantage is $Adv_A^B(\lambda) = \varepsilon$, so the probability that $\mathcal{A}$ distinguishes between challengers in Games 3 and 2 is $\frac{\varepsilon}{K}$. Since $Pr[GW_3] = \frac{1}{2}$, we have $|Pr[GW_2] - \frac{1}{2}| = |Pr[GW_2] - Pr[GW_3]| = \frac{\varepsilon}{K}$. Combining this with $Pr[GW_0] = Pr[GW_1] = Pr[GW_2]$, we obtain $|Pr[GW_0] - \frac{1}{2}| = \frac{\varepsilon}{K}$. The proof is complete.

5 Analysis and Comparison

We have compared and analyzed the proposed scheme here. Specifically, Table 1 presents the comparison of characteristics between this scheme and other schemes, while Table 2 shows the comparison of communication costs between them.

In Table 1, we compare the features of our scheme with those of other Boolean search schemes. As shown, schemes such as Chen [16], Li [17], Li [15], and Xu [24] lack quantum security(QS), and some schemes also have shortcomings in efficient certificate management(ECM) and multi-user support(MU). In contrast, the scheme we propose achieves coverage in all four dimensions: no need for a key field(NKF), quantum security, certificate management efficiency, and multi-user support.

In Table 2, we compare the communication costs of our scheme with other lattice-based keyword search schemes [25–28]. $\mathbb{Z}$ represents the size of an element in $\mathbb{Z}_q$, ℓ

denotes the length of the keyword, l' is the length of the ID, and k is a parameter that combines the string length and security parameters. The comparison results show that our scheme can maintain a low communication cost for a key and ciphertext while enabling more complex Boolean searches.

Table 1. Comparison of features

Scheme	NKF	QS	ECM	MU
Chen[16]	√	×	√	√
li[17]	√	×	×	×
li[15]	×	×	√	√
xu[24]	√	×	×	√
Ours	√	√	√	√

Table 2. Comparison of communication cost

Scheme	Secret key	Ciphertext
Zhang [25]	m^2Z	$(\ell + m\ell + m)Z$
Yu [26]	m^2Z	$(2Nm + 1)Z$
Liu [27]	$m(m + 1)Z + kn$	$k(2mZ + Z + 1)$
Fan [28]	$m(m + 1)Z + kn$	$2(1 + m)Z + l'$
Ours	m^2Z	$(2m + 1)Z$

6 Conclusions

This paper addresses the contradiction between data encryption and efficient searching in cloud storage. It focuses on keyword field dependency, lack of quantum security, and storage and computing overhead increasing with the number of keywords in existing solutions for Boolean keyword searching. It proposes a lattice-based post-quantum secure identity-based Boolean keyword searchable encryption scheme. This scheme is based on an improvement of Lin's transformation method, which converts signed keyword sets into vectors and combines inner product encryption structures for efficient set inclusion verification. It supports complex Boolean searches without the need for keyword fields, and the overhead of ciphertext is not affected by the number of keywords. Based on the LWE hard problem over lattices, the scheme has been proven to satisfy IND-ID-CPA security under the random oracle model, effectively resisting quantum computing attacks and addressing the vulnerabilities of traditional schemes that rely on bilinear pairings, susceptible to quantum attacks. Comparative analysis with existing schemes shows that this scheme achieves comprehensive coverage across four key dimensions: no need for a keyword field, quantum security, efficient certificate management, and multi-user support. Additionally, the communication costs for keys and ciphertext are both at a very low level. Future work will further explore the integration of range search technology to expand functionality while maintaining efficiency, enabling more granular access control and search permission management, and enhancing the scheme's applicability in complex cloud environments.

References

1. Song, D.X., Wagner, D., Perrig, A.: Practical techniques for searches on encrypted data. In: Proceeding 2000 IEEE Symposium on Security and Privacy, SP 2000, pp. 44–55. IEEE (2000)
2. Boneh, D., Di Crescenzo, G., Ostrovsky, R., Persiano, G.: Public key encryption with keyword search. In: Cachin, C., Camenisch, J.L. (eds.) EUROCRYPT 2004. LNCS, vol. 3027, pp. 506–522. Springer, Heidelberg (2004). https://doi.org/10.1007/978-3-540-24676-3_30
3. Xu, P., et al.: Generating searchable public-key ciphertexts with hidden structures for fast keyword search. IEEE Trans. Inf. Forensics Secur. **10**(9), 1993–2006 (2015)
4. Chen, R., Mu, Y., Yang, G., Guo, F., Wang, X.: Dual-server public-key encryption with keyword search for secure cloud storage. IEEE Trans. Inf. Forensics Secur. 789—798 (2016)
5. Lu, Y., et al.: Constructing certificateless encryption with keyword search against outside and inside keyword guessing attacks. IEEE China Commun. **16**(7), 156–173 (2019)
6. Shamir, A.: Identity-based cryptosystems and signature schemes. In: Blakley, G.R., Chaum, D. (eds.) CRYPTO 1984. LNCS, vol. 196, pp. 47–53. Springer, Heidelberg (1985). https://doi.org/10.1007/3-540-39568-7_5
7. Liu, J., Lai, J., Huang, X.: Dual trapdoor identity-based encryption with keyword search. Soft. Comput. **21**(10), 2599–2607 (2015). https://doi.org/10.1007/s00500-015-1960-6
8. Noroozi, M., Karoubi, I., Eslami, Z.: Designing a secure designated server identity-based encryption with keyword search scheme: still unsolved. Ann. Telecommun. **73**(11–12), 769–776 (2018)
9. Li, H., Huang, Q., Shen, J., Yang, G., Susilo, W.: Designated-server identity-based authenticated encryption with keyword search for encrypted emails. Inf. Sci. **481**, 330–343 (2019)
10. Lin, Z., et al.: Identity-based encryption with disjunctive, conjunctive and range keyword search from lattices. IEEE Trans. Inf. Forensics Secur. **19**, 8644–8657 (2024)
11. Cash, D., Jarecki, S., Jutla, C., Krawczyk, H., Roşu, M.-C., Steiner, M. : Highly-scalable searchable symmetric encryption with support for Boolean queries. In: IACR Cryptology ePrint Archive 8042, pp. 353–373 (2013)
12. Wang, Y., Wang, J., Sun, S. :Towards multi-user searchable encryption supporting Boolean query and fast decryption. J. Univ. Comput. Sci. **2017**, 24–38 (2017)
13. Sun, S.-F., Liu, J.K., Sakzad, A., Steinfeld, R., Yuen, T.H.: An efficient non-interactive multi-client searchable encryption with support for Boolean queries. In: Askoxylakis, I., Ioannidis, S., Katsikas, S., Meadows, C. (eds.) ESORICS 2016. LNCS, vol. 9878, pp. 154–172. Springer, Cham (2016). https://doi.org/10.1007/978-3-319-45744-4_8
14. Kamara, S., Moataz, T.: Boolean searchable symmetric encryption with worst-case sub-linear complexity. In: Coron, J.-S., Nielsen, J.B. (eds.) EUROCRYPT 2017. LNCS, vol. 10212, pp. 94–124. Springer, Cham (2017). https://doi.org/10.1007/978-3-319-56617-7_4

15. Yuan, L., Haiyan, W., Shulan, W., Ding, Y.: Attribute-based searchable encryption scheme supporting efficient range search in cloud computing. In: Data Science in Cyberspace, pp. 1–8 (2021)
16. Zehong, C., Fangguo, Z., Peng, Z., Hanbang, Z.: Multi-user boolean searchable encryption supporting fast ranking in mobile clouds. Comput. Commun. **164**, 100–113 (2020)
17. Feng, L., et al.: Towards efficient verifiable Boolean search over encrypted cloud data. IEEE Trans. Cloud Comput. **11**(1), 839–853 (2023)
18. Regev, O.: New lattice-based cryptographic constructions. J. ACM **51**(6), 899–942 (2004)
19. Regev, O.: On lattices, learning with errors, random linear codes, and cryptography. J. ACM **56**(6), 1–40 (2009)
20. Gentry, C., Peikert, C., Vaikuntanathan, V. : Trapdoors for hard lattices and new cryptographic constructions. In: IACR Cryptology ePrint Archive, pp. 197–206 (2008)
21. Ajtai, M. : Generating hard instances of the short basis problem. In: Automata, Languages and Programming, pp. 1–9 (1999)
22. Agrawal, S., Boneh, D., Boyen, X.: Efficient lattice (H)IBE in the standard model. In: Gilbert, H. (ed.) EUROCRYPT 2010. LNCS, vol. 6110, pp. 553–572. Springer, Heidelberg (2010). https://doi.org/10.1007/978-3-642-13190-5_28
23. Agrawal, S., Boneh, D., Boyen, X.: Lattice basis delegation in fixed dimension and shorter-ciphertext hierarchical IBE. In: Rabin, T. (ed.) CRYPTO 2010. LNCS, vol. 6223, pp. 98–115. Springer, Heidelberg (2010). https://doi.org/10.1007/978-3-642-14623-7_6
24. Peiming, X., Shaohua, T., Peng, X., Qianhong, W., Honggang, H., Willy, S.: Practical multi-keyword and boolean search over encrypted e-mail in cloud server. IEEE Trans. Serv. Comput. **14**(6), 1877–1889 (2019)
25. Zhang, X., Xu, C., Wang, H., Zhang, Y., Wang, S.: FS-PEKS: lattice-based forward secure public-key encryption with keyword search for cloud-assisted industrial internet of things. IEEE Trans. Dependable Secur. Comput. **18**(3), 1019–1032 (2021)
26. Yu, X., Xu, C., Xu, L., Mei, L.: Hardening secure search in encrypted database: a KGA-resistance conjunctive searchable encryption scheme from lattice. Soft. Comput. **26**(21), 11139–11151 (2022)
27. Liu, Z., Tseng, Y., Tso, R., Mambo, M., Chen, Y. : Public-key authenticated encryption with keyword search: cryptanalysis, enhanced security, and quantum-resistant instantiation. In: IACR Cryptology ePrint Archive, pp. 423–436 (2022)
28. Yajun, F., Baodong, Q., Dong, Z.: A lattice-based designated-server public-key authenticated encryption with keyword search. J. Syst. Architect. **145** (2023)

Privacy Protection

Temporal Knowledge Graph Representation Learning with Evolutional Representation and Attention Mechanism

Yue Wang[1], Yanli Chen[2(✉)], Wei Ke[3], Yanxia Zhou[1], and Zhicheng Dong[1(✉)]

[1] College of Information Science and Technology, Xizang University, Lhasa, China
dongzc666@163.com

[2] School of Big Data and Computer Science, Guizhou Normal University, Guiyang, China

[3] Faculty of Electronic and Information Engineering, Xi'an Jiaotong University, Xi'an, China

Abstract. Temporal knowledge graphs (TKGs) play an increasingly prominent role in dynamic scenarios such as social network evolution, financial forecasting, and smart city operations, where capturing evolving relationships is critical. Temporal knowledge graph representation learning holds significant practical value by modeling dynamic semantic evolution. However, significant challenges, which is reflected in effectively encoding temporal patterns and integrating structural dependencies across evolving graph states, exists in current approaches. To meet these challenges, we propose a novel framework named as temporal knowledge graph representation and learning with evolutional representation and attention mechanism that synergizes two complementary innovations within an encoder-decoder architecture. The first innovation involves evolutionary representation learning, which dynamically captures the evolution of entities and relations by integrating structural dependencies within temporal snapshots through RGCN operations. The second innovation focuses on modeling sequential patterns across adjacent states. The time-aware attention mechanism adaptively weights historical contexts through attention layers, amplifying salient temporal correlations for query-specific prediction. Through these components, the proposed TKGRAM can improve key issues in existing methods. Extensive experiments on ICEWS benchmarks demonstrate TKGRAM's consistent superiority, outperforming state-of-the-art baselines, and further, ablation studies validate both innovations as critical drivers of performance gains.

Keywords: Temporal Knowledge Graph · Representation learning · Temporal evolutional representation · Attention mechanism

1 Introduction

Knowledge graphs (KGs) have emerged as a fundamental framework for representing structured information in various domains, such as social networks, finan-

J. Yu and Y. Chen (Eds.): FCS 2025, CCIS 2805, pp. 135–149, 2027.
https://doi.org/10.1007/978-981-95-9514-3_9

cial transactions, and smart city operations, where data naturally forms complex relational structures [26]. Typically, KGs represent semantic information using triples $(subject, relation, object)$, enabling applications like knowledge base completion (KBC), recommendation systems, and semantic search. Large-scale KGs such as DBpedia, FreeBase, and WordNet [25] have been widely adopted, yet they suffer from inherent incompleteness due to data sparsity. Knowledge graph representation learning addresses this by mapping entities and relations into continuous low-dimensional vector spaces, effectively capturing underlying semantic patterns and facilitating tasks like entity prediction [27]. However, a critical limitation of traditional KG representation learning models is their static nature–they ignore the temporal dimension, treating all facts as timeless. This oversight leads to inaccurate semantic representations in dynamic real-world scenarios, where relationships evolve over time, rendering static models inadequate for capturing the fluidity of knowledge.

The emergence of temporal knowledge graphs (TKGs) addresses this gap by incorporating timestamps into quadruples $(subject, relation, object, time)$, thereby modeling the dynamic evolution of relationships. And temporal knowledge graph representation learning has gained prominence, aiming to embed temporal features into low-dimensional vectors while preserving structural integrity, thereby enhancing practical applications like event prediction and anomaly detection [28]. Despite their potential, existing TKG representation learning models face persistent challenges in effectively encoding temporal patterns and integrating structural dependencies across evolving graph states [29].

First, encoding temporal patterns remains problematic: TKGs exhibit dynamic topologies where entity features change over time, necessitating time encoding functions that respect inherent properties like periodicity and relative time significance [30]. Current approaches, such as simple feed-forward neural networks or recurrent models (e.g., LSTM in TA-TransE [12]), lack theoretical depth and struggle to model complex temporal distributions, particularly for sparse or irregular event sequences. Second, integrating structural dependencies is insufficient: TKGs consist of sequences of static snapshots, but existing methods often treat quadruples independently, failing to capture global topological features [31]. Third, different relations exert varying influences on entities over time, and current attention mechanisms assign uniform weights, ignoring temporal-specific importance.

To overcome these challenges, we propose TKGRAM, a novel framework that synergizes two complementary innovations within an encoder-decoder architecture. First, our evolutional representation learning module dynamically captures entity and relation evolution by integrating structural dependencies within each temporal snapshot using relational graph convolutional networks. This approach ensures that entity states evolve coherently over time, preserving both local graph structures and long-term dependencies. Second, the time-aware attention mechanism adaptively weights historical contexts through attention layers, amplifying salient temporal correlations for query-specific prediction. Integrated through a unified framework, TKGRAM effectively encodes temporal patterns while enriching structural representations, overcoming the fragmentation in prior works.

Our contributions in this paper are threefold:

- We introduce TKGRAM, a novel temporal knowledge graph representation learning framework that synergizes evolutional representation learning with time-aware attention. This enables simultaneous modeling of structural dependencies and temporal evolution, outperforming state-of-the-art baselines.
- We develop an evolutional representation learning module using RGCN for intra-snapshot structural aggregation and GRU for inter-snapshot sequential modeling, capturing entity and relation dynamics with theoretical rigor.
- We design a time-aware attention mechanism that dynamically assigns weights to historical contexts based on query relevance, significantly improving prediction accuracy across three benchmark datasets.

Extensive experiments on ICEWS benchmarks demonstrate TKGRAM's superiority, with consistent gains in mrr and hits@n metrics. Ablation studies confirm both innovations as critical drivers. The rest of this paper is organized as follows. Section 2 reviews related works on static and temporal KG representation learning. Section 3 introduces preliminaries, including TKG definitions and key notations. Section 4 details the TKGRAM model architecture. Section 5 presents experimental results and analysis, and Sect. 6 summarizes the entire text and discusses future development.

2 Related Work

This section introduces traditional static knowledge graph (KG) representation learning models and their temporal knowledge graph (TKG) representation learning models.

2.1 Static Knowledge Graph Representation Learning

Currently, most existing knowledge graph representation learning models are designed for static knowledge graphs (KGs). These models can be broadly classified into three main categories.

Translation-Based Methods: This category enforces translation constraints defined by relations between head and tail entities. They assess the plausibility of triples by computing the euclidean distance between the translated head entity vector and the tail entity vector. Representative models include TransE [1], TransH [2], and TransR [3]. Notably, the simplicity and efficiency of TransE have inspired numerous subsequent extensions.

Semantic Matching Methods: Models in this category evaluate fact plausibility by aligning the latent semantic representations of entities and relations within the vector space. Prominent and widely adopted models include RESCAL [4], DistMult [5], ComplEx [6], and SimplE [7].

Neural Network-Based Methods: This group primarily leverages the strong feature extraction and non-linear fitting capabilities of neural networks to model KG structures. Representative examples include ConvE [8], ConvKB [9], and RGCN [10].

However, a critical limitation of all these approaches is their oversight of temporal information. Consequently, they fail to capture the dynamic nature inherent in real-world knowledge. This limitation inevitably leads to lower accuracy when these models are applied to temporal knowledge graphs (TKGs).

2.2 Temporal Knowledge Graph Representation Learning

Compared with static KGs, TKGs explicitly incorporate temporal timestamps alongside facts. To effectively capture this temporal dimension, temporal knowledge graph representation learning (TKGRL) methods are designed to account for timestamps during model construction. These methods can be broadly categorized into transformation-based methods, decomposition-based methods and graph neural networks-based methods.

Transformation-Based Methods: Within the transformation-based approach, temporal information is incorporated by modeling either timestamps or relations as transformations operating between entities. Building upon existing TKGRL methods, TKGRL is achieved by either integrating temporal information directly or by projecting entities and relations onto temporal hyperplanes. TTransE [11] incorporates temporal information by concatenating it to relations. It converts the quadruple (s, r, o, t) into the triple $(s, [r|t], o)$, thereby enforcing a time-aware translation constraint. Unlike TTransE's approach, which simply concatenates temporal information to relations, TA-TransE [12] generates more expressive synthetic relations. It achieves this by treating relations as sequences incorporating temporal information and utilizing LSTM networks to learn time-aware representations of relation types. HyTE [13] jointly learns representations of entities, relations, and temporal information. This is achieved by projecting the temporal knowledge graph onto distinct timestamp-specific hyperplanes, effectively decomposing the TKG into multiple static subgraphs corresponding to each timestamp. Other representative papers include RotatE [14], Tero [15], ChronoR [16] and RotateQVS [17].

Decomposition-Based Methods: The primary objective of knowledge graph representation learning is to derive low-dimensional vector representations for the graph's elements. Tensor decomposition offers three key capabilities that are inherently well-suited to this objective: dimension reduction, missing data completion, and implicit relation mining. T-SimplE [18] models the temporal knowledge graph as a 4th-order tensor and applies tensor decomposition to learn embeddings of entities, relations, and timestamps. TLT-KGE [19] integrates temporal representations into the embeddings of entities and relations, verifying the effectiveness of this integration in both complex and hyper complex spaces. TuckER [20] leverages the tucker decomposition for knowledge graph link prediction; it represents the knowledge graph as a 3rd-order tensor and decomposes it into a core tensor, entity subject embeddings, relation embeddings, and entity object embeddings.

Graph Neural Networks-Based Methods: Graph neural networks (GNNs) possess powerful capabilities for structural modeling. Through GNNs, entities can learn enriched representations by integrating both their own attribute features and the global topological features of the graph. Furthermore, multiple specialized variants of graph neural networks exist, including relational graph convolutional networks and graph attention networks. DEGAT [21] proposes a dynamic embedding graph attention network. It employs graph attention networks to derive static entity representations by aggregating features from neighboring entities and relations. T^2TKG [22] presents a novel approach specifically designed to overcome the limitation of existing temporal knowledge graph methods in explicitly capturing both intra-time and inter-time latent dependencies, enabling more accurate prediction of future facts. TEA-GNN [23] learns entity representations via a graph attention network. This model incorporates both relational and temporal information directly into the GNN architecture.

3 Preliminaries

In this section, we present the preliminaries of our work, including the definition of temporal knowledge graph and representation learning for temporal knowledge graph. Then, we briefly introduce the main symbols used in this paper for easy reference. The symbols are shown in Table 1.

Table 1. The main notations.

Symbol	Explanation
$\mathcal{G}$	A TKG
$\mathcal{G}_t$	A KG snapshot at time t in a TKG
$\mathcal{E}, \mathcal{R}, \mathcal{T}$	Set of entities, relations and times in $\mathcal{G}$
$\|\mathcal{E}\|, \|\mathcal{R}\|, \|\mathcal{T}\|$	The numbers of entities, relations and times
m	The numbers of adjacent snapshot
h	Initialization of embedding
s^t, o^t , r	Entity s, o , r appears at time t
$h_{s,t}, h_{r,t}$	Entity and relation embedding at time t

Definition 1: Temporal Knowledge Graph. Formally, a temporal knowledge graph (TKG) constitutes a multi-relational directed graph $\mathcal{G} = (\mathcal{E}, \mathcal{R}, \mathcal{T})$,

where temporal edges (s, r, o, t) are timestamped interactions between entities. Here, $\mathcal{E}$, $\mathcal{R}$ and $\mathcal{T}$ denote discrete sets of entities, relations, and timestamps. The TKG is dynamically structured as a sequence of static snapshots: $\{\mathcal{G}_1, \mathcal{G}_2, \ldots, \mathcal{G}_T\}$, with each snapshot $\mathcal{G}_t$ corresponding to time $t \in \mathcal{T}$. For a fact in a subgraph $\mathcal{G}_t$, it can be represented as a quadruple (s, r, o, t) which indicats that entity $s \in \mathcal{E}$ is related to entity $o \in \mathcal{E}$ via relation $r \in \mathcal{R}$ at time $t \in \mathcal{T}$. According to the time range, TKGs has two settings: interpolation and extrapolation, corresponding to historical and future temporal reasoning tasks respectively. Given a TKG with timestamps from t_0 to t_T, interpolation mainly aimed at inferring missing facts that occur at time $t < t_T$, on the contrary, extrapolation attempts to predict facts that occur at time $t > t_T$. More formally, the purpose of this study is to predict either the subject in a given query $(s, r, ?, t)$ or the relation in a given query $(s, ?, o, t)$.

Definition 2: Representation Learning for Temporal Knowledge Graph. Given a TKG $\mathcal{G} = (\mathcal{E}, \mathcal{R}, \mathcal{T})$, the objective of TKG representation learning is to learn mapping functions, $F_e : \mathcal{E} \rightarrow \mathbb{R}^{\mathbf{d}}$ and $F_r : \mathcal{R} \rightarrow \mathbb{R}^{\mathbf{d}}$ that project entities and relations, respectively. Where $\mathbf{d}$ is the dimensionality of the vector representations. These mappings are designed to capture the temporal dynamics of the knowledge graph through the learned representations.

4 Our Approach

Figure 1 depicts the overall architecture of TKGRAM model. As shown, the model adopts an encoder-decoder framework. The encoder module maps entities and relations into a continuous low-dimensional vector space while simultaneously fusing structural information and temporal features. The encoder consists of two components: evolutionary representation learning module and time-aware attention module. The evolutionary representation learning module integrates structural dependencies at each timestamp in the KG, sequential patterns across temporally adjacent facts, and the static attributes of entities into evolutionary representations of entities and relations. The time-aware attention module improves encoding accuracy by assigning distinct weights to different timestamps based on their varying contributions to the final results.

4.1 Evolutional Representation Learning

Evolutional representation learning module focuses on acquiring states from the recent m historical knowledge graph snapshots adjacent to the query time. It aggregates structural dependencies within each snapshot and employs gated recurrent unit to learn temporal patterns of entities and relations, thereby enhancing the model's capacity to capture temporal dynamics across evolving knowledge graphs. First, to account for periodic patterns such as recurring

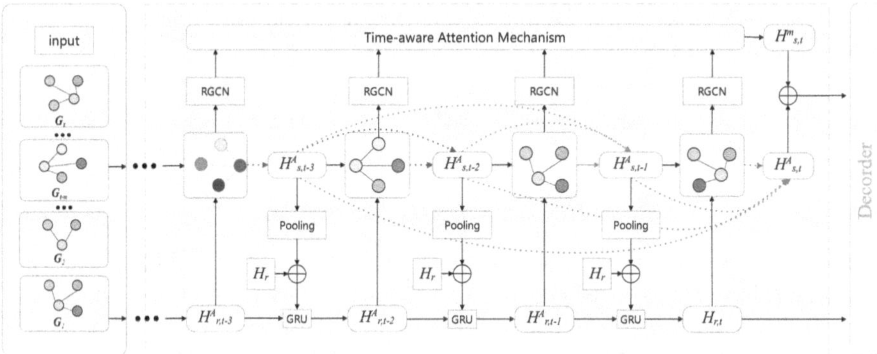

Fig. 1. Overall architecture of TKGRAM model.

events, this study employs Fourier Transform to convert time domain information into the frequency domain, formally expressed as follows:

$$\varphi(d) = \frac{1}{\sqrt{k}}\left[\cos(dw_1+b_1), \sin(dw_1+b_1), \ldots, cos(dw_k+b_k), \sin(dw_k+b_k)\right], \quad (1)$$

where $d = t_q - t_i$ is the time interval that is modeled by rescaling a learnable time w_t with a time bias b, $cos(\cdot)$ and $sin(\cdot)$ are periodic activation functions, k is the number of periods. Then, for KG snapshots on each timestamp, we leverage a relation-aware graph convolutional network (RGCN) to capture structural dependencies within each temporal snapshot. The specific operation is as follows:

$$g = \sigma\left(W_g\left[h_{\|}\varphi(d)\right]\right) \quad (2)$$

$$\boldsymbol{h} = g \cdot h + (1-g) \cdot h \quad (3)$$

$$\boldsymbol{h}_{s,t}^{(l+1)} = f\left(\frac{1}{|\mathcal{N}_{s,t}|}\sum_{o^t\in\mathcal{N}_{s,t}} W_1^l\left(\boldsymbol{h}_{o,t}^{(l)} + \boldsymbol{h}_{r,t}^{(l)}\right) + W_2^l\boldsymbol{h}_{s,t}^{(l)}\right) \quad (4)$$

where W_g is a trainable linear transformation matrix, $\|$ indicates vector concatenation operation, $\sigma(\cdot)$ is RReLU activation function and g_t is trainable attention weight function. The information aggregation of entities and relations is defined as Eq. 4. Where $\boldsymbol{h}_{s,t}^{(l)}$ and $\boldsymbol{h}_{r,t}^{(l)}$ denote the embedding of entities and relations in the l-th layers of the relational graph neural network in each KG snapshot at t-th timestamp, separately. For simplicity, W_1^l and W_2^l are trainable parameters for aggregating features and self-loop in the l-th layer, $f(\cdot)$ is the RReLU activation function, and $|\mathcal{N}_{s,t}|$ is the set of neighbors of entity s in time t as a normalization constant.

To progressively update the evolving representations of entities and relations in temporal order, we employ Gated Recurrent Unit (GRU) to integrate temporal dependencies from historical subgraphs for each query. This framework explicitly captures temporal evolution patterns of entities and relations, formally formulated as follows:

$$H_{s,t} = GRU\left(H_{s,t-1}, H_{s,t-1}^{A}\right) \tag{5}$$

$$r_t' = \left[f_{ave}\left(H_{s,t-1}, \mathcal{H}_{r,t}\right); r\right] \tag{6}$$

$$H_{r,t} = GRU\left(H_{r,t-1}, R'\right) \tag{7}$$

where $H_{s,t-1}^{A}$ denotes the embedding matrix of entity s after RGCN information aggeration. $H_{s,t}$ consists of all $h_{s,t}$ of entity s at time t. For relation, in order to maintain the consistency with the update of the entity embedding in the subgraph sequence, GRU is also used for update. Where $\mathcal{H}_{r,t}$ is all entities connected to relation r at time t, r_t' is obtained by $H_{s,t-1}$ and $\mathcal{H}_{r,t}$ using f_{ave}, the f_{ave} is mean pooling operation, and R' consists of r_t of all relations. $H_{r,t}$ is finally embedding of relation at time t.

4.2 Time-Aware Attention Mechanism

After following temporal processing steps, we derive dynamic evolutionary representations of entities and relations across the most recent m-step period. However, these representations fail to capture the relative contribution of individual historical timestamps to query prediction. To address this limitation, we introduce a time-aware attention mechanism that dynamically evaluates the impact of different knowledge graph snapshots specifically for the query. The mechanism operates as follows:

$$h_{s,t}' = W_3\left[h_{s,t} \| f_{ave}(h_{r,t})\right] \tag{8}$$

$$\alpha_i = f_2\left(W_4\left(h_{t-m+i}^{A} + h_{s,t}'\right)\right) \quad i \in [1, m-1] \tag{9}$$

$$\boldsymbol{h}_{s,t}^{m} = h_{s,t} + \sum_{i=2}^{m} \alpha_i h_{s,t-m+i} \tag{10}$$

where α_i denotes the attention score for each historical timestamp, W_3 and W_4 are the learnable weight matrices, $f_2(\cdot)$ is the sigmoid activate function,$h_{s,t}^{m}$ is the final representation after time-aware attention mechanism. Ultimately, the final entity embeddings are derived by dynamically integrating outputs from both the evolutionary representation learning module and the time-aware attention mechanism. The operation is as follows:

$$h_{s,t} = \lambda(\mathbf{g}) \odot h_{s,t}^{m} + (1 - \lambda(\mathbf{g})) \odot h_{s,t}^{A} \tag{11}$$

where $\lambda(\cdot)$ is the sigmoid function, aiming to restrict the range of each element to $[0, 1]$. The $\mathbf{g}$ is the gate parameter to adjust the weights. The $\odot$ is element-wise multiplication. And the $h_{s,t}^{A}$ is the result of the Eq. 5.

4.3 Decoder

To evaluate prediction performance for inferred entities and relations, we employ ConvTransE [24] - a widely adopted decoder in temporal knowledge graph reasoning - as our scoring function. This approach effectively captures dependencies by integrating convolutional networks with translational embedding techniques. The scoring function is formulated as follows:

$$p_t(o \mid s, r, \mathcal{G}_{<t}) = \sigma\Big(h_{o,t} \cdot ConvTransE\big(h_{s,t}, h_{r,t}\big)\Big), \tag{12}$$

$$p_t(r \mid s, o, \mathcal{G}_{<t}) = \sigma\Big(h_{r,t} \cdot ConvTransE\big(h_{s,t}, h_{o,t}\big)\Big), \tag{13}$$

where $\sigma(\cdot)$ is the sigmoid function. $h_{s,t}$, $h_{o,t}$ and $h_{r,t}$ are the embedding representations of entities s, o and relation r at timestamp t. $\mathcal{G}_{<t}$ represents the temporal subgraphs that occurred before the time point t. Following previous studies, the loss function during training is defined as:

$$\mathcal{L}_e = -\sum_{t=0}^{T} \sum_{(s,r,o,t)\in\mathcal{G}_t} \log p_t(o \mid s, r, \mathcal{G}_{<t}) \tag{14}$$

$$\mathcal{L}_r = -\sum_{t=0}^{T} \sum_{(s,r,o,t)\in\mathcal{G}_t} \log p_t(r \mid s, o, \mathcal{G}_{<t}) \tag{15}$$

where Eq. 14 and Eq. 15 respectively represent the loss functions for entity prediction and relation prediction.The $(s, r, o, t) \in \mathcal{G}_t$ represents all quadruples at time t. The final loss is $\mathcal{L} = \lambda\mathcal{L}_e + (1-\lambda)\mathcal{L}_r$, λ is the hyperparameter balancing different tasks.

5 Experiments

This section presents an empirical evaluation of the proposed model on three public datasets. We first detail the experimental setup, covering datasets, evaluation metrics, and baselines, then analyze the results. Additionally, ablation studies assess the contribution of each core component to model performance.

5.1 Experimental Setup

Datasets. To establish robust evaluation frameworks, we conducted experiments using three widely recognized event-centric temporal knowledge graphs (TKGs) within computational social science. The benchmark datasets–namely, ICEWS14, ICEWS18, and ICEWS05-15–are derived from the Integrated Crisis Early Warning System (ICEWS), which captures geopolitical event data. Adhering to temporal evaluation protocols, we employed a strict chronological partitioning scheme: training (80%), validation (10%), and testing (10%) splits. The statistical details of these datasets are summarized in Table 2.

Table 2. Statistics of ICEWS14, ICEWS18 and ICEWS05-15 .

Dateset	Entities $\mathcal{E}$	Relations $\mathcal{R}$	Time Interval	Train Facts	Valid Facts	Test Facts
ICEWS14	7,218	230	24 h	74,845	8,514	7,371
ICEWS18	23,033	256	24 h	373,018	45,995	49,545
ICEWS05-15	10,488	251	14 h	368,868	46,302	46,159

Evaluation Metrics. In this study, performance is evaluated using the Mean Reciprocal Rank (MRR) and Hits@1, 3, 10 metrics. MRR is computed as the average reciprocal rank of the correct predictions. Hits@1, Hits@3, and Hits@10 measure the proportion of cases where the correct answer appears among the top 1, 3, and 10 ranked results, respectively. The mathematical formulations for MRR and Hits@n are provided below:

$$\text{MRR} = \frac{1}{N} \sum_{(s,r,o,t)\in\mathcal{G}} \left(\frac{1}{\text{rank}(s,r,o,t)}\right) \tag{16}$$

$$\text{Hits@n} = \frac{1}{N} \sum_{(s,r,o,t)\in\mathcal{G}} \mathbb{I}\left[\text{rank}(s,r,o,t) \leq n\right] \tag{17}$$

In these formulations, $\text{rank}(s,r,o,t)$ denotes the ordinal position of the correct quadruple (s,r,o,t) within the ranked predictions, where N represents the total instances in the evaluation set. The indicator function $\mathbb{I}$ yields 1 when the condition is true and 0 otherwise, with n taking integer values $\{1,3,10\}$.

Baselines. To validate the superiority of our model, we conduct comprehensive comparisons against both static and temporal knowledge graph representation learning methods. For static knowledge graphs , baseline methods include DistMult [5], ComplEx [6], and RotatE [14], etc. For temporal approaches, our model is compared against the time-aware method described in Sect. 2.2, such as RE-GCN [31], CRAFT [33], and STDN [32], etc.

5.2 Results and Analysis

Table 3 and Table 4 present experimental results on the ICEWS14, ICEWS05-15, and ICEWS18 datasets, revealing consistent underperformance of static knowledge graph models compared to temporal knowledge graph approaches across most evaluation metrics. This performance gap primarily stems from static models' inherent limitation: they generate time-invariant representations for entities and relations, thereby failing to capture evolving temporal dynamics.

Experimental results demonstrate that our model achieves state-of-the-art performance, surpassing baseline methods. Compared with STDN, TKGRAM improves the MRR by 3.08 and 1.95% points on ICEWS14 and ICEWS18, respectively. The results show the effectiveness of the TKGRAM model in predicting entities. Notably, across the ICEWS datasets, our model demonstrates the most significant performance improvement on ICEWS14. In addition, the results show that the model performance on the ICEWS14 and ICEWS05-15 dataset are better than those on the ICEWS18 dataset. The primary reason is that at the same quantity level of relations, the number of entities of ICEWS18 dataset is excessive. Thus, this imbalance triggers heightened complexity in entity interactions, which presents a significant challenge for extracting meaningful information from such intricate data patterns.

Table 3. Results on ICEWS14 and ICEWS05-15 . The time-filtered MRR, H@1, H@3, and H@10 metrics are multiplied by 100. The best results are boldfaced.

Dataset	ICEWS14				ICEWS05-15			
	MRR	H@1	H@3	H@10	MRR	H@1	H@3	H@10
Distmult	15.44	10.91	17.24	23.92	29.73	13.93	36.19	47.54
ComplEx	28.21	18.83	28.54	49.71	30.35	14.72	37.47	47.21
R-GCN	28.03	19.42	31.95	44.83	30.24	14.21	36.48	43.16
RotatE	21.31	10.26	24.35	44.75	24.71	13.22	29.04	48.16
CyGNet	35.05	25.73	39.01	53.55	36.81	26.61	41.63	56.22
RE-GCN	40.39	30.66	44.96	59.21	48.03	37.33	53.85	68.27
TiRGN	44.04	33.83	48.95	63.84	50.04	39.25	56.13	70.71
CENET	39.02	29.62	43.23	57.49	41.95	32.17	46.93	60.43
CRAFT	45.71	35.05	51.83	65.21	**50.14**	**39.56**	56.18	70.09
TFCE	43.70	33.79	48.69	64.31	48.84	38.21	56.57	**71.67**
STDN	43.26	33.71	48.16	61.47	-	-	-	-
Ours	**46.34**	**36.54**	**52.46**	**70.26**	48.57	38.85	**56.84**	71.03

5.3 Ablation Study

The superior performance of TKGRAM is primarily attributed to its dual capabilities: evolutional representation learning and temporal-aware attention mechanisms, fundamentally transforming temporal knowledge modeling. To further validate the contribution of each core component, we conducted rigorous ablation studies on the ICEWS14 and ICEWS05-15 datasets. As shown in table 5. We systematically constructed reduced variants of the TKGRAM model: one variant removed the temporal-aware attention mechanism (TKGRAM-w/o-AR), preserving only evolutional representation updates; another variant disabled the evolutional representation learning module (TKGRAM-w/o-ER).

Detailed results, as presented in Table 5, unequivocally demonstrate the critical necessity of both proposed mechanisms. Removal of the temporal-aware attention mechanism (TKGRAM-w/o-AR) caused a pronounced degradation in performance, evident in significant decreases in MRR. This decline underscores the indispensable role of the temporal attention module in dynamically integrating and weighting relevant historical contexts across adjacent timesteps for

Table 4. Results on ICEWS05-15. The time-filtered MRR, H@1, H@3, and H@10 metrics are multiplied by 100. The best results are boldfaced.

Dataset	ICEWS18			
	MRR	H@1	H@3	H@10
Distmult	17.95	13.12	20.71	29.32
ComplEx	17.95	13.12	20.71	29.32
R-GCN	18.04	8.57	15.68	20.32
RotatE	12.78	4.01	14.89	31.91
CyGNet	24.93	15.90	28.28	42.61
RE-GCN	30.58	21.01	34.34	48.75
TiRGN	33.66	23.19	37.99	54.22
CENET	27.85	18.15	31.63	46.98
CRAFT	34.21	23.96	38.53	54.11
TFCE	33.09	22.88	37.32	53.65
STDN	32.67	22.80	37.32	53.65
Ours	**34.62**	**24.08**	**39.03**	**57.26**

effective dependency modeling. Similarly, disabling the evolutive representation learning module (TKGRAM-w/o-ER) resulted in substantial performance loss. This degradation highlights that the explicit modeling of entity state evolution through sequential updates is crucial for effectively capturing and propagating historical information. This starkly illustrates that the synergistic interplay between evolutive representation learning and temporal-aware attention is fundamental to TKGRAM's state-of-the-art effectiveness.

Table 5. Performance comparison of TKGRAM variants on ICEWS datasets

Dataset	ICEWS14			ICEWS05-15		
	MRR	H@1	H@10	MRR	H@1	H@10
TKGRAM	46.34	36.54	70.26	48.57	38.85	71.03
TKGRAM-w/o-ER	42.45	33.74	64.39	43.72	34.83	63.38
TKGRAM-w/o-AR	44.75	34.26	67.03	46.08	36.05	65.87

6 Conclusion

This paper introduced TKGRAM, a novel framework for temporal knowledge graph representation learning that synergizes evolutional representation learning and a time-aware attention mechanism. By dynamically capturing entity/relation evolution through RGCN-based structural aggregation and GRU-driven sequential modeling, while adaptively weighting historical contexts via attention, TKGRAM overcomes key limitations in existing methods. Extensive experiments on ICEWS benchmarks demonstrate TKGRAM's consistent superiority over state-of-the-art baselines in mrr and hits@n metrics. Ablation studies confirm both modules as critical drivers. Future work will explore generating time-aware discriminative negative samples to further enhance reasoning robustness.

Acknowledgments. This study was funded by the National Key Research and Development Program of China(2023YFC3304504); the Science and Technology Major Project of Tibetan Autonomous Region of China(XZ202201ZD0006G); 2024 Xizang Autonomous Region Central Guided Local Science and Technology Development Fund Project(XZ202401YD0015).

References

1. Bordes, A., Usunier, N., Garcia-Duran, A., Weston, J., Yakhnenko, O.: Translating embeddings for modeling multi-relational data. In: Proceedings of the 26th International Conference on Neural Information Processing Systems, pp. 2787–2795. NIPS'13, Curran Associates Inc. (2013)
2. Wang, Z., Zhang, J., Feng, J., Chen, Z.: Knowledge graph embedding by translating on hyperplanes. In: Proceedings of the Twenty-Eighth AAAI Conference on Artificial Intelligence, pp. 1112–1119. AAAI'14, AAAI Press (2014)
3. Lin, Y., Liu, Z., Sun, M., Liu, Y., Zhu, X.: Learning entity and relation embeddings for knowledge graph completion. In: Proceedings of the Twenty-Ninth AAAI Conference on Artificial Intelligence, pp. 2181–2187. AAAI'15, AAAI Press (2015)
4. Nickel, M., Tresp, V., Kriegel, H.P.: A three-way model for collective learning on multi-relational data. In: Proceedings of the 28th International Conference on Machine Learning, pp. 809–816. ICML'11, Omnipress (2011)
5. Yang, B., Yih, S.W., He, X., Gao, J., Deng, L.: Embedding entities and relations for learning and inference in knowledge bases. In: Proceedings of the third International Conference on Learning Representations, pp. 809–816. ICLR'15 (2015)
6. Trouillon, T., Welbl, J., Riedel, S., Gaussier, E., Bouchard, G.: Complex embeddings for simple link prediction. In: Proceedings of the 33rd International Conference on International Conference on Machine Learning, pp. 2071–2080. ICML'16, JMLR.org (2016)
7. Kazemi, S.M., Poole, D.: Simple embedding for link prediction in knowledge graphs. In: Proceedings of the 32nd International Conference on Neural Information Processing Systems, pp. 4284–4295. NeurIPS'18, Curran Associates, Inc. (2018)
8. Dettmers, T., Minervini, P., Stenetorp, P., Riedel, S.: Convolutional 2D knowledge graph embeddings. In: Proceedings of the Thirty-Second AAAI Conference on Artificial Intelligence, pp. 1811–1818. AAAI'18, AAAI Press (2018)
9. Nguyen, D.Q., Nguyen, T.D., Nguyen, D.Q., Phung, D.: A novel embedding model for knowledge base completion based on convolutional neural network. In: Proceedings of the 2018 Conference of the North American Chapter of the Association for Computational Linguistics: Human Language Technologies, pp. 327–333. NAACL'18, Association for Computational Linguistics (2018)

10. Schlichtkrull, M., Kipf, T.N., Bloem, P., van den Berg, R., Titov, I., Welling, M.: Modeling Relational Data with Graph Convolutional Networks. In: Gangemi, A., et al. (eds.) ESWC 2018. LNCS, vol. 10843, pp. 593–607. Springer, Cham (2018). https://doi.org/10.1007/978-3-319-93417-4_38
11. Leblay, J., Chekol, M.W.: Deriving validity time in knowledge graph. In: Companion Proceedings of the The Web Conference 2018, International World Wide Web Conferences Steering Committee, Republic and Canton of Geneva, CHE, pp. 1771–1776 (2018). https://doi.org/10.1145/3184558.3191639
12. García-Durán, A., Dumančić, S., Niepert, M.: Learning sequence encoders for temporal knowledge graph completion. In: Proceedings of the 2018 Conference on Empirical Methods in Natural Language Processing, Association for Computational Linguistics, Stroudsburg, PA, USA, pp. 4816–4821 (2018). https://doi.org/10.18653/v1/D18-1516
13. Dasgupta, S.S., Ray, S.N., Talukdar, P.: Hyte: Hyperplane-based temporally aware knowledge graph embedding. In: Proceedings of the 2018 Conference on Empirical Methods in Natural Language Processing, pp. 2001–2011. EMNLP'18, Association for Computational Linguistics (2018)
14. Sun, Z., Deng, Z., Nie, J., Tang, J.: Rotate: Knowledge graph embedding by relational rotation in complex space. In: 7th International Conference on Learning Representations, ICLR 2019, OpenReview.net, New Orleans, LA, USA (2019). https://openreview.net/forum?id=HkgEQnRqYQ
15. Xu, C., Nayyeri, M., Alkhoury, F., Shariat Yazdi, H., Lehmann, J.: TeRo: A time-aware knowledge graph embedding via temporal rotation. In: Proceedings of the 28th International Conference on Computational Linguistics, Association for Computational Linguistics, Stroudsburg, PA, USA. pp. 1583–1593 (2020). https://aclanthology.org/2020.colingmain.139
16. Sadeghian, A., Armandpour, M., Colas, A., Wang, D.Z.: Chronor: Rotation based temporal knowledge graph embedding. In: Proceedings of the AAAI Conference on Artificial Intelligence, AAAI2021, AAAI Press, Palo Alto, CA, USA. pp. 6471–6479 (2021). https://ojs.aaai.org/index.php/AAAI/article/view/16802
17. Chen, K., Wang, Y., Li, Y., Li, A.: RotateQVS: Representing temporal information as rotations in quaternion vector space for temporal knowledge graph completion. In: Proceedings of the 60th Annual Meeting of the Association for Computational Linguistics (Volume 1: Long Papers), Association for Computational Linguistics, Stroudsburg, PA, USA. pp. 5843–5857 (2022). https://doi.org/10.18653/v1/2022.acl-long.402
18. Lin, L., She, K.: Tensor decomposition-based temporal knowledge graph embedding. In: 32nd IEEE International Conference on Tools with Artificial Intelligence, ICTAI 2020, Baltimore, MD, USA, November 9–11, 2020, IEEE, Baltimore, MD, USA. pp. 969–975 (2020). https://doi.org/10.1109/ICTAI50040.2020.00151
19. Zhang, F., Zhang, Z., Ao, X., Zhuang, F., Xu, Y., He, Q.: Along the time: Timeline-traced embedding for temporal knowledge graph completion. In: Proceedings of the 31st ACM International Conference on Information & Knowledge Management, Association for Computing Machinery, New York, NY, USA. pp. 2529–2538 (2022). https://doi.org/10.1145/3511808.3557233
20. Balazevic, I., Allen, C., Hospedales, T.: TuckER: Tensor factorization for knowledge graph completion. In: Proceedings of the 2019 Conference on Empirical Methods in Natural Language Processing and the 9th International Joint Conference on Natural Language Processing (EMNLP-IJCNLP), Association for Computa-

tional Linguistics, Stroudsburg, PA, USA. pp. 5185–5194 (2019). https://doi.org/10.18653/v1/D19-1522, https://aclanthology.org/D19-1522
21. Wang, J., Zhu, C., Zhu, W.: Dynamic embedding graph attention networks for temporal knowledge graph completion. In: Memmi, G., Yang, B., Kong, L., Zhang, T., Qiu, M. (eds.) Knowledge Science, pp. 722–734. Engineering and Management, Springer International Publishing, Cham (2022)
22. Zhang, M., Xia, Y., Liu, Q., Wu, S., Wang, L.: Learning latent relations for temporal knowledge graph reasoning. In: Proceedings of the 61st Annual Meeting of the Association for Computational Linguistics (Volume 1: Long Papers), pp. 12617–12631 (2023a)
23. Xu, C., Su, F., Lehmann, J.: Time-aware graph neural network for entity alignment between temporal knowledge graphs. In: Proceedings of the 2021 Conference on Empirical Methods in Natural Language Processing, Association for Computational Linguistics, Stroudsburg, PA, USA. pp. 8999–9010 (2021). https://doi.org/10.18653/v1/2021.emnlp-main.709, https://aclanthology.org/2021.emnlp-main.709
24. Shang, C., Tang, Y., Huang, J., Bi, J., He, X., Zhou, B.: End-to-end Structure Aware Convolutional Networks for Knowledge Base Completion (2018). https://doi.org/10.48550/arXiv.1811.04441, http://arxiv.org/abs/1811.04441
25. Zhang, F., Wang, X., Li, Z., Li, J.: Transrhs: A representation learning method for knowledge graphs with relation hierarchical structure. In: Proceedings of the Twenty-Ninth International Joint Conference on Artificial Intelligence, pp. 2987-2993. IJCAI'20, International Joint Conferences on Artificial Intelligence Organization (2020)
26. Bu, Y., Lu, Y., Wang, Y., Huang, B., Tao, Y., Shi, Y.: A survey of temporal knowledge graph reasoning. In: 2024 7th International Conference on Pattern Recognition and Artificial Intelligence (PRAI), pp. 657–661 (2024). https://doi.org/10.1109/PRAI62207.2024.10826769, https://ieeexplore.ieee.org/abstract/document/10826769
27. Cai, Z., He, Z., Guan, X., Li, Y.: Collective data-sanitization for preventing sensitive information inference attacks in social networks. IEEE Trans. Dependable Secure Comput. **15**, 577–590 (2018)
28. Zhou, W.T., Kang, Z., Tian, L., Su, Y.: Intensity-free convolutional temporal point process: incorporating local and global event contexts. Inf. Sci. **646**, 119318 (2023)
29. Lacroix, T., Obozinski, G., Usunier, N.: Tensor decompositions for temporal knowledge base completion. In: International Conference on Learning Representations, pp. 1–12. ICLR'20 (2020)
30. Jiang, T., et al.: Encoding temporal information for time-aware link prediction. In: Proceedings of the 2016 Conference on Empirical Methods in Natural Language Processing, pp. 2350–2354. EMNLP'16, Association for Computational Linguistics (2016)
31. Li, Z.: Temporal knowledge graph reasoning based on evolutional representation learning. In: Proceedings of the 44th International ACM SIGIR Conference on Research and Development in Information Retrieval, pp. 408–417 (2021)
32. Li, X., et al.: Intricate spatiotemporal dependency learning for temporal knowledge graph reasoning. ACM Trans. Knowl. Discov. Data **18**(6), 1–19 (2024)
33. Zhang, S., Wei, W., Huang, R., Xie W., Chen, D.Z.: Modeling historical relevant and local frequency context for representation-based temporal knowledge graph forecasting. In: Findings of the Association for Computational Linguistics: EMNLP 2024, Miami, Florida, USA: Association for Computational Linguistics, pp. 7675–7686 (2024). https://doi.org/10.18653/v1/2024.findings-emnlp.451

A Robust Audio Watermarking Algorithm Based on Adaptive Quantization Strategy and Feature Classification

Pei Zhu[1], Rongsheng Li[2(✉)], Zhenghui Liu[1], Zhipeng Yang[1], and Song Zhang[1]

[1] School of Computer and Information Technology, Xinyang Normal University, Xinyang 464000, China
yangzp@xynu.edu.cn, zhangsong@xynu.edu.cn

[2] School of Marxism Studies, Xinyang Normal University, Xinyang 464000, China
rslixy@163.com

Abstract. With the rapid development of the Internet and digital signal processing technologies, digital audio has become an important medium for information dissemination. However, there are some lawbreakers who infringe upon the legitimate rights and interests of creators through illegal means such as recapturing. To address this issue, this paper proposes a robust audio watermarking algorithm based on an adaptive quantization strategy driven by a Sigmoid classifier. The method first segments the audio and extracts DWT-CLM features to construct intra-frame vectors. Then, the Sigmoid classifier determines the frame structure and selects either a fixed or adaptive quantization step size for embedding and extraction. Experimental results demonstrate that the proposed algorithm achieves a lower watermark bit error rate while maintaining good audio quality, exhibiting strong robustness and adaptability.

Keywords: audio watermarking · adaptive quantization · recapturing attack · robustness

1 Introduction

With digital audio content widely used across music platforms, short video apps, podcasts, and online education, the copyright protection of audio works has become increasingly pressing. Users can easily download, transcode, or re-upload audio assets, and even bypass platform copyright-detection mechanisms through technical means, leading to unauthorized duplication, distribution, and misuse. These actions not only violate the legitimate rights of content creators and copyright holders, but also impose administrative and legal pressures on hosting platforms [1]. Against this backdrop, digital watermarking technology–thanks to its embeddable, imperceptible, and traceable nature, has become a key means of protecting audio copyrights.

J. Yu and Y. Chen (Eds.): FCS 2025, CCIS 2805, pp. 150–164, 2027.
https://doi.org/10.1007/978-981-95-9514-3_10

Digital audio watermarking is an information hiding technique that embeds copyright information, identity labels, or traceability data into audio signals. Its core objective is to enable covert information transmission, copyright protection, and content authentication without compromising the perceptual quality of the audio. At present, a variety of approaches have been proposed to enhance the robustness and transparency of audio watermarking. Several studies remain focused on transform-domain embedding, such as Fourier, DCT, or DWT-based methods, which demonstrate strong performance against common attacks like compression and filtering [2–5]. To overcome the limitations of fixed embedding strategies, researchers have introduced fuzzy inference, adaptive quantization, and mode modulation mechanisms to improve resilience against nonlinear distortions [4,6,7]. In parallel, hybrid frameworks and inter-frame correlation analysis have been explored to strengthen synchronization robustness, leading to significant reductions in bit error rates [6,8]. With the rapid advancement of deep learning, intelligent schemes leveraging local invariant points, adaptive embedding strength, and invertible neural networks have also been applied to audio watermarking [9,10], further advancing the field toward adaptive and intelligent solutions. Reference [5] proposes an audio watermarking technique based on discrete wavelet transform that embeds watermark information into the low-frequency sub-band to enhance resistance against common attacks such as compression and filtering. This method achieves a high embedding capacity and a low bit error rate while preserving audio quality, confirming the potential of the DWT domain for audio security applications. However, the method still employs a fixed embedding strategy, failing to dynamically adjust embedding strength according to local frame characteristics, and thus offers limited adaptability to complex channel environments such as recapturing, which provides an important impetus for subsequent research on adaptive watermarking strategies.

To overcome the fixed embedding strategies, absence of frame-level discrimination, and insufficient robustness against nonlinear attacks that plague existing audio watermarking schemes, this paper introduces a robust anti-recapturing audio watermarking algorithm that integrates adaptive quantization with feature-based classification. It markedly lowers the bit error rate under recapturing and similar attacks while preserving audio imperceptibility and embedding flexibility, thereby offering a practical enhancement over traditional fixed-embedding approaches.

The contributions of this paper are summarized as follows:

1) A frame-level Sigmoid classifier is designed to discriminate audio characteristics using statistical features.
2) An adaptive quantization mechanism dynamically switches between fixed and variable step sizes for watermark embedding.

Experiments show consistently lower bit error rates and better robustness against common signal processing and recapturing attacks compared with existing methods.

2 Related Work

2.1 Discrete Wavelet Transform

Discrete Wavelet Transform(DWT) [5,11–13] is a method for joint time-frequency analysis of signals. Unlike traditional transforms, DWT simultaneously reveals the local characteristics of a signal in both the time and frequency domains. It is a localized analysis technique with adaptable time and frequency windows, and thus offers excellent multiresolution capabilities. Thanks to its superior timefrequency localization, DWT has been widely adopted in digital signal processing, demonstrating significant advantages in audio compression, feature extraction, and watermark embedding.

Let $\mathcal{A} = \{a_l \mid 1 \leq l \leq L\}$ be an audio signal of length L, and let a_l denote the l-th sample. Performing a D-level DWT on $\mathcal{A}$ yields wavelet coefficients $F_A^D, G_A^D, G_A^{D-1}, \ldots, G_A^1, F_A^D$, where F_A^D represents the approximation coefficients at the D-th level of wavelet decomposition, where most of the signal's energy is concentrated. G_A^j represents the detail coefficients at the j-th level of decomposition ($j = 1, 2, 3, \ldots, D$). The recursive computation formula for the D-level DWT is as follows:

$$F_A^0 = A \tag{1}$$

$$F_A^j[n] = \sum_k F_A^{j-1}[k] \cdot h[2n-k] \tag{2}$$

$$G_A^j[n] = \sum_k F_A^{j-1}[k] \cdot g[2n-k] \tag{3}$$

where $h[n]$ represents the low-pass filter, $g[n]$ represents the high-pass filter, $j = 1, 2, 3, \ldots, D$, and each level of decomposition includes downsampling operations. Figure 1 is a structural diagram of a 3-level DWT.

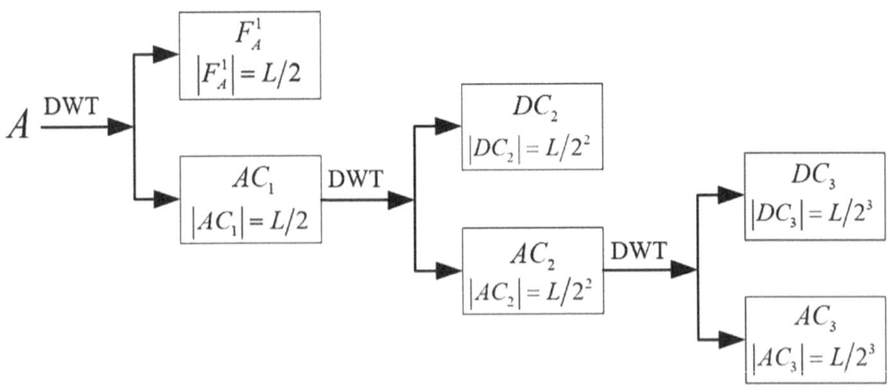

Fig. 1. Structural diagram of a 3-level DWT for audio signal A.

2.2 DWT-CLM Features

The CLM feature used in this paper is conceptually related to the cepstral trend statistics previously introduced in patchwork-based audio watermarking schemes to enhance robustness against desynchronization and recapturing attacks [11]. In that framework, the cepstral linear modulation descriptor characterizes the linear variation of cepstral coefficients across frames. The formulation in Eq. 4 follows a similar principle of capturing cepstral dynamics but incorporates two key modifications: (1) the feature is derived from the approximation coefficients of a multi-level discrete wavelet transform (DWT) rather than directly from the raw cepstrum, thereby exploiting the energy compaction and multi-resolution properties of wavelets; and (2) the statistic is normalized by frame length, which improves stability against amplitude variations. Consequently, the proposed DWT-CLM feature can be regarded as a wavelet-domain adaptation of the cepstral linear modulation measure. This adaptation preserves the robustness advantages of the original feature while improving its suitability for frame-level classification and adaptive quantization in the present watermarking pipeline.

Definition of DWT-CLM Features. DWT-CLM feature is an audio feature extraction method that combines Discrete Wavelet Transform (DWT) and Cepstral Linear Modulation (CLM). DWT captures multi-scale timefrequency information of the audio signal, while CLM describes the linear variation trends of cepstral coefficients, effectively reflecting the structural characteristics of the signal. This feature is highly robust against distortions such as recapturing and compression, making it suitable for security applications like audio watermarking.

Let $A = \{a_l \mid 1 \leq l \leq L\}$ be an audio signal of length L, and upon perform a D-level DWT on A , a set of wavelet coefficients can be obtained, including an approximation component$F_A^D$and multiple detail components $G_A^D, G_A^{D-1}, \ldots, G_A^1$. Since the energy of the signal is primarily concentrated in the approximation component after DWT, this paper selects the D-th-level approximation component F_A^D as the basis for feature extraction to construct the proposed DWT-CLM features ($F_A^D = \{f_1, f_2, \ldots, f_N\}$). The method is shown in Eq. 4, and its length is $N = L/2^D$.

$$D = \frac{\left| \sum_{n=1}^{N} \log_2 \left(|f_n| / \lambda \right) \right|}{N} \tag{4}$$

As defined by Eq. 4, the more coefficients with large magnitudes in the approximation component F_A^D , the greater the obtained feature value; conversely, the smaller the feature value.This formulation differs from the cepstral linear modulation feature defined in [12] in two important respects: (1) the feature is extracted from the approximation coefficients of a multi-level DWT rather than directly from the raw cepstral coefficients, thereby leveraging the energy compaction property of the wavelet domain; (2) the proposed expression employs a logarithmic ratio with respect to the normalization factor L, instead of computing a linear slope across cepstral indices. These modifications yield a wavelet-domain variant that preserves the robustness advantages of cepstral trend analysis while offering improved stability under amplitude variations and better suitability for frame-level adaptive quantization.

Robustness Analysis. This paper extracts DWT-CLM features from the audio signal and embeds the watermark information into these features to generate watermarked audio. After signal processing and recapturing attacks, the DWT-CLM features in the watermarked signal should remain stable, ensuring that the embedded watermark information can be accurately extracted. When the watermarked audio is illegally distributed, the watermark information can be extracted by detecting the DWT-CLM features in the target audio, providing reliable evidence for copyright authentication. To verify the robustness of this feature, systematic experiments will be conducted for analysis in the following sections.

Randomly select an audio signal with a sampling frequency of 44.1 kHz (containing L samples) from the sample library, as shown in Fig. 2.

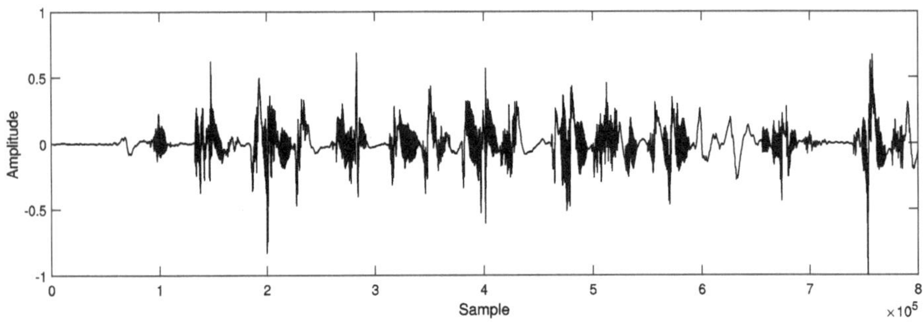

Fig. 2. Original speech signal.

Step 1. Divide A into P frames, denote the i-th frame as A_i, with each frame having a length of N.

Step 2. Perform DWT on A_i (here, a 3-level transform is chosen) and calculate its DWT-CLM features F_i .

Subject the selected audio signal to a re-recording attack, and denote the recapturing signal as A', as shown in Fig. 3.

Repeat steps 1 and 2, segment A' into frames, and calculate the DWT-CLM features $F_i{}'$ for each frame. Figure 4 provides a comparison of the DWT-CLM features for each frame of the audio signal A and the signal after a recapturing attack A', with each frame length set to 2200.

It can be observed that the DWT-CLM features of each audio frame after recapturing have changed little compared to the original signal, which to some

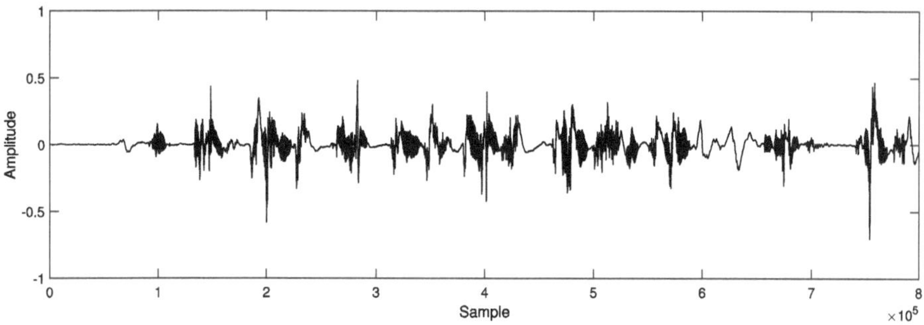

Fig. 3. Duplicated speech signal.

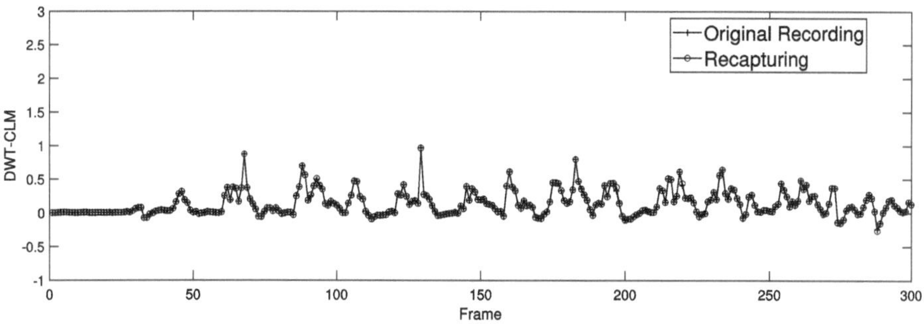

Fig. 4. DWT-CLM features of the original speech signal and the duplicated speech signal.

extent indicates the robustness of this feature against recapturing attacks. If the method of quantizing the DWT-CLM features is used to embed the watermark, it can ensure that the watermarked signal, after signal processing, has a high probability of correctly extracting the watermark information.

2.3 Sigmoid Classifier-Driven Adaptive Quantization Mechanism

In digital audio watermarking, the setting of the quantization step size has a decisive impact on the transparency of watermark embedding and the robustness of watermark extraction [14]. If a uniform fixed step size is used for all frames, it can lead to an overly strong embedded signal in low-energy frames, causing noticeable distortion. Conversely, an overly weak signal in high-energy frames can result in loss of the watermark. Therefore, this paper proposes a frame-level adaptive quantization mechanism driven by a Sigmoid classifier, which automatically determines whether to use a fixed or adaptive step size based on frame characteristics [15], thereby balancing robustness and transparency.

Sigmoid Classifier. The Sigmoid function is a common S-shaped activation function, and its curve is shown in Fig. 5.

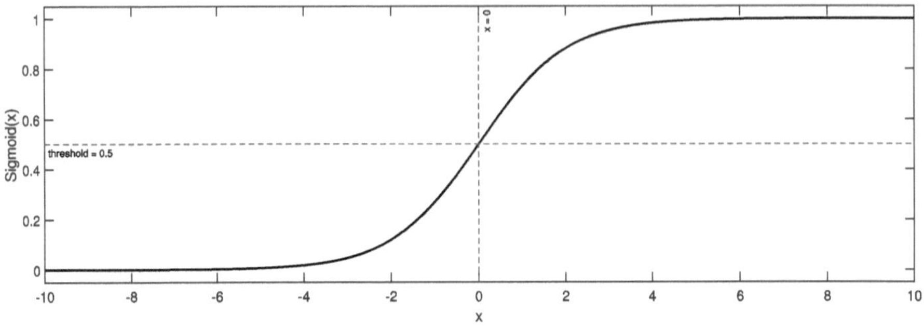

Fig. 5. Sigmoid function curve.

It can be known that the output value of the Sigmoid function always lies within the (0,1) interval. Near the threshold of 0.5, the Sigmoid function has a distinct non-linear boundary capability. And the larger the input value, the closer the output is to 1; conversely, the smaller the input value, the closer the output is to 0. Therefore, this model is suitable for binary classification decisions and is well-suited for mapping features that change continuously. It is defined as follows:

$$\sigma(x) = \frac{1}{1 + e^{-x}} \tag{5}$$

In this algorithm, we extract feature vectors based on frames, and construct a four-dimensional frame feature vector f_i consisting of the zero-crossing rate ZCR_i, variance Var_i , and the sum of wavelet energy E_i of each frame. We then use a sigmoid classifier to determine whether to use a quantization step size. The structure of the classifier is given by the following:

$$P_i = \frac{1}{1 + \exp\left(-\left(W \cdot f_i + b\right)\right)} \tag{6}$$

where W is the well-trained weight vector, b is the bias term, $f_i \in (0,1)$ represents the probability that the current frame uses a fixed step size. When P is above a certain threshold T, it is determined that the current frame structure is stable and suitable for using a fixed quantization step size; otherwise, an adaptive quantization step size is used. Here, T is the threshold setting.

Strategy Decision and Step Size Control Mechanism. Based on the output probability P_i from the classifier, the system employs the following decision-making strategy:

$$\delta_i = \begin{cases} \Delta, P_i \geq T \\ \lambda_{adapt} \cdot \sqrt{E_i}, P < T \end{cases} \tag{7}$$

where Δ represents the preset fixed step size, λ_{adapt} represents an adaptive factor used to dynamically control the size of the step, ensuring the maintenance of embedding quality and perceptual balance in frames susceptible to interference.

After the step size determination is completed, the embedding features are quantized, and the corresponding watermark bits are embedded into the signal structure. The entire process is shown in Fig. 6, which ensures that each frame can adaptively adjust embedding parameters.

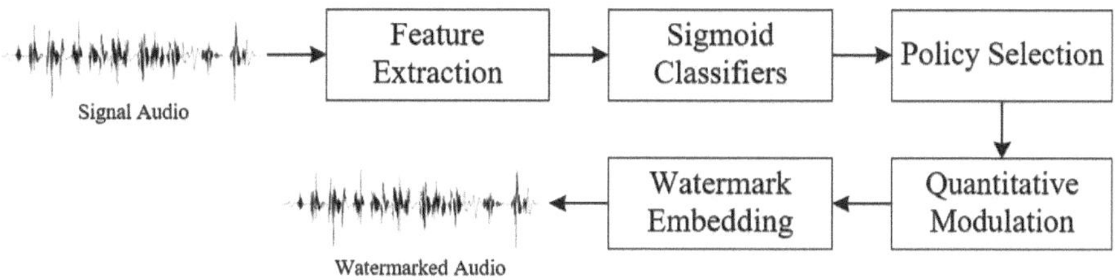

Fig. 6. Classifier-Driven embedding control flowchart.

This mechanism enables the system to perform targeted watermark embedding control based on the local characteristics of audio frames, balancing robustness and stealthiness. Especially in scenarios such as nonlinear distortion and recapturing attacks where it has significant advantages.

3 Proposed Method

3.1 Watermark Embedding

Let's assume $A = \{a_l | 1 \leq l \leq L\}$ represents the carrier audio signal, the symbol a_l denotes the i-th sample point of audio A, and L represents the number of sample points contained in the audio signal A.The general process of watermark embedding is shown in Fig. 7.

Step 1. Frame segmentation. Divide the carrier signal A uniformly into frames of N points each, resulting in a frame set $\{A_i\}_{i=1}^{M}$, where $M = \lfloor L/N \rfloor$ is the total number of frames. The i-th frame is denoted as A_i, with a length of N. Divide A_i into two segments, front and back, denoted as $A_{i,1}$ and $A_{i,2}$, each with a length of $N/2$.

Step 2. Calculate the DWT-CLM features of the front and back segments of the i-th frame, denoted as $A_{i,1}$ and $A_{i,2}$, using Eq. 4, and denote these features as $D_{i,1}$ and $D_{i,2}$, respectively.

Step 3. Let ω_i be the watermark information to be embedded. Here, ω_i is embedded into A_i by quantizing $D_{i,1}$ and $D_{i,2}$.

(1) Calculate the difference feature $R_i = |D_{i,1} - D_{i,2}|$, and simultaneously compute the zero-crossing rate ZCR_i, variance Var_i, and energy E_i, to construct a four-dimensional feature vector $f_i = [R_i, ZCR_i, Var_i, E_i]$.

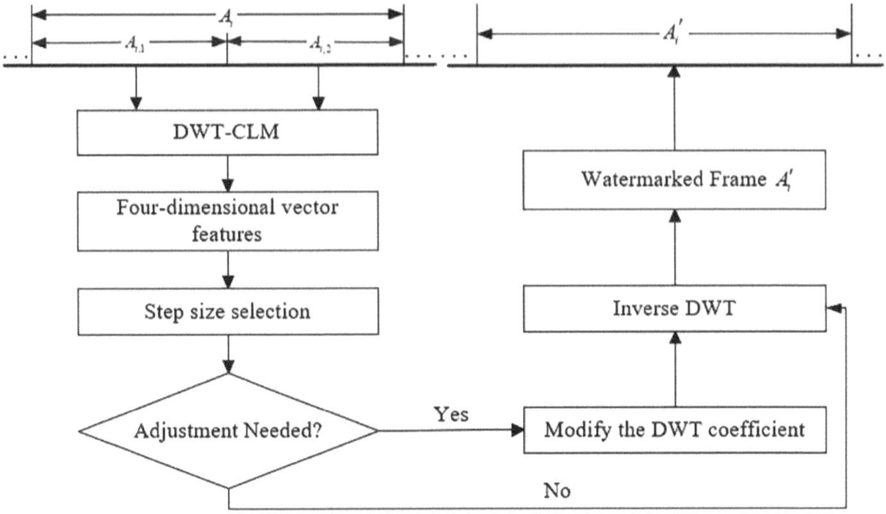

Fig. 7. Block diagram of the watermark embedding process.

(2) Based on the weights W and bias b of the sigmoid classifier, calculate the classification probability p_i for the current frame using Eq. 6. Compare the probability value p_i with the threshold θ, and select the quantization step size Δ_i using Eq. 7.
(3) Given the current quantization result $\hat{\omega}_i = \text{mod}\left(round\left(R_i/\Delta_i\right), 2\right)$, if $\hat{\omega}_i \neq \omega_i$, then adjust $D_{i,1}$ to satisfy the condition $D_{i,1}^{new} = \lceil R_i/\Delta_i \rceil \cdot \Delta_i$. Calculate the scaling factor $\alpha_i = D_{i,1}^{new}/D_{i,1}$, and update the low-frequency approximation coefficients. Otherwise, if $\hat{\omega}_i = \omega_i$, then the embedding condition is already satisfied, and no modifications are made.
(4) Perform the inverse DWT with the modified approximation component and the original detail coefficients to complete the embedding of ω_i, obtaining the watermarked signal A_i' for the frame i.

Step 4. By applying the above method and steps to embed the watermark signal into other audio frames, we can obtain the watermarked audio signal.

3.2 Watermark Extraction

The watermark extraction process is similar to the watermark embedding process. Assume the watermarked signal is denoted as A', with a length of L'. Perform frame segmentation and division on the watermarked signal A_i', and denote the first audio frame as A_1'. Take the extraction of watermark information from the first frame as an example, and introduce the method of watermark extraction. The general process of watermark extraction is shown in Fig. 8.

Step 1. Divide A_1' into two segments, front and back, denoted as $A_{1,1}'$ and $A_{1,2}'$ respectively.

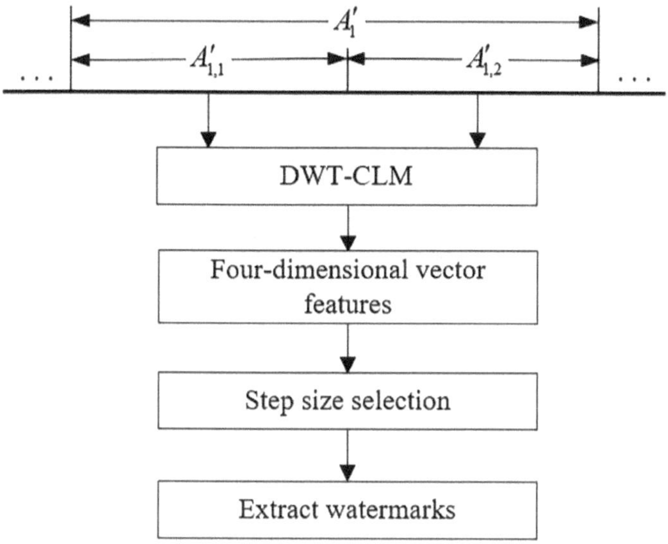

Fig. 8. Watermark extraction process flowchart.

Step 2. Calculate the DWT-CLM features of $A_{1,1}'$ and $A_{1,2}'$ using Eq. 4, denoted as $D_{1,1}'$ and $D_{1,2}'$ respectively. Calculate the residual $R_1' = |D'_{1,1} - D'_{1,2}|$ between $D_{1,1}'$ and $D_{1,2}'$.

Step 3. Calculate the feature vector $f_1' = [R'_1, ZCR'_1, Var'_1, E'_1]$ of the current frame and determine the step size Δ_1' based on the sigmoid classifier.

Step 4. Extract the watermark ω from R'_1 and Δ_1'. $\omega = \text{mod}(round(R'_1/\Delta'_1), 2)$.

By applying the aforementioned steps to extract watermarks from the remaining audio frames, the full watermark information can be obtained.

3.3 Classifier Training Protocol

To implement the frame-level classification required for adaptive quantization, a logistic regression (Sigmoid) classifier was trained using four frame-level features: autocorrelation coefficient, zero-crossing rate, variance of the DWT-CLM feature, and short-time energy. The feature vectors were extracted from all frames in the experimental dataset and stored in a structured format. Each frame was labeled as either fixed or adaptive according to the embedding strategy applied during preliminary analysis, where frames marked as "fixed" were assigned label 1 (stable) and those marked as "adaptive" were assigned label 0 (unstable).

Prior to training, the features were normalized using z-score standardization:

$$X_{norm} = \frac{X - \mu}{\sigma} \tag{8}$$

where μ and σdenote the mean and standard deviation of the training set. Logistic regression was then performed via maximum likelihood estimation, yielding the model

$$y = \sigma(Wx + b) \tag{9}$$

where W is the weight vector, b the bias term, and $\sigma(\cdot)$ the sigmoid activation function.

The classifier output was thresholded at 0.5 to obtain binary predictions. Accuracy was evaluated on the training dataset, and the final model achieved a classification accuracy of 92.3%, indicating reliable separation between stable and unstable frames. The learned weights, bias, and normalization parameters were saved in for subsequent use in watermark embedding and extraction.

4 Experimental Results

In practical application scenarios, the watermarked audio signal may undergo various signal processing due to user operations or external environmental factors, such as compression, filtering, and may even face targeted attack behaviors. These processes may lead to changes in the number of audio sample points or misalignment of frame content, thereby causing synchronization mismatch issues between watermark embedding and extraction, affecting the effective extraction and verification of the watermark.

To test the performance of the algorithm presented in this paper, the following experiments are conducted using MATLAB for simulation. Randomly select 200 segments of different types of audio signals from the sample library, which are 16-bit quantized and have a sampling frequency of 44.1 kHz. First, embed the watermark into the selected audio signals, then use different recording devices to perform recapturing attacks on the watermarked signals (three different recording devices: Sony PCM-D100 recorder, Huawei P40, and iPhone 14), to obtain the attacked signals.

The recapturing experiments were conducted in a standard office environment with a measured background noise level of 35 dB SPL (A-weighted) and a reverberation time (RT60) of approximately 0.4 s. The playback system consisted of a Focusrite Scarlett 2i2 (3rd Gen) sound card connected to a PreSonus Eris E3.5 studio monitor. The speaker was placed 1 m from the recording device at a 0-degree azimuth angle. Playback volume was calibrated to 75 dB SPL (A-weighted) at the recording position, measured with a calibrated B&K Type 2250 sound level meter.

The recording devices–Sony PCM-D100, Huawei P40, and iPhone 14–were placed at the designated position using a fixed stand. Their built-in microphones were used with all automatic gain control and noise reduction features disabled. The recording levels were manually set to avoid clipping while maintaining an adequate signal-to-noise ratio. This setup aims to provide a reproducible baseline for evaluating robustness against typical consumer-grade recapturing scenarios.

4.1 Perceptual Transparency

In audio watermarking technology, perceptual transparency is one of the important indicators to measure the quality of a watermark embedding algorithm, representing the degree to which the watermark embedding affects the original perceptual quality of the audio. Common evaluation methods include subjective and objective approaches. Subjective methods involve providing both the original audio signal and the watermarked signal to a group of listeners, who then distinguish between the two signals based on their subjective impressions. Listeners then rate the differences between the two signals according to the Subjective Difference Grades (SDG) [16], with the average score serving as the result of the subjective listening quality test. The meanings of the SDG ratings are shown in Table 1.

Table 1. SDG Evaluation Criteria

SDG Score	Description of Impairments	Quality Assessment
0.0	Imperceptible	Excellent
-1.0	Perceptible but not annoying	Good
-2.0	Slightly annoying	Fair
-3.0	Annoying	Poor
-4.0	Very annoying	Bad

Common objective evaluation methods include the Signal-to-Noise Ratio (SNR). It is a method that treats the embedded watermark information as noise added to the original audio and is used to measure the extent to which the watermark embedding alters the original signal. The calculation formula for SNR is shown in Eq. 10 [16].

$$S_{NR} = 10\lg\left(\frac{\sum_{l=1}^{N} a(l)^2}{\sum_{l=1}^{N} \left(a(l) - a'(l)\right)^2}\right) \tag{10}$$

where $a(l)$ represents the l-th sample point of the original audio signal, $a'(l)$ represents the l-th sample point of the watermarked audio signal,N is the length of the audio signal. To meet the requirement of perceptual transparency, the SNR value should be greater than 20 dB.

4.2 Robustness

To verify the robustness of the proposed audio watermarking algorithm against attacks such as recapturing, this paper uses the Bit Error Rate (BER) [17] as the core evaluation metric to measure the accuracy of watermark extraction after an attack. The definition of BER is shown in Eq. 11.

$$B_{ER} = \frac{1}{M}\sum_{m=1}^{M} w(m) \oplus w'(m) \tag{11}$$

where M represents the length of the embedded watermark, $w(m)$ and $w'(m)$ represent the original watermark information and the extracted watermark information, respectively, $\oplus$ denotes the XOR operation. It is easy to understand that the lower the BER value, the lower the error rate of the extracted watermark, the stronger the robustness of the algorithm, and the greater the resistance to attacks.

Table 2 lists the statistical mean values of the BER for watermark extraction from 200 watermarked audio signals after undergoing some common signal processing (MP3 compression at rates of 64 and 128 kb/s, resampling with cutoff frequencies of 11 kHz and 16 kHz, and low-pass filtering with a cutoff frequency of 16 kHz) as well as recapturing attacks, and compares them with references [18,19].

Table 2. BER values of watermarked signals after different signal processing and recapturing attacks.

Signal Processing and Recapturing	BER		
	Reference [18]	Reference [19]	This Paper
MP3 Compression (64 kb/s)	15	19	2
MP3 Compression (128 kb/s)	12	18	4
Resampling (44.1→20.05→44.1 kHz)	13	18	4
Low-pass Filtering (16 kHz)	0	0	0
Recapturing	12	18	5
Jittering (1/100)	7	8	2
Jittering (1/10)	2	3	0
Time-scaling (90%)	13	15	7
Time-scaling (110%)	12	13	6

The experimental results demonstrate that the watermark embedded by the proposed algorithm can still be completely and accurately extracted from the watermarked signal after undergoing a low-pass filtering attack with a cutoff frequency of 16 kHz and jittering attack with a rate of 1/1000. For other types of signal processing operations, although the watermark cannot be extracted completely accurately and there are some extraction errors, the number of extracted errors is less than that of the algorithms provided in references [18,19].

5 Conclusion

This paper proposes a robust audio watermarking algorithm based on an adaptive quantization strategy and feature classification, aiming to effectively resist signal distortion caused by various signal processing and recapturing operations. By introducing statistical features such as zero-crossing rate, variance, and energy to build a classifier, dynamic analysis of the frame content structure is achieved, which in turn allows for the adaptive selection of either fixed or variable step sizes for watermark embedding. The results demonstrate that the proposed algorithm exhibits good inaudibility and strong resistance to signal processing operations, and it also shows a certain level of robustness against recapturing attacks.

Acknowledgments. This paper is supported by Henan Province Key Research and Development Project (No. 241111212200). We would like to thank the anonymous reviewers for their constructive suggestions.

Competing Interest. The authors have no competing interests to declare that are relevant to the content of this article.

References

1. Hua, G., Huang, J., Shi, Y.Q., Goh, J., Thing, V.L.: Twenty years of digital audio watermarking-a comprehensive review. Signal Process. **128**, 222–242 (2016)
2. Yamni, M., et al.: An efficient watermarking algorithm for digital audio data in security applications. Sci. Rep. **13**(1), 18432 (2023)
3. Salah, E., Amine, K., Redouane, K., Fares, K.: A fourier transform based audio watermarking algorithm. Appl. Acoust. **172**, 107652 (2021)
4. Mosleh, M., Setayeshi, S., Barekatain, B., Mosleh, M.: A novel audio watermarking scheme based on fuzzy inference system in DCT domain. Multimedia Tools Appl. **80**(13), 20423–20447 (2021)
5. Prabhu, V., Sundar, L., Drishya, V., Rajeshwari, G.R.A., Thomas, D.R.: A discrete wavelet transform-based audio watermarking technique for digital security. Inte. J. Nonlin. Dynam. Control **2**(2), 183–198 (2022)
6. Li, Q., Xing, Z., Wang, J., Huang, G., Yuan, X.: Qlaw: An improved quantization-based local audio watermarking scheme using inter-frame correlation. IEEE Access (2025)
7. Mushtaq, S., Mehraj, S., Parah, S.A.: Mmrwfas: mode modulation technique based robust watermarking framework for audio signals. Appl. Acoust. **239**, 110835 (2025)
8. Sijabat, P.P., Budiman, G., Sa'idah, S.: A resilient hybrid audio watermarking framework against signal processing attacks. In: 2025 IEEE International Conference on Artificial Intelligence for Learning and Optimization (ICoAILO), pp. 74–80. IEEE (2025)
9. Wu, S., Guan, H., Liu, J., Zeng, Z., Huang, Y., Zhang, S.: Imperceptible audio watermarking with local invariant points and adaptive embedding strength. Chin. J. Electron. **34**(2), 495–509 (2025)
10. Zhu, J., et al.: Robust audio watermarking based on invertible neural network. Acad. J. Comput. Inf. Sci. **8**(3), 10–17 (2025)
11. Raave, M.: Watermarking time-series data using DWT. Ph.D. thesis, Delft University of Technology (2024)
12. Natgunanathan, I., Xiang, Y., Hua, G., Beliakov, G., Yearwood, J.: Patchwork-based multilayer audio watermarking. IEEE/ACM Trans. Audio Speech Lang. Process. **25**(11), 2176–2187 (2017). https://doi.org/10.1109/TASLP.2017.2749001

13. He, J., Liu, Z., Lin, K., Qian, Q.: A novel audio watermarking algorithm robust against recapturing attacks. Multimedia Tools Appl. **82**(12), 18599–18616 (2023)
14. Jiang, W., Huang, X., Quan, Y.: Audio watermarking algorithm against synchronization attacks using global characteristics and adaptive frame division. Signal Process. **162**, 153–160 (2019). https://doi.org/10.1016/j.sigpro.2019.04.017, https://www.sciencedirect.com/science/article/pii/S0165168419301446
15. Luo, Y., Peng, D., Sang, Y., Xiang, Y.: Dual-domain audio watermarking algorithm based on flexible segmentation and adaptive embedding. IEEE Access **7**, 10533–10545 (2019). https://doi.org/10.1109/ACCESS.2019.2890972
16. Alshathri, S., Hemdan, E.E.D.: An efficient audio watermarking scheme with scrambled medical images for secure medical internet of things systems. Multimedia Tools Appl. **82**(13), 20177–20195 (2023)
17. Liu, X., Li, X., Shi, C., Niu, X., Xiong, L.: A novel svd-based adaptive robust audio watermarking algorithm. Multimedia Tools Appl. **83**(27), 69443–69465 (2024)
18. Nadeau, A., Sharma, G.: An audio watermark designed for efficient and robust resynchronization after analog playback. IEEE Trans. Inf. Forensics Secur. **12**(6), 1393–1405 (2017)
19. Li, J., Wang, H.X., Wu, T., Sun, X.m., Qian, Q.: Norm ratio-based audio watermarking scheme in dwt domain. Multimedia Tools Appl. **77**, 14481–14497 (2018)

Temporal Knowledge Graph Reasoning Based on Path-Aware Temporal Gating

Ruobing Li[1(✉)], Zhicheng Dong[1], and Yanli Chen[2]

[1] College of Information Science and Technology, Xizang University, Lhasa, China
lrbydsa66@163.com

[2] School of Big Data and Computer Science, Guizhou Normal University, Guiyang, China

Abstract. Temporal knowledge graphs have garnered significant attention for their ability to represent and reason over dynamic relational structures in evolving environments. However, irrelevant path interference and inadequate use of temporal information in temporal knowledge graph (TKG) reasoning becomes the main challenges. To improve these issues, we propose a new path-aware temporal gating reasoning model (PA-Temp). In the proposed model, we uses query relational path extraction and attention aggregation mechanism to dynamically select semantically-related multi-hop paths and avoid irrelevant path noise. Meanwhile, a time-aware gating module is designed to fuse explicit time coding and dynamic entity embedding to capture the sudden events and periodic evolution characteristics of entity states. In addition, a skip information flow mechanism is introduced to adaptively filter redundant information in historical snapshots by combining time decay and path similarity constraints. Experiments results on benchmark dataset supplemented with GDELT show that performance of PA-Temp is significantly superior to existing models in link prediction tasks, especially in MRR and Hits@1/3 indicators, verifying the effectiveness of its path-aware and time-gating collaborative optimization.

Keywords: Temporal knowledge graph reasoning · Path perception · Time gating mechanism · Attention aggregation

1 Introduction

As a structured semantic network, knowledge graphs can clearly depict the correlation between entities and are widely applied in fields such as information retrieval and intelligent question answering [19]. However, the entity relationships in the real world are not static and unchanging but continuously evolve over time. For instance, the relationship between Russia and Ukraine has undergone a dramatic transformation from cooperation to confrontation over the past decade or so. This series of dynamic changes is difficult to be accurately expressed through a static knowledge graph. Therefore, the Temporal Knowledge Graph (TKG) emerged as an extension of the static knowledge graph. Its core feature lies in the dynamic evolution of the states of entities and relations over time. By introducing timestamps (such as (s,r,o,t)) for each fact triplet, Realize fine-grained modeling of historical event sequences to more accurately depict complex dynamic scenes in the real world [21].

J. Yu and Y. Chen (Eds.): FCS 2025, CCIS 2805, pp. 165–178, 2027.
https://doi.org/10.1007/978-981-95-9514-3_11

In recent years, TKG reasoning has become a research hotspot in the field of knowledge graphs. Its core task is to predict missing information based on existing temporal facts, which is specifically divided into interpolation and extrapolation. The Interpolation setting seeks to fill in missing facts within a specified range of timestamps, while extrapolation attempts to predict future events based on historical knowledge [5]. However, the existing TKG reasoning models are limited by their insufficient collaborative modeling ability for "semantic paths" and "temporal information" when dealing with extrapolation tasks, which severely restricts the reasoning accuracy and generalization ability. Specifically, two key challenges need to be addressed urgently:

Firstly, the absence of a semantic path filtering mechanism leads to interference from irrelevant paths. Although current mainstream methods (such as RE-NET [5], REGCN [6], etc.) use sequential graph neural networks (GNNs) to process historical snapshots to capture implicit temporal information, their aggregation mechanisms are usually based on neighborhood expansion of the entire graph or local subgraphs, and do not explicitly extract semantly related multi-hop paths for the subject - relation pairs in the current query. This leads to a large amount of semantically irrelevant or time-disconnected path information being indiscriminately incorporated into the coding process, seriously interfering with the accuracy of prediction. For example, for predicting future events (?, Request for Aid, Ukraine, 2025-03-10), based on the historical snapshots that are related to Ukraine (US, Approved Military Aid, Ukraine, 2024-09-15), (EU, Freeze Funds to Ukraine, Internal Disputes of the EU, 2024-11-30), (China, Propose Peace Initiative, United Nations, 2024-10-05), (India, Neutral Statement, Russian-Ukrainian Conflict, 2024-12-18), if the model aggregates these paths indiscriminately, it may mistakenly interpret "China proposes a Peace Initiative" or "India remains neutral" as potential aid signals, thereby underestimating the possibility that Ukraine will continue to seek assistance from the West.

Secondly, there is a lack of path-level time-aware mechanisms. Most models (such as TANGO [8] and CEN [10]) only rely on the implicit temporal information of snapshots and do not fuse the explicit features of timestamps with path information. Even if some models (such as HiSMatch) attempt to combine entity subgraphs with query subgraphs, they fail to dynamically filter out time-related effective paths through attention mechanisms or similarity constraints, resulting in the improper utilization of information from distant irrelevant paths or semantically disconnected paths. The utilization of historical paths by existing methods is "undifferentiated", without considering the time attenuation effect and time consistency of the paths, resulting in the inability of time information to guide the effective screening of paths.

In response to the above challenges, this paper proposes a novel path-aware temporal gated knowledge graph reasoning model. Firstly, semantic paths are extracted based on query relations, and path representations are aggregate through the attention mechanism. Meanwhile, explicit time coding and time-aware gating mechanisms are utilized to capture the temporal evolution characteristics of entities. Finally, the path-enhanced representation is fused with dynamic temporal embedding to achieve precise temporal knowledge graph reasoning. The main contributions of this article are as follows:

1) A new reasoning model combining path enhancement and temporal gating mechanism is proposed, which can effectively capture the deep semantic relationships between entities and their changes over time.
2) A dedicated time-aware gating module was designed, which enhanced the model's ability to handle emergencies and improved its support for extrapolated prediction tasks.
3) Extensive experimental verification was conducted on multiple public datasets, and the results show that the proposed model significantly outperforms existing methods in prediction tasks.

2 Related Work

Temporal knowledge graph reasoning technology originated from static knowledge graph reasoning and has been continuously developing as the understanding of the importance of the time dimension deepens. This section reviews the foundation of path reasoning in static knowledge graphs, and then focuses on sorting out the existing methods of temporal knowledge graph reasoning, and classifies and discusses them according to their technical routes.

2.1 Static KG Reasoning

Path reasoning in static knowledge graphs lays the foundation for TKG reasoning. Its main idea is to capture potential logical rules by using multi-hop relationship paths between entities, which has good interpretability. Research in this field has undergone an evolution from traditional probabilistic models to neuro-symbolic ensembles and then to applications of reinforcement learning.

The early working path ranking algorithm (PRA [7]) generates paths through random walks and learns path features, and uses these paths as features to derive logical rules based on paths using probabilistic graphical models, laying the foundation for path reasoning. However, PRA is confronted with the problem of sparse feature. To this end, ProPPR effectively alleviates sparsity and enhances the generalization ability of the model by introducing text content and integrating spatial similarity heuristic methods. With the rise of neural networks, the powerful representation capabilities of neural networks have begun to be utilized to enhance the semantic combination and reasoning ability of paths. RNNPath introduces recurrent neural networks to encode path sequences, enhancing the semantic representation ability of paths. LogSumExp designs a logical combination method across all elements, featuring an attention mechanism for multiple reasoning. Structured methods based on variational reasoning and reinforcement learning have further promoted the development of path reasoning. DIVA [22] has proposed a unified variational reasoning framework, dividing multi-hop reasoning into two steps, namely pathfinding and path reasoning. DeepPath [8] innovatively employs reinforcement learning agents to search for information rich paths, significantly enhancing the efficiency and quality of path selection.

2.2 Temporal KG Reasoning

To overcome the limitations of static path reasoning, it began to explore how to integrate temporal information into path reasoning, thereby achieving dynamic path reasoning. The existing TKG reasoning methods can be roughly classified into the following categories:

Dynamic Embedding-Based Methods: These methods learn the representations of entities and relations that evolve over time. RE-Net utilizes recurrent neural networks to aggregate the historical neighbor information of entities and capture their dynamic evolution. RLEE [4] introduces a time decay mechanism to assign higher weights to more recent historical facts. Re-Temp [2] filters irrelevant history through the dynamic information flow mechanism, enhancing the model's ability to focus on key time points. Such methods usually have stronger representational power, but their interpretability is relatively weak.

Method Based on Time Series Rule Mining: This type of method focuses on extracting logical rules with time constraints from the historical data of TKG. TLogic [6] extracts rules through temporal random walks and calculates confidence levels for prediction. It has good robustness against sparse data. ALRE-IR [5] is implemented by analyzing the relationship paths in historical subgraphs, and then aligning these paths with current events to formulate effective rules for prediction.

Dynamic Path Search Method Based on Reinforcement Learning: This type of method regards temporal reasoning as a sequential decision-making problem on historical snapshots and seeks meaningful reasoning paths through agents. TITer [3] designed an agent capable of perceiving time span, which directly models time evolution by walking between consecutive historical snapshots to find evidence paths. TPath [15]: Through time path constraints and reward functions, it guides the agent to search for paths in historical and possible future states, while performing link prediction and completion of the time series knowledge graph.

Method Based on Temporal Graph Neural Networks: Utilizing graph neural networks to aggregate information on continuous snapshots or event streams. DaeMon [1] abandons the traditional entity embedding and instead directly extracts features from the temporal subgraph structure, making predictions by analyzing the patterns of the subgraph sequence. EvoKG [16] captures the constantly changing structure and temporal dynamics in TKG by conducting cyclic modeling of events, and builds the interaction between entities based on the temporal neighborhood aggregation framework.

3 Problem Formulation

Temporal knowledge graph is a timestamp directed graph G = (E,R,T), where E is the entity set, R is the relation set, and T is the timestamp set. Temporal knowledge graphs are snapshots arranged in time series, i.e. G = $\{G_0, G_1, G_2, \ldots, G_t\}$, each snapshot consists of a series of quadruples (s,r,o,t) to store facts occurring at the current moment, $s,o \in E$ are subject and object entities, $r \in R$ represents relationships, $t \in T$ is a timestamp.

This paper focuses on the extrapolation task, i.e., given a future timestamp tq and the corresponding query q = (s,r,?, t) or q = (?, r,o,t), the model predicts the missing entity (subject or object) based on the factual information contained in the historical snapshot before t.

4 Methodology

In this chapter, we will explain the proposed model in detail, as shown in the Fig. 1. The model consists of three units: path awareness mechanism, dynamic time embedding and skip information flow. Next, we begin to describe in detail the architecture of the model and its learning and reasoning processes.

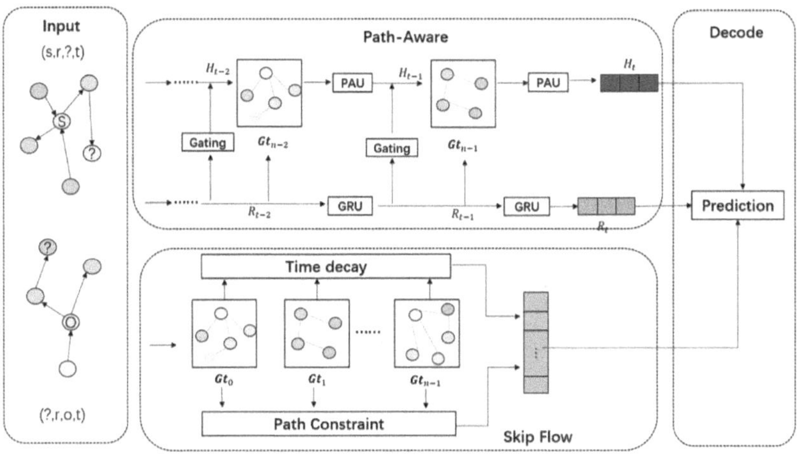

Fig. 1. Framework Overview.

4.1 Model Overview

The main idea of PA-Temp is to extract multi-hop relational paths Ps-oi from query subject s to candidate object oi from historical snapshot according to query, the representation of each path is generated by the concatenation of relational embeddings, and then the generated paths are weighted and aggregated through the attention mechanism for related paths. Dynamic temporal embedding is to obtain dynamic embedding of entities through GRU, and a time-aware gating is designed to obtain the event embedding at the current time, and the dynamic embedding of entities is obtained by adjusting static and temporal embedding through gating. The path representation and dynamic embedding of entities are fused to obtain the final entity representation. Skipping information flow is to assign different weights to each historical subgraph to determine the fusion of historical information.

4.2 Dynamic Time Embedding

In order to make entities flexible to respond to unexpected events, a gating mechanism of dynamic time embedding is designed, which integrates static and dynamic embedding through event aware gating and explicit time coding. Static embeddings represent features of entities that do not change over time; dynamic embeddings capture features of entities that change over time.

Dynamic embedding is composed of trend components and seasonal components, where the trend component is a linear transformation over t and the seasonal component is a periodic function of t to represent periodic changes and long-term trends of entities over time.

$$h_{s,t}^{S} = h_{s,t}^{S} \tag{1}$$

$$h_{s,t}^{D} = w_a t + \sin(2\pi w_b t) \tag{2}$$

$h_{s,t}^{S}$, $h_{s,t}^{D}$ in Eqs. (1) and (2), represent the static and dynamic embeddings of the subject entity at time t respectively, w_a and w_b are learnable parameters.

Event embedding is obtained from the frequency of relationships in snapshots, mainly to obtain the characteristics of unexpected events of entities. Specifically, the frequency of each relationship type in the snapshot is count, and the relationship type is obtained by weighted summation.

$$h_{\text{event}}^{t} = \sum_{r_i \in R_t} \alpha_i \cdot h_{r_i} \tag{3}$$

$$\alpha_i = \frac{f(r_i)}{\sum_j f(r_j)} \tag{4}$$

where the embedding of the event at time t is represented h_{event}^{t} in Eq. (3), α_i is the attention weight, r_j and r_i is the relational embedding.

Event-aware gating enables the model to adaptively adjust the sensitivity to event information at different points in time to better capture and exploit the impact of time dynamics on entity relationships. Gated computation depends primarily on static representations of events and entities. When events are important, g indicating that entity representations are more inclined to dynamic embedding of entities. When events are not important, g indicating that entity representations are more inclined to static embedding of entities.

$$g = \sigma\left(w_g [h_{\text{event}}; h_{s,t}^{S}]\right) \tag{5}$$

$$h_{s,t}' = g \odot h_{s,t}^{D} + (1 - g) \odot h_{s,t}^{S} \tag{6}$$

4.3 Path-Aware Mechanism

Because it is difficult to capture the complex semantic association hidden in context only by iteration focusing on entity's own characteristics, and multi-hop path between entities can reveal this association, path-aware mechanism is adopted to inject semantic information contained in path into entity representation, and the dynamic change of this association with time is modeled.

Path perception is mainly to capture the local features of entities. First, multiple paths between entities are extracted by query, each path is composed of a series of relationships, for example, $s \rightarrow^{r_1} o_1 \rightarrow^{r_2} o_2$ it is a 2-hop path, and the path is represented as a relationship sequence p = $[r_1, r_2]$. The path is aggregated by attention mechanism, and the obtained path representation is fused with the entity to make the entity representation contain semantic information. Path extraction and aggregation are performed on each historical subgraph to update the entity representation step by step to better capture complex relationships and temporal dynamics between entities. Detailed calculations are as follows:

$$h_{pi} = \oplus_{j=1}^{k} h_{r_j} \tag{7}$$

$$\alpha_i = soft\max(w_c[h_{pi}; h_{s,r}]) \tag{8}$$

$$h_{\text{agg}} = \sum_i \alpha_i h_{p_i} \tag{9}$$

$$h_{s,t} = h'_{s,t} + w_d h_{\text{agg}} \tag{10}$$

Note that the output of each timestamp is also the input to the next timestamp. Equation (7) shows the path representation, which embeds relationships for integration operations. Equation (8) calculates the similarity between the path and the query to obtain the attention weight of each path. Equation (9) is a weighted sum of all path representations to obtain an aggregate path representation. Equation (10) combines the aggregate path representation h_{agg} with entity dynamic embedding $h'_{s,t}$ to obtain a new entity representation.

4.4 Skip Information Flow

Path-aware mechanisms focus on capturing local features using multi-hop paths between entities, while skipping information flow focuses on filtering and selecting important global features in time series. Skipping the information flow mechanism is to filter unnecessary historical snapshot information according to entity dependencies in the query, thus focusing on those snapshots useful for the current prediction task.

For skipping irrelevant historical snapshots, time decay mechanism and path similarity constraint are adopted. The time decay mechanism is determined according to the time difference between the current timestamp and the historical timestamp. Since the longer the time difference, the less the impact on the query. Therefore, the historical information closer to the current time point will receive higher weight, while the earlier information will be given lower weight. It can be expressed as:

$$\beta_j^{\text{decay}} = e^{-\lambda \cdot (t_{\text{current}} - t_j)} \tag{11}$$

Path similarity constraint is implemented by calculating similarity between query path and historical path, and then adjusting the weight of historical information in the final entity representation according to this similarity. The goal is to make the model focus more on path information relevant to the current query and ignore paths that are irrelevant.

$$\beta_{\text{j}}^{path} = soft\max(\cos(h_r, h_{agg})) \tag{12}$$

Time decay mechanism and path similarity constraint jointly determine the weights of different historical snapshots from time dimension and semantic dimension respectively. Thus the final entity representation is a weighted sum of the different historical subgraphs. Formally:

$$\beta_j = \alpha \beta_j^{\text{decay}} + (1-\alpha)\beta_{\text{j}}^{path} \tag{13}$$

$$h_{s,t}^{final} = h_{s,t} + \sum_{j=1}^{m} \beta_j \cdot h_{s,t-j} \tag{14}$$

4.5 Decode

The design of the decoding process and loss function is critical to accurately predict missing entities from encoded entity and relationship representations. Therefore, ConvTransE is used as the decoder to evaluate the score of candidate answers; the loss function uses cross-entropy loss to measure the difference between the predicted score and the true label output by the decoder.

$$L = -\sum_{e_c \in \mathcal{E}} z_c \cdot \log(\text{softmax}(s(e_s, r, e_c, t))) \tag{15}$$

If the classification is correct, z_c it is 1, otherwise it is 0. The training goal is to minimize the total loss of all queries.

5 Experiments

5.1 Experiment Setup

Datasets: Experiments were performed on three typical TKG data sets: ICEWS14, ICEWS18, GDEL. ICEWS14 and ICEWS18 are derived from the Integrated Crisis Early Warning System, which stores political events in 2014 and 2018 respectively. GDELT comes from a global event, language and mood database that records long-term, high-frequency events in multiple domains around the world. Overall statistics for each dataset are shown in Table 1. All data sets are divided into Training, Validation, and Test sets in chronological order.

Table 1. Statistics of Datasets (N_{train}, N_{valid}, N_{test} are the numbers of facts in training, validation, and test sets.)

Datasets	E	R	N_{train}	N_{valid}	N_{test}	N_{time}
ICEWS14	7128	230	63685	13823	13222	365
ICEWS18	23033	256	373018	45995	49545	304
GDELT	7691	240	1734399	238765	305241	2976

Evaluation Criteria: We use MRR, hits@1, hits@3 and hits@10 as evaluation criteria to evaluate the performance of the model. In order to make the evaluation result more accurate, the existing quadruples are excluded from the candidate set.

Base Model: Since our inference task is for extrapolation settings, we do not compare models under interpolation settings, only models under extrapolation settings. So we chose RE-GCN, xERTE, TiTer, CEN, HiSMatch, TiRGN, Re-Temp, DaeMon as our benchmarks.

Implementation Details: The embedding dimension is set to 200, which is also the hidden dimension of the graph model and the decoder hidden dimension. The number of layers of the neural network is 3 and the packet loss rate is set to 0.3. Adam has a learning rate of 1e-3 for optimization and a sample size of 10 for path aggregation. The model is trained on a training set of up to 30 epochs, and we stop training when validation performance does not improve within 10 consecutive epochs. The trained model is then used to evaluate the test set.

5.2 Experimental Result

The experimental results of this model are shown in Table 2. It can be seen that PA-Tem results are slightly lower than Re-Temp only at H@10 of ICEWS14 and better than the baseline model on all other data sets. Since ICEWS14 and ICEWS18 have a time granularity of one day, the modeling ability of short-term time series relationship of the model is high, and the performance of the model on these two data sets is better, which indicates that the model can capture dense time series relationship in the neighborhood more accurately; while GDELT covers multiple fields, the events are scattered, and the cross-domain generalization ability of the model is high, and the effect of model improvement on this data set is not obvious. Therefore, it is necessary to improve the performance of the model under sparse and multi-source data. The experimental results show that PA-Temp outperforms DaeMon model on all data sets, and path extraction and aggregation exist in both models, while PA-Temp outperforms DaeMon model, indicating the feasibility of filtering irrelevant paths by path constraint. However, the performance difference between PA-Temp and Re-Temp is small, so more efficient path aggregation strategies need to be explored to further improve PA-Temp performance.

Table 2. Performance(%) with Baseline models. The highest value is bold and the second highest is underlined.

Model	ICEWS14				ICEWS18			
	MRR	H@1	H@3	H@10	MRR	H@1	H@3	H@10
RE-GCN	41.50	30.86	46.60	62.47	30.55	20.00	34.73	51.46
xERTE	40.12	32.11	44.73	56.25	29.31	21.03	33.51	46.48
TiTer	41.73	32.74	46.46	58.44	29.96	22.06	33.41	44.96
CEN	42.20	32.08	47.46	61.31	31.50	21.70	35.44	50.59
HiSMatch	46.42	35.91	51.63	66.84	33.99	23.91	37.90	53.94
TiRGN	44.04	33.83	48.95	63.84	33.66	23.19	37.99	54.22
Re-Temp	48.04	37.32	53.60	**68.90**	35.82	25.02	40.36	57.30
DaeMon	45.32	34.96	50.07	63.72	31.85	22.67	35.92	49.80
TECHS	43.88	34.59	49.36	61.95	30.85	21.81	35.39	49.82
PA-Temp	**48.39**	**37.53**	**54.27**	68.85	**35.89**	**25.13**	**40.42**	**57.62**

Model	GDELT			
	MRR	H@1	H@3	H@10
RE-GCN	19.31	11.99	20.61	33.59
xERTE	18.07	12.31	20.05	30.32
TiTer	17.02	11.23	19.81	26.92
CEN	19.89	12.61	21.16	34.09
HiSMatch	22.01	14.45	23.80	36.61
TiRGN	21.67	13.63	23.27	37.60
Re-Temp	25.05	15.70	27.14	44.16
DaeMon	20.73	13.65	22.53	34.23
TECHS	-	-	-	-
PA-Temp	**25.12**	**15.75**	**27.24**	**44.23**

5.3 Analysis of Historical Length

To study the performance variation trend of the model under different historical lengths, we conducted experiments of different historical lengths on the ICEWS14, ICEWS18 and GDELT datasets. The experiment tested the variation of the historical length from 1 to 7. The experimental results which illustrate the performance variation across these history lengths are shown in Fig. 2.

Through the analysis of different historical lengths, we found that when the historical length ranges from 1 to 5, the performance will improve as the historical length increases. On the ICEWS14 and ICEWS18 datasets, performance will decline when the historical length exceeds 5, and the performance improvement is more significant when the historical length is 3. However, as the historical length of GDELT increases, although its performance has improved, the effect is not significant. This indicates that the recent few historical snapshots can assist the model in reasoning, while in a longer history, irrelevant information will increase the model's computational load and does not help the model's performance.

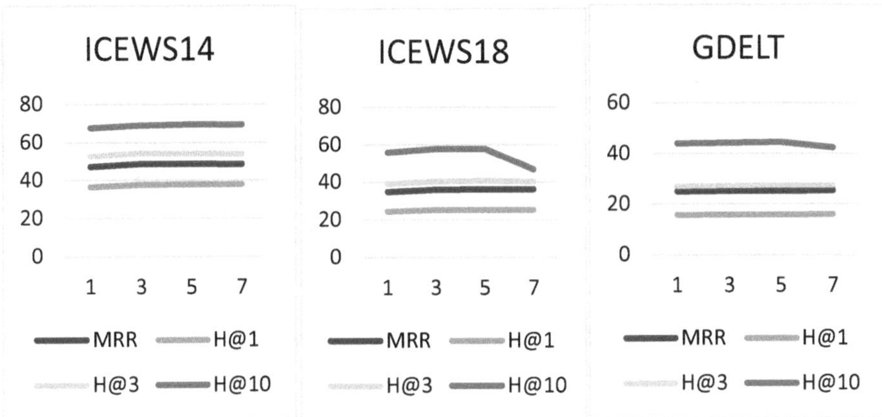

Fig. 2. Performance of Different History Length.

5.4 Ablation Experiment

To evaluate the effectiveness of the three core modules in the PA-Temp model—the path-aware mechanism, dynamic time embedding, and skipping information flow—we performed a series of ablation studies on the ICEWS14, ICEWS18, and GDELT datasets. The results are presented in Table 3.

Removing the Path-Aware Mechanism: Through the experimental results, we observed a significant decline in model performance. This indicates that explicitly extracting and aggregating multi-hop relationship paths related to queries not only injects deep semantic information into the model but also helps it understand complex causal chains or indirect associations, thereby compensating for the deficiency of relying only on local neighborhood aggregation.

Remove Time-Aware Gating: Experimental results show that through the gating mechanism, explicit time signals and sudden events can be more effectively integrated, thereby more accurately modeling the dynamic evolution of entities. Simply mixing static and dynamic information cannot deal with unexpected events and will lead to a weakened temporal sensitivity of the model. Therefore, the effectiveness of this module has been proved.

Remove the Skip Information Flow Mechanism: The model fuses historical snapshots using average weighting, without considering time decay and path similarity constraints. The impact of the skip information flow mechanism on model performance is most significant. This indicates that the skip information flow mechanism intelligently screens historical information through time decay and path similarity constraints, effectively avoiding the interference of irrelevant historical information.

Table 3. Ablation Study on ICEWS14, ICEWS18 and GDELT

Model	ICEWS14				ICEWS18			
	MRR	H@1	H@3	H@10	MRR	H@1	H@3	H@10
PA-Temp	48.39	37.53	54.27	68.85	35.89	25.13	40.42	57.62
-path	35.46	26.41	38.83	53.27	23.03	15.23	25.39	38.37
-time	35.83	26.89	39.45	52.99	23.16	15.37	25.48	38.53
-skip	27.97	20.96	30.39	41.18	20.12	13.19	22.13	33.99

Model	GDELT			
	MRR	H@1	H@3	H@10
PA-Temp	25.12	15.75	27.24	44.23
-path	21.67	13.00	23.09	39.30
-time	21.23	12.57	22.55	38.89
-skip	17.12	10.94	17.81	28.96

6 Conclusion

This paper proposes a new reasoning model combining path perception and temporal gating, which can effectively capture the deep semantics and temporal variation of entities. A path perception module is designed to extract and aggregate relational paths, so that entity representation contains rich semantic information. A time decay factor is designed to control the use of historical information automatically according to the passage of time, so as to improve the performance and adaptability of the model on time-sensitive tasks.

Acknowledgments. Thanks to the National Key Research and Development Program (2023YFC3304504), the 2024 Central Guidance Fund Project for Local Science and Technology Development of the Xizang Autonomous Region (XZ202401YD0015), and the Natural Science Basic Research Program of Guizhou Province (QIANKEHEJICHU-ZD [2025]-043) Financial support for this research.

References

1. Dong, H., Ning, Z., Wang, P., et al.: Adaptive path-memory network for temporal knowledge graph reasoning. arXiv preprint arXiv:2304.12604 (2023)

2. Wang, K., Han, S.C., Poon, J.: Re-temp: relation-aware temporal representation learning for temporal knowledge graph completion. arXiv preprint arXiv:2310.15722 (2023)
3. Sun, H., Zhong, J., Ma, Y., et al.: Timetraveler: reinforcement learning for temporal knowledge graph forecasting. arXiv preprint arXiv:2109.04101 (2021)
4. Liu, L., Xiao, C., Wang, S., et al.: Relation logical reasoning and relation-aware entity encoding for temporal knowledge graph reasoning. In: Proceedings of the 31st International Conference on Computational Linguistics, pp. 1323–1332 (2025)
5. Mei, X., Yang, L., Cai, X., et al.: An adaptive logical rule embedding model for inductive reasoning over temporal knowledge graphs. In: Proceedings of the 2022 Conference on Empirical Methods in Natural Language Processing, pp. 7304–7316 (2022)
6. Liu, Y., Ma, Y., Hildebrandt, M., et al.: Tlogic: temporal logical rules for explainable link forecasting on temporal knowledge graphs. In: Proceedings of the AAAI Conference on Artificial Intelligence, vol. 36, no. 4, pp. 4120–4127 (2022)
7. Lao, N., Cohen, W.W.: Relational retrieval using a combination of path-constrained random walks. Mach. Learn. **81**(1), 53–67 (2010)
8. Xiong, W., Hoang, T., Wang, W.Y.: Deeppath: a reinforcement learning method for knowledge graph reasoning. arXiv preprint arXiv:1707.06690 (2017)
9. Li, Y., Sun, S., Zhao, J.: TIRGN: time-guided recurrent graph network with local-global historical patterns for temporal knowledge graph reasoning. In: IJCAI, pp. 2152–2158 (2022)
10. Niu, G., Li, B.: Logic and commonsense-guided temporal knowledge graph completion. In: Proceedings of the AAAI Conference on Artificial Intelligence, vol. 37, no. 4, pp. 4569–4577 (2023)
11. Li, Z., Jin, X., Li, W., et al.: Temporal knowledge graph reasoning based on evolutional representation learning. In: Proceedings of the 44th International ACM SIGIR Conference on Research and Development in Information Retrieval, pp. 408–417 (2021)
12. Zhang, M., Xia, Y., Liu, Q., et al.: Learning latent relations for temporal knowledge graph reasoning. In: Proceedings of the 61st Annual Meeting of the Association for Computational Linguistics, vol. 1: Long Papers, pp. 12617–12631 (2023)
13. Niu, G., Zhang, Y., Li, B., et al.: Rule-guided compositional representation learning on knowledge graphs. In: Proceedings of the AAAI Conference on Artificial Intelligence, vol. 34, no. 03, pp. 2950–2958 (2020)
14. Wu, J., Cao, M., Cheung, J.C.K., et al.: Temp: temporal message passing for temporal knowledge graph completion. arXiv preprint arXiv:2010.03526 (2020)
15. Li, X., Zhou, H., Yao, W., et al.: Sequential and repetitive pattern learning for temporal knowledge graph reasoning. In: Proceedings of the 2024 Joint International Conference on Computational Linguistics, Language Resources and Evaluation (LREC-COLING 2024), pp. 14744–14754 (2024)
16. Chen, T., Long, J., Yang, L., et al.: HTCCN: temporal causal convolutional networks with hawkes process for extrapolation reasoning in temporal knowledge graphs. In: Proceedings of the 2024 Conference of the North American Chapter of the Association for Computational Linguistics: Human Language Technologies, vol. 1: Long Papers, pp. 4056–4066 (2024)
17. Zong, L., Xie, Z., Ma, C., Liu, X., Zhang, X., Xu, B.: RENN: a rule embedding enhanced neural network framework for temporal knowledge graph completion (2024)
18. Chen, W., Wan, H., Wu, Y., et al.: Local-global history-aware contrastive learning for temporal knowledge graph reasoning. In: 2024 IEEE 40th International Conference on Data Engineering (ICDE), pp. 733–746. IEEE (2024)
19. Wang, J., Wang, B., Qiu, M., et al.: A survey on temporal knowledge graph completion: taxonomy, progress, and prospects. arXiv preprint arXiv:2308.02457 (2023)
20. Jin, W., Qu, M., Jin, X., et al.: Recurrent event network: autoregressive structure inference over temporal knowledge graphs. arXiv preprint arXiv:1904.05530 (2019)

21. Liang, K., Meng, L., Liu, M., et al.: A survey of knowledge graph reasoning on graph types: static, dynamic, and multi-modal. IEEE Trans. Pattern Anal. Mach. Intell. **46**(12), 9456–9478 (2024)
22. Chen, W., Xiong, W., Yan, X., Wang, W.: Variational knowledge graph reasoning. arXiv preprint arXiv:1803.06581 (2018)

Multimodal Fake News Detection via Evidence Retrieval with Vision-Language Model

Liwei Dong[1], Yanli Chen[1(✉)], Yongrong Wang[1], and Hanzhou Wu[2]

[1] Guizhou Normal University, Guiyang, China
liwei.dong@gznu.edu.cn, yanli_027@163.com
[2] Shanghai University, Shanghai, China
hanzhou@shu.edu.cn

Abstract. Fake news poses a serious threat to public trust and social stability, especially in the age of social media. Existing approaches to multimodal fake news detection have increasingly turned to Large Vision-Language Models (LVLMs) for their strong reasoning ability across text and images. However, current methods still face two major challenges: LVLMs are limited by the information available within the input, and their predictions often lack grounding in external evidence, leading to unreliable or weakly interpretable results. To improve these issues, we propose MERF (Multimodal Evidence Retrieval Framework), a unified approach that augments LVLMs with web-based evidence. MERF performs content search and reverse image retrieval to construct an enriched evidence set, which is then incorporated into the LVLM for evidence-aware reasoning. By grounding veracity assessment in external textual and visual context, MERF improves both accuracy and interpretability. Experiments on the Weibo and MR2 datasets demonstrate that MERF consistently outperforms state-of-the-art baselines, highlighting the importance of retrieval-augmented reasoning in robust multimodal fake news detection.

Keywords: Multimodal Fake News Detection · Vision-Language Models · Evidence Retrieval

1 Introduction

The pervasive dissemination of fake news via social media platforms has raised significant societal concerns, undermining public confidence and misrepresenting reality [1]. In particular, the viral nature of such content–often amplified through emotional appeal, sensational imagery, or coordinated inauthentic behavior–has made timely and accurate detection increasingly urgent. Consequently, the field of fake news detection (FND) has gained substantial attention, aiming to automatically flag deceptive or inaccurate information using algorithmic techniques [2–4]. Broadly speaking, FND techniques can be grouped into two main

J. Yu and Y. Chen (Eds.): FCS 2025, CCIS 2805, pp. 179–187, 2027.
https://doi.org/10.1007/978-981-95-9514-3_12

paradigms: content-oriented and propagation-based strategies. Content-based methods analyze intrinsic characteristics of news items, including lexical style, emotional tone, sentiment, and linguistic patterns [5], while propagation-based methods focus on external cues, such as how news spreads through social networks, who shares it, and in what temporal patterns [6,7]. In this study, we adopt a content-centric approach, as it remains applicable even in the early stages of dissemination when propagation signals are sparse. Earlier FND models typically relied on traditional machine learning or deep learning pipelines in combination with handcrafted features [8,9]. Although these approaches capture some useful patterns, they often struggle with generalizability and require domain-specific feature engineering. With the advent of Large Language Models (LLMs), the field has witnessed substantial improvements in factual reasoning, semantic understanding, and text classification [10,11]. However, LLMs are inherently unimodal, processing only textual inputs, and thus are unable to capture the rich visual signals often present in fake news. To overcome this limitation, researchers have proposed Large Vision-Language Models (LVLMs) [12], which are trained on massive image-text datasets and capable of multimodal reasoning. These models have shown remarkable performance in tasks such as visual question answering, captioning, and commonsense inference [13]. Nonetheless, their reasoning is largely constrained by the information within the input, and their performance deteriorates when claims are ambiguous, incomplete, or context-dependent, which is common in real-world misinformation.To address these challenges, we propose MERF (Multimodal Evidence Retrieval Framework), a novel architecture designed to augment LVLMs with external evidence. MERF retrieves both textual and visual context from the web, using a parallel cross-modal retrieval process that provides relevant background information to support informed reasoning. By grounding the model's decisions in real-world evidence, MERF enhances both the accuracy and explainability of multimodal fake news detection.

Our key contributions include:

We introduce a lightweight framework that guides LVLMs through multi-perspective veracity assessment with external evidence.

We design a multimodal retrieval module that collects relevant textual and visual information from the web for cross-checking.

We evaluate MERF on benchmark datasets, showing improved performance over existing models.

2 Related Work

Multimodal fake news detection has attracted increasing attention as misinformation in social media often includes both text and visual content. Early approaches typically fused features from text and image via concatenation or co-attention mechanisms. For instance, MCAN employs stacked co-attention layers to capture interactions between modalities, demonstrating strong baseline performance on Twitter and Weibo datasets [14]. To improve semantic alignment across modalities, MVAE adopts a multimodal variational autoencoder to learn latent representations that reconstruct both modalities [15]. SEPM further enhances multimodal fusion via progressive feature refinement and hierarchical semantic alignment, showing improved fine-grained understanding [16]. Meanwhile, HFMFN explores hierarchical fusion techniques to balance inter-modal and intra-modal feature learning, aiming to avoid modality dominance and ensure stable predictions [17]. More recently, COOLANT introduces a cross-modal contrastive learning framework to explicitly align vision and language features. It integrates contrastive learning objectives with an auxiliary consistency task and an attention-guided aggregation module, achieving state-of-the-art performance on benchmark datasets [18]. This method demonstrates that fine-grained contrastive alignment combined with interpretable modality re-weighting significantly enhances robustness and accuracy. Despite progress, current models often rely on static unimodal encoders and lack external contextual support. Our work builds upon these foundations by introducing a retrieval-augmented architecture [19] that incorporates dynamic web-based evidence to guide large vision-language models, achieving more interpretable and context-aware misinformation detection.

3 Proposed Method

We propose MERF (Multimodal Evidence Retrieval Framework), a lightweight pipeline that enhances LVLMs' reasoning by incorporating external evidence. By lightweight, we specifically refer to the fact that MERF does not involve retraining or fine-tuning of LVLMs. Instead, it introduces an evidence retrieval step and prompt restructuring, which add negligible computational overhead while significantly enhancing interpretability. As shown in Fig. 1, MERF includes a web-based retrieval module that gathers relevant textual and visual information to supplement the original news input. This evidence is then structured and fed into the LVLM, enabling more informed and explainable fake news detection.

3.1 Multimodal Retrieval

To enrich reasoning, MERF incorporates a multimodal retrieval module that collects relevant evidence from the web [20]. For textual retrieval, the full news text is used as a query to preserve semantic completeness. A Google-based crawler fetches the top-K related webpages, which are then filtered based on semantic similarity and parsed into structured information (e.g., title, time, content) using a lightweight LLM pipeline. For visual retrieval, MERF performs reverse image search to trace the source and previous usage of the input image. Retrieved results are categorized according to content type. Text-based pages follow the same processing as textual retrieval, while video content is briefly summarized to extract essential metadata. This process enables the system to identify reused or misleading images and enhances the model's ability to detect visual-textual inconsistencies. By combining both semantic textual context and visual provenance cues, the module strengthens MERF's ability to reason beyond surface-level features. It is particularly useful for detecting subtle manipulations or

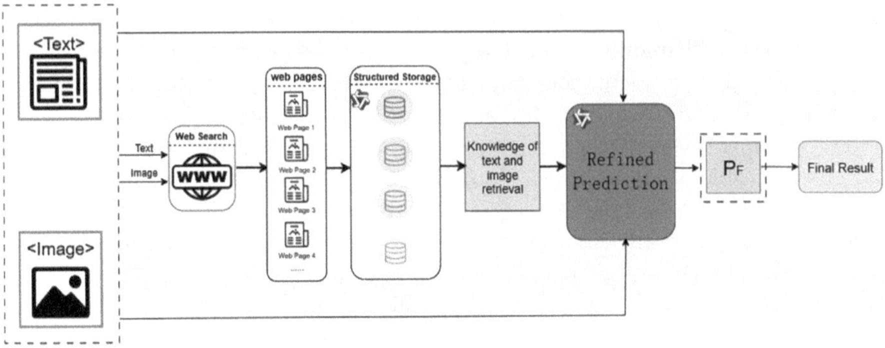

Fig. 1. The MERF framework consists of two main stages. In Stage 1 (Multimodal Retrieval), the system collects relevant textual and visual evidence from the web through content search and reverse image lookup. In Stage 2 (Refined Prediction), the retrieved evidence is structured and integrated into instruction-guided prompts, allowing the LVLM to perform contextual and cross-modal reasoning. This two-stage pipeline enhances detection accuracy, interpretability, and robustness against multimodal misinformation

recontextualized media. The entire retrieval output is structured into a unified evidence set and stored for downstream reasoning, as shown in Fig. 2.

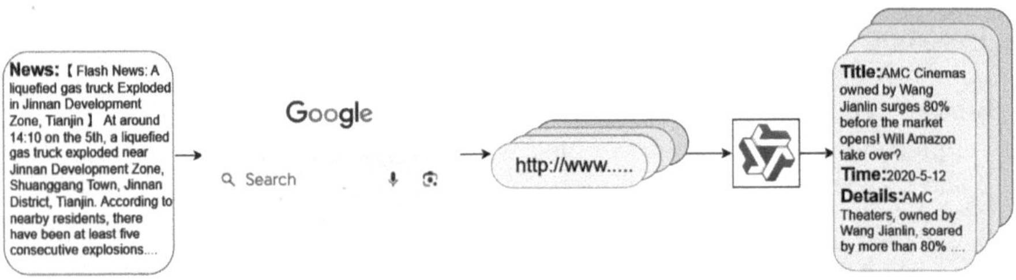

Fig. 2. Retrieval of the textual content of N relevant URLs obtained from an Internet search, and then storage by LVLM according to the specified structure.

3.2 Refined Prediction

In the final stage, MERF constructs an instruction-guided prompt that consolidates all previously collected information, including the original news content, associated image, and retrieved textual and visual evidence. This comprehensive prompt enables the LVLM to conduct contextual and multimodal reasoning, guiding it to examine semantic alignment, detect cross-modal inconsistencies, and incorporate background knowledge from external sources. Unlike conventional classification approaches, MERF encourages the model to engage in evidence-aware decision-making by considering both supporting and contradicting evidence. Importantly, MERF does not introduce an additional handcrafted scoring system. Instead, it leverages instruction-guided prompting to explicitly present different types of evidence and to direct the LVLM to compare them, assess their credibility, and generate a rationale that reflects how they are weighted. For example, prompts instruct the model to "analyze whether retrieved evidence supports or contradicts the claim, and base the final judgment on the quality rather than the quantity of evidence." This design minimizes the risk of being misled when most retrieved texts support misinformation, and ensures that the LVLM grounds its prediction in credibility and cross-modal consistency.

4 Experiments

4.1 DATA

To evaluate the proposed framework, we adopt the Weibo dataset [21], we obtain a refined set of 7,853 instances, making it well-suited for evaluating multimodal misinformation detection models. In addition to the Weibo dataset, we evaluate MERF on a benchmark from the 2023 IKCEST Belt and Road Big Data Competition. This dataset targets multimodal fake news detection, with text-image pairs labeled as real or fake. It covers multiple domains (e.g., news, health, finance) and includes multilingual content. The dataset was officially released by the competition and used by over 3,800 teams from 19 countries. We adopt the official data split for evaluation (Table 1).

Table 1. Statistics of the pre-processed Weibo and MR2 datasets.

Dataset	Train		Test		Total
	Fake	Real	Fake	Real	
Weibo	3783	3748	996	1000	9527
MR2	4765	3561	766	853	9945

4.2 Experiment Settings

Large Visual-Language Models. In this work, we adopt Qwen-VL-Max [22], the most capable model in Alibaba Cloud's Qwen vision-language model family. It is accessed via an OpenAI-compatible API and builds on the foundation of Qwen-VL-Plus with further improvements in visual understanding and instruction-following. The model supports ultra-high-resolution images (up to 12 million pixels) and utilizes a Vision Transformer with windowed attention, enabling it to efficiently process images of varying sizes while retaining fine-grained visual detail. Thanks to these enhancements, Qwen-VL-Max demonstrates strong performance on complex multimodal tasks, particularly those involving precise image-text alignment, visual grounding, and semantic consistency across modalities. These capabilities make it highly suitable for our framework, which relies on accurate cross-modal reasoning for fake news detection.

4.3 Baselines

MVAE [23]: A multimodal variational autoencoder that learns shared latent representations from text and image pairs. It performs fake news classification based solely on post content, without considering propagation signals.

SpotFake [4]: A fusion-based model that combines text features from XLNet and image features from VGG19 for multimodal fake news detection.

MoPeD [24]: A modality-perception learning framework that uses CLIP and CNN-based dual encoders, along with cross-modality fusion and aggregation modules, to identify key features across modalities.

Table 2. Performance comparison of different methods on the Weibo and MR2 datasets.

Dataset	Method	Accuracy	Rumor			Non-Rumor		
			Precision	Recall	F1	Precision	Recall	F1
Weibo	MVAR	0.824	0.854	0.769	0.809	0.802	0.875	0.837
	SpotFake	0.870	0.855	0.892	0.873	0.769	0.807	0.787
	MoPeD	0.883	0.879	0.887	0.883	0.887	0.878	0.883
	MERF	**0.912**	**0.978**	**0.897**	**0.936**	**0.892**	**0.977**	**0.932**
MR2	MVAR	0.807	0.826	0.742	0.782	0.790	0.861	0.824
	SpotFake	0.862	0.848	0.879	0.863	0.776	0.802	0.789
	MoPeD	0.881	0.867	0.882	0.874	0.887	0.876	0.881
	MERF	**0.904**	**0.958**	**0.893**	**0.924**	**0.888**	**0.952**	**0.919**

4.4 Ablation Studies

To quantify the contribution of the multimodal retrieval module, we compare MERF with a variant that bypasses retrieval and directly feeds the post text and image into the LVLM, using only instruction-guided Chain-of-Thought (CoT) prompting. Removing the retrieval step causes notable performance degradation: accuracy drops by 8.1% points on Weibo and 7.9% points on MR2. This confirms that external evidence is the primary driver of MERF's gains, enabling the model to resolve context-dependent or ambiguous claims that are difficult to judge from the post content alone. We observe the largest declines on cases involving reused or recontextualized images, where reverse image search and web content retrieval provide decisive cues (Table 3).

Table 3. Ablation on multimodal retrieval. Accuracy (Acc) is reported; Δ indicates the drop without retrieval.

Method	Weibo Acc	MR2 Acc
MERF (full)	0.912	0.904
w/o Retrieval	0.831 (Δ–0.081)	0.825 (Δ–0.079)

4.5 Result and Analysis

Table 2 presents the performance comparison between MERF and several representative baselines on the Weibo and MR2 datasets. Results are evaluated using accuracy, precision, recall, and F1-score for both rumor and non-rumor categories. On the Weibo dataset, MERF achieves the highest accuracy (0.912) and the best F1-scores for both rumor (0.936) and non-rumor (0.932), indicating its strong balance and reliability across categories. Compared to traditional models like MVAE and SpotFake, MERF demonstrates consistent improvements in all metrics. Although MoPeD performs competitively (accuracy: 0.883), its rumor F1-score (0.883) still falls short, highlighting MERF's advantage in detecting subtle misinformation through evidence-driven reasoning. On the more challenging MR2 dataset, MERF maintains its superiority, achieving the best accuracy (0.904) and outperforming all baselines in both rumor F1 (0.924) and non-rumor F1 (0.919). Notably, even the strongest competitor, MoPeD, shows lower F1 scores (0.874/0.881), which suggests MERF's robustness and generalization across domains. Overall, MERF consistently delivers state-of-the-art performance on both datasets, confirming the effectiveness of integrating external evidence and cross-modal analysis in multimodal fake news detection. Beyond quantitative improvements, MERF provides clear qualitative advantages. On Weibo, the gain in rumor F1 (+5.3% over MoPeD) is largely due to the ability of reverse image retrieval to detect reused or recontextualized images, which conventional fusion models often misclassify. On MR2, which covers multilingual and domain-diverse content, MERF consistently outperforms baselines, highlighting its robustness to heterogeneous distributions. Furthermore, the ablation study confirms that removing retrieval leads to a drop of over 8% in accuracy, underscoring that external evidence is the key driver of MERF's improvements. Case analyses also show that MERF's rationales explicitly reference both corroborating and contradicting evidence, demonstrating that the framework enhances interpretability in addition to accuracy.

5 Conclusion

In this work, we proposed MERF, a unified framework for multimodal fake news detection that enhances the reasoning capabilities of LVLMs through explicit evidence retrieval. While LVLMs demonstrate strong performance in processing multimodal content, they often fall short when accurate verification requires external context. MERF addresses this challenge by integrating web-based textual and visual retrieval, allowing the model to generate more accurate and explainable predictions. Experimental results on the Weibo and MR2 datasets show that MERF consistently outperforms existing baselines across multiple evaluation metrics, achieving higher accuracy, F1-scores, and greater robustness. Ablation studies further validate the effectiveness of each module, especially the contribution of retrieved evidence in guiding LVLM-based reasoning.

Acknowledgements. This work was supported by the National Key Research and Development Program of China (Grant No. 2023YFC3304504), the 2024 Xizang Autonomous Region Central Guided Local Science and Technology Development Fund Project (Grant No. XZ202401YD0015), and the Basic Research Program for Natural Science of Guizhou Province (Grant No. QIANKEHEJICHU-ZD[2025]-043).

Disclosure of Interests. The authors declare that they have no competing interests relevant to the content of this paper.

Future Directions. Despite its effectiveness, MERF has several limitations. Its dependence on online retrieval introduces latency and may lead to reproducibility issues due to the dynamic nature of web content. The quality and relevance of retrieved evidence can also be affected by search engine ranking noise. Furthermore, the current framework focuses solely on text and static images, while real-world misinformation increasingly involves other modalities such as video and audio. In future work, we plan to extend MERF to support these modalities, investigate more efficient and reliable retrieval strategies, and explore lightweight deployment solutions to enable real-time misinformation detection in practical applications.

References

1. Aïmeur, E., Amri, S., Brassard, G.: Fake news, disinformation and misinformation in social media: a review. Soc. Netw. Anal. Min. **13**(1), 30 (2023)
2. Capuano, N., Fenza, G., Loia, V., Nota, F.D.: Content-based fake news detection with machine and deep learning: a systematic review. Neurocomputing **530**, 91–103 (2023)
3. Mridha, M.F., Keya, A.J., Hamid, M.A., Monowar, M.M., Rahman, M.S.: A comprehensive review on fake news detection with deep learning. IEEE Access **9**, 156151–156170 (2021)
4. Singhal, S., Shah, R.R., Chakraborty, T., Kumaraguru, P., Satoh, S.I.: Spotfake: a multi-modal framework for fake news detection. In: 2019 IEEE Fifth International Conference on Multimedia Big Data (BigMM), pp. 39–47. IEEE (2019)
5. Zhou, X., Zafarani, R.: A survey of fake news: fundamental theories, detection methods, and opportunities. ACM Comput. Surv. (CSUR) **53**(5), 1–40 (2020)
6. Qi, P., et al.: FAKESV: a multimodal benchmark with rich social context for fake news detection on short video platforms. In: Proceedings of the AAAI Conference on Artificial Intelligence, vol. 37, no. 12, pp. 14444–14452 (2023)
7. Gravanis, G., Vakali, A., Diamantaras, K., Karadais, P.: Behind the cues: a benchmarking study for fake news detection. Expert Syst. Appl. **128**, 201–213 (2019)

8. Ahmed, H., Traore, I., Saad, S.: Detection of online fake news using N-Gram analysis and machine learning techniques. In: Traore, I., Woungang, I., Awad, A. (eds.) ISDDC 2017. LNCS, vol. 10618, pp. 127–138. Springer, Cham (2017). https://doi.org/10.1007/978-3-319-69155-8_9
9. Sahoo, S.R., Gupta, B.B.: Multiple features based approach for automatic fake news detection on social networks using deep learning. Appl. Soft Comput. **100**, 106983 (2021)
10. Chen, C., Shu, K.: Combating misinformation in the age of LLMs: opportunities and challenges. AI Mag. **45**(3), 354–368 (2024)
11. Hu, B., et al.: Bad actor, good advisor: exploring the role of large language models in fake news detection. In: Proceedings of the AAAI Conference on Artificial Intelligence, vol. 38, no. 20, pp. 22105–22113 (2024)
12. Achiam, J., et al.: GPT-4 Technical report. arXiv preprint arXiv:2303.08774 (2023)
13. Du, Y., et al.: Guiding pretraining in reinforcement learning with large language models. In: International Conference on Machine Learning, pp. 8657–8677. PMLR (2023)
14. Wu, Y., Zhan, P., Zhang, Y., Wang, L., Xu, Z.: Multimodal fusion with co-attention networks for fake news detection. In: Findings of the Association for Computational Linguistics: ACL-IJCNLP 2021, pp. 2560–2569 (2021)
15. Khattar, D., Goud, J.S., Gupta, M., Varma, V.: MVAE: multimodal Variational autoencoder for fake news detection. In: Proceedings of the World Wide Web Conference (WWW), pp. 2915–2921 (2019)
16. Wang, H., Guo, J., Liu, S., Chen, P., Li, X.: SEPM: multiscale semantic enhancement-progressive multimodal fusion network for fake news detection. Expert Syst. Appl. **2025**, 127741 (2025)
17. Wu, F., Chen, S., Gao, G., Ji, Y., Jing, X.Y.: Balanced multi-modal learning with hierarchical fusion for fake news detection. Pattern Recogn. **164**, 111485 (2025)
18. Wang, L., Zhang, C., Xu, H., Xu, Y., Xu, X., Wang, S.: Cross-modal contrastive learning for multimodal fake news detection. In: Proceedings of the 31st ACM International Conference on Multimedia (ACM MM), pp. 5696–5704 (2023)
19. Lewis, P., et al.: Retrieval-augmented generation for knowledge-intensive NLP tasks. Adv. Neural. Inf. Process. Syst. **33**, 9459–9474 (2020)
20. Jin, Z., Cao, J., Guo, H., Zhang, Y., Luo, J.: Multimodal fusion with recurrent neural networks for rumor detection on microblogs. In: Proceedings of the 25th ACM International Conference on Multimedia, pp. 795–816 (2017)
21. Jiang, Y., Wang, Y.: Large visual-language models are also good classifiers: a study of in-context multimodal fake news detection. arXiv preprint arXiv:2407.12879 (2024)
22. Bai, S., et al.: Qwen2.5-VL technical report. arXiv preprint arXiv:2502.13923 (2025)
23. Khattar, D., Goud, J.S., Gupta, M., Varma, V.: MVAE: multimodal variational autoencoder for fake news detection. In: The World Wide Web Conference, pp. 2915–2921 (2019)
24. Wang, B., Wu, G., Li, X., Gao, J., Hu, Y., Yin, B.: Modality perception learning-based determinative factor discovery for multimodal fake news detection. IEEE Trans. Neural Netw. Learn. Syst. (2024)

Security Analysis of LWE/RLWE-Based Diffie-Hellman-Like Post-quantum Authenticated Key Exchange Protocols in the eCK Model

Wentao Li[1], Yu Zhou[2(✉)], Jinhua Wang[2], Xiangyu Pan[2], Yanyan Liu[2], Yu Li[2], and Wenhui Ye[2]

[1] The 30th Research Institute of China Electronics Technology Group Corporation, Chengdu, China
[2] National Key Laboratory of Security Communication, Chengdu, China
zhouyu.zhy@qq.com

Abstract. In the post-quantum era, LWE/RLWE-based Diffie-Hellman-like post-quantum authenticated key exchange (PQ-AKE) protocols have gained considerable attention due to their advantages with parameter flexibility, forward secrecy and key independence. At the same time, in terms of the security analysis of the PQ-AKE, the eCK model is widely recognized as one of the strong and comprehensive security models for PQ-AKE protocols. However, we discovered lots of known LWE/RLWE-based Diffie-Hellman-like PQ-AKE protocols are insecure in the eCK model. In this paper, We conduct security analysis of two classic Diffie-Hellman-like PQ-AKE protocols and found them to be insecure in the eCK model. Based on above analysis, we abstract a generalized framework from these protocols. Then we prove that protocols following this framework are insecure against eCK adversaries. In response to these security shortcomings, we propose a new framework for constructing Diffie-Hellman-like PQ-AKE protocols that are provably secure in the eCK model. The framework provides guidelines for future secure protocol constructions.

Keywords: Authenticated key exchange · eCK model · Lattice-based cryptography · Post-quantum protocol

1 Introduction

With the rapid development of information technology, network communication has become deeply integrated into every aspect of modern society. A secure and

This research in part was supported by National Key Laboratory of Security Communication Foundations (2023, 6142103042303; 2024, 6142103042404; 2024, 6142103042405), and Stability Program of National Key Laboratory of Security Communication (2024, WD202411).

J. Yu and Y. Chen (Eds.): FCS 2025, CCIS 2805, pp. 188–204, 2027.
https://doi.org/10.1007/978-981-95-9514-3_13

efficient communication infrastructure serves as a cornerstone of today's digital world. Authenticated key exchange (AKE) protocols play a fundamental role in ensuring secure communication, enabling two parties to mutually authenticate each other and establish a shared session key over an untrusted channel. However, the rapid advancement of quantum computing technologies and the emergence of quantum algorithms such as Shor's algorithm [1] and Grover's algorithm [2] have posed significant threats to traditional authenticated key exchange protocols.

To solve this critical security challenge, there is an urgent need to design post-quantum authenticated key exchange protocol (PQ-AKE) that remain secure even against quantum adversaries. Currently, there are two primary approaches to designing such protocols: 1) PQ-AKE protocols based on key encapsulation mechanisms; 2) Diffie-Hellman-like PQ-AKE protocols. Among these, the Diffie-Hellman-like PQ-AKE protocols have the following advantages: separation of key exchange and authentication, parameter flexibility, forward secrecy and key independence. Therefore, this protocols consitutes a fundamental research direction in the post-quantum protocol.

From a security analysis perspective, it is essential to account for the enhanced capabilities of quantum adversaries. In the post-quantum context, the eCK model [3] provides a robust and comprehensive security model. The eCK model endows adversaries with substantial capabilities, effectively characterzing the attack scenarios that post-quantum protocols must confront. Consequently, analyzing Diffie-Hellman-like PQ-AKE protocols in the eCK model is of great importance for ensuring their security.

In the paper, we conducts a security analysis of two classic Diffie-Hellman-like PQ-AKE protocols in the eCK model, then summarizing a design frameworks of these protocols. We concludes that the protocols designed under this framework are insecure in the eCK model. Therefore, this paper proposes a framework for secure Diffie-Hellman-like PQ-AKE protocols in the eCK model.

1.1 Related Work

Diffie-Hellman-like PQ-AKE protocols are derived from the design principle of the classical Diffie-Hellman key exchange protocol. In contrast to traditional Diffie-Hellman-like AKE protocols relying on classical hard problems (such as integer factorization and discrete logarithm problem), PQ-AKE replace these problems with lattice-based problems (such as LWE [4], RLWE [5], and SIS [6]), thereby aiming to resist attacks from quantum adversaries.

Lindner and Peikert [7] first proposed a LWE-based Diffie-Hellman-like key exchange framework in 2011. Their construction introduced a public random matrix A, and the parties held secret vector $\mathbf{s}$ and error vector $\mathbf{e}$. Through computing messages $\mathbf{b} = A\mathbf{s} + \mathbf{e}$, the parties exchanged the messages to establish a shared key. This scheme naturally mimics the classical Diffie-Hellman protocol. However, the inherent "error vector" introduced by LWE lead to the issue of inconsistent session keys.

To solve this issue, Ding et al. [8] in 2012 proposed a Ding's error mechanism, which solved the issue of inconsistent keys caused by errors vector in

LWE. Later, Peikert [9] in 2014 introduced a more efficient Peikert's error reconciliation mechanism, which simplified the reconciliation process and improved protocol performance. To further enhance communication efficiency, Bos et al. [10] proposed a multi-bits Peikert's error reconciliation mechanism.

Building upon these theoretical foundations, several concrete post-quantum Diffie-Hellman-like key exchange protocols have been proposed. Bos et al. [11] introduced the key exchange protocol BCNS15 based on the RLWE problem and Peikert's error reconciliation mechanism. In 2015, Alkim et al. [12] improved upon BCNS15 and proposed the well-known NewHope protocol. In 2016, Bos et al. [13] developed the Frodo protocol, based on the LWE and Peikert's error reconciliation mechanism. However, all of these protocols lack built-in authentication and therefore are vulnerable to active adversaries.

To address the lack of authentication issue in the aforementioned protocols, Bos et al. [11] integrated BCNS15 into the TLS protocol, combining the key exchange protocol with digital signatures to create a PQ-AKE protocol. Zhao et al. [14] proposed a PQ-AKE protocol using a combination of RLWE-based key exchange and message authentication codes. Zhao et al. [15] proposed a biometric-based authentication protocol combining biometric credentials with post-quantum key exchange protocols, in client-server communication settings.

The security analysis of PQ-AKE protocols crucially depends on the choice of security models. Currently, the primary security models include the BR model [16], CK model [17], CK+ model [18], and eCK model [3]. Table 1 outlines the comparative features of these models.

Table 1. Comparison of Security Models: BR, CK, CK+, and eCK

Model	Long-term Key Leakage	Ephemeral Key Leakage	Session Key Leakage	Concurr-ency	Combination Attacks
BR	Yes	No	Yes	Yes	No
CK	Yes	Partial	Yes	Yes	No
CK+	Yes	Partial	Yes	Yes	No
eCK	Yes	Yes	Yes	Yes	Yes

From the Table 1, it is evident that the eCK model is one of the strong standardized security models. Therefore, the eCK model is considered the strong and comprehensive model for analyzing and constructing robust AKE protocols, especially in the post-quantum setting.

1.2 Motivation and Contributions

We found that lots of classic LWE/RLWE-based Diffie-Hellman-style PQ-AKE protocols [11,14,15,19] are insecure in the eCK model. This is a critical security gap in current protocol designs. In response to this issue, we makes the following contributions:

- We conduct a security analysis of two classic Diffie-Hellman-like PQ-AKE protocols in the eCK model. Based on the analysis, we abstract their framework, and prove that protocols following this framework are insecure against eCK adversaries.
- We propose a design framework for constructing Diffie-Hellman-like PQ-AKE protocols that are secure in the eCK model, providing guidelines for future secure protocol constructions.

1.3 Organization of the Paper

Section 2 presents lattice-based problems, Diffie-Hellman-like authenticated key exchange protocols and eCK model. In the Sect. 3, we give the role of the eCK model in PQ-AKE and security analysis of classic two LWE/RLWE-based Diffie-Hellman-like PQ-AKE protocols in the eCK model. Security analysis of the protocol frameworks in the eCK model is shown in Sect. 4. In the end, we conclude this paper in the Sect. 5.

2 Preliminaries

2.1 Notation

The symbols and their descriptions are shown in Table 2.

Table 2. Notation

Symbol	Description
$\mathbb{R}^m$	m-dimensional Euclidean space
χ_σ	a discrete Gaussian distribution with variance σ
$\mathbb{Z}_q$	a ring of residues modulo q
$x \leftarrow \chi_\sigma$	a random sample from χ_σ
$\boldsymbol{b}$	a vector
$\boldsymbol{A}$	a matrice
$\boldsymbol{u} \leftarrow \mathbb{Z}_q^n$	a vector $\boldsymbol{u}$ randomly sampled from $\mathbb{Z}_q^n$
$\boldsymbol{u} \leftarrow \chi_\sigma^n$	a vector $\boldsymbol{u}$ sampled from distribution χ_σ

2.2 Lattice-Based Problems

Definition 1 (Lattice). *A lattice Λ is a discrete additive subgroup of a linear space $\mathbb{R}^m$, defined as the set of all integer linear combinations of a basis $\boldsymbol{B} = (\boldsymbol{b}_1, \boldsymbol{b}_2, \ldots, \boldsymbol{b}_n)$ of linearly independent vectors. If the basis consists of n vectors in $\mathbb{R}^m$, then n is the rank of the lattice, and Λ is expressed as*

$$\Lambda(\boldsymbol{B}) = \{\sum_{i=1}^{n} \lambda_i \boldsymbol{b}_i \mid\mid \lambda_i \in Z\}. \tag{1}$$

Λ is a full-rank lattice if $n = m$.

Definition 2 (Search-LWE Problem). *Let n, m, q be the positive integers. the Search-LWE problem is to recover $\boldsymbol{s}$ given $\boldsymbol{A}$ and $\boldsymbol{b}$, where the vector $\boldsymbol{b} = \boldsymbol{A}\boldsymbol{s} + \mathbf{e} \mod q$, the matrix $\boldsymbol{A} \leftarrow \mathbb{Z}_q^{n\times m}$, the secret vector $\boldsymbol{s} \leftarrow \chi_\sigma^n$, and the error vector $\mathbf{e} \leftarrow \chi_\sigma^m$.*

Definition 3 (Decision-LWE Problem). *Let n, m, q be the positive integers. Decision-LWE involves distinguishing between the following two distributions:*

Distributions 1 is $(\boldsymbol{A}, \boldsymbol{b})$, where $\boldsymbol{A} \leftarrow \mathbb{Z}_q^{n\times m}$, $\boldsymbol{b} = \boldsymbol{A}\boldsymbol{s} + \mathbf{e} \mod q$, $\boldsymbol{s} \leftarrow \chi_\sigma^n$ and $\mathbf{e} \leftarrow \chi_\sigma^m$.

Distributions 2 is $(\boldsymbol{A}, \mathbf{c})$, where $\boldsymbol{A} \leftarrow \mathbb{Z}_q^{n\times m}$ and $\mathbf{c} \leftarrow \mathbb{Z}_q^m$.

2.3 Diffie-Hellman-Like Authenticated Key Exchange Protocols

Diffie-Hellman-like AKE protocols structurally emulate the classical Diffie-Hellman mechanism. These protocols are designed to perform secure key agreement, provide mutual authentication, and satisfy formal security guarantees in well-established security models such as BR, CK, or CK+.

Currently, lots of known Diffie-Hellman-like PQ-AKE protocols are constructed by combining post-quantum key exchange schemes (typically based on LWE or RLWE) with classical authentication mechanisms (such as digital signatures and message authentication codes).

Typically, these protocols exhibit the following three structural features:

- Bilateral Key Contribution. Both parties generate their own secret keys, which are exchanged and combined through a certain arithmetic structure (similar to the group operation of traditional DH) to generate a session key.
- Key Mixing Function. A deterministic function (such as hash functions and KDFs) is used to combine own keys and the other party's key informations to derive a consistent and uniform session key.
- Authentication mechanism integration. On the basis of the key exchange protocol, cryptographic schemes (such as digital signatures and MAC) are introduced to perform identity authentication, finally to realize the AKE protocols.

2.4 eCK Model

The eCK model [3] is an extension of the classical Canetti-Krawczyk (CK) security model [17], designed to provide stronger and more comprehensive security guarantees for AKE protocols. The eCK model plays a pivotal role in security analysis of AKE protocols. Especially in the post-quantum era., the eCK model serves as a practical design guideline for developing robust PQ-AKE protocols to reist quantum adversaries. The eCK model primarily consists of the following three components.

Participants. In the eCK model, multiple protocol participants are typically denoted by capital letters such as A, B, C etc. When A and B execute the AKE protocol normally, two AKE sessions will be generated respectively. These two sessions match each other.

Each session is identified by a unique session identifier (sid), often denoted as:

$$sid = (role, ID, ID^*, comm_1, \ldots, comm_n),$$

where $ID, ID^* \in \{A, B, C, \ldots\}$ represents the session participants. $role \in \{owner, peer\}$ indicates whether ID is the session initiator ($owner$) or the session responder ($peer$), and $comm_i$ denotes the $i - th$ communication message. For example, if A initiates a two-round AKE protocol with B, the session identifier for A is $sid = (owner, A, B, comm_1, comm_2)$, and for B it is $sid^* = (peer, B, A, comm_1, comm_2)$.

Adversary Model. In the eCK framework, the adversary M is considered a powerful entity with full control over the communication network. The adversary can intercept, delay, modify, replay, and inject arbitrary messages. The model grants the adversary access to the following types of queries:

- $Send(A, B, m)$. Simulates sending message m from party A to B, and returns the response of party A.
- $Long-term\ Key\ Reveal(A)$. Reveals the long-term key of party A.
- $Ephemeral\ Key\ Reveal(sid)$. Reveals the ephemeral key of a given session sid.
- $Session\ Key\ Reveal(sid)$. Reveals the session key of the specified session sid.

Security Game. After making the above queries, the adversary selects a fresh session sid and issues a $Test(sid)$ query.

1) The notion of fresh session sid is defined as follows.
   Let $sid = (Owner, A, B, comm_1, comm_2)$ be a session initiated by A with peer B. If a matching session exists, then the matching session is $sid^* = (Peer, B, A, comm_1, comm_2)$. Then, if session sid is considered fresh, none of the following conditions hold:

- The adversary has issued a *Session Key Reveal*(sid) query for sid, or for its matching session sid^*, if it exists.
- If a matching session exists, and the adversary has issued either: $Long - Term\ Key\ Reveal(A)$ and $Ephemeral\ Key\ Reveal(sid)$, or $Long - Term\ Key\ Reveal(B)$ and $Ephemeral\ Key\ Reveal(sid^*)$.
- If no matching session exists, and the adversary has either: $Long - Term\ Key\ Reveal(B)$, or $Long - Term\ Key\ Reveal(A)$ and $Ephemeral\ Key\ Reveal(sid)$.

2) $Test(sid)$ is shown below.
Oracle selects a random bit $b \in \{0, 1\}$. If $b = 0$, the oracle returns the real session key of sid. If $b = 1$, the oracle returns a random value of the same length. Finally, the adversary outputs a guess bit b'. The adversary wins the game if $b' = b$, and fails otherwise.
The advantage of adversary M in the eCK security game is defined as:

$$Adv(M) = \|P[M wins] - \frac{1}{2}\|. \tag{2}$$

An AKE protocol is considered secure in the eCK model if it satisfies the following two conditions.

1) Correctness: Both parties in the protocol can calculate the same session key.
2) Security: The advantage $Adv(M)$ of the adversary M is a negligible function.

3 Security Analysis of Classic Protocols in the eCK Model

3.1 The Role of the eCK Model in Post-quantum Authenticated Key Exchange Protocols

Quantum adversaries, equipped with Shor's [1] and Grover's [2] algorithms, can potentially break many classical hard problems (such as integer factorization and discrete logarithm problem). Therefore, PQ-AKE protocols must satisfy stronger security guarantees. In the posr-quantum setting, the eCK model can well describe the attack scenarios that post-quantum protocols need to face, so playing acritical role. Its advantages can be summarized in three key aspects in the security analysis of PQ-AKE protocols.

Definition of Adversary's Capabilities. The eCK model assumes that the adversary has full control over all communication channels. Therefore, the adversary can conduct active attacks. At the same time, the adversary is given three abilities: long-term key reveal, ephemeral key reveal and session key reveal. Table 3 summarizes the adversary's capabilities and their corresponding attack scenarios.

Table 3. Adversary's Capabilities and Corresponding Attack Scenarios in the eCK Model

Adversary's Capability	Corresponding Attack Scenario
Active attacks	Man-in-the-middle (MitM), replay, impersonation
Long-term key reveal	Exposure of long-term key (forward secrecy)
Ephemeral key reveal	Ephemeral key compromise due to weak randomness
Session key reveal	Session key compromise due to poor encryption algorithms

Adversary's capabilities closely align with those of quantum adversaries, who may exploit quantum algorithms to extract secret information. By simulating such powerful attackers, the eCK model enables us to evaluate how well a PQ-AKE protocol maintains residual security even when critical keys are partially compromised.

Definition of Fresh Session. A fundamental component of the eCK model is the definition of fresh session, which defines the boundary of adversarial capability. This is particularly crucial in PQ-AKE protocols, as it ensures the following properties.

- Indistinguishability of session keys: As long as the adversary has not simultaneously compromised both the long-term keys and ephemeral keys of a party, the session key must remain computationally indistinguishable from a random value.
- Key leakage tolerance: Leakage of partial secret information in one session or its matching session does not compromise the entire protocol.

The concept accurately evaluates important post-quantum properties such as forward secrecy and key compromise impersonation (KCI) resilience, which are considered essential in post-quantum protocol.

Constructive Security Proofs in the eCK Model. The eCK model supports the construction and proof of secure PQ-AKE protocols. In the eCK model, when building PQ-AKE protocols based on lattice problems (such as SIS, LWE, or RLWE), designers can reduce the protocol's security to the corresponding lattice problem through security games.

In summary, the eCK model serves as a rigorous and comprehensive model for analyzing the security of PQ-AKE protocols.

3.2 Security Analysis of the TLS Protocol with RLWE-Based Key Exchange and Digital Signature in the eCK Model

In the work of Bos et al. [11], the BCNS15 key exchange protocol is introduced, as illustrated in Fig. 1.

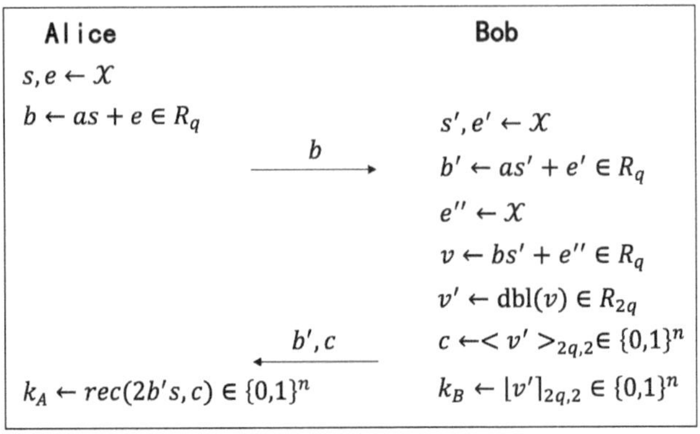

Fig. 1. BCNS15 Key Exchange Protocol.

The protocol is constructed based on the RLWE and Peikert's error reconciliation mechanism to realize a Diffie-Hellman-like post-quantum key exchange protocol. Then the BCNS15 was integrated into the TLS framework, as depicted in Fig. 2.

In practice, the BCNS15 protocol is designed to be compatible with the existing TLS infrastructure. It combines BCNS15 protocol with classical identity authentication schemes to construct a Diffie-Hellman-like PQ-AKE protocol. In Fig. 2, The Certificate, CertificateRequest and other functions in the box represent TLS's standard authentication components (e.g., RSA or ECDSA digital signatures).

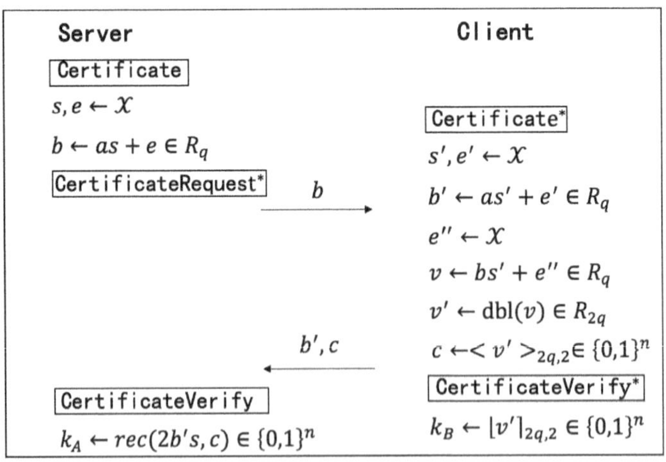

Fig. 2. The TLS Protocol with RLWE-based Key Exchange and Digital Signature.

Security Analysis. In the eCK model, the adversary is granted powerful capabilities. In particular, the adversary is allowed to perform *Long − term Key Reveal* and *Ephemeral Key Reveal* queries.

For the TLS protocol, the adversary can proceed with the following attack strategy in the eCK model:

1) The adversary captures the exchanged public messages $b^{'}$ and c from the communication channel.
2) The adversary issues an *Ephemeral Key Reveal* query to obtain the server's ephemeral key s.
3) Using the known ephemeral key s, the adversary computes an intermediate value $u = b^{'}s$.
4) The adversary invokes the Peikert's error reconciliation mechanism for u, such as $k = rec(2u, c)$, thus successfully deriving the session key $k = k_A = k_B$.

This attack clearly demonstrates that the protocol is insecure in the eCK model. Since the adversary only obtains any party's ephemeral key, it can compute the session key.

Remark. Through above analysis, it is found that the TLS protocol is designed by combining the BCNS15 protocol with classical identity authentication schemes. The session key derivation relies solely on ephemeral keys, without long-term keys. So, if the temporary keys is leaked, the session key becomes insecure. To solve this weakness, subsequent protocols have adopted a new design principle: the session key should depend on both long-term keys and ephemeral keys. A classic example is the lattice-based authenticated key exchange protocol proposed by Zhao et al. [14], where the session key generation process incorporates both the long-term secret keys and ephemeral keys. This design enhances resilience against ephemeral key leakage. In the next section, we present a detailed security analysis of Zhao et al.'s protocol.

3.3 Security Analysis of the Lattice-Based Authenticated Key Exchange Protocol in the eCK Model

Zhao et al. [14] proposed a lattice-based authenticated key exchange protocol relying on the RLWE and Peikert's error reconciliation mechanisms, as illustrated in Fig. 3.

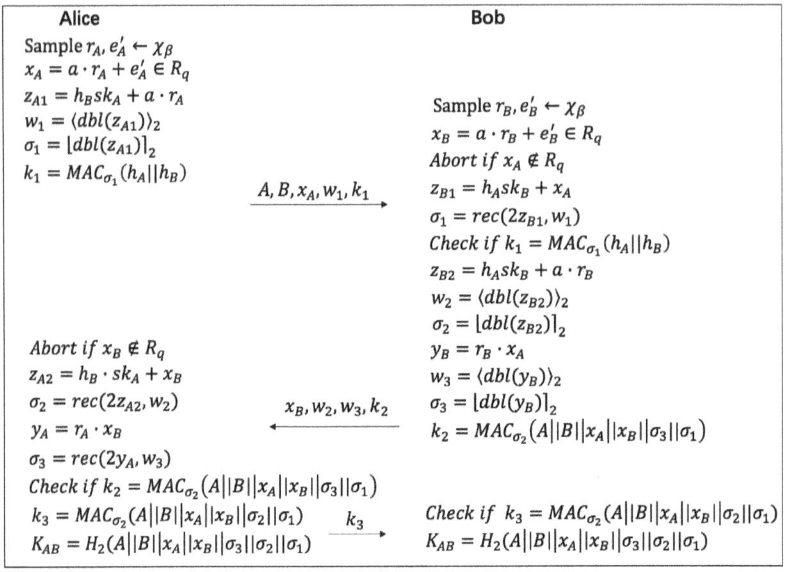

Fig. 3. the Lattice-Based Authenticated Key Exchange Protocol.

Security Analysis. Compared to the previously discussed schemes, a major improvement of this protocol is that the session key generation requires the combined the long-term keys and ephemeral keys. Although this protocol has been proven secure in the CK+ model, it can still be broken by an adversary in the eCK model. The concrete attack proceeds as follows:

1) The adversary captures the exchanged public messages $x_A, w_1, k_1, x_B, w_2, w_3, k_2$ from the communication channel.
2) The adversary issues a $Long{-}term\ Key\ Reveal$ query to obtain the long-term key sk_A of Alice
3) Using Alice's long-term key sk_A, the adversary computes intermediate values $z'_{A1} = h_B \cdot sk_A + x_A$ and $z_{A2} = h_B \cdot sk_A + x_B$, where $z'_{A1} \approx z_{A1} \approx z_{B1}$.
4) Because of $z'_{A1} \approx z_{B1}$, the adversary invokes Peikert's error reconciliation mechanism to get the authenticated keys $\sigma_1 = rec(2z'_{A1}, w_1)$ and $\sigma_2 = rec(2z_{A2}, w_2)$.
5) The adversary issues a *Ephemeral Key Reveal* query to acquire the ephemeral key r_B of Bob.
6) Using Bob's ephemeral key sk_A , the adversary computes intermediate value $y_B = x_A \cdot r_B$.
7) The adversary invokes Peikert's error reconciliation mechanism to get the negotiated keys $\sigma_3 = \lfloor dbl(y_B) \rceil_2$.
8) Finally, the adversary combines the authenticated keys σ_1, σ_2 and the negotiated key σ_3 by applying the hash function H_2 to derive the final session key $K_{AB} = H_2(A||B||x_A||x_B||\sigma_1||\sigma_2||\sigma_3)$, thereby compromising the protocol.

Through the above analysis, we demonstrate that this PQ-AKE protocol is insecure in the eCK model.

4 Security Analysis of the Protocol Framework in the eCK Model

4.1 A Framework Based on the Separation of Long-Term Keys and Ephemeral Keys

In the previous section, we analyzed two classic Diffie-Hellman-like PQ-AKE protocols in the eCK model. We abstracts a common design features of the protocols: separation of long-term keys and ephemeral keys to design PQ-AKE protocols. More comprehensively, long-term keys are used for identity authentication, and temporary keys are used for key exchange. Then, PQ-AKE protoclos are designed by combining identity authentication with key exchange.

Through the above analysis, we abstract the framework of such protocols shown in Fig. 4.

Protocol Framework Process. Let the communicating parties be Alice and Bob. Alice hodes the long-term key lk_A, and generates ephemeral key ek_A for the session. Meanwhile, Bob holds lk_B, and generates ephemeral key ek_B for the session. In the framework, the functions f_1, F_1 represent the identity authentication processes. The functions f_2, F_2 typically rely on specific hard problem such as LWE or RLWE.

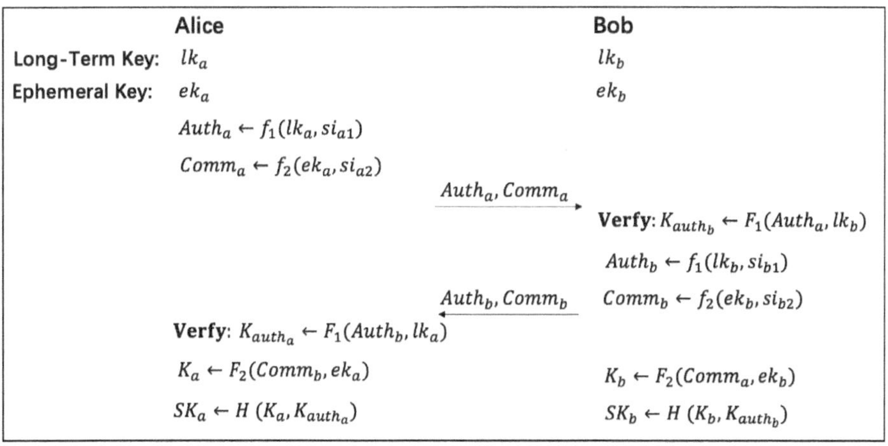

Fig. 4. The Framework Based on the Separation of Keys Long-term and Ephemeral Keys.

The protocol proceeds as follows:

1) Alice initiates the session.
   - Alice computes an authentication message: $Auth_a \leftarrow f_1(lk_a, si_{a1})$, and a key exchange message: $Comm_a \leftarrow f_2(ek_a, si_{a2})$, where si_{a1} and si_{a2} are public parameters.
   - Alice sends the message $(Auth_a,\ Comm_a)$ to Bob.
2) Bob responses the session.
   - Upon receiving $(Auth_a,\ Comm_a)$, Bob uses his long-term key lk_a to verify Alice's identity:$K_{auth_b} \leftarrow F_1(Auth_a, lk_b)$. If the authentication fails, the session is aborted.
   - Bob computes his own authentication message $Auth_b \leftarrow f_1(lk_b, si_{b1})$, and generates a key exchange message $Comm_b \leftarrow f_2(ek_b, si_{b2})$, where si_{b1} and si_{b2} are public parameters.
   - Bob replies with $(Auth_b,\ Comm_b)$.
   - Bob computes the key: $K_b \leftarrow F_2(Comm_a, ek_b)$.
   - Bob derives the session key: $SK_b \leftarrow H(K_b, K_{auth_b})$.
3) Alice completes the session.
   - Upon receiving $(Auth_b,\ Comm_b)$, Alice uses her long-term private key lk_b to verify Bob's identity:$K_{auth_a} \leftarrow F_1(Auth_b, lk_a)$. If the authentication fails, the session is aborted.
   - Alice computes the key :$K_a \leftarrow F_2(Comm_b, ek_a)$.
   - Alice derives the session key: $SK_a \leftarrow H(K_a, K_{auth_a})$.

In the end, both parties derive the same session key $SK_a = SK_b$.

Security Analysis. In the eCK model, the attack strategy proceeds as follows:

1) The adversary gets authentication message $Auth_b$ and key exchange message $Comm_a$ from the public channnel.
2) The adversary performs a $Long-term\ Key\ Reveal$ query to obtain Alice's long-term secret key lk_a, and an *Ephemeral Key Reveal* query to obtain Bob's ephemeral secret key ek_b.
3) Using lk_a, the adversary is able to compute the authentication key $K_{auth_a} \leftarrow F_1(Auth_b, lk_a)$.
4) With Bob's ephemeral key ek_b, the adversary computes the key: $K_b \leftarrow F_2(Comm_a, ek_b)$.
5) Combining the K_{auth_a} and $K_b = K_a$, the adversary can compute the session key $SK = H(K_{auth_a}, K_b)$, thereby breaking the session security.

As a result, this protocol framework is insecure in the eCK model.

4.2 A Framework Based on the Combination of Long-Term Keys and Ephemeral Keys

In order to solve the above problems, we draw inspiration from the original article [3] that proposed the eCK model. In this article, the NAXOS authenticated key exchange protocol is proposed, which is provably secure in the eCK model. Based on the design features of the NAXOS protocol, we propose a framework based on combination of long-term keys and ephemeral keys. The framework is illustrated in Fig. 5.

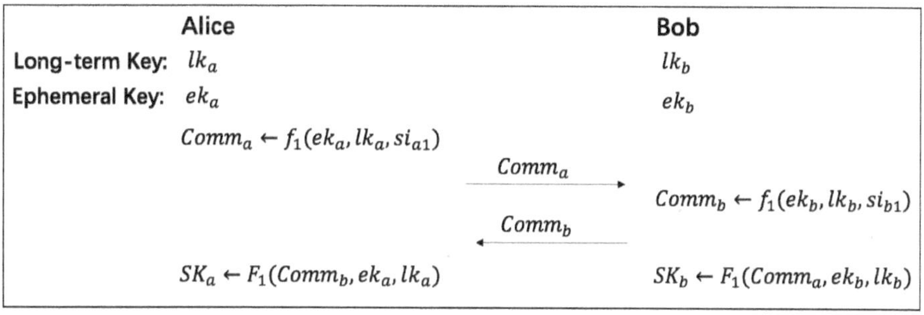

Fig. 5. The Framework Based on the Combination of Long-term Keys and Ephemeral Keys.

Protocol Framework Process. Let the two communicating parties be Alice and Bob. Alice hodes the long-term key lk_A, and generates ephemeral key ek_A for the session. Meanwhile, Bob holds long-term key lk_B, and generates ephemeral key ek_B for the session. The functions f_1 and F_1, typically rely on specific hard problem such as LWE or RLWE.

The protocol proceeds as follows:

1) Alice initiates the session:
   - Alice uses a function that combines her ephemeral key ek_a and long-term key lk_a to compute her key exchange message $Comm_a \leftarrow f_1(ek_a, lk_a, si_{a1})$, where si_{a1} is public parameter.
   - Alice sends the message $Comm_a$ to Bob.
2) Bob responds the session:
   - Upon receiving $Comm_a$, combining his ephemeral key ek_b and long-term key lk_b, Bob computes his own key exchange message $Comm_b \leftarrow f_1(ek_b, lk_b, si_{b1})$, where si_{b1} is public parameter.
   - Bob replies to Alice with $Comm_b$.
   - Bob derives the session key $SK_b \leftarrow F_2(Comm_a, ek_b, lk_b)$.
3) Alice completes the session:
   - Upon receiving $Commm_b$, Alice computes the session key: $SK_a \leftarrow F_2(Comm_b, ek_a, lk_a)$.

In the end, both Alice and Bob compute the same session key: $SK_a = SK_b$.

Remark. Moreover, the above framework uses implicit authentication, where identity verification is achieved as a byproduct of successful key exchange. If we want to further improve the authentication security of the protocol, we can also add authentication components such as MAC, digital signature, etc. This extension transforms the protocol from one that offers implicit authentication to one that provides explicit authentication.

The core strength of this framework lies in its combination of long-term and ephemeral keys, which enables authentication and key exchange to be carried out simultaneously within a unified process. By tightly coupling the long-term keys and ephemeral keys, this framework can effectively resist attacks from adversaries in the eCK model. Therefore, if we wants to design a secure Diffie-Hellman-like PQ-AKE exchange protocol in the eCK model, we can design it based on this framework.

When constructing LWE/RLWE-based PQ-AKE protocols in this framework, we cannot simply substitute LWE/RLWE with the classical hard problems (e.g., integer factorization and discrete logarithm problem) in the traditional Diffie-Hellman-like AKE protocol. Since the specific properties of LWE, especially its secret and error terms, require careful handling. If we designes the LWE-based protocol in this framework, then f_1 is designed based on the LWE and $Comm_a$ may take the form:

$$Comm_a = \boldsymbol{A} \cdot \boldsymbol{s} + \boldsymbol{e}. \tag{3}$$

To ensure security in the eCK model, both $\boldsymbol{s}$ and $\boldsymbol{e}$ must be derived by combining the long-term and ephemeral keys through secure functions, e.g.: $\boldsymbol{s} \leftarrow g_1(lk_a, ek_a)$ and $\boldsymbol{e} \leftarrow g_2(lk_a, ek_a)$. If either $\boldsymbol{s}$ or $\boldsymbol{e}$ is derived solely from a single secret (either the long-term key or ephemeral key), the protocol becomes vulnerable. Specifically:

- Case 1: Both $\boldsymbol{s}$ and $\boldsymbol{e}$ are derived from the long-term key or ephemeral key only. E.g., $\boldsymbol{s} \leftarrow g_1(lk_a)$, $\boldsymbol{e} \leftarrow g_2(ek_a)$ then an adversary can perform a $Long-term\ Key\ Reveal$ to compute $\boldsymbol{s}$, compromising the LWE assumption.
- Case 2: $\boldsymbol{s}$ is derived solely from the long-term key or ephemeral key. E.g., $\boldsymbol{s} \leftarrow g_1(ek_a)$, $\boldsymbol{e} \leftarrow g_2(lk_a, ek_a)$, then an adversary can perform a $Ephemeral\ Key\ Reveal$ to compute $\boldsymbol{s}$, compromising the LWE assumption.
- Case 3: $\boldsymbol{e}$ is derived solely from the long-term key or ephemeral key. E.g., $\boldsymbol{s} \leftarrow g_1(lk_a, ek_a)$, $\boldsymbol{e} \leftarrow g_2(lk_a)$, then an adversary can perform a $Long-term\ Key\ Reveal$ to compute $\boldsymbol{e}$, compromising the LWE assumption.

In traditional hard problems, such concerns are often absent. For instance, in the discrete logarithm problem, assume the form of $Comm_a$ is $Comm_a = g^{lk_a \cdot ek_a}$. Even if an adversary obtains the long-term secret key lk_a and $Comm_a$, the problem of recovering ek_a from $g^{lk_a \cdot ek_a}$ remains computationally hard due to self-reducibility. However, LWE lacks such self-reducibility, making the protocol more fragile under partial key exposure. Based on this framework, Li et al. [20] developed a provably secure Diffie-Hellman-like PQ-AKE protocol based on RLWE in the eCK model.

5 Conclusion

In this paper, we conducted security analysis of classic LWE/RLWE-based Diffie-Hellman-like PQ-AKE protocols in the eCK model. We summarize the frameworks based on separation of long-term keys and ephemeral keys for these protocols. Then we conclude that the PQ-AKE protocols based on this framework is insecure in the eCK model. To address above issue, we propose a framework based on combination of long-term keys and ephemeral keys. Based on this framework, the PQ-AKE can be secure in the eCK model. Our results contribute to the design of future post-quantum protocols that are robust in strong security models such as eCK. In the future works, we will focus on designing secure lattice-based PQ-AKE.

References

1. Shor, P.W.: Algorithms for quantum computation: discrete logarithms and factoring. In: 35th Annual Symposium on Foundations of Computer Science (FOCS), pp. 124–134. IEEE, Santa Fe (1994)
2. Grover, L.K.: A fast quantum mechanical algorithm for database search. In: 28th Annual ACM Symposium on Theory of Computing (STOC), pp. 212–219. ACM, Philadelphia (1996)
3. LaMacchia, B., Lauter, K., Mityagin, A.: Stronger security of authenticated key exchange. In: Susilo, W., Liu, J.K., Mu, Y. (eds.) ProvSec 2007. LNCS, vol. 4784, pp. 1–16. Springer, Heidelberg (2007). https://doi.org/10.1007/978-3-540-75670-5_1
4. Regev, O.: On lattices, learning with errors, random linear codes, and cryptography. J. ACM **56**(6), 34:1–34:40 (2009). https://doi.org/10.1145/1568318.1568324
5. Lyubashevsky, V., Peikert, C., Regev, O.: On ideal lattices and learning with errors over rings. In: Gilbert, H. (ed.) EUROCRYPT 2010. LNCS, vol. 6110, pp. 1–23. Springer, Heidelberg (2010). https://doi.org/10.1007/978-3-642-13190-5_1
6. Micciancio, D., Regev, O.: Worst-case to average-case reductions based on Gaussian measures. SIAM J. Comput. **37**(1), 267–302 (2007). https://doi.org/10.1137/S0097539705447360
7. Lindner, R., Peikert, C.: Better key sizes (and attacks) for LWE-based encryption. In: Kiayias, A. (ed.) CT-RSA 2011. LNCS, vol. 6558, pp. 319–339. Springer, Heidelberg (2011). https://doi.org/10.1007/978-3-642-19074-2_21
8. Ding, J., Xie, X., Lin, X.: A simple provably secure key exchange scheme based on the LWE problem. IACR Cryptol. ePrint Arch. 2012/688 (2012)
9. Perlner, R.: Optimizing information set decoding algorithms to attack cyclosymmetric MDPC codes. In: Mosca, M. (ed.) PQCrypto 2014. LNCS, vol. 8772, pp. 220–228. Springer, Cham (2014). https://doi.org/10.1007/978-3-319-11659-4_13
10. Bos, J.W., Costello, C., Naehrig, M., Stebila, D.: Post-quantum key exchange for the TLS protocol from RLWE. In: IEEE Symposium on Security and Privacy, pp. 553–570. IEEE, San Jose (2015). https://doi.org/10.1109/SP.2015.40
11. Gaborit, P., Hauteville, A., Tillich, J.-P.: RankSynd a PRNG based on rank metric. In: Takagi, T. (ed.) PQCrypto 2016. LNCS, vol. 9606, pp. 18–28. Springer, Cham (2016). https://doi.org/10.1007/978-3-319-29360-8_2

12. Alkim, E., Ducas, L., Pöppelmann, T., Schwabe, P.: Post-quantum key exchange - a new hope. In: USENIX Security Symposium 2016, pp. 327–343. USENIX Association, Austin (2016)
13. Bos, J.W., Costello, C., Ducas, L.: Frodo: take off the ring practical, quantum-secure key exchange from LWE. In: ACM Conference on Computer and Communications Security (CCS), pp. 1006–1018. ACM, Vienna (2016). https://doi.org/10.1145/2976749.2978425
14. Zhao, Z., Lian, H., Shen, J.: An identity-based authenticated key exchange protocol based on lattices. J. Cryptol. Res. **11**(2), 444–454 (2024)
15. Zhao, Z., Huang, L., Tang, Y.: Biometric authenticated key exchange protocol based on RLWE. Appl. Res. Comput. **37**(11), 3437–3440 (2020)
16. Daemen, J., Govaerts, R., Vandewalle, J.: Weak keys for IDEA. In: Stinson, D.R. (ed.) CRYPTO 1993. LNCS, vol. 773, pp. 224–231. Springer, Heidelberg (1994). https://doi.org/10.1007/3-540-48329-2_20
17. Canetti, R., Krawczyk, H.: Analysis of key-exchange protocols and their use for building secure channels. In: Pfitzmann, B. (ed.) EUROCRYPT 2001. LNCS, vol. 2045, pp. 453–474. Springer, Heidelberg (2001). https://doi.org/10.1007/3-540-44987-6_28
18. Nguyen, P.Q., Pointcheval, D.: Analysis and improvements of NTRU encryption paddings. In: Yung, M. (ed.) CRYPTO 2002. LNCS, vol. 2442, pp. 210–225. Springer, Heidelberg (2002). https://doi.org/10.1007/3-540-45708-9_14
19. Seyhan, K., Akleylek, S.: A new lattice-based password authenticated key exchange scheme with anonymity and reusable key. PeerJ Comput. Sci. **10**, e1791 (2024)
20. Li, W., Zhou, Y.: A quantum-resistant authenticated key exchange protocol based on RLWE and error reconciliation mechanism. In: 10th International Conference on Cloud Computing and Big Data Analytics (ICCCBDA), pp. 443–450. IEEE, Chengdu (2025)

MD-GCN: Multi-view Dependency-Enhanced Graph Convolutional Network for Aspect-Based Sentiment Analysis

Chuan Bai, Yanli Chen(✉), Jietao Zheng, Xing Zhang, and Jie Su

Guizhou Normal University, Guiyang, China
nauhc@gznu.edu.cn, yanli_276@163.com

Abstract. With the rapid development of artificial intelligence technologies, sentiment analysis has become a fundamental task in intelligent information processing. In particular, for the detection of misinformation, sentiment consistency serves as a crucial indicator. Since sentiment is often influenced by multiple factors, assessing consistency requires consideration from different aspects. However, traditional document-level or single-aspect sentiment analysis methods focus on limited perspectives, making it difficult to accurately capture genuine sentiment consistency. To address this issue, this paper proposes a novel Multi-view Dependency-enhanced Graph Convolutional Network (MD-GCN) to advance the application of aspect-based sentiment analysis (ABSA) in multi-aspect consistency evaluation. MD-GCN constructs three complementary syntactic graphs: the Token-to-Token (T2T) graph preserves global syntactic dependencies, the Token-to-Aspect (T2A) graph strengthens aspect-centered local context, and the Aspect-to-Opinion (A2O) graph explicitly links aspect terms with opinion words. These heterogeneous views are integrated through a multi-view attention fusion mechanism, which dynamically balances their contributions to highlight the most informative structural features. Experiments on three widely used benchmark datasets demonstrate that MD-GCN not only achieves significant improvements over strong baselines in terms of accuracy and macro-F1, but also substantially enhances the modeling of aspect-level sentiment consistency. The results highlight the potential of multi-aspect sentiment analysis as a reliable indicator for verifying misinformation.

Keywords: Sentiment Consistency · Aspect-based Sentiment Analysis · Graph Convolutional Network · Multi-view Fusion · Misinformation Detection

1 Introduction

With the rapid advancement of artificial intelligence and natural language processing, sentiment analysis has emerged as a fundamental task in intelligent

J. Yu and Y. Chen (Eds.): FCS 2025, CCIS 2805, pp. 205–220, 2027.
https://doi.org/10.1007/978-981-95-9514-3_14

information processing. As sentiment often involves multiple aspects that are interrelated, relying solely on holistic or single-level analysis is insufficient to uncover the genuine sentiment tendencies expressed in text. Aspect-based Sentiment Analysis (ABSA) has therefore attracted considerable attention, as it enables fine-grained identification of sentiment polarity across different aspects. For example, while a news headline may convey a positive emotion, the details in the body may implicitly reveal a negative inclination, such inconsistencies provide critical cues for downstream applications. Compared with traditional document-level or sentence-level sentiment analysis, ABSA captures the dependencies between sentiments and specific aspects with greater precision, thereby offering more authentic sentiment representations and enabling a more reliable assessment of sentiment consistency. This capability holds significant value for tasks such as misinformation detection, public opinion monitoring, and product review analysis. Most existing sentiment analysis approaches, however, remain limited to a single perspective or perform only shallow multi-view integration, making them less effective in capturing the complex, cross-aspect sentiment consistency present in real-world texts. To address this issue, we propose a novel Multi-view Dependency-enhanced Graph Convolutional Network (MD-GCN). The model constructs three complementary syntactic views: a Token-to-Token (T2T) graph that preserves global syntactic dependencies, a Token-to-Aspect (T2A) graph that strengthens aspect-centered local context, and an Aspect-to-Opinion (A2O) graph that explicitly models dependencies between aspect terms and opinion words. On this basis, a multi-view attention fusion mechanism is employed to adaptively balance the contributions of different views, enabling more comprehensive modeling of multi-aspect sentiment consistency. Extensive experimental results on multiple benchmark datasets demonstrate that MD-GCN significantly outperforms strong baselines, validating its effectiveness and generalizability in aspect-level sentiment consistency modeling and providing reliable technical support for misinformation detection. The main contributions of this paper are summarized as follows:

- We propose a novel Multi-view Dependency-enhanced Graph Convolutional Network (MD-GCN), which provides a new technical perspective for applying aspect-based sentiment analysis (ABSA) to misinformation detection.
- We design three complementary syntactic views—Token-to-Token (T2T), Token-to-Aspect (T2A), and Aspect-to-Opinion (A2O)—together with a multi-view attention fusion mechanism, enabling the effective capture of sentiment relationships and latent contradictions across different aspects.
- We conduct extensive experiments on three standard benchmark datasets, and the results demonstrate that the proposed method significantly outperforms existing models in terms of both accuracy and macro-F1, validating the practicality and generalizability of multi-aspect sentiment analysis for misinformation detection.

The remainder of the paper is structured as follows: Sect. 2 presents related work, reviewing literature relevant to aspect-level sentiment analysis models, with particular emphasis on graph neural networks and attention mechanisms. Section 3 comprehensively elaborates the model framework and each component, including methods used and computational formulas. Section 4 validates model effectiveness through extensive experimental studies, compares our approach with multiple ABSA baseline models, and discusses results, comparative analyses, and ablation studies. Section 5 addresses the advantages and limitations of the proposed model. Finally, Sect. 6 summarizes the study and briefly outlines future research directions.

2 Related Work

2.1 Attention-Based Approaches

Early ABSA models exploited attention mechanisms to extract semantic features associated with aspect words [8]. Zou et al. [35] proposed LBSA, a lexicon-based supervised attention model that integrates sentiment lexicons into a recurrent neural network, improving interpretability and performance in sentiment classification tasks. Tay et al. [20] proposed AF-LSTM, an Aspect Fusion LSTM that integrates word-aspect relationships using circular convolution and correlation to improve aspect-specific sentiment analysis, outperforming state-of-the-art models. Li et al. [9] developed the HAPN model with hierarchical attention to derive sentence-level sentiment. Tang et al. [19] enabled spontaneous learning of attention supervision signals. Wu et al. [28] proposed RA-CNN and RAO-CNN models for ABSA, using residual attention and aspect terms modules to enhance performance, achieving state-of-the-art results. Yang et al. [29] incorporated interactive aspect-attention to align opinions and aspects. Feng et al. [4] constructed an attention-assisted graph module for semantic enrichment. Zhu et al. [34] proposed a deformable convolutional network for aspect-level sentiment analysis, using adaptive receptive fields and cross-correlation attention to outperform CNN-based models on multiple datasets.

2.2 GNN-Based Approaches

Graph neural networks (GNNs), especially GCNs [22], are widely used to capture syntactic structure. Sun et al. [16] introduced CDT using GCN over BiLSTM embeddings along dependency trees. Huang and Carley [5] proposed TD-GAT for direct aspect-opinion connections via dependency graphs. Tian et al. [21] proposed A-GCN, a dependency-driven approach for relation extraction that uses attention mechanisms on dependency trees and includes dependency type information. Liang et al. [10] proposed BiSyn-GAT+, a Bi-Syntax aware Graph Attention Network that uses constituent tree syntax to model sentiment relations and outperforms state-of-the-art methods. Zhou and Law [33] proposed SRGN, a Semantic Relatedness-enhanced Graph Network that integrates semantic relatedness through an Edge-gated Graph Convolutional Network (EGCN) and an aspect-context attention module, achieving superior performance on multiple datasets. [15] designed dual-GCNs for aspect and sentence level semantics. Zhang et al. [31] proposed CLF-TrLSTM, a framework that combines

Tree-Structured LSTM and contrastive learning to improve aspect-based sentiment analysis, achieving state-of-the-art performance. Wang et al. [26] proposed SVCCL-GCN, a hybrid model that combines a self-attention variational autoencoder with capsule-based contrastive learning to jointly capture syntactic and semantic dependencies, achieving superior performance on multiple ABSA benchmarks. Phukon et al. [13] developed FakeDetectionGCN, an ABSA-based GCN framework for fake review detection that integrates aspect–sentiment relations with external semantic knowledge to identify deceptive content. Wan et al. [23] introduced KA-GCN, which integrates external knowledge and a position-aware attention mechanism into GCNs to address segmentation inconsistencies and relation strength modeling, achieving state-of-the-art results on public datasets.

3 Methodology

This section provides a detailed explanation of the proposed model and its key components. We first define the task of aspect-based sentiment analysis (ABSA), and then present our proposed model, which adopts a multi-view dependency-enhanced architecture to comprehensively capture syntactic and semantic information for fine-grained sentiment classification.

3.1 Problem Definition of ABSA

Given a sentence sequence $S = \{w_1, w_2, \ldots, w_n\}$ consisting of n words, where each w_i represents a word in the sentence, and an aspect term $A = \{a_1, a_2, \ldots, a_m\}$, where each a_j is a word forming an aspect term and $A \subseteq S$ represents a contiguous subsequence of S. A sentence may contain multiple aspect terms, each associated with an independent sentiment label. The goal of Aspect-Based Sentiment Analysis (ABSA) is to predict the sentiment polarity label $P \in \{\text{positive}, \text{negative}, \text{neutral}\}$ for each given aspect term in the sentence.

3.2 The Overall Framework of the Model

We propose a multi-view dependency-enhanced graph convolutional network, which aims to model the fine-grained interactions between aspect and opinion words by leveraging multi-view syntactic structures and multilayer graph convolution operations. As illustrated in Fig. 1, the architecture of the proposed model consists of four core modules: Contextual Embedding Module, Multi-view Graph Construction and Fusion Module, Graph Convolutional Encoding Module, Sentiment Classification Module. Specifically, **first**, the raw input sentence is processed by the contextual embedding module, which employs a Bi-LSTM encoder to capture contextual representations. **Second**, aspect-aware and self-attention mechanisms are applied to generate enhanced semantic matrices. **Third**, dependency-type matrices are constructed based on syntactic dependency trees from three views: T2A, T2T, and A2O. **Fourth**, these view-specific matrices are integrated into a unified dependency-type matrix. **Finally**, a multi-view attention mechanism is applied to aggregate the dependency matrices and produce the final multi-view attention matrix, which is then passed into the GCN layers for further representation learning.

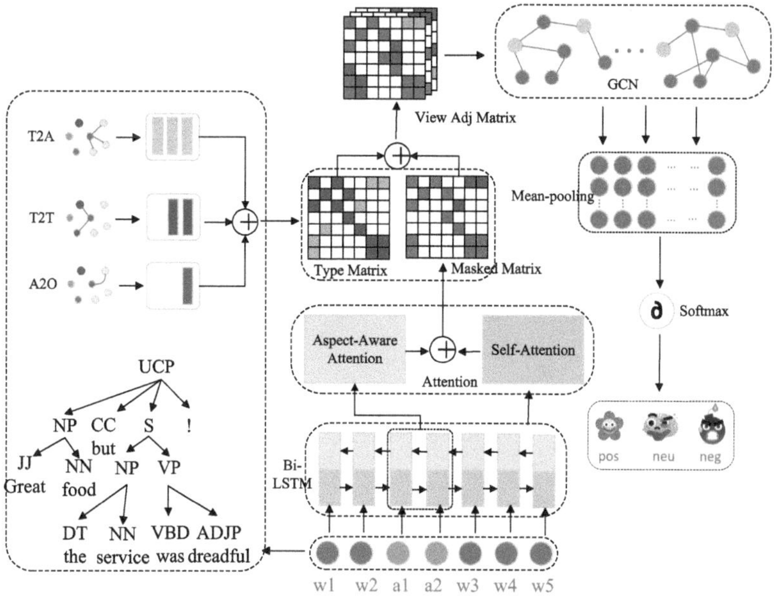

Fig. 1. The architecture of the proposed MD-GCN model.

Contextual Embedding Module. Each word is mapped to a pre-trained static embedding (e.g., GloVe or word2vec), combined with part-of-speech (POS) and position embeddings. For an input sequence $S = \{w_1, \ldots, w_n\}$, the embedding of word w_i is represented as:

$$e_i = \left[e_i^{\text{word}}; e_i^{\text{pos}}; e_i^{\text{post}}\right] \tag{1}$$

where e_i^{word}, e_i^{pos}, and e_i^{post} represent word, POS, and position embeddings, respectively. To capture contextual dependencies within the input sequence, we employ a bidirectional Long Short-Term Memory (BiLSTM) network. The BiLSTM processes the input sequence in both forward and backward directions to generate comprehensive contextual representations. The computations are defined as follows:

$$\overrightarrow{H}_{\text{lstm}} = \overrightarrow{\text{Bi-LSTM}}(H), \tag{2}$$

$$\overleftarrow{H}_{\text{lstm}} = \overleftarrow{\text{Bi-LSTM}}(H), \tag{3}$$

where H denotes the input sequence representation, typically constructed from word embeddings combined with part-of-speech and position embeddings. $\overrightarrow{H}_{\text{lstm}}$ and $\overleftarrow{H}_{\text{lstm}}$ represent the hidden states computed by the forward and backward LSTM layers, respectively. These two directional outputs are concatenated to obtain the final contextual embedding:

$$H_{\text{lstm}} = [\overrightarrow{H}_{\text{lstm}}; \overleftarrow{H}_{\text{lstm}}], \tag{4}$$

This bidirectional structure enables the model to effectively encode both preceding and succeeding contextual information for each token, which is essential for accurately capturing sentiment polarity, especially in complex aspect-based sentiment analysis scenarios.

Semantic Feature Construction. To enrich the semantic features, both aspect-aware attention and self-attention mechanisms are employed. The resulting attention matrices are:

$$A^i_{\text{aspect}} = \begin{bmatrix} a^{(i)}_{11} & a^{(i)}_{12} & \cdots & a^{(i)}_{1N} \\ a^{(i)}_{21} & a^{(i)}_{22} & \cdots & a^{(i)}_{2N} \\ \vdots & \vdots & \ddots & \vdots \\ a^{(i)}_{N1} & a^{(i)}_{N2} & \cdots & a^{(i)}_{NN} \end{bmatrix}, \quad A^i_{\text{self}} = \begin{bmatrix} s^{(i)}_{11} & s^{(i)}_{12} & \cdots & s^{(i)}_{1N} \\ s^{(i)}_{21} & s^{(i)}_{22} & \cdots & s^{(i)}_{2N} \\ \vdots & \vdots & \ddots & \vdots \\ s^{(i)}_{N1} & s^{(i)}_{N2} & \cdots & s^{(i)}_{NN} \end{bmatrix} \tag{5}$$

Equation (5) explicitly represents the aspect-aware attention matrix A^i_{aspect} and the self-attention matrix A^i_{self} for the i-th token in the sentence. Both matrices are of size $N \times N$, where N denotes the number of tokens in the input sequence. A^i_{aspect} captures the attention distribution between the i-th token and all other tokens in the sentence with respect to the given aspect. Each element $a^{(i)}_{ij}$ reflects the relevance between token w_i and token w_j under aspect-guided attention for token w_i. A^i_{self} models the semantic interactions among all tokens using self-attention, independent of aspect information. The element $s^{(i)}_{ij}$ denotes the self-attention weight between token w_i and token w_j from the perspective of token w_i. These matrices are later fused to form a unified semantic attention matrix that enriches the input representation by incorporating both global contextual cues and aspect-specific dependencies.

$$A^i_{\text{sem}} = A^i_{\text{aspect}} + A^i_{\text{self}} \tag{6}$$

$$A^i_{\text{mask}} = \text{softmax}(A^i_{\text{sem}} + M^i). \tag{7}$$

where each $A^i_{\text{sem}} \in \mathbb{R}^{N \times N}$ denotes the semantic enhancement for token i. The enhanced semantic matrix A^i_{sem} is then integrated with syntactic graph structures for multi-view representation learning. Here, we use the softmax operator to apply the distance mask M^i to the semantic feature matrix A^i_{sem}. The entries with value $-\infty$ in M^i will suppress the corresponding entries in A^i_{sem} to zero after softmax. As i increases, A^i_{mask} can capture more global information.

Multi-view Syntactic Graph Construction. To better capture syntactic structures and aspect-related semantics in aspect-based sentiment analysis (ABSA), we design three distinct syntactic graph views to guide graph neural network learning. These include: Token-to-Token (T2T) graph, Token-to-Aspect (T2A) graph, and Aspect-Opinion (A2O) graph. These views provide complementary structural priors from multiple linguistic perspectives.

1) Token-to-Token (T2T) Graph. The T2T graph is constructed based on dependency parsing, which analyzes the grammatical structure of a sentence by establishing binary dependency relations between words, where one word serves as the head and the other as the dependent. If a dependency arc exists between two tokens, an edge is created between them in the graph, and the adjacency matrix is defined as shown in Eq. (8). For example, in the sentence "Great food but the service was dreadful!", the word "Great" modifies "food", indicating a dependency relation. Therefore, the corresponding entry A_{ij}^{T2T} in the adjacency matrix is set to 1, reflecting a direct syntactic connection between the two tokens. The adjacency matrix is defined as:

$$A_{ij}^{\mathrm{T2T}} = \begin{cases} 1, & \text{if a dependency arc exists between } w_i \text{ and } w_j \\ 0, & \text{otherwise} \end{cases} \tag{8}$$

This graph captures the basic syntactic dependency structure within the sentence.

2) Token-to-Aspect (T2A) Graph. The T2A graph explicitly connects context words to aspect terms. Each non-aspect token is linked to its nearest aspect term within a predefined distance threshold d (e.g., 2 or 3). The adjacency matrix is defined as shown in Eq. (9). For example, in the sentence *"Great food but the service was dreadful!"*, suppose "service" is the aspect term. The word "was", which is not part of the aspect but lies within the distance threshold d, is connected to "service" in the graph. Consequently, the adjacency entry A_{ij}^{T2A} is set to 1, where w_i = "was" and w_j = "service", indicating a local contextual relationship between the non-aspect word and the aspect. The adjacency matrix is given by:

$$A_{ij}^{\mathrm{T2A}} = \begin{cases} 1, & w_j \in A,\ w_i \notin A, \text{ and } \mathrm{distance}(w_i, w_j) \leq d \\ 0, & \text{otherwise} \end{cases} \tag{9}$$

This view emphasizes the syntactic and positional importance of aspect terms by simulating an attention distribution centered on them as the semantic anchors of a sentence.

3) Aspect-to-Opinion (A2O) Graph. To directly model the link between sentiment expressions and aspect terms, we introduce the Aspect-to-Opinion (A2O) graph. Sentiment words are identified using a predefined sentiment lexicon (e.g., SentiWordNet), and are directly connected to aspect terms. The adjacency matrix is defined as shown in Eq. (10). For example, in the sentence *"Great food but the service was dreadful!"*, suppose "service" is the aspect term and "dreadful" is recognized as a sentiment word. An edge is then created from "dreadful" to "service", and the corresponding adjacency matrix entry A_{ij}^{A2O} is set to 1, where w_i = "dreadful" and w_j = "service". This connection explicitly captures the sentiment-aspect relationship for downstream sentiment classification. The adjacency matrix is defined as:

$$A_{ij}^{\mathrm{A2O}} = \begin{cases} 1, & w_i \in O,\ w_j \in A \\ 0, & \text{otherwise} \end{cases} \tag{10}$$

This structure bridges long-distance sentiment-aspect dependencies and mitigates the impact of sentiment ambiguity. Given the four adjacency matrices $\{A^{(T2T)}, A^{(T2A)}, A^{(A2O)}\}$, each view is processed independently through a GCN encoder to obtain the node representations $\{H^{(1)}, H^{(2)}, H^{(3)}\}$. A multi-view attention fusion mechanism is employed to compute the final node representation H:

$$A_{\mathrm{type}} = \sum_{v=1}^{3} \alpha_v A^{(v)} \tag{11}$$

where α_v denotes the attention weight for each view, learned dynamically from the input using a two-layer feedforward network. This enables the model to assign appropriate importance to each syntactic perspective during learning.

Adjacency Matrix Composition. At this point, we have obtained the semantic feature matrices $\{A_{\mathrm{sem}}^1, \ldots, A_{\mathrm{sem}}^P\}$ and the dependency type information matrix A_{type}. We then aggregate the semantic and syntactic information to construct the i-th view adjacency matrix for the GNN as follows:

$$A^i = A_{\mathrm{mask}}^i + A_{\mathrm{type}}, \tag{12}$$

Finally, we obtain P distinct view-specific adjacency matrices $\{A^1, \ldots, A^P\}$, each reflecting different combinations of semantic and syntactic information.

GNN and Multi-view Attention Mechanism. After obtaining the sentence embedding H and P distinct adjacency matrices $\{A^1, \ldots, A^P\}$, the output embedding of the l-th GNN layer is computed as:

$$H^l = \sigma\left(\mathrm{AGG}_{i=1}^{P}(A^i H^{l-1}) W^l\right), \quad H^0 = H, \tag{13}$$

where $H^l \in \mathbb{R}^{N \times D}$ denotes the output of the l-th layer, $W^l \in \mathbb{R}^{D \times D}$ is the learnable weight matrix for the l-th layer, and σ is the activation function. The operator $\mathrm{AGG}_{i=1}^{P}$ refers to the proposed multi-view attention mechanism, which is defined as follows:

$$H_{\mathrm{avg}}^i = \frac{1}{N \times D} \sum_{r=1}^{N} \sum_{s=1}^{D} \left(A^i H^{l-1}\right)_{r,s}, \tag{14}$$

$$\alpha_{\text{view}} = W_2 \left(\sigma \left(W_1 H_{\text{avg}} \right) \right), \tag{15}$$

$$\text{AGG}_{i=1}^{P}(A^i H^{l-1}) = \sum_{i=1}^{P} \alpha_{\text{view}}^{i} A^i H^{l-1}, \tag{16}$$

where $H_{\text{avg}}^{i} \in \mathbb{R}$ is the average value of the i-th view, $H_{\text{avg}} \in \mathbb{R}^{P \times 1}$ collects all view-wise averages, and $\alpha_{\text{view}} \in \mathbb{R}^{P \times 1}$ denotes the view weights. W_1 and W_2 are learnable parameters, and α_{view}^{i} is the i-th element of α_{view}.

The multi-view attention mechanism adaptively extracts useful information from various views, leveraging their diversity and complementarity to enhance the representation capability for downstream tasks. Furthermore, it reduces the noise introduced by incorporating excessive views [32].

Output Layer. After obtaining the final output representation, it is passed through a fully connected layer, followed by a softmax classifier. This layer maps the features into a three-dimensional sentiment classification space (positive, negative, neutral) to predict polarity and obtain the sentiment probability distribution corresponding to each aspect term. The mapping is defined as:

$$Y(a) = \text{softmax}(WU + b) \tag{17}$$

4 Experiments

In this section, comprehensive experiments are conducted on three benchmark datasets, accompanied by detailed demonstrations. First, various experimental settings are introduced, including discussions on the datasets and evaluation metrics. Next, the baseline models used for comparison with the proposed model are presented. Then, the experimental results are reported and thoroughly analyzed. Finally, the impact of different factors on the model's performance is discussed in detail.

4.1 Experimental Setup

Datasets. This study evaluates the proposed model using three publicly available benchmark datasets. The **Restaurant14** and **Laptop14** datasets are obtained from the SemEval-2014 task [14]. Each instance in these datasets contains one or more aspect terms, with many sentences incorporating implicit sentiment information. The third dataset, **Twitter** [2], consists of user-generated posts from the Twitter social media platform, where each sentence contains only a single aspect. The statistical distribution of sentiment polarities (i.e., positive, negative, and neutral) across all three datasets is summarized in Table 1.

Table 1. Sentiment distribution for training and testing sets across datasets

Dataset	Positive		Neutral		Negative	
	Train	Test	Train	Test	Train	Test
Twitter	1561	173	3127	346	1560	173
Restaurant14	2164	728	637	196	807	196
Laptop14	994	341	464	169	870	128

Parameter Settings. In the experiments, token embeddings are initialized with 300-dimensional pre-trained GloVe vectors [12], which are concatenated with 30-dimensional part-of-speech and position embeddings. The combined embeddings are fed into a BiLSTM encoder with a dropout rate of 0.7. The batch size is set to 16, and the number of GCN layers is fixed at 2. The Adam optimizer is used with a learning rate of 0.002.

4.2 Baselines

To comprehensively assess the effectiveness of our proposed model, we compare it against a variety of strong baseline methods, categorized into two groups:

1) Attention-Based Methods

- **TD-LSTM** [17]: Employs two target-specific LSTM networks to separately encode the left and right contexts of the aspect term.
- **ATAE-LSTM** [27]: Uses attention-based LSTM and incorporates aspect information into the input for better sentiment alignment.
- **MemNet** [18]: Introduces multi-hop attention mechanisms to model contextual word importance with respect to the target.
- **AOA** [6]: Applies attention-over-attention to jointly model aspects and contextual representations.
- **RAM** [1]: Utilizes multiple memory slots and attention hops to capture interactions between opinion words and aspects.
- **IAN** [11]: Designs interactive attention mechanisms between context and aspect representations to enhance mutual learning.

2) GNN-Based Methods

- **MGAN** [3]: Introduces fine- and coarse-grained attention networks to enhance word-level interactions.
- **ASCNN** [30]: Uses convolutional layers over dependency structures to extract sentiment features.
- **ASGCN** [30]: Combines LSTM and GCN over syntactic dependency trees to integrate syntactic features.

- **R-GAT** [24]: Incorporates relational attention to regulate edge importance in the dependency graph.
- **CRF-GCN** [7]: Combines conditional random fields and GCN, introducing global nodes for sentiment prediction.
- **IDGNN** [25]: Proposes dual GCNs to fuse semantic and syntactic information for complex sentence modeling.
- **SVCCL-GCN** [26]: proposed SVCCL-GCN, a hybrid GCN that uses variational autoencoder and contrastive learning to improve sentiment classification.

4.3 Experimental Results and Analysis

The comparative experimental results of all models are shown in Table 2. In the table, the best performance on each dataset is highlighted in bold, while the second-best is underlined. Missing results are indicated by a dash "-".

Table 2. Performance comparison on three benchmark datasets (ACC and F1).

Model	Restaurant14		Laptop14		Twitter	
	ACC	F1	ACC	F1	ACC	F1
Attention						
TD-LSTM	78.00	66.73	71.83	68.43	70.80	69.00
ATAE-LSTM	78.60	67.02	68.88	69.93	68.84	66.60
MemNet	78.16	65.83	70.33	64.09	69.65	67.68
AOA	79.94	70.42	72.60	67.50	72.30	70.20
RAM	80.23	70.80	74.49	71.35	69.36	67.30
IAN	79.26	70.09	72.05	67.38	72.50	70.81
GNN						
MGAN	81.49	71.48	76.21	71.42	74.62	73.53
ASCNN	81.73	73.10	72.62	66.72	71.05	69.45
ASGCN	80.77	72.02	75.55	71.05	72.15	70.40
R-GAT	83.30	76.08	77.42	73.76	75.57	73.82
CRF-GCN	82.71	73.87	75.83	74.78	–	–
IDGNN	83.40	76.37	78.07	74.17	**75.94**	74.37
SVCCL-GCN	**84.05**	74.56	77.70	74.96	75.39	73.70
Ours	83.63	**77.25**	**78.67**	**75.39**	75.68	**74.57**

From Table 2, it can be observed that **attention-based models** (e.g., TD-LSTM, ATAE-LSTM, MemNet, AOA, RAM, IAN) generally achieve lower performance compared with graph-based models. On the **Restaurant14** dataset, their F1 scores mostly range from 65.83 (MemNet) to 70.80 (RAM), and on **Laptop14**, from 64.09 (MemNet) to 71.35 (RAM). Although AOA and IAN slightly improve the performance, they still fall short of graph-based methods. On the **Twitter** dataset, their F1 scores fluctuate between 66.60 (ATAE-LSTM) and 70.81 (IAN), reflecting their limited ability to handle informal and noisy text. These results confirm that attention-based architectures, though capable of modeling contextual interactions, struggle to effectively capture syntactic dependencies crucial for fine-grained sentiment reasoning.

In contrast, **graph-based models** exhibit significant and consistent improvements across all datasets. Models such as MGAN, ASGCN, and R-GAT demonstrate that integrating syntactic dependency graphs enhances the ability to capture structural sentiment relations. For example, R-GAT achieves strong results with F1 scores of 76.08, 73.76, and 73.82 on Restaurant14, Laptop14, and Twitter, respectively. Similarly, SVCCL-GCN delivers competitive and stable performance across datasets, with F1 scores of 74.56, 74.96, and 73.70. IDGNN attains the highest accuracy on Twitter (75.94) and second-best F1 (74.37), verifying the advantage of graph-enhanced dependency reasoning.

Our proposed model achieves the **highest F1 scores on all three datasets**, reaching 77.25 on Restaurant14, 75.39 on Laptop14, and 74.57 on Twitter, which surpasses all existing baselines. In terms of accuracy, it ranks **first on Laptop14 (78.67)** and **second on both Restaurant14 (83.63 vs. 84.05 of SVCCL-GCN)** and **Twitter (75.68 vs. 75.94 of IDGNN)**, showing a near-optimal balance between precision and recall. These consistent gains demonstrate that the proposed model not only excels on well-structured reviews (Restaurant14, Laptop14) but also maintains robustness in noisy, user-generated text (Twitter).

The performance improvements can be attributed to the proposed **multi-view syntactic modeling framework**. Specifically, the Token-to-Token (T2T) view captures global grammatical relations, the Token-to-Aspect (T2A) view reinforces local contextual relevance around aspect terms, and the Aspect-to-Opinion (A2O) view directly links opinion expressions to their target aspects. The attention-fusion mechanism adaptively weighs these views, allowing the model to emphasize the most informative syntactic cues during sentiment reasoning.

Overall, the experimental results verify that our model effectively integrates multi-view syntactic dependencies to enhance sentiment prediction. It achieves consistent improvements in both accuracy and F1 across datasets of varying linguistic complexity, demonstrating strong robustness, adaptability, and generalization ability in aspect-based sentiment analysis tasks.

5 Conclusion

This paper proposes a novel Multi-view Dependency-enhanced Graph Convolutional Network (MD-GCN), designed to enhance the capability of aspect-based sentiment analysis (ABSA) in sentiment consistency modeling and to explore its application in content security tasks such as misinformation detection. The proposed framework constructs three complementary syntactic graphs—Token-to-Token (T2T), Token-to-Aspect (T2A), and Aspect-to-Opinion (A2O)—and integrates them through a multi-view attention fusion mechanism, effectively capturing sentiment dependencies across different aspects. Experimental results on benchmark datasets, including Restaurant14, Laptop14, and Twitter, demonstrate that MD-GCN consistently achieves significant improvements over strong baselines in terms of both accuracy and macro-F1, highlighting the advantages of multi-view modeling in capturing complex sentiment relationships. These findings not only underscore the reliability of ABSA as an important indicator for verifying the authenticity of misinformation but also provide a practical solution for misinformation detection in broader contexts of social governance and information security.

6 Limitations and Future Work

Although MD-GCN demonstrates strong performance in modeling aspect-level sentiment and offers promising potential for sentiment consistency analysis in fake news detection, several limitations remain. First, the model is highly dependent on the accuracy of syntactic parsers and sentiment lexicons. Parsing errors or incomplete coverage of sentiment-bearing words may negatively affect the reliability of the T2T and A2O views. Second, the current framework primarily focuses on sentence-level syntactic structures, without incorporating higher-order discourse relations or long-range semantic dependencies that are often critical in deceptive or manipulative narratives. Future work will proceed in several directions. One priority is the integration of syntactic error correction mechanisms and lexicon expansion strategies to improve robustness in noisy and adversarial environments. Another avenue is to combine the proposed multi-view framework with large-scale pre-trained language models (e.g., BERT, GPT-4) to capture deeper semantic cues and contextual signals relevant for detecting subtle inconsistencies in fake news. Furthermore, extending MD-GCN to multilingual and domain-specific corpora—such as cross-lingual news articles, legal documents, or medical records—will be essential for evaluating the model's adaptability in diverse security-critical applications. Finally, we envision coupling sentiment consistency analysis with causal reasoning and multi-modal data (e.g., text-image pairs in misinformation) to build more comprehensive and trustworthy systems for combating online misinformation and enhancing information security.

Acknowledgements. This work was supported by the National Key Research and Development Program of China (No. 2023YFC3304504), the 2024 Xizang Autonomous Region Central Guided Local Science and Technology Development Fund Project (No. XZ202401YD0015), and the Basic Research Program for Natural Science of Guizhou Province (No. QIANKEHEJICHU-ZD[2025]-043).

References

1. Chen, P., Sun, Z., Bing, L., Yang, W.: Recurrent attention network on memory for aspect sentiment analysis. In: Proceedings of the 2017 Conference on Empirical Methods in Natural Language Processing, pp. 452–461 (2017)
2. Dong, L., Wei, F., Tan, C., Tang, D., Zhou, M., Xu, K.: Adaptive recursive neural network for target-dependent twitter sentiment classification. In: Proceedings of the 52nd Annual Meeting of the Association for Computational Linguistics (volume 2: Short papers), pp. 49–54 (2014)
3. Fan, F., Feng, Y., Zhao, D.: Multi-grained attention network for aspect-level sentiment classification. In: Proceedings of the 2018 Conference on Empirical Methods in Natural Language Processing, pp. 3433–3442 (2018)
4. Feng, S., Wang, B., Yang, Z., Ouyang, J.: Aspect-based sentiment analysis with attention-assisted graph and variational sentence representation. Knowl.-Based Syst. **258**, 109975 (2022)
5. Huang, B., Carley, K.M.: Syntax-aware aspect level sentiment classification with graph attention networks arXiv preprint arXiv:1909.02606 (2019)
6. Huang, B., Ou, Y., Carley, K.M.: Behavioral-cultural modeling and prediction and behavior representation in modeling and simulation. In: International Conference on Social Computing, Aspect level sentiment classification with attention-over-attention neural networks, pp. 197–206. Springer, Cham (2018)
7. Huang, B., Zhang, J., Ju, J., Guo, R., Fujita, H., Liu, J.: CRF-GCN: an effective syntactic dependency model for aspect-level sentiment analysis. Knowl.-Based Syst. **260**, 110125 (2023)
8. Huang, X., Li, J., Wu, J., Chang, J., Liu, D., Zhu, K.: Flexibly utilizing syntactic knowledge in aspect-based sentiment analysis. Inf. Process. Manage. **61**(3), 103630 (2024)
9. Li, L., Liu, Y., Zhou, A.: Hierarchical attention based position-aware network for aspect-level sentiment analysis. In: Proceedings of the 22nd Conference on Computational Natural Language Learning, pp. 181–189 (2018)
10. Liang, S., Wei, W., Mao, X.L., Wang, F., He, Z.: Bisyn-gat+: Bi-syntax aware graph attention network for aspect-based sentiment analysis. arXiv preprint arXiv:2204.03117 (2022)
11. Ma, D., Li, S., Zhang, X., Wang, H.: Interactive attention networks for aspect-level sentiment classification. arXiv preprint arXiv:1709.00893 (2017)
12. Pennington, J., Socher, R., Manning, C.D.: Glove: global vectors for word representation. In: Proceedings of the 2014 Conference on Empirical Methods in Natural Language Processing (EMNLP), pp. 1532–1543 (2014)
13. Phukon, P., Potikas, P., Potika, K.: Detecting fake reviews using aspect-based sentiment analysis and graph convolutional networks. Appl. Sci. Basel **15**(7) (2025). https://doi.org/10.3390/app15073771
14. Pontiki, M., Galanis, D., Pavlopoulos, J., Papageorgiou, H., Androutsopoulos, I., Manandhar, S.: SemEval-2014 task 4: aspect based sentiment analysis. In: Nakov, P., Zesch, T. (eds.) Proceedings of the 8th International Workshop on Semantic Evaluation (SemEval 2014), pp. 27–35. Association for Computational Linguistics, Dublin, Ireland (2014). https://doi.org/10.3115/v1/S14-2004, https://aclanthology.org/S14-2004/
15. Shang, W., et al.: Aspect-level sentiment analysis based on aspect-sentence graph convolution network. Inf. Fusion **104**, 102143 (2024)

16. Sun, K., Zhang, R., Mensah, S., Mao, Y., Liu, X.: Aspect-level sentiment analysis via convolution over dependency tree. In: Proceedings of the 2019 Conference on Empirical Methods in Natural Language Processing and the 9th International Joint Conference on Natural Language Processing (EMNLP-IJCNLP), pp. 5679–5688 (2019)
17. Tang, D., Qin, B., Feng, X., Liu, T.: Effective LSTMs for target-dependent sentiment classification. arXiv preprint arXiv:1512.01100 (2015)
18. Tang, D., Qin, B., Liu, T.: Aspect level sentiment classification with deep memory network. arXiv preprint arXiv:1605.08900 (2016)
19. Tang, J., et al.: Progressive self-supervised attention learning for aspect-level sentiment analysis. arXiv preprint arXiv:1906.01213 (2019)
20. Tay, Y., Tuan, L.A., Hui, S.C.: Learning to attend via word-aspect associative fusion for aspect-based sentiment analysis. In: Proceedings of the AAAI Conference on Artificial Intelligence, vol. 32 (2018)
21. Tian, Y., Chen, G., Song, Y., Wan, X.: Dependency-driven relation extraction with attentive graph convolutional networks. In: Proceedings of the 59th Annual Meeting of the Association for Computational Linguistics and the 11th International Joint Conference on Natural Language Processing (Volume 1: Long Papers), pp. 4458–4471 (2021)
22. Veličković, P., Cucurull, G., Casanova, A., Romero, A., Lio, P., Bengio, Y.: Graph attention networks. arXiv preprint arXiv:1710.10903 (2017)
23. Wan, B., Wu, P., Han, P., Li, G.: Aspect-based sentiment analysis by knowledge and attention integrated graph convolutional network. Appl. Soft Comput. **171** (2025). https://doi.org/10.1016/j.asoc.2025.112763
24. Wang, K., Shen, W., Yang, Y., Quan, X., Wang, R.: Relational graph attention network for aspect-based sentiment analysis. arXiv preprint arXiv:2004.12362 (2020)
25. Wang, P., Tao, L., Tang, M., Wang, L., Xu, Y., Zhao, M.: Incorporating syntax and semantics with dual graph neural networks for aspect-level sentiment analysis. Eng. Appl. Artif. Intell. **133**, 108101 (2024)
26. Wang, X., Liu, L., Chen, Z., Wang, H., Yu, B.: Graph convolutional network based on self-attention variational autoencoder and capsule contrastive learning for aspect-based sentiment analysis. Expert Syst. Appl. **279**, 127172 (2025)
27. Wang, Y., Huang, M., Zhu, X., Zhao, L.: Attention-based LSTM for aspect-level sentiment classification. In: Proceedings of the 2016 Conference on Empirical Methods in Natural Language Processing, pp. 606–615 (2016)
28. Wu, C., et al.: Residual attention and other aspects module for aspect-based sentiment analysis. Neurocomputing **435**, 42–52 (2021)
29. Yang, B., Han, D., Zhou, R., Gao, D., Wu, G.: Aspect opinion routing network with interactive attention for aspect-based sentiment classification. Inf. Sci. **616**, 52–65 (2022)
30. Zhang, C., Li, Q., Song, D.: Aspect-based sentiment classification with aspect-specific graph convolutional networks. arXiv preprint arXiv:1909.03477 (2019)
31. Zhang, Q., Wang, S., Li, J.: A contrastive learning framework with tree-LSTMs for aspect-based sentiment analysis. Neural Process. Lett. **55**(7), 8869–8886 (2023)
32. Zhang, Z., Zhou, Z., Wang, Y.: SSEGCN: syntactic and semantic enhanced graph convolutional network for aspect-based sentiment analysis. In: Proceedings of the 2022 conference of the North American Chapter of the association for computational linguistics: human language technologies, pp. 4916–4925 (2022)
33. Zhou, T., Law, K.M.: Semantic relatedness enhanced graph network for aspect category sentiment analysis. Expert Syst. Appl. **195**, 116560 (2022)

34. Zhu, C., Yi, B., Luo, L.: Base on contextual phrases with cross-correlation attention for aspect-level sentiment analysis. Expert Syst. Appl. **241**, 122683 (2024)
35. Zou, Y., Gui, T., Zhang, Q., Huang, X.J.: A lexicon-based supervised attention model for neural sentiment analysis. In: Proceedings of the 27th International Conference on Computational Linguistics, pp. 868–877 (2018)

Towards Security and Efficient Cloud Resource Scheduling with Double Deep Q-Networks

Minglong He, Ling Xiong, and Zhicai Liu(✉)

School of Computer and Software Engineering, Xihua University, Chengdu, China
liuzc.xhu@foxmail.com

Abstract. Resource scheduling in cloud computing environments is a highly complex task. On the one hand, the complexity stems from the diverse optimization challenges posed by various cloud infrastructures; on the other hand, user requests in these environments are highly dynamic. Ineffective scheduling strategies can lead to increased task processing times, which in turn may result in a higher number of Service Level Agreement (SLA) violations. Such performance degradation directly undermines cloud security by creating resource starvation for security agents (e.g., intrusion detection systems) and prolonging task exposure windows, thereby increasing the attack surface. Although heuristic algorithms have been widely used, they often exhibit significant limitations when dealing with large-scale systems, thereby diminishing their practical effectiveness. To address these challenges, this paper proposes a scheduling method based on the Double Deep Q-Network (DDQN) framework. The proposed method is designed for a discrete action space and incorporates a dynamic reward function capable of adaptively reflecting the quality of current scheduling decisions. While directly optimizing for performance metrics, this function inherently promotes a more stable and predictable resource environment, which is a fundamental prerequisite for robust cloud security. Experiments are conducted in a simulation environment built using SimPy, evaluating the proposed approach under low, medium, and high workload scenarios. The method is benchmarked against FirstMax, Random, A2C, PPO, and DQN algorithms. Performance is assessed based on metrics including makespan, average task completion time, average weighted turnaround time, and number of SLA violations. Experimental results demonstrate that the DDQN-based scheduling algorithm consistently outperforms all baseline methods across all workload levels. Compared to the second-best performing method, FirstMax, the proposed approach reduced makespan by an average of 1.15%, decreased average task completion time by an average of 1.24%, improved average weighted turnaround time by an average of 6.36%, and reduced SLA violations by 24.87% on average.

Keywords: Cloud Computing · Cloud Security · Deep Reinforcement Learning · Resource Scheduling · Double Deep Q-Networks

J. Yu and Y. Chen (Eds.): FCS 2025, CCIS 2805, pp. 221–236, 2027.
https://doi.org/10.1007/978-981-95-9514-3_15

1 Introduction

Cloud computing provides users with convenient and cost-effective computational resources, enabling them to access a wide range of cloud services via the Internet without the need to purchase or maintain physical hardware [1]. Cloud computing offers diverse services, including computing, storage, software, and function invocation. Two key technologies underpin the powerful capabilities of cloud computing: virtualization and distributed computing [2]. Virtualization technology allows physical resources to be abstracted into multiple independent units, each of which can operate like a physical machine while sharing the underlying hardware. Distributed computing technology enables multiple independent computing nodes within a system to collaborate over a network, thereby improving system throughput. These technologies have empowered the development of cloud computing but have also significantly increased the complexity of resource management in such environments. Furthermore, the dynamic nature of user demands and the constraints imposed by cloud service-level agreements (SLAs) further complicate cloud resource management.

Heuristic resource scheduling approaches, which leverage flexible and efficient approximation methods, have been successfully applied to large-scale resource scheduling optimization problems in cloud computing environments [3]. Common heuristic algorithms include greedy algorithms, genetic algorithms, and ant colony optimization. These methods are well-suited for dynamic cloud environments, capable of handling multi-objective optimization problems, and demonstrate strong fault tolerance. However, heuristic scheduling algorithms also exhibit certain limitations. For example, they may become trapped in local optima during the scheduling process, leading to inefficient resource utilization. In addition, the complexity of these algorithms increases with the number of optimization objectives [4].

The advent of deep learning has led to an expansion of fields now leveraging deep learning algorithms to address numerous challenges in real-world applications [5]. In the context of cloud computing resource scheduling optimization, deep reinforcement learning algorithms utilize historical data monitored by the cloud computing center for training, thereby generating scheduling strategies tailored to specific cluster environments and optimization objectives [6]. The deep reinforcement learning-based resource scheduling process comprises three stages: information collection, decision making, and feedback learning. During the information collection phase, the scheduling center gathers user task requests and the resource status of the current server cluster. In the decision-making phase, the deep reinforcement learning algorithm employs deep neural networks to process the collected data and formulate scheduling decisions. Finally, in the feedback learning phase, the scheduling center executes the scheduling according to the final decision, calculates the reward for this decision, and updates the parameters of the neural network based on the reward feedback. The implementation of a deep reinforcement learning algorithm for cloud computing resource scheduling is illustrated in Fig. 1.

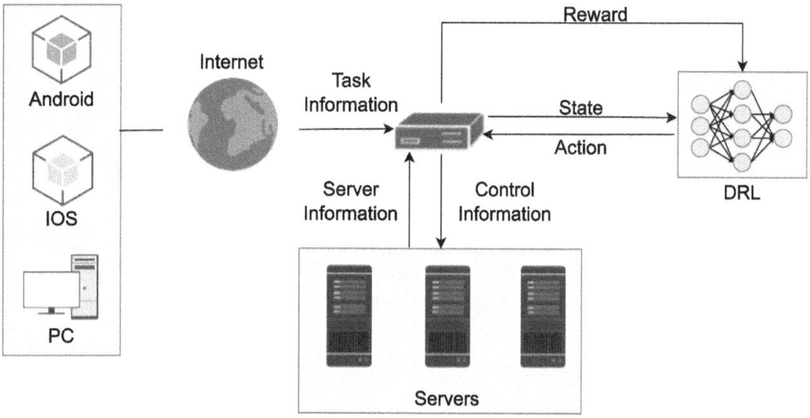

Fig. 1. The scheduling center is responsible for the collection of task and server information. Subsequently, the scheduling center utilizes a deep neural network to formulate a decision. Following the decision's execution by the scheduling center, the neural network receives a reward to facilitate its continual improvement.

Deep reinforcement learning (DRL) algorithms consist of three key elements: state, action space, and reward. The action space can be categorized into discrete and continuous types. In a discrete action space, resource scheduling decisions are made by assigning tasks to VMs or selecting specific VMs. Conversely, continuous action spaces enable dynamic system-wide performance tuning through the allocation of specific CPU or memory resources. A commonly used DRL algorithm for discrete action spaces is Deep Q-Network (DQN) [8], which leverages deep neural networks to process raw state inputs directly, addressing the limitations of traditional Q-learning that requires handcrafted features. Additionally, DQN enhances data efficiency and algorithmic stability by storing historical samples and employing random sampling for training. However, DQN suffers from an overestimation issue: the target network selects actions based on the maximum Q-value, and over time, the discrepancy between the target and the original networks increases, potentially leading the scheduling policy into local optima. To mitigate this problem, the Double Deep Q-Network (DDQN) algorithm [9] decouples action selection and evaluation within the target network. In DDQN, the target network focuses solely on evaluating actions, while all action selections are made by the original network. This design enables the original network to more accurately reflect the current scheduling state during updates, thereby improving responsiveness to environmental changes and addressing the overestimation inherent in DQN. Therefore, this paper proposes a cloud resource scheduling method based on a discrete action space integrated with the DDQN algorithm. To enable DDQN to dynamically assess the quality of scheduling decisions during training, a reward function is designed to reflect the effectiveness of current scheduling decisions in real time. The code for the cloud resource scheduling method based on DDQN is available at https://github.com/HelloDVA/DDQN. The main contributions of this work are as follows:

1) A cloud computing resource scheduling method based on a discrete action space is designed using the DDQN algorithm. This method leverages deep reinforcement learning to address the challenges of resource management and security in cloud computing environments.
2) A virtual cloud computing simulation environment is constructed using SimPy to emulate the resource scheduling process. The design of the reward function encapsulates the critical value of each decision in upholding SLA compliance and mitigating security risks and reinforces the overall security posture by ensuring critical services and security mechanisms operate within their required resource boundaries.
3) The proposed scheduling algorithm is evaluated under three different task arrival scenarios. Makespan, average task completion time, average weighted turnaround time, and number of SLA violations are used as performance metrics. Comparative experiments are conducted against several baseline scheduling algorithms, including FirstMax, Random, A2C, PPO, and DQN.

2 Related Work

2.1 Traditional Resource Scheduling Methods

Traditional resource scheduling approaches include rule-based and heuristic-based algorithms. Rule-based scheduling methods encompass First-Come-First-Served (FCFS), Random Scheduling, and Shortest Job First (SJF), among others. Heuristic-based methods include Genetic Algorithms (GA) [11], Ant Colony Optimization (ACO) [13], and Particle Swarm Optimization (PSO) [12]. While these algorithms are effective in addressing resource scheduling problems in cloud computing environments, they often exhibit limitations when faced with complex and dynamic scenarios.

To address these challenges, researchers have proposed improvements tailored to specific environments, leading to enhanced performance in particular use cases. To mitigate resource contention in cloud environments, the authors in [10] proposed a weighted random scheduling strategy that assigns varying weights to virtual machines (VMs) based on their performance levels, thereby balancing uneven resource allocation. To address the tendency of GAs to converge to local optima in cloud environments, the authors in [14] introduced an improved GA-based scheduling strategy. Similarly, in [15], the author proposed an improved PSO (IPSO) algorithm to enhance the global search performance and avoid premature convergence, demonstrating higher efficiency in locating global optima. In order to reduce the computational complexity of ACO in cloud environments, the author in [16] introduced the Path-Critical-Point Ant Colony Optimization (PCP-ACO) algorithm. This method first performs a topological sort of tasks and then allocates resources using ACO, thereby accelerating convergence.

2.2 Resource Scheduling Based on Deep Reinforcement Learning

Reinforcement learning (RL) methods can be broadly categorized into value-based and policy-based approaches based on their strategy iteration mechanisms. DRL integrates deep neural networks into traditional RL, enabling the handling of high-dimensional and complex state spaces, which is crucial for solving complex environmental challenges. Both value-based and policy-based methods have been extensively applied in cloud resource scheduling tasks. Value-based RL approaches, such as Q-learning [17], SARSA [18], and Deep Q-Network [19], estimate a value function to evaluate state-action pairs and assist agents in selecting optimal policies. In contrast, policy-based methods, including Advantage Actor-Critic (A2C) [20], Asynchronous Advantage Actor-Critic (A3C) [21], and Proximal Policy Optimization (PPO) [22], directly optimize the policy that guides the agent's actions.

In DRL-based resource scheduling, specific optimization objectives are often defined to guide learning. Common objectives include task completion time, energy consumption, and monetary cost. Among these, task completion time is the most prevalent, as it reflects the system's responsiveness and is critical to user satisfaction. Energy consumption and cost metrics are more relevant from a sustainability and long-term efficiency perspective. For example, in [23,24], task completion time is used as the primary optimization objective, yielding favorable results. In [25,26], energy consumption is adopted to investigate the impact of DRL on sustainability, with results showing that DRL algorithms effectively reduce energy usage in cloud data centers.

Value-based methods are particularly suited for discrete action spaces; however, DQN suffers from overestimation bias, which can cause the agent to converge to suboptimal solutions in large-scale task environments. To address this issue, this paper adopts a DDQN-based resource scheduling method, using task completion time as the optimization goal. To enhance the agent's sensitivity to the quality of its decisions, a reward function is designed based on the chosen objective, enabling the agent to dynamically assess the effectiveness of each scheduling action.

3 Method

3.1 Cloud Resource Scheduling Method Based on DDQN

The process of the DDQN-based cloud computing resource scheduling algorithm is depicted in Fig. 2. The input state information for the scheduling method includes the resource status of the currently available virtual machines and the resource demand of the first task in the waiting queue. Based on the current state information, the agent performs a scheduling decision by allocating virtual machines to the current task. Each scheduling instance is recorded in a replay buffer. When the buffer contains a sufficient number of transitions to satisfy training criteria, the agent randomly samples a batch of experiences from the buffer to compute gradients and update its neural network parameters accordingly.

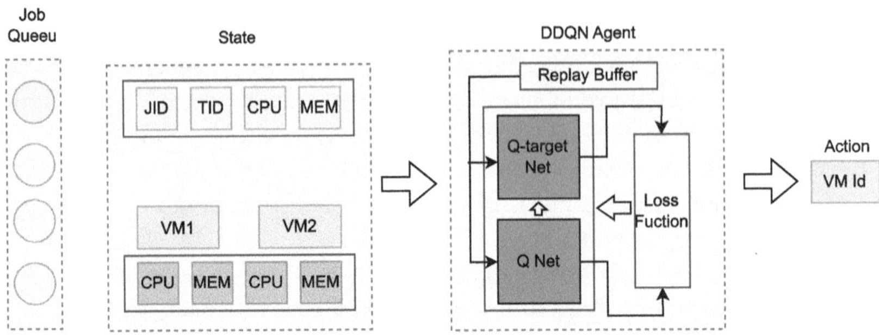

Fig. 2. DDQN-based Cloud Resource Scheduling Method.

3.2 Problem Definition

Table 1.

Table 1. Notation Definitions

job_i	The i-th job in the environment, where $i \in 0, 1, 2, ..., n$
$task_{ij}$	The j-th subtask of the i-th job, where $j \in 0, 1, 2, ..., m$
$task_{cpu}$	CPU resources required by the subtask
$task_{mem}$	Memory resources required by the subtask
$machines$	Set of all virtual machines in the environment
$machine_k$	The k-th virtual machine in the environment, where $k \in 0, 1, 2, ..., o$
$machine_{cpu}$	Current CPU resource availability of a virtual machine
$machine_{mem}$	Current memory resource availability of a virtual machine

State Space. The state space characterizes the external environment perceived by the agent, forming the foundation upon which decisions are made. In a cloud computing resource scheduling environment, the state comprises information on both the virtual machines and the pending tasks. A user-submitted job is defined as $job_i = task_{i1}, task_{i2}, ..., task_{ij}$, where each job contains multiple subtasks. Each subtask encapsulates the required computational resources, specifically $task_{cpu}$ and $task_{mem}$. Each machine records its current resource status, including $machine_{cpu}$ and $machine_{mem}$, and all virtual machines are denoted as $machines$. The formal definition of the state space is as follows:

$$state = \{task_{cpu}, task_{mem}, machines\} \tag{1}$$

Action Space. The action space defines the set of possible actions available to the agent and can be categorized into discrete or continuous types based on the nature of the actions. In the context of cloud resource scheduling, the agent must assign each pending task to an appropriate virtual machine according to the current system state. Accordingly, the action space is defined as a discrete set consisting of the identifiers of all available virtual machines:

$$action = \{machine_1, machine_2, ..., machine_k\} \tag{2}$$

Reward Function. The reward function is a critical component in deep reinforcement learning-based scheduling algorithms, serving as the primary mechanism to drive the learning process of the agent. It significantly influences the agent's policy and overall performance [28]. The design of the reward function needs to encapsulate the critical value of each decision in upholding SLA compliance and mitigating security risks. To intrinsically align the agent's objectives with these security and reliability concerns, the reward is strategically defined as the difference between the available CPU resources of the selected VM and the CPU requirements of the scheduled task:

$$reward = machine_{cpu} - task_{cpu} \tag{3}$$

This formulation creates a direct, quantitative link between scheduling decisions and their impact on system state: A positive reward is granted when the allocated VM has ample resources to not only meet the task's demands but also maintain a healthy buffer. This signifies a high-quality, security-aware decision that prevents resource contention, ensures stable performance, and upholds SLA guarantees—a foundational aspect of a secure cloud environment. A negative reward (penalty) is incurred when the chosen VM has insufficient resources, immediately flagging a decision that induces resource strain. This directly correlates to an increased risk of SLA violation and its associated security ramifications, such as service degradation and expanded attack surfaces. The magnitude of the penalty reflects the severity of the resource shortfall.

Thus, by maximizing cumulative reward, the agent learns a policy that proactively avoids high-risk scheduling choices, prioritizes resource efficiency, and reinforces the overall security posture by ensuring critical services and security mechanisms operate within their required resource boundaries.

3.3 Loss Function

Both DDQN and DQN utilize deep neural networks to calculate the Q-values associated with different actions, and make action selections and network updates based on these Q-values. The agent calculates the Q-values corresponding to different actions for the current state s_t using the Q-network, and selects the action a_t with the highest Q-value. After executing the action, the agent receives the corresponding reward r_t and the updated state s_{t+1}. The entire experience tuple $\{s_t, a_t, r_t, s_{t+1}\}$ is stored in the Replay Buffer for future network updates. During the network update phase, historical experiences are first sampled from the Replay Buffer, and the Q-network is used to estimate the Q of the best action a_t in state s_t. Similarly, the Q-target network is used to estimate the Q_{target} of the best action a_{t+1} in the next state s_{t+1}. The Q, Q_{target}, and the reward r_t are then used to update the deep neural network. The update formula for the DQN deep network is as follows:

$$Loss = (r_t + \gamma \max Q_{target}(s_{t+1}, a_{t+1}; \theta') - Q(s_t, a_t; \theta))^2 \tag{4}$$

In the DQN, the Q-target network selects the maximum estimated value in state s_{t+1} for computation. However, the Q-target network does not participate in the scheduling decision-making process, which can lead to errors in the Q-network update, causing the scheduling method to fail to select the optimal solution for the current state. DDQN addresses the overestimation issue inherent in DQN. As shown in Fig. 2, before calculating Q_{target}, the Q-network first computes the best action a'_{t+1} for state s_{t+1}, and then the Q-target network directly computes the estimated value Q_{target} corresponding to the action a'_{t+1} in state s_{t+1}. Since the Q-network is involved in the computation of Q_{target}, this effectively mitigates the overestimation problem present in DQN. The deep network update formula for DDQN is as follows:

$$a'_{t+1} = \arg\max Q(s_{t+1}; \theta) \tag{5}$$

$$Loss = \left(r_t + \gamma Q_{target}(s_{t+1}, a'_{t+1}; \theta') - Q(s_t, a_t; \theta)\right)^2 \tag{6}$$

4 Experiment

4.1 Experiment Conditions

Simulation Environment. CloudsimPy is a cloud simulation framework implemented in Python and based on the discrete-event simulation library SimPy. Its compatibility with Python-based deep learning libraries makes it highly suitable for research on machine learning and deep learning-based resource management techniques [27]. In the simulation environment, the `Job` and `Machine` modules represent job and virtual machine modeling, respectively. The `Simulation` module simulates the cloud environment and contains the `Cluster` (for managing VMs), `Broker` (for submitting task requests), and `Scheduler` (for task scheduling).

Dataset. The Alibaba Cluster Trace 2017 dataset, which contains 900 jobs and 5103 subtasks, each associated with specific CPU and memory requirements. To evaluate the scheduling performance under different workload conditions, three scenarios with low, medium, and high task arrival rates are defined by varying the number of VMs and tasks. Each VM in the simulation has 5 CPU cores and 1 GB of memory.

Baselines. The comparison includes traditional scheduling methods such as FirstMax and Random, as well as deep reinforcement learning approaches: A2C [29], PPO [30], and DQN [8]. In the experiment, all DRL-based methods use the same hyperparameters, reward function, state, and action space for fairness. This comparison highlights the effectiveness of DRL-based scheduling algorithms over both traditional and alternative DRL approaches.

4.2 Hyperparameter Configuration.

In deep learning, hyperparameters are manually specified prior to the execution of an algorithm and are critical in influencing the model's behavior and performance. These hyperparameters affect various aspects such as training speed, convergence, model capacity, and generalization ability. The hyperparameter settings utilized in this study for model training are summarized in Table 2.

Table 2. Hyperparameter settings used in the experiment

Hyperparameter	Value
Batch Size	5
Epochs	200
Learning Rate	1×10^{-4}
Hidden Dim	64
Activation	ReLU
Optimizer	Adam
Gamma	0.99
Epsilon	1
Epsilon Decay	0.05
Epsilon End	0.05

4.3 Evaluation Metrics

Average Task Completion Time. This metric quantifies the average time taken for a job from its submission to the system until its execution is completed. It reflects both queuing delay and execution time, indicating the efficiency of the scheduling strategy on a per-task basis:

$$T_{\text{avg}}^{\text{completion}} = \frac{1}{n} \sum_{i=1}^{n} \left(job_i^{\text{finished}} - job_i^{\text{start}} \right) \tag{7}$$

Average Weighted Turnaround Time. This metric is defined as the ratio of a job's completion time to its actual execution time. It is used to assess fairness and responsiveness, particularly in scenarios involving jobs with varying durations:

$$T_{\text{avg}}^{\text{weighted}} = \frac{1}{n}\sum_{i=1}^{n}\frac{job_i^{\text{finished}} - job_i^{\text{start}}}{job_i^{\text{duration}}} \tag{8}$$

Makespan. Makespan is defined as the total time taken from the arrival of the first job to the completion of the last job. It reflects the overall throughput efficiency of the scheduling method:

$$T_{\text{makespan}} = \max_i(job_i^{\text{finished}}) - \min_i(job_i^{\text{start}}) \tag{9}$$

SLA Violations. An SLA violation occurs when the resources available on the virtual machine allocated to a task cannot meet the resource requirements of the task, leading to task allocation failure. Let machine_k denote the virtual machine allocated to the current task, and $task_{ij}$ represent the j-th resource requirement of the i-th task. The number of SLA violations is defined as:

$$N_{\text{SLA}} = \sum_{i=1}^{n}\sum_{j=1}^{m}\mathbb{I}\left(machine_k^{\text{cpu}} < task_{ij}^{\text{cpu}} \;||\; machine_k^{\text{mem}} < task_{ij}^{\text{mem}}\right) \tag{10}$$

where $\mathbb{I}(\cdot)$ is the indicator function returning 1 if the condition is true, and 0 otherwise. A higher N_{SLA} indicates lower quality of service and poorer adherence to user-defined deadlines.

4.4 Experimental Results

The experiments consist of three different traffic levels. Low-load: 100 jobs with 1059 subtasks handled by 10 VMs. Medium-load: 200 jobs with 1848 subtasks handled by 15 VMs. High-load: 300 jobs with 2486 subtasks handled by 20 VMs.

Makespan. The makespan results for different methods under three scenarios are shown in Table 3. In all three scenarios, low, medium, and high, deep reinforcement learning methods consistently outperform the random scheduling method in terms of makespan. The DQN and DDQN, which are better equipped for handling discrete action spaces, consistently achieve lower makespan than FirstMax and Random methods across all three scenarios. The performance of A2C and PPO methods is generally worse than that of FirstMax in most scenarios. Among them, DDQN demonstrates the lowest makespan in all three scenarios, thus achieving the best overall performance.

Table 3. Makespan

	FirstMax	Random	A2C	PPO	DQN	DDQN
Low	1746	1855	1774	1836	1722	**1712**
Medium	2426	2559	2476	2540	2404	**2402**
High	2792	2889	2836	2884	2780	**2777**

Average Task Completion Time. The results for the average task completion time under different methods and scenarios are presented in Table 4. The overall performance of the methods follows a similar trend to that of makespan. Deep reinforcement learning-based scheduling methods still outperform the random scheduling approach, with both DDQN and DQN demonstrating superior performance compared to FirstMax and Random. Among these methods, DDQN achieves the optimal performance.

Table 4. Average Task Completion Time

	FirstMax	Random	A2C	PPO	DQN	DDQN
Low	812	872	830	864	806	**800**
Medium	1202	1296	1241	1282	1186	**1185**
High	1535	1610	1569	1600	1524	**1522**

Average Weighted Turnaround Time. The results for the average weighted turnaround time under different methods and scenarios are presented in Table 5. A higher average weighted turnaround time indicates that the task completion time is closer to the actual execution time required by the task, suggesting that the scheduling method is more effective in managing tasks of varying lengths. The experimental results show that DQN and DDQN exhibit superior scheduling capabilities for both long and short tasks compared to the other methods, with DDQN achieving the best performance.

Table 5. Average Weighted Turnaround Time

	FirstMax	Random	A2C	PPO	DQN	DDQN
Low	0.876	0.849	0.879	0.857	0.889	**0.894**
Medium	0.574	0.551	0.567	0.556	0.589	**0.590**
High	0.456	0.434	0.447	0.441	0.462	**0.463**

SLA Violations. The results for the SLA violations under different methods and scenarios are presented in Table 6. Among the methods evaluated, DDQN and DQN outperform the other comparison methods, showing significantly fewer SLA violations across all three different scheduling scenarios. This indicates that value-based deep reinforcement learning methods are more suitable for tasks involving a discrete action space. Furthermore, DDQN achieves the best performance, demonstrating that addressing the issue of overestimation in Q-values improves the scheduling method's overall effectiveness.

As the user demand increases, all methods show a decline in SLA adherence. This could be attributed to the increase in the number of virtual machines, which reduces the difficulty of making scheduling decisions. The increased resource availability allows for better allocation, thereby lowering the incidence of SLA violations.

Table 6. Number of SLA Violations

	FirstMax	Random	A2C	PPO	DQN	DDQN
Low	357	488	443	505	357	**321**
Medium	129	250	155	260	106	**101**
High	42	137	86	142	29	**24**

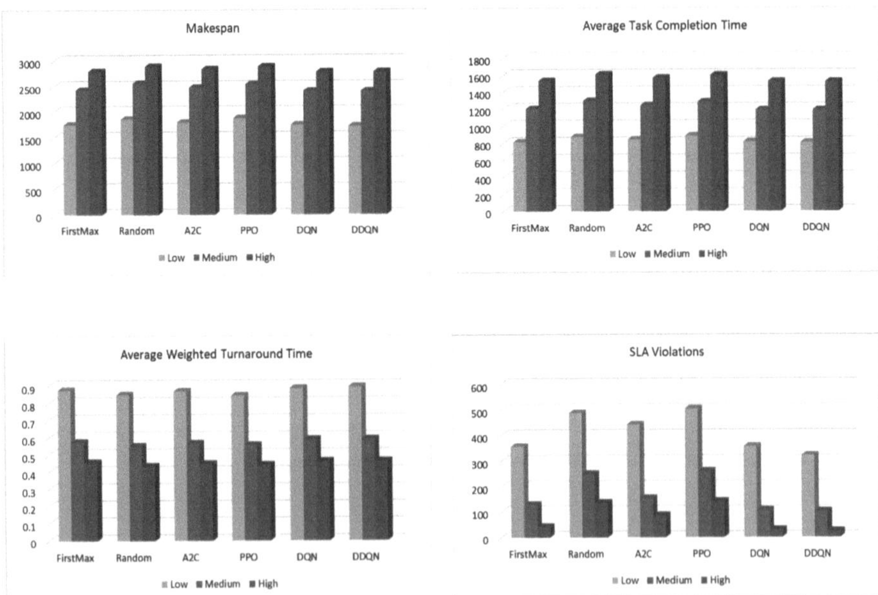

Fig. 3. Visualization information of makespan, average task completion time, average weighted turnaround time, and SLA violations for different scheduling methods in different low, medium, and high scenarios.

Visualization Analysis. The bar charts for makespan, average task completion time, average weighted turnaround time, and SLA violations under different methods and scenarios are shown in Fig. 3. Upon examining the bar charts, it is evident that DQN and DDQN outperform the other comparison methods, with DDQN achieving the best performance. However, as the user demand increases, the performance gap between DQN and DDQN and the suboptimal method FirstMax begins to narrow.

There are two potential reasons for this phenomenon. The first reason may be the increased number of long tasks, which reduces the frequency of scheduling failures. Additionally, with a slight increase in the number of virtual machines, the scheduling pressure for short tasks is alleviated, further reducing scheduling failures. The second reason relates to hardware limitations and the need for fairness in comparison. The training period for deep reinforcement learning algorithms is set to 200 epochs. As user demand grows, it may be necessary to extend the training period to achieve better performance.

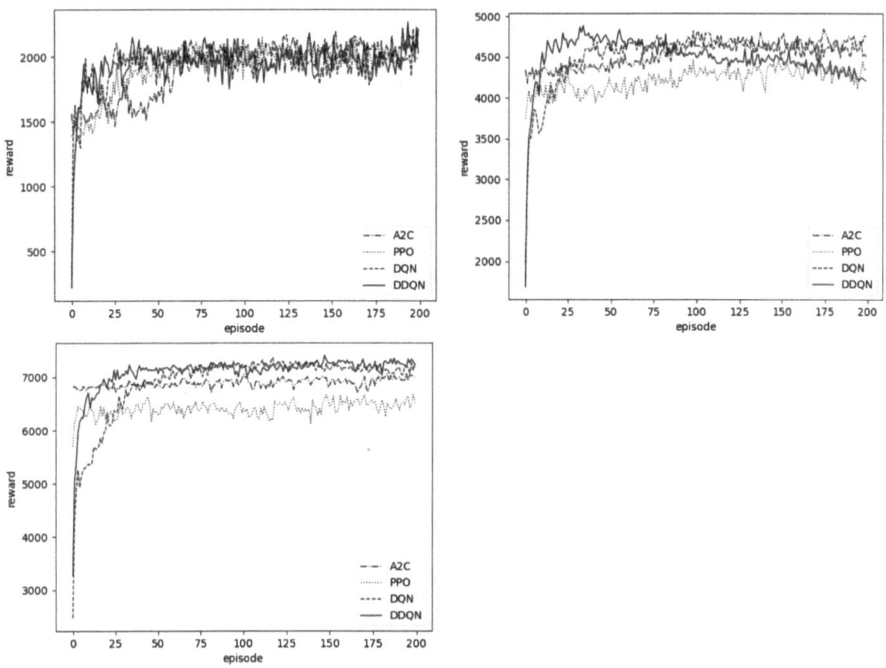

Fig. 4. Reward accumulation of different deep reinforcement learning methods in low, medium, and high scenarios.

Cumulative Reward Convergence Analysis. The cumulative reward is the total reward accumulated by the deep reinforcement learning method after completing one round of scheduling. When the reward remains within a specific range, it indicates the stability of the method. The cumulative rewards for reinforcement learning methods under three different scenarios are shown in Fig. 4. A comparative analysis reveals that DQN and DDQN exhibit good adaptability in environments with discrete action spaces. These methods are explored for a longer period before reward convergence, leading to the highest final accumulated reward. On the other hand, A2C and PPO did not explore sufficiently during policy updates, which may be attributed to the fact that A2C and PPO are not well-suited for problems with discrete action spaces. Comparing DQN and DDQN, it can be observed that DDQN, with its modified deep network update approach, successfully addresses the overestimation problem. This allows DDQN to reduce the likelihood of the scheduling method getting stuck in local optima and to converge to higher rewards more quickly.

5 Conclusion

This study demonstrates that high-performance resource scheduling is not merely an efficiency problem but also a crucial aspect of cloud security hygiene. Our DRL-based approach provides a foundational step towards building secure-by-design cloud infrastructures. A resource scheduling strategy is developed based on the Double Deep Q-Network framework, designed for a discrete action space. In addition, a dynamically adaptive reward function is introduced to reflect the quality of scheduling decisions in real time, thereby facilitating more effective learning by the agent. Experiments are conducted in a simulation environment built using SimPy in Python, with workload scenarios categorized into low, medium, and high traffic levels based on the Alibaba 2017 cluster trace dataset. Experimental results demonstrate that the DDQN-based scheduling method achieves the best overall performance. Compared with the second-best method, FirstMax, the proposed approach reduces makespan by an average of 1.15%, decreases average task completion time by 1.24%, improves average weighted turnaround time by 6.36%, and reduces the number of SLA violations by 24.87% on average. Most importantly, the significant reduction in SLA violations (24.87% on average) and task completion times indicates that our approach can effectively mitigate the performance-induced security risks. By ensuring timely task completion and stable resource provisioning, our scheduler helps maintain the operational integrity of security-critical services and reduces the system's vulnerability window. However, the model has yet to be validated in real-world deployment scenarios. Furthermore, while the custom-designed reward function used in the experiments effectively reduces SLA violations, it may potentially lead to resource fragmentation. For future work, we plan to explicitly incorporate security-specific metrics (e.g., security alert severity, compliance status of nodes) into the reward function to develop a security-aware scheduler that can directly respond to active threats.

References

1. Wang, L., et al.: Cloud computing: a perspective study. N. Gener. Comput. **28**, 137–146 (2010)
2. Gong, C., Liu, J., Zhang, Q., Chen, H., Gong, Z.: The characteristics of cloud computing. In: 2010 39th International Conference on Parallel Processing Workshops, pp. 275–279. IEEE (2010)
3. Singh, P., Dutta, M., Aggarwal, N.: A review of task scheduling based on metaheuristics approach in cloud computing. Knowl. Inf. Syst. **52**(1), 1–51 (2017). https://doi.org/10.1007/s10115-017-1044-2
4. Zhan, Z.H., Liu, X.F., Gong, Y.J., Zhang, J., Chung, H.S.H., Li, Y.: Cloud computing resource scheduling and a survey of its evolutionary approaches. ACM Comput. Surv. (CSUR) **47**(4), 1–33 (2015)
5. Mathew, A., Amudha, P., Sivakumari, S.: Deep learning techniques: an overview. In: Hassanien, A.E., Bhatnagar, R., Darwish, A. (eds.) AMLTA 2020. AISC, vol. 1141, pp. 599–608. Springer, Singapore (2021). https://doi.org/10.1007/978-981-15-3383-9_54
6. Zhou, G., Tian, W., Buyya, R., Xue, R., Song, L.: Deep reinforcement learning-based methods for resource scheduling in cloud computing: a review and future directions. Artif. Intell. Rev. **57**(5), 124 (2024)
7. Botvinick, M., Ritter, S., Wang, J.X., Kurth-Nelson, Z., Blundell, C., Hassabis, D.: Reinforcement learning, fast and slow. Trends Cogn. Sci. **23**(5), 408–422 (2019)
8. Mnih, V., Kavukcuoglu, K., Silver, D., et al.: Human-level control through deep reinforcement learning. Nature **518**(7540), 529–533 (2015)
9. Van Hasselt, H., Guez, A., Silver, D.: Deep reinforcement learning with double q-learning. In: Proceedings of the AAAI Conference on Artificial Intelligence, vol. 30, no. 1 (2016)
10. Xin, Y., Xie, Z.Q., Yang, J.: A load balance oriented cost efficient scheduling method for parallel tasks. J. Netw. Comput. Appl. **81**, 37–46 (2017)
11. Lin, Y.K., Chong, C.S.: Fast GA-based project scheduling for computing resources allocation in a cloud manufacturing system. J. Intell. Manuf. **28**, 1189–1201 (2017)
12. Kumar, M., Sharma, S.C.: PSO-based novel resource scheduling technique to improve QoS parameters in cloud computing. Neural Comput. Appl. **32**(16), 12103–12126 (2020)
13. Gupta, A., Garg, R.: Load balancing based task scheduling with ACO in cloud computing. In: 2017 International Conference on Computer and Applications (ICCA), pp. 174–179. IEEE (2017)
14. Kuang, L., Gong, T., OuYang, S., Gao, H., Deng, S.: Offloading decision methods for multiple users with structured tasks in edge computing for smart cities. Futur. Gener. Comput. Syst. **105**, 717–729 (2020)
15. Yu, H.: Evaluation of cloud computing resource scheduling based on improved optimization algorithm. Complex Intell. Syst. **7**(4), 1817–1822 (2021)
16. Shobeiri, P., Akbarian Rastaghi, M., Abrishami, S., Shobiri, B.: PCP-ACO: a hybrid deadline-constrained workflow scheduling algorithm for cloud environment. J. Supercomput. **80**(6), 7750–7780 (2024)
17. Uma, J., Vivekanandan, P., Shankar, S.: Optimized intellectual resource scheduling using deep reinforcement Q-learning in cloud computing. Trans. Emerg. Telecommun. Tech. **33**(5), e4463 (2022)
18. Alfakih, T., Hassan, M.M., Gumaei, A., Savaglio, C., Fortino, G.: Task offloading and resource allocation for mobile edge computing by deep reinforcement learning based on SARSA. IEEE Access **8**, 54074–54084 (2020)

19. Mangalampalli, S., Karri, G.R., Kumar, M., Khalaf, O.I., Romero, C.A.T., Sahib, G.M.A.: DRLBTSA: deep reinforcement learning based task-scheduling algorithm in cloud computing. Multimedia Tools Appl. **83**(3), 8359–8387 (2024)
20. Lu, J., Yang, J., Li, S., Li, Y., Jiang, W., Dai, J., Hu, J.: A2C-DRL: dynamic scheduling for stochastic edge-cloud environments using A2C and deep reinforcement learning. IEEE Internet Things J. (2024)
21. Krishna, M.S.R., Mangalampalli, S.S.: PWSA3C: prioritized workflow scheduler in cloud computing using asynchronous advantage actor critic (A3C) algorithm. IEEE Access (2024)
22. Bilgaiyan, S., Pandey, A., Roy, S., Shukla, S., Gourisaria, M.K.: PPO-based autoscaling for CPU efficiency in edge computing. In: 2024 Second International Conference on Networks, Multimedia and Information Technology (NMITCON), pp. 1–6. IEEE (2024)
23. Huang, Y., et al.: Deep adversarial imitation reinforcement learning for QoS-aware cloud job scheduling. IEEE Syst. J. **16**(3), 4232–4242 (2021)
24. Chen, X., Yang, L., Chen, Z., Min, G., Zheng, X., Rong, C.: Resource allocation with workload-time windows for cloud-based software services: a deep reinforcement learning approach. IEEE Trans. Cloud Comput. **11**(2), 1871–1885 (2022)
25. Kang, K., Ding, D., Xie, H., Yin, Q., Zeng, J.: Adaptive DRL-based task scheduling for energy-efficient cloud computing. IEEE Trans. Netw. Serv. Manage. **19**(4), 4948–4961 (2021)
26. Zhao, J., Rodríguez, M.A., Buyya, R.: A deep reinforcement learning approach to resource management in hybrid clouds harnessing renewable energy and task scheduling. In: 2021 IEEE 14th International Conference on Cloud Computing (CLOUD), pp. 240–249. IEEE (2021)
27. Li, F., Hu, B.: Deepjs: job scheduling based on deep reinforcement learning in cloud data center. In: Proceedings of the 4th International Conference on Big Data and Computing, pp. 48–53 (2019)
28. Ladosz, P., Weng, L., Kim, M., Oh, H.: Exploration in deep reinforcement learning: a survey. Inf. Fusion **85**, 1–22 (2022)
29. Mnih, V., et al.: Asynchronous methods for deep reinforcement learning. In: International Conference on Machine Learning, pp. 1928–1937. PMLR (2016)
30. Schulman, J., Wolski, F., Dhariwal, P., Radford, A., Klimov, O.: Proximal policy optimization algorithms arXiv preprint arXiv:1707.06347 (2017)

AI Security

Certificateless Offline/Online Signcryption for Lightweight Anonymous Authentication in VANETs

Abdullahi Yahya Imam[1,2] and Fagen Li[1(✉)]

[1] School of Computer Science and Engineering, University of Electronic Science and Technology of China, Chengdu 611731, Sichuan, China
fagenli@uestc.edu.cn

[2] Department of Information Technology, Faculty of Computing, Bayero University, Kano, Nigeria
ayimam.it@buk.edu.ng

Abstract. Vehicular ad hoc networks (VANETs) play a vital role in intelligent transportation systems by facilitating communication between vehicles and roadside infrastructure. Various cryptographic solutions have been proposed to ensure the security of the communicating data with vehicular privacy. Signcryption, a primitive that combines encryption and signature in a single logical operation, has been widely employed to achieve security with efficiency. However, many of the existing schemes rely on computationally expensive bilinear pairings, making them unsuitable for critical applications in VANETs especially for onboard units (OBUs) that generally have limited computing resources. Although recently proposed pairing-free schemes offer improved efficiency, various security weaknesses such as public key replacement attacks make them insecure. In this paper, we propose a lightweight signcryption scheme, where OBUs perform the time-consuming operations offline. Our scheme avoids the use of bilinear pairings while eliminating the security weaknesses present in the existing pairing-free constructions. Formal proofs, based on hardness of elliptic curve computational Diffie-Hellman (ECDH) and discrete logarithm (ECDL) problems, have shown that our scheme achieves confidentiality and unforgeability. Furthermore, performance analysis demonstrates that our scheme significantly reduced computational overhead particularly for resource-constrained OBUs in comparison to existing schemes.

Keywords: certificateless signcryption · conditional privacy preserving · anonymous authentication

1 Introduction

The rapid advancement of the internet of things (IoT) has greatly revolutionized road traffic management and safety. Vehicular ad hoc networks (VANETs)

J. Yu and Y. Chen (Eds.): FCS 2025, CCIS 2805, pp. 239–256, 2027.
https://doi.org/10.1007/978-981-95-9514-3_16

emerged as a key enabler of intelligent transportation systems, facilitating real-time communication between vehicles and road infrastructure (V2I) [1]. Through such networks, vehicles can share information about traffic conditions, accidents or road hazards, thereby enhancing road safety, traffic efficiency, and overall driving experience [2]. In a typical VANET, vehicles are equipped with cameras, sensors and onboard unit (OBUs) that communicate with roadside units (RSUs) via dedicated short range communication (DSRC). Serving as edge node, an RSU processes vehicular data and forwards the information to the cloud server (CS) for long-term references and analysis. However, the communications in VANET occur over open wireless channels, making them inherently vulnerable to threats such as eavesdropping, data tampering, and identity spoofing [3]. Addressing these challenges requires robust cryptographic solutions that guarantee authentication, integrity, confidentiality and user privacy. The highly dynamic feature of the VANET environment and the limited computing resources in OBUs are the major constraints, which demand lightweight cryptographic solutions that balance efficiency with robust security [4]. Traditional approaches that independently apply encryption and digital signature result in excessive computational and communication costs, rendering them impractical for VANETs. To overcome these drawbacks, Zheng [5] introduced signcryption, a cryptographic primitive that logically combines encryption and signature into a single operation.

Over the years, various signcryption schemes [6,7] have been proposed to enhance VANET authentication. Early solutions employed bilinear pairings, which are highly time-consuming operations that make the schemes unsuitable for VANETs. Although the works of [10,22] proposed pairing-free schemes that improved performance, their computational overheads are still too high for critical real-time communications. To address this challenge, yet pairing-free schemes by Cui et al. [8], Xie et al. [9], Zhang et al. [18] introduced offline computations of message-independent operations in advance, significantly reducing the overhead of resource-intensive operations before the real-time online communication phase. However, weaknesses that include the failure to satisfy the essential security requirements make the schemes vulnerable to various threats. For instance, Wang et al. [26] recently revealed critical security vulnerability of public key replacement attacks in the scheme of Xie et al. [9]. Although Wang et al. [26] later proposed a more secure variant, employing resource-consuming bilinear pairings, which negatively affects its efficiency in VANETs.

These limitations in the existing schemes highlight the need for a lightweight pairing-free scheme that robustly ensures security and practical efficiency especially for resource-constrained OBUs. Motivated by the above challenges, this work makes the following contributions:

1. We propose a certificateless signcryption scheme that remedies the ciphertext forgeability (in the Xie et al. [9]'s scheme) revealed by Wang et al. [26], while avoiding the resource-intensive bilinear pairing operations.
2. We formally prove that our scheme achieves confidentiality and ciphertext unforgeability based on the hardness of the computational DiffieHellman (CDH) and elliptic curve discrete logarithm (ECDL) problems.

3. Performance evaluation we conducted has shown that our scheme significantly reduces computational overhead compared to the related schemes, making it especially suitable for resource-constrained OBUs in VANET environments.

2 Related Works

Early designs of authentication in VANETs such as Raya and Hubaux [11] largely relied on traditional public key infrastructure (PKI). The burden of certificate management inherent to PKI paradigm constituted excessive computational and communication overheads, which make them inefficient for VANETs. After Shamir [12] introduced identity-based cryptography (IBC), which eliminates the need for certificates, Yu and Yang [13] proposed an IBC-based signcryption scheme using ECC. While IBC resolved certificate management, it brought in the key escrow problem, leading to the risk of public key disclosure and message repudiation. In 2003, Al-Riyami and Paterson [14] proposed certificateless cryptography (CLC) that mitigates key escrow problem while eliminating certificate management. In this paradigm, the key generation center (KGC) provides users with partial private keys, while users generate their own secret values. Utilizing this paradigm, Liu et al. [15] proposed a certificateless signcryption (CLSC) scheme. However, Weng et al. [17] demonstrated that Lius's scheme was insecure against both distinguishability and forgery attacks. Xie and Zhang [19] later put forward a new CLSC scheme with formal proofs of security, but its reliance on bilinear pairings made it computationally expensive and unsuitable for resource-constrained VANETs. With the introduction of edge computing framework by [20] that optimizes messages delivery between users and the edge nodes, Basudan et al. [7] presented a scheme for roads condition monitoring with vehicular anonymity. In this approach, vehicles deliver messages to the edge nodes, which process them and forward the information to the cloud servers. To further enhance the efficiency, Cui et al. [8] proposed an edge-based authentication scheme with signcryption algorithm split into offline and online phases. Although this approach reduced real-time computational overhead by precomputing message-independent operations ahead of online signcryption, Cui et al.'s did not support traceability and unlinkabilty, essential requirements for anonymous authentication. Recently, Xie et al. [9] proposed an online/offline signcryption scheme that performed most time-intensive computations offline, while only the lightweight operations need to be carried out online. However, Wang et al. [26] revealed a critical security vulnerability of public key replacement attacks in Xie et al.'s scheme. Wang et al. [26] later proposed a new scheme but using the expensive bilinear pairings, which result in high computational overhead and make it unsuitable for time-critical VANETs.

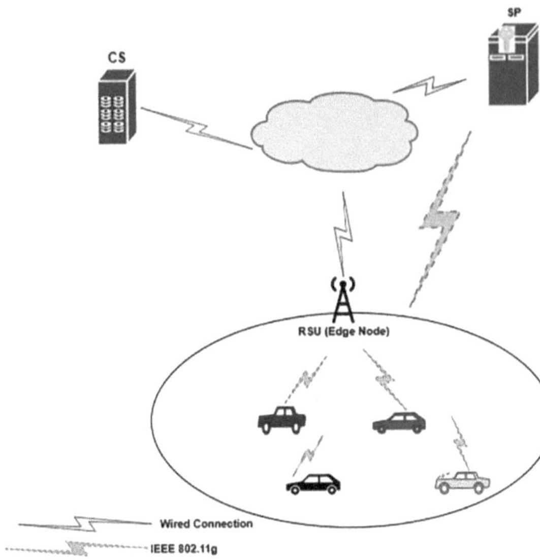

Fig. 1. System Model of the VANETs.

3 System Model

The system is modeled in Fig. 1 with components that include service provider (SP), roadside unit (RSU) as edge node, cloud server (CS) and onboard units (OBUs) as vehicular entities. These components are described as follows:

1. *Service provider (SP)*: This is a trusted entity that serves as KGC for CLC as well as tracing authority (TA) in the system. As KGC, it generates partial private keys for OBUs and RSUs. In addition, it can trace any malicious OBU or RSU in the network. SP is assumed to have abundant computing and storage capacities.
2. *Roadside unit (RSU)*: This entity serving as edge node is generally placed across the road. An RSU has less computing capacity than the SP but more resources than an OBU; however, it is not a trusted entity. This entity receives ciphertexts from OBUs, unsigncrypts them, processes the information, and then sends the information to CS for storage and future analysis.
3. *Cloud server (CS)*: Having huge storage capacity, this entity serves as data center. It receives the authenticated information from the edge nodes and then stores them for future references.
4. *Onboard unit (OBU)*: This component is usually embedded in every a vehicle (V). OBU generally has limited computing power, which necessitates its urge for protocols with light computational operations. A vehicle gathers real-time traffic data and signcrypts them before transmitting the ciphertext to a nearby edge node.

3.1 Security Requirements

The design of our proposed scheme takes into consideration the following security requirements.

1. *Confidentiality*: This requirement ensures only authorized edge can retrieve the plaintext from a ciphertext.
2. *Message authentication and integrity*: This property ensures that a message originates from a legitimate vehicle and has not been altered during transmission.
3. *Anti-replay attacks*: This property ensures the freshness of messages, preventing any entity from maliciously retransmitting a previously sent message without detection.
4. *Unlinkability*: This property guarantees that multiple ciphertexts generated by the same vehicle cannot be linked to each other, and that an edge node cannot associate any message with a unique vehicle in the system.
5. *Anonymity*: This privacy-preserving property ensures that the real identity of a vehicle remains hidden, such that the receiver (edge node) cannot infer the vehicle's original identity from the ciphertext.
6. *Traceability*: This property enables the TA to disclose the real identity of any malicious vehicle from a given ciphertext. Although the ciphertext does not reveal the real identity of a vehicle, the TA can conditionally trace it when necessary.

4 Proposed Offline/Online Signcryption Scheme for Anonymous Authentication

This section describes our proposed scheme for anonymous authentication in VANETs. The scheme consists of the following algorithms.

Algorithm 1: System Setup

Input: k
Output: s, $params$
SS(k)
  Select G whose generator is P
  randomly select $s \in \mathbb{Z}_p^*$
  $S_{pub} \leftarrow sP$
  select $H_1, H_3 : \{0,1\}^* \rightarrow \mathbb{Z}_p^*$
  select $H_2 : \{0,1\}^* \rightarrow \{0,1\}^{|m|}$
  $params \leftarrow \{G, p, P, S_{pub}, H_1, H_2, H_3\}$
return s, $params$

1. *System Setup*: On the security parameter k, the TA serving as KGC runs the steps as detailed in Algorithm 1. KGC selects an additive group G with a generator P. Then, KGC selects a master secret s and computes a master public

key S_{pub}. Afterwards, KGC selects three secure hash functions H_1, H_2, H_3 where $|m|$ is the plaintext length, and outputs a master secret s and system parameters $params$. The s is kept secret, but the $params$ are publicly announced in the system.

2. *Generate Secret Value and Precomputed Commitment*: A vehicle V_i in the IoV inputs its real identity ID_i to the Algorithm 2, which generates a one-time set of (a_i, A_i, c_i, C_i). The a_i is a self-chosen secret value, A_i will be part of its public key, while c_i and C_i respectively serve as precomputed committal and commitment. The value c_i would be used as a random seed in the offline signcryption phase. The pre-computation of the c_i and C_i reduces the computational cost ahead of the real-time message signcryption.

Algorithm 2: Generate Secret Value and Precomputed Commitment

Input: ID_i
Output: (a_i, A_i, c_i, C_i)
GSVPC(ID_i)
  randomly select $a_i, c_i \in \mathbb{Z}_p^*$
  $A_i \leftarrow a_i P$
  $C_i \leftarrow c_i P$
return (a_i, A_i, c_i, C_i)

3. *Extract Partial Private Key*: The KGC uses the master secret s, together with the real identity ID_i and A_i submitted by a vehicle V_i, to execute the Algorithm 3. The outputs issued to the V_i are the pseudonym PID_i for the corresponding real identity ID_i and the partial private key $ppk_i = (d_i, B_i)$.
4. *Private Key Generation*: Upon obtaining a pseudonym and its partial private key (PID_i, d_i, B_i) from the KGC, a V_i that has already generated the self-chosen secret value a_i in Algorithm 2, outputs its private key SK_i as defined in Algorithm 4.

Algorithm 3: Extract Partial Private Key

Input: s, ID_i, A_i
Output: (PID_i, d_i, B_i)
EPPK(s, ID_i, A_i)
  choose randomly $y_i \in \mathbb{Z}_p^*$
  $Y_i \leftarrow y_i P$
  $PID_i \leftarrow ID_i \oplus H_1(sA_i)$
  $\theta_1^i \leftarrow H_1(PID_i, A_i, S_{pub})$
  $d_i \leftarrow (y_i + \theta_1^i s) \bmod p$
  $B_i \leftarrow A_i + Y_i$
return (PID_i, d_i, B_i)

Algorithm 4: Generate Private Key

Input: a_i, d_i
Output: SK_i

GPK(a_i, d_i)
    $SK_i \leftarrow (a_i, d_i)$
return SK_i

Algorithm 5: Generate Public Key

Input: A_i, B_i
Output: PK_i

GPUK(A_i, B_i)
    $PK_i \leftarrow (A_i, B_i)$
return PK_i

5. *Generate Public Key*: A V_i runs Algorithms 5 to generate the public key PK_i corresponding to its private key SK_i. Similarly, an edge node with identity ID_e generates its private-public keys pair $SK_e = (a_e, d_e)$ and $PK_e = (A_e, B_e)$. Unlike vehicles, the edge node does not need to generate any precomputed commitment in Algorithm 2.

Algorithm 6: Offline Signcrypt

Input: ID_e, PK_e, c_i
Output: U_i

OFSCR(c_i, PK_e)
    $\theta_1^e \leftarrow H_1(ID_e, A_e, S_{pub})$
    $U_i \leftarrow c_i(B_e + \theta_1^e S_{pub})$
return U_i

Algorithm 7: Online Signcrypt

Input: $PID_i, SK_i, PK_i, ID_e, PK_e, c_i, C_i, U_i, m_i, T_s^i$
Output: σ_i

ONSC($PID_i, SK_i, PK_i, ID_e, PK_e, c_i, C_i, U_i, m_i, T_s^i$)
    $\theta_2^i \leftarrow H_2(U_i, ID_e, PK_e)$
    $x_i \leftarrow m_i \oplus \theta_2^i$
    $\theta_3^i \leftarrow H_3(m_i, T_i, U_i, PID_i, PK_i, \theta_2^i)$
    $\delta_i \leftarrow c_i + (a_i + d_i)\theta_3^i \bmod p$
    $\sigma_i \leftarrow (x_i, \delta_i, C_i, T_s^i)$
return σ_i

Note: To achieve the security goal of ciphertexts unlinkability, a unique set (PID_i, SK_i , PK_i, c_i, C_i) is used for each message signcryption. Thus, for every new message, V_i employs a fresh tuple (PID_i, SK_i, PK_i, c_i, C_i). For simplicity, a vehicle may generate and store a multiple of such tuples in advance.

6. *Offline Signcrypt*: Vehicle V_i executes Algorithm 6 using the edge node's identity ID_e, its public key PK_e, and the precomputed committal value c_i. This algorithm outputs a partial ciphertext U_i.
7. *Online Signcrypt*: This is the real-time signcryption phase, typically performed by a vehicle while in motion. Upon generating a message m_i, the V_i employs a tuple $(PID_i, SK_i\ , PK_i, c_i, C_i)$ togather with the edge node's identity ID_e and public key PK_e to execute Algorithm 7. The output is the ciphertext $\sigma_i = (x_i, \delta_i, C_i, T_s^i)$, where T_s^i denotes the sending timestamp.

Algorithm 8: Unsigncrypt

Input: $ID_e, SK_e, PK_e, PID_i, PK_i, \sigma_i$
Output: m_i or $\perp$

UNSCR($ID_e, SK_e, PID_i, PK_i, \sigma_i$)
  if $\Delta T_i \leq T_r^i - T_s^i$
    $U_i \leftarrow (a_e + d_e)C_i$
    $\theta_1^i \leftarrow H_1(PID_i, A_i, S_{pub})$
    $\theta_2^i \leftarrow H_2(U_i, ID_e, PK_e)$
    $m_i \leftarrow x_i \oplus \theta_2^i$
    $\theta_3^i \leftarrow H_3(m_i, T_s^i, U_i, PID_i, PK_i, \theta_2^i)$
    if $\delta_i P \neq C_i\ +\ \theta_3^i(B_i + \theta_1^i S_{pub})$
      return $\perp$
return m_i

Algorithm 9: Aggregate Unsigncrypt

Input: $ID_e, SK_e, PK_e,\ (PID_i, PK_i, \sigma_i)_{i=1}^n$
Output: $\{m_i\}_{i=1}^n$ or $\perp$

UNSCR($ID_e, SK_e,\ (PID_i, PK_i, \sigma_i)_{i=1}^n$)
  for $i = 1, \ldots, n$
    $\delta_{agg} \leftarrow \Sigma_{i=1}^n \delta_i \bmod p$
  for $i = 1, \ldots, n$
    if $\Delta T_i \leq T_r^i - T_s^i$
      $U_i \leftarrow (a_e + d_e)C_i$
      $\theta_1^i \leftarrow H_1(PID_i, A_i, S_{pub})$
      $\theta_2^i \leftarrow H_2(U_i, ID_e, PK_e)$
      $m_i \leftarrow x_i \oplus \theta_2^i$
      $\theta_3^i \leftarrow H_3(m_i, T_s^i, U_i, PID_i, PK_i, \theta_2^i)$
  if $\delta_{agg} P \neq \Sigma_{i=1}^n C_i\ +\ \Sigma_{i=1}^n \theta_3^i(B_i + \theta_1^i S_{pub})$
    return $\perp$
return $\{m_i\}_{i=1}^n$

---

8. *Unsigncrypt*: Upon receiving a ciphertext σ_i, the edge node first records a receiving timestamp T_r^i and checks time validity ΔT_i by computing the difference between the receiving T_r^i and the sending T_s^i timestamps. If ΔT_i is within the valid range, the edge node uses its private $SK_e = (a_e, d_e)$ to unsigncrypt the ciphertext as described in Algorithm 8.

9. *Aggregate Unsigncrypt*: The edge node typically receives multiple ciphertexts from different vehicles within short time intervals. To improve efficiency, it performs aggregate unsigncrypt rather than processing each ciphertext individually. After receiving n ciphertexts from n vehicles, the edge node checks the time validity of each ciphertext, and then recovers the corresponding plaintext messages $m_1, ..., m_n$ as provided in Algorithm 9.

Proof of Correctness
The decryption of the ciphertext holds as follows:

$$\begin{aligned} U_i &= c_i(B_e + \theta_1^e S_{pub}) \\ &= c_i(A_e + Y_e + \theta_1^e S_{pub}) \\ &= c_i(a_e P + y_e P + \theta_1^e sP) \\ &= c_i(a_e + y_e + \theta_1^e s)P \\ &= (a_e + y_e + \theta_1^e s)c_i P \\ &= (a_e + d_e)C_i \end{aligned}$$

The signature verification is established as:

$$\begin{aligned} \delta_i P &= c_i P + (a_i + d_i)\theta_3^i P \\ &= c_i P + (a_i + (y_i + \theta_1^i s))\theta_3^i P \\ &= c_i P + \theta_3^i (a_i P + y_i P + \theta_1^i sP) \\ &= C_i + \theta_3^i ((A_i + Y_i) + \theta_1^i S_{pub}) \\ &= C_i + \theta_3^i (B_i + \theta_1^i S_{pub}) \end{aligned}$$

The aggregate signature verification is established as:

$$\begin{aligned} \delta_{agg} P &= \Sigma_{i=1}^n \delta_i P = \Sigma_{i=1}^n c_i + \Sigma_{i=1}^n (a_i + d_i)\theta_3^i P \\ &= \Sigma_{i=1}^n c_i + \Sigma_{i=1}^n (a_i + (y_i + \theta_1^i s))\theta_3^i P \\ &= \Sigma_{i=1}^n c_i P + \Sigma_{i=1}^n \theta_3^i (a_i P + y_i P + \theta_1^i sP) \\ &= \Sigma_{i=1}^n C_i + \Sigma_{i=1}^n \theta_3^i ((A_i + Y_i) + \theta_1^i S_{pub}) \\ &= \Sigma_{i=1}^n C_i + \Sigma_{i=1}^n \theta_3^i (B_i + \theta_1^i S_{pub}) \end{aligned}$$

5 Analysis of Security

In this section, we formally prove the security of our scheme in the random oracle model (ROM) against ADV_I and ADV_{II} adversaries. The adversary

ADV_I represents an outsider attacker who can substitute the public key of any vehicle, but has no access to the system's master secret key. In contrast, ADV_{II} denotes a malicious SP who has access to the master secret and can generate the partial private key of any vehicle, but cannot substitute any public key. We then demonstrate that the scheme satisfies the other essential security requirements for anonymous authentication.

5.1 Formal Security Proofs of the Scheme

Here, we prove ciphertext unforgeability and confidentiality of the scheme against any probabilistic polynomial time ($\mathcal{PPT}$) adversaries ADV_I and ADV_{II}. The proofs are structured as a challenge-response game between a challenger $\mathcal{C}$ and an adversary.

Theorem 1. *(Ciphertext Confidentiality): The scheme achieves indistinguishability under adaptive chosen-ciphertext attacks (IND-CCA2) against ADV_I, based on the assumption of CDH hardness in the random random oracle model (ROM).*

Proof: On the input of a CDH instance $(G, P, \alpha P, \beta P)$, if the adversary ADV_I can break IND-CCA2 of the scheme with non-negligible advantage ϵ, then the challenger $\mathcal{C}$ can solve the CDH problem with the following probability: $\epsilon' \geq \epsilon\left(1 - \frac{1}{q_{H_1}}\right)^{q_{ppk}}\left(1 - \frac{q_{UNSC}}{2^k}\right)\left(\frac{1}{q_{H_1}q_{H_2}}\right)$, where q_{H_1}, q_{H_2}, q_{ppk} and q_{UNSC} respectively denote the maximum number of queries to H_1, H_2, *Partial Private Key* and *Unsigncryption* oracles.

1. *System Setup*: The challenger $\mathcal{C}$ sets $S_{pub} = \alpha P$ and generates the system parameters $params = \{G, p, P, S_{pub}, H_1, H_2, H_3\}$. Then $\mathcal{C}$ forwards the $params$ to ADV_I.
2. *Phase-I Queries*: After $\mathcal{C}$ selects the target identity ID_I, ADV_I can adaptively query any of the following:
   (a) *Create Vehicle*: $\mathcal{C}$ maintains an initially empty list L_{CV} containing tuples $(ID_i, a_i, A_i, ppk_i, Y_i, B_i, \theta_1^i)$. Upon receiving a query on ID_i from ADV_I, $\mathcal{C}$ replies with $PK_i = (A_i, B_i)$ if the corresponding tuple exists in L_{CV}. Otherwise, if $ID_i \neq ID_I$, $\mathcal{C}$ randomly generates $a_i, ppk_i, \theta_1^i \in \mathbb{Z}_p^*$ and computes $A_i = a_i P$, $Y_i = ppk_i P - \theta_1^i S_{pub}$, sets $y_i = \perp$ and $Y_i = A_i + B_i$. If $ID_i = ID_I$, $\mathcal{C}$ sets $ppk_i = \perp$, randomly generates $a_i, y_i, \theta_1^i \in \mathbb{Z}_p^*$, computes $A_i = a_i P$, $Y_i = y_i P$ and $B_i = A_i + Y_i$. Then, $\mathcal{C}$ sets $PK_i = (A_i, B_i)$ and updates the tuple $(ID_i, a_i, A_i, ppk_i, Y_i, B_i, \theta_1^i)$ in L_{CV} and replies $PK_i = (A_i, B_i)$ to ADV_I as public key.
   (b) *Extract Partial Private Key*: For a query on ID_i, if $ID_i \neq ID_I$, $\mathcal{C}$ retrieves (ppk_i, B_i) from L_{CV} and replies same to ADV_I. Otherwise, $\mathcal{C}$ aborts this query.
   (c) *Set Secret Value*: For a query on ID_i, $\mathcal{C}$ retrieves a_i and returns it to ADV_I.

(d) *Replace Public Key*: For a query (ID_i, PK_i'), $\mathcal{C}$ sets $A_i = A_i', B_i = B_i', a_i = \perp$ in L_{CV}.
(e) H_1-*Oracle*: For a query on (ID_i, A_i, S_{pub}), $\mathcal{C}$ retrieves the tuple $(ID_i, a_i, A_i, ppk_i, Y_i, B_i, \theta_1^i)$ from L_{CV} and returns θ_1^i to ADV_I .
(f) H_2-*Oracle*: $\mathcal{C}$ maintains an initially empty list L_{H_2} of tuples $(U_i, ID_e, PK_e, \theta_2^i)$. For a query on (U_i, ID_e, PK_e), if the tuple exists in L_{H_2}, $\mathcal{C}$ returns θ_2^i to ADV_I. Otherwise, $\mathcal{C}$ randomly generates $\theta_2^i \in \mathbb{Z}_p^*$, updates the L_{H_2}, and returns θ_2^i.
(g) H_3-*Oracle*: $\mathcal{C}$ maintains an initially empty list L_{H_3} of tuples $(m_i, T_s^i, U_i, ID_i, PK_i, \theta_2^i, \theta_3^i)$. For a query on $(m_i, T_s^i, U_i, ID_i, PK_i, \theta_2^i)$, $\mathcal{C}$ returns θ_3^i to ADV_I if the tuple exists in the L_{H_3}; otherwise, $\mathcal{C}$ randomly generates $\theta_3^i \in \mathbb{Z}_p^*$, updates L_{H_3} and returns θ_3^i.
(h) *Signcryption Query*: Upon a query on (ID_i, ID_e, m_i, T_s^i) or $(ID_i, a_i', ID_e, m_i, T_s^i)$ for the situation $PK_i = (A_i, B_i)$ has been substituted with PK_i', $\mathcal{C}$ obtains the matching tuples for ID_i and ID_e from L_{CV}. If $ID_i \neq ID_I$, $\mathcal{C}$ signcrypts m_i and replies the result to ADV_I. But, if $ID_i = ID_I$, $\mathcal{C}$ randomly generates $\delta_i \in \mathbb{Z}_p^*$ and retrieves θ_1^i and θ_3^i from L_{H_1} and L_{H_3} respectively. Next, $\mathcal{C}$ computes $C_i = \delta_i P - \theta_3^i(A_i + \theta_3^i S_{pub})$, $U_i = (a_e + d_e)C_i$, retrieves θ_2^i from L_{H_2} and generates $x_i = m_i \oplus \theta_2^i$. Then, $\mathcal{C}$ updates L_{H_3} with $((m_i, T_s^i, U_i, ID_i, PK_i, \theta_2^i)$ and returns $\sigma_i = (x_i, \delta_i, C_i, T_s^i)$.
(i) *Unsigncryption Query*: Upon a query on $(ID_i, ID_e, \sigma_i, T_s^i)$ or $(ID_i, ID_e, a_e, \sigma_i, T_s^i)$ in case PK_e has been substituted with PK_e', $\mathcal{C}$ retrieves ID_i and ID_e from L_{CV}. If $ID_e \neq ID_I$, $\mathcal{C}$ performs unsigncryption and returns the result to ADV_I. If $ID_e = ID_I$, $\mathcal{C}$ retrieves θ_3^i and θ_1^i from L_{H_3} and L_{H_1} respectively. Then, $\mathcal{C}$ checks if $\delta_i P = C_i + \theta_3^i(B_i + \theta_1^i S_{pub})$; if true, it accepts and returns m_i; otherwise, it rejects σ_i.

3. *Challenge Phase*: ADV_I generates two messages of equal length m_0^* and m_1^* and sends $\{ID_i^*, m_0^*, m_1^*, ID_e^*\}$ to $\mathcal{C}$. $\mathcal{C}$ randomly chooses $j \in \{0, 1\}$ and sets $C_i^* = \beta P$. Then, $\mathcal{C}$ randomly chooses $x_i^*, \delta_i^* \in \mathbb{Z}_p^*$, sets $\sigma_i^* = (x_i^*, \delta_i^*, C_i^*, T_s^{i*})$, and sends it to ADV_I.
4. *Phase-II Queries*: ADV_I can repeat the same queries as in *Phase-I*, and $\mathcal{C}$ replies accordingly. However, ADV_I cannot query either *Partial Private Key* on ID_e^* or the *Unsigncrypt* on m_i^* with ID_i^* and ID_e^*.
5. *Guess Phase*: ADV_I outputs a guess bit j' of j and wins if $j' = j$. If $ID_e^* = ID_I$, and ADV_I queries H_2 oracle on (U_i^*, ID_e^*, PK_e^*), $\mathcal{C}$ can compute $U_i^* = (a_e^* + d_e^*)C_i^* = a_e^* C_i^* + y_e^* C_i^* + \theta_1^{e*}\beta S_{pub} = a_e^* C_i^* + y_e^* C_i^* + \theta_1^{e*}\alpha\beta P$. Thus, $\mathcal{C}$ outputs the CDH solution as $\alpha\beta P = (\theta_1^{e*})^{-1}(C_i^* - (a_e^* + y_e^*)U_i^*)$.

The advantage of $\mathcal{C}$ in the game is defined as follows: probability $\mathcal{C}$ does not abort during *Partial-Private-Key* and *Unsigncryption* queries in the game is at least $\left(1 - \frac{1}{q_{H_1}}\right)^{q_{ppk}}(1 - \frac{q_{UNSC}}{2^k})$. The probability that ID_I has been selected by ADV_I in the *Challenge Phase* is at least $\frac{1}{q_{H_1}}$. In the *Guess Phase*, probability that ADV_I specifically queries the H_2 oracle is $\frac{1}{q_{H_2}}$. Since

ADV_I breaks IND-CCA2 security of the scheme with non-negligible advantage ϵ, $\mathcal{C}$ solves the given CDH instance with non-negligible probability $\epsilon' \geq \epsilon\left(1 - \frac{1}{q_{H_1}}\right)^{q_{ppk}} \left(1 - \frac{q_{UNSC}}{2^k}\right)\left(\frac{1}{q_{H_1}q_{H_2}}\right)$. ■

Theorem 2. *(Ciphertext Confidentiality): The scheme achieves indistinguishability under adaptive chosen-ciphertext attacks (IND-CCA2) against ADV_{II}, based on the assumption of CDH hardness in the random oracle model (ROM).*

Proof: On the input of a CDH instance $(G, P, \alpha P, \beta P)$, if the adversary ADV_{II} can break IND-CCA2 of the scheme with non-negligible advantage ϵ, then the challenger $\mathcal{C}$ can solve the CDH problem with probability: $\epsilon' \geq \epsilon\left(1 - \frac{1}{q_{H_1}}\right)^{q_{ssv}}$ $\left(1 - \frac{q_{UNSC}}{2^k}\right)\left(\frac{1}{q_{H_1}q_{H_2}}\right)$, where q_{H_1}, q_{H_2}, q_{ssv} and q_{UNSC} respectively denote the maximum number of queries to H_1, H_2, *Set Secret Value* and *Unsigncryption* oracles.

This proof proceeds in a manner similar to the proof of Theorem 1 except that ADV_{II} cannot replace the public key of any V_i, nor does it need to query the partial private key. In addition, ADV_{II} is not allowed to query *Set Secret Value* for the challenge identity. ■

Theorem 3. *(Unforgeability) The scheme is existentially unforgeable under chosen message attacks (EUF-CMA) against ADV_I, based on the assumption of ECDL hardness in the random oracle model.*

Proof: On the input of an ECDL instance $(G, P, \alpha P)$, if the adversary ADV_I can break EUF-CMA of the scheme with non-negligible advantage ϵ, then $\mathcal{C}$ can solve the ECDL problem with probability $\epsilon' \geq \epsilon\left(1 - \frac{1}{q_{H_1}}\right)^{q_{ppk}}\left(\frac{1}{q_{H_1}}\right)$, where q_{H_1} and q_{ppk} respectively denote the maximum number of queries to H_1 and *Partial Private Key*.

1. *System Setup and Queries Phases*: These are performed in a similar manner to the corresponding queries in the proof of Theorem 1.
2. *Forgery Phase*: ADV_I chooses ID_i^* and ID_e^* as identities of sender and recipient respectively. For this recipient ID_e^*, the ADV_I forges a ciphertext σ_i^* on m_i^* for the ID_i^*. Although ADV_I never extracts the partial private key for ID_i^* or queries the *Signcrypt* on (ID_i^*, ID_e^*, m_i^*) to obtain $\sigma_i^* = (x_i^*, \delta_i^*, C_i^*, T_s^{i*})$, the ciphertext remains valid. According to the Forking Lemma [24], ADV_I can produce another ciphertext $\sigma_1' = (x_i^*, \delta_i', C_i^*, T_s^{i*})$ with same commitment but with $\theta_3^{i*} \neq \theta_3^{i'}$. Therefore, from the resulting equations:

$$\delta_i^* P = C_i^* + \theta_3^{i*}(B_i^* + \theta_1^{i*} S_{pub}) \quad (1)$$
$$\delta_i' P = C_i^* + \theta_3^{i'}(B_i^* + \theta_1^{i*} S_{pub}) \quad (2)$$
$$\delta_i^* P = C_i^* + \theta_3^{i*}(A_i^* + Y_i^* + \theta_1^{i*} \alpha P) \quad (3)$$
$$\delta_i' P = C_i^* + \theta_3^{i'}(A_i^* + Y_i^* + \theta_1^{i*} \alpha P) \quad (4)$$

Then, $\mathcal{C}$ successfully solves the ECDL problem as:

$$\alpha = \left((\delta_i^* - \delta_i') (\theta_3^{i*} - \theta_3^{i'})^{-1} - a_i^* - y_i^*\right) (\theta_1^*)^{-1} \bmod p$$

The advantage of $\mathcal{C}$ in the game is defined as follows: the probability $\mathcal{C}$ does not abort during *Partial-Private-Key* queries is at least $\left(1 - \frac{1}{q_{H_1}}\right)^{q_{ppk}}$ in the game. While, the probability that $\mathcal{C}$ does not abort in the forgery phase is at least $\frac{1}{q_{H_1}}$. Therefore, $\mathcal{C}$ solves the ECDL problem with non-negligible probability $\epsilon' \geq \epsilon \left(1 - \frac{1}{q_{H_1}}\right)^{q_{ppk}} \frac{1}{q_{H_1}}$. ■

Theorem 4. *(Unforgeability) The scheme is existentially unforgeable under chosen message attacks (EUF-CMA) against ADV_{II}, based on the assumption of ECDL hardness in the random oracle model.*

Proof: On the input of an ECDL instance $(G, P, \alpha P)$, if the adversary ADV_{II} can break EUF-CMA of the scheme with non-negligible advantage ϵ, then $\mathcal{C}$ can solve the ECDL problem with probability $\epsilon' \geq \epsilon \left(1 - \frac{1}{q_{H_1}}\right)^{q_{ssv}} \left(\frac{1}{q_{H_1}}\right)$, where q_{H_1} and q_{ssv} respectively denote queries to H_1 oracle and S*Set Secret Value.*

This proof proceeds in a manner similar to the proof of Theorem 3 except that ADV_{II} cannot replace public key of any V_i, nor does it need to query the partial private key. In addition, ADV_{II} is not allowed to query *Set Secret Value* for the challenge identity. ■

Table 1. Comparison of Security Properties

Schemes	Confidentiality	Unforgeability	Traceability	Unlinkability	Anonymity
CHW [8]	√	√	×	×	√
WGZ [10]	√	√	×	×	×
ZLLCFD [18]	√	√	√	–	√
WC [22]	√	√	√	unproven	√
Ours	√	√	√	√	√

5.2 Security Requirements Analysis

In addition to the formal security proofs in the previous subsection, analysis of the security requirements is as follows.

1. *Confidentiality*: The formal security proofs in Theorems 1 and 2 have proved that our scheme meets confidentiality requirements.
2. *Message Authentication and Integrity*: The formal security proofs of unforgeability in Theorems 3 and 4 have shown that our scheme provides message authentication and integrity.

3. *Anti-Replay Attacks*: A fresh timestamp attached to every ciphertext based on which a recipient can determine the freshness of a received message makes our scheme replay-proof. Thus, the timestamp enables an edge node to uncover any replay message.
4. *Unlinkability*: Since a V_i applies a unique $(PID_i, SK_i, PK_i, c_i, C_i)$ to every new message, this results to a unique and randomized ciphertext. Consequently, any two or more ciphertexts originating from the same sender V_i remain unlinkable.
5. *Anonymity:* A pseudonym PID_i in every signcryption phase hides the real identity ID_i of a vehicle since $PID_i = ID_i \oplus H_1(sA_i)$. This technique prevents any receiver (edge node) from discovering the ID_i in the ciphertext. Thus, without knowing the master secret key s, disclosing ID_i from a PID_i requires solving the ECDL problem in sA_i.
6. *Traceability:* Since the trusted authority (TA) owns the master secret s and is assigned to conditionally trace a malicious vehicle V_i from a PID_i, TA alone can recover that vehicle's real identity as $ID_i = PID_i \oplus H_1(sA_i)$.

Table 2. Running Times of ECC Operations

Operation	Notation	Running time (ms)
Scalar multiplication	T_s	1.0254
Point addition	T_{pa}	0.0369
Hashing to $\mathbb{Z}_p^*$	T_h	0.00696

Security Comparison. The result illustrated in Table 1 compares the security requirements between these related schemes [8,10,18,22] and ours. The symbol $\surd$ in the table denotes that a scheme satisfies the requirement, while $\times$ indicates the scheme does not satisfy the requirement. Moreover, the $-$ symbol signifies the scheme does not demonstrate the feature. The result has shown that only our scheme meets all the essential properties for anonymous authentication in VANETs.

6 Performance Analysis

This section analyses the performances of our scheme and the relevant schemes [8,10,18,22] from the perspective of costs of computations and communications. The analysis is carried out based on the experimental setting that ensures 80 bits security level using JPBC library from [25]. The setting uses an additive group G on the curve $E : y^2 = (x^3 + ax + b) \bmod p$, where $a, b \in \mathbb{F}_p$. The prime number p is 20 bytes long. The experiments have been run on a computer with Intel i5-6200U CPU 2.30 GHz(4 CPUs) and 8 GB memory.

6.1 Cost of Computation Analysis

Table 2 provides the execution times of the fundamental operations on the elliptic curve for this experiment. Based on that, the execution times for signcryption, single unsigncryption and aggregate unsigncryption algorithms of the schemes [8,10,18,22] are presented in Table 3. The signcryption algorithms in [8,18] and our scheme employ offline operations as a mechanism of improving the computational performance. This approach facilitates pre-computation of message-independent operations ahead of real-time communication that applies online phase. The [8] specifically incurs $T_s + 4T_h = 1.0532$ ms in online signcryption, while the [18] requires $T_s + 3T_h = 1.0463$ ms. However, our scheme needs $2T_h = 0.0139$ ms only. On the other hand, the schemes [10,22] which rather support online computations only, their signcryption operations incur $3T_s + 2T_{pa} + 2T_h = 3.1639$ ms and $3T_s + 12_{pa} + 4T_h = 3.1778$ ms respectively. For the unsigncryption algorithms, a single ciphertext [8] costs $5T_s + 2T_{pa} + 3T_h = 5.2217$ ms, while [10] and [18] respectively require $4T_s + 3T_{pa} + 3T_h = 4.2332$ ms and $5T_s + 2T_{pa} + 4T_h = 5.2286$ ms. However, the scheme [22] incurs $5T_s + 3T_{pa} + 4T_h = 5.2655$ ms which is the largest, while our scheme requires $4T_s + 2T_{pa} + 3T_h = 4.1963$ ms clearly the lowest time. Furthermore, it can be deduced from the table that our scheme still has the least cost of unsigncrypting multiple ciphertexts through aggregate unsigncryption. To better understand the results, Fig. 2 clearly displays the costs of computations for signcryption algorithms in the compared schemes. Meanwhile, costs of single unsigncryption for 10, 20 and 30 ciphertexts are presented in Fig. 3. Last but not the least, for dense traffic scenarios, Fig. 4 illustrates the costs of both single and aggregate unsigncryption for 300 ciphertexts. Therefore, having the lowest costs of computation for online signcryption, single and aggregate unsigncryption algorithms, our scheme outperforms the other related schemes.

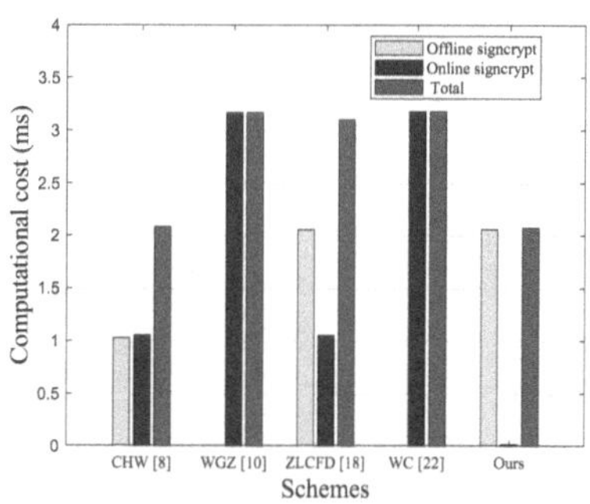

Fig. 2. Costs for signcrypting a message.

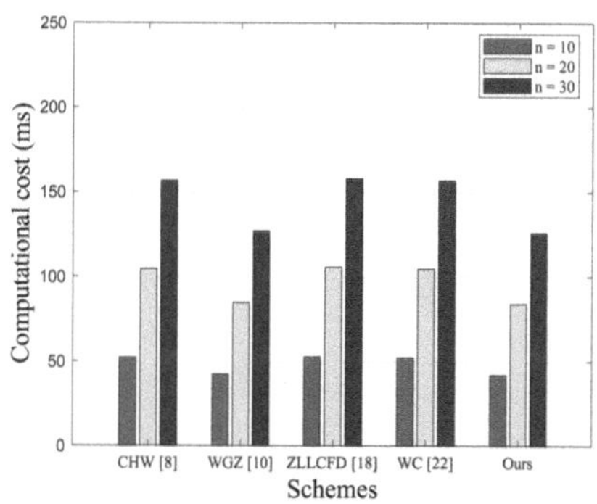

Fig. 3. Costs for single unsigncryption on 10, 20 and 30 ciphertexts.

Table 3. Computational Costs Comparison

Schemes	Single Signcryption (ms)	Single Unsigncryption (ms)	Aggregate Unsigncryption
CHW [8]	offline (T_s) + online ($T_s + 4T_h$)	$5T_s + 2T_{pa} + 3T_h$	$(4n+1)T_s + (3n-1)T_{pa} + (2n+1)T_h$
WGZ [10]	online ($3T_s + 2T_{pa} + 2T_h$)	$4T_s + 3T_{pa} + 3T_h$	$(3n+3)T_s + (4n+1)T_{pa} + (3n+1)T_h$
ZLLCFD [18]	offline (T_s) + online ($T_s + 3T_h$)	$5T_s + 2T_{pa} + 4T_h$	$(4n+1)T_s + (3n-1)T_{pa} + (3n+1)T_h$
WC [22]	online ($3T_s + 2T_{pa} + 4T_h$)	$5T_s + 3T_{pa} + 4T_h$	$5nT_s + 3nT_{pa} + 4nT_h$
Ours	offline ($2T_s + T_h$) + online ($2T_h$)	$4T_s + 2T_{pa} + 3T_h$	$(3n+1)T_s + (3n-1)T_{pa} + 3nT_h$

6.2 Cost of Communications Analysis

Here, we analyze the ciphertext sizes of the schemes [8,10,18,22] and ours. Over the finite field $|\mathbb{F}_p|$ where p is 20 bytes long, an element in the additive ECC group $|G|$ costs 2×20 bytes $= 40$ bytes. The $|m|$, as the length of a single plaintext message, is assumed to be 20 bytes, while the $|T|$ marks a timestamp size as 4 bytes. In the simulation, each ciphertext comprises $(2|G|)$ as a public key, $|\mathbb{Z}_p^*|$ as a sender's identity or pseudonym, and $(|G|, 2|\mathbb{Z}_p^*|, |T|)$ as the actual signcryption result with its timestamp. Combining a sender's public key, its identity or pseudonym and the signcryption result, the total data a vehicle V_i transmits per session is $3|G| + 3|Z_p^*| + |m| + |T|$. A ciphertext in all the schemes incurs the same communication cost in the experiment.

Therefore, maintaining a relatively equivalent ciphertext size in addition to achieving all the security requirements with the lowest computational cost for both online signcryption and unsigncryption, our scheme outperforms the relevant existing schemes.

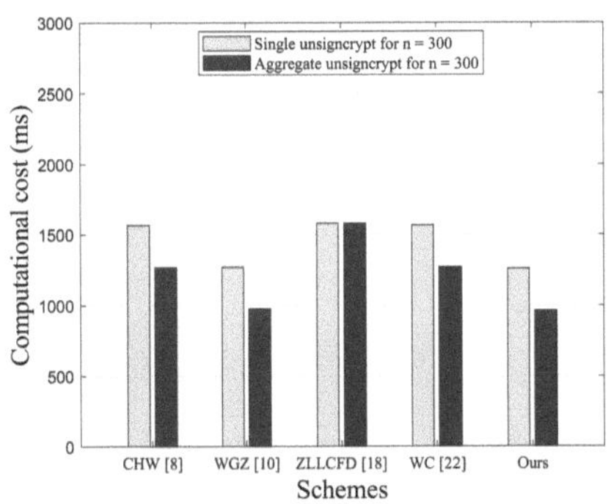

Fig. 4. Comparison of computational costs for single unsigncryption and aggregate unsigncryption on 300 ciphertexts.

7 Conclusion

In this paper, we propose a lightweight signcryption scheme for anonymous authentication in VANETs. While eliminating bilinear pairings, our scheme shifts the time-consuming computational operations to offline as well as supporting aggregate unsigncryption, an important feature in dense traffic scenarios. Formal security proofs of confidentiality and unforgeability have demonstrated that our scheme is secure, in addition to satisfying all the other essential security requirements for anonymous authentication. Performance evaluation has shown that our scheme significantly improves the efficiency compared to the related schemes. Therefore, these desirable properties make our proposed scheme suitable for anonymous authentication in VANETs.

Acknowledgments. This work was supported by the Sichuan Science and Technology Program (Grant No. 2025YFHZ0320).

Disclosure of Interests. The authors have no competing interests to declare that are relevant to the content of this article.

References

1. Imam, A.Y., Li, F.: Efficient lightweight anonymous authentication scheme with certificateless aggregation for VANETs. J. Inf. Secur. Appl. **91**, 104053–104053 (2025)
2. Lee, M., Atkison, T.: VANET applications: past, present, and future. Veh. Commun. **28**, 100310 (2021)
3. Ali, I., Chen, Y., Li, J., Wakeel, A., Pan, C., Ullah, N.: Efficient offline/online heterogeneous-aggregated signcryption protocol for edge computing-based internet of vehicles. IEEE Trans. Intell. Transp. Syst. **24**(12), 14506–14519 (2023)
4. Elkhalil, A., Zhang, J., Elhabob, R., Eltayieb, N.: An efficient signcryption of heterogeneous systems for Internet of Vehicles. J. Syst. Architect. **113**, 101885 (2021)
5. Zheng, Y.: Digital signcryption or how to achieve cost(signature & encryption) << cost(signature) + cost(encryption). In: Proceedings of Annual International Cryptology Conference, pp. 165–179. Springer, Cham (1997)
6. Lu, H., Xie, Q.: An efficient certificateless aggregate signcryption scheme from pairings. In: Proceedings of 2011 International Conference on Electronics, Communications and Control (ICECC), pp. 132-135. Ningbo, China (2011)
7. Basudan, S., Lin, X., Sankaranarayanan, K.: A privacy-preserving vehicular crowdsensing-based road surface condition monitoring system using fog computing. IEEE Internet Things J. **4**(3), 772–782 (2017)
8. Cui, M., Han, D., Wang, J.: An efficient and safe road condition monitoring authentication scheme based on fog computing. IEEE Internet Things J. **6**(5), 9076–9084 (2019)
9. Xie, Z., Chen, Y., Ali, I., Pan, C., Li, F., He, W.: Efficient and secure certificateless signcryption without pairing for edge computing-based internet of vehicles. IEEE Trans. Veh. Technol. **72**(5), 5642–5653 (2023)
10. Wang, B., Guo, X., Zhang, S.: A secure and efficient pairing-free certificateless aggregate signcryption scheme for V2G networks. IEEE Syst. J. **18**(1), 282–293 (2024)
11. Raya, M., Hubaux, J.-P.: Securing vehicular ad hoc networks. J. Comput. Secur. **15**(1), 39–68 (2007)

12. Shamir, A.: Identity-based cryptosystems and signature schemes. In: Blakley, G.R., Chaum, D. (eds.) CRYPTO 1984. LNCS, vol. 196, pp. 47–53. Springer, Heidelberg (1985). https://doi.org/10.1007/3-540-39568-7_5
13. Yu, H.F., Yang, B.: Identity-based hybrid signcryption scheme using ECC. J. Softw. **26**(12), 3174–3182 (2015)
14. Al-Riyami, S.S., Paterson, K.G.: Certificateless public key cryptography. In: Laih, C.-S. (ed.) ASIACRYPT 2003. LNCS, vol. 2894, pp. 452–473. Springer, Heidelberg (2003). https://doi.org/10.1007/978-3-540-40061-5_29
15. Liu, Z., Hu, Y., Zhang, X., Ma, H.: Certificateless signcryption scheme in the standard model. Inf. Sci. **180**(3), 452–464 (2010)
16. Chen, Z., Jin, C., Chen, G., Jin, Y., Zong, H.: A heterogeneous online/offline signcryption scheme for Internet of Vehicles. Veh. Commun. **43**, 100635–100635 (2023). https://doi.org/10.1016/j.vehcom.2023.100635
17. Weng, J., Yao, G., Deng, R.H., Chen, M.-R., Li, X.: Cryptanalysis of a certificateless signcryption scheme in the standard model. Inf. Sci. **181**(3), 661–667 (2011)
18. Zhang, W., Liu, S., Liu, Y., Cao, J., Fu, B., Du, Y.: A certificateless online/offline aggregate signcryption scheme against collusion attacks based on fog computing. Electronics **12**(23), 4747 (2023)
19. Xie, W., Zhang, Z.: Efficient and provably secure certificateless signcryption from bilinear maps. In: Proceedings of IEEE International Conference on Wireless Communication, Network Information Secure, pp. 558–562 (2010)
20. Bonomi, F., Milito, R., Zhu, J., Addepalli, S.: Fog computing and its role in the Internet of Things. In: Proceedings of 1st, Ed., MCC Workshop Mobile Cloud Computing, pp. 13–16 (2012)
21. Yu, H., Ren, R.: Certificateless elliptic curve aggregate signcryption scheme. IEEE Syst. J. **16**(2), 2347–2354 (2022)
22. Wu, W., Chen, J.: A security-enhanced certificateless aggregate signcryption scheme for vehicular ad hoc networks. Veh. Commun. **53**, 100897 (2025)
23. Chen, Y., Lu, Z., Xiong, H., Xu, W.: Privacy-preserving data aggregation protocol for fog computing-assisted vehicle-to-infrastructure scenario. Secur. Commun. Netw. **2018**, 1–14 (2018)
24. Pointcheval, D., Stern, J.: Security arguments for digital signatures and blind signatures". J. Cryptol. **13**(3), 361–396 (2000)
25. De Caro, A., Iovino, V.: jPBC: java pairing based cryptography. In: IEEE Symposium on Computers and Communications (ISCC). Kerkyra, Greece, vol. 2011, pp. 850–855 (2011)
26. Wang, Y., Jia, X., Bao, Y., Cao, Y., Wen, J.: Efficient and provably secure offline/online heterogeneous Signcryption scheme for VANETs. IEEE Internet Things J. **11**(24), 41248–41260 (2024)

A Multi-modal Data Recommendation Model Based on CNN

Xu Zhang[1], Xu Chen[1], Yijie Gao[2], Junxi Zhu[1], Yanjin Kong[1], and Yilei Wang[1(✉)]

[1] School of Computer Science, Qufu Normal University, Rizhao 276827, China
wang_yilei2019@qfnu.edu.cn

[2] School of Computer Science, Yantai Engineering Vocational and Technical College, Yantai 264006, China

Abstract. Against the background of the increasing popularity of digital reading and the explosive growth of book resources, traditional recommender systems face significant challenges in dealing with the cold-start and sparse data problems. To this end, this paper proposes a book recommendation model that integrates multimodal features, combines convolutional neural networks (CNN) with structured and unstructured data, and synthesizes book cover images, text profiles, numerical indicators, and categorical information for modeling. EfficientNetB0 is used to extract image features, TF-IDF to extract text features, combined with embedded coding and standardized processing of numerical features, in order to improve the feature expression ability. Finally, the Web terminal is constructed by Flask, and the system supports three modes of ISBN similarity map recommendation, preference matching recommendation and popular book recommendation to realize multi-scene adaptation. Through multimodal feature fusion, the system effectively improves the accuracy and practicability of recommendation, and has good popularization value.

Keywords: Book recommendation · Convolutional neural network (CNN) · EfficientNet · TF-IDF · Hybrid model

1 Introduction

With the rapid development of internet technology and the widespread adoption of digital reading, book resources are growing at an exponential rate. Faced with an overwhelming amount of book data, users encounter the problem of "information overload" when selecting books that align with their interests and needs. Personalized recommendation technology has emerged as a solution. Recommendation systems analyze user behavior and book characteristics to provide users with books that better match their preferences, significantly improving information retrieval efficiency and reading experience.

Traditional recommendation methods primarily include collaborative filtering and content-based recommendation [1, 2]. While these methods have improved recommendation effectiveness to some extent, they exhibit significant shortcomings in areas such as data sparsity, cold start issues [3, 4], and feature dimension expansion. In recent years,

J. Yu and Y. Chen (Eds.): FCS 2025, CCIS 2805, pp. 257–264, 2027.
https://doi.org/10.1007/978-981-95-9514-3_17

convolutional neural networks (CNNs) have been increasingly adopted in the field of recommendation systems due to their outstanding performance in image recognition [5] and natural language processing [6, 7]. CNNs excel at extracting deep semantic information from unstructured data, making them particularly suitable for processing multimodal data such as book covers and content summaries, thereby offering new insights for optimizing book recommendation systems [8].

In response to the above challenges, this paper proposes a CNN book recommendation system that integrates images, text, and structured information. The main contributions include:

1) Proposing a multi-modal recommendation framework: integrating book cover images, text features such as book titles and author biographies, and structured data such as book recommendation scores and positive review rates to construct a hybrid feature input, thereby enhancing the system's recommendation diversity and robustness.
2) Introducing the EfficientNetB0 model: Utilizing a lightweight CNN model to extract cover image features, improving feature extraction efficiency and reducing system deployment costs.
3) Implementing multiple recommendation strategies: Building three recommendation services based on ISBN input, category preferences, and global popularity, adapting to different user scenarios and enhancing the system's practicality and flexibility.
4) Developing an interactive web platform: Using the Flask framework to develop a frontend-backend integrated web application, enabling visualization of recommendation functions and user interaction, and validating the system's feasibility.
5) This study not only expands the application pathways of deep learning-based recommendation systems theoretically but also provides effective solutions for improving user experience and optimizing resource allocation on book platforms in practice.

2 Preliminaries

Before constructing a book recommendation system based on convolutional neural networks, it is important to understand the following three key theoretical foundations for the design and implementation of the system. This paper selects convolutional neural networks (CNN), Term Frequency-Inverse Document Frequency (TF-IDF) text feature extraction methods [9], and collaborative filtering and cold start problems as the core theoretical foundations supporting this study.

2.1 Convolutional Neural Network

Convolutional neural networks (CNNs) are a typical deep learning architecture widely used in image processing and visual recognition tasks. Their core mechanisms include local connections, weight sharing, and multi-layer feature abstraction, enabling effective extraction of both local and global features from images. The main components include:

1) Convolution layers: Scan the input image using convolution kernels to extract local spatial features;

2) Pooling layers: Reduce the dimensions of feature maps to enhance the model's translation invariance;
3) Fully connected layer: maps the extracted high-level features to classification or regression tasks;
4) Activation function: commonly uses the ReLU activation function to introduce nonlinear expressive capabilities.

This system adopts the lightweight EfficientNetB0 network [10] as the image feature extraction module. This model achieves a balance between parameter count and model accuracy through a composite scaling strategy, making it particularly suitable for efficient inference tasks in resource-constrained scenarios.

2.2 TF-IDF (Term Frequency-Inverse Document Frequency)

It is a classic text weighting statistical method widely used in information retrieval and text mining [11–13]. Its core idea is that the more frequently a word appears in a document and the less frequently it appears in the entire corpus, the higher its importance. The calculation formula is as follows:

$$TF - IDF(t, d) = TF(t, d) \times log\left(\frac{N}{DF(t)}\right)$$

In this study, the system uses TF-IDF to perform vectorization processing on text information such as book descriptions, titles, and publishers. These text features are then fused with image features to jointly participate in the construction of the book recommendation model [14], thereby enhancing the system's ability to understand semantic information.

2.3 Collaborative Filtering and the Cold Start Problem

Collaborative filtering is a recommendation technique based on user historical behavior or preference data [15, 16], widely used in various recommendation systems. It is mainly divided into two categories:

1) User-based collaborative filtering: recommends items that other users like by analyzing the similarity between users [17];
2) Item-based collaborative filtering: recommends items based on user ratings or behavioral similarity [18].

Although collaborative filtering methods have high recommendation accuracy, they also face two key issues:

1) Data sparsity: Most users interact with only a small number of items, resulting in a severely sparse user-item rating matrix [19];
2) Cold start problem: For new users or new items, the system lacks sufficient historical information to provide effective recommendations [20].

To address these issues, this paper introduces CNN to extract image features, combines TF-IDF to obtain text semantic information, and constructs multi-modal feature inputs to overcome the limitations of collaborative filtering in cold start and data sparsity scenarios, thereby achieving a "content-driven" recommendation mechanism.

3 Model Introduction

Given the diversity of book information, this paper adopts four types of feature representation methods to construct four feature submodules: image, text, numerical, and category (Table 1).

1) The image feature branch uses the pre-trained EfficientNetB0 network to extract deep features from book covers. This network was pre-trained on the ImageNet dataset and can efficiently capture the visual content of images. This branch further reduces the dimension to 256 dimensions using a fully connected layer, employs the ReLU activation function, and combines L2 regularization to prevent overfitting.
2) The text feature branch handles book title, author, and keyword info. It converts text to 500-D sparse vectors via TF-IDF, then extracts features through a 128-D fully connected layer. TF-IDF retains key lexical features, and the output matches the image branch's dimension for feature fusion.
3) The numerical feature branch handles four standardized book features (recommendation rate, positive review rate, price, publication year) via a 64-D fully connected layer to capture nonlinear relationships. These numerical metrics provide objective balance to subjective text and image features.
4) The category feature branch employs an embedding layer to convert discrete book category labels into 8-D continuous vectors, capturing semantic relationships so similar categories are closer in space. These vectors are then flattened and combined with other features for further processing.

Table 1. Key feature processing method.

Feature type	Handling method	Dimension
Image features	EfficientNetB0 Global average pooling	1280
Text features	TF-IDF vectorization	500
Numerical features	MinMaxScaler standardization	4
Category features	LabelEncoder + Embedding	8

The core component of the model is the feature fusion layer, which concatenates the outputs of the four branches along the channel dimension. Image features are passed through a fully connected layer with an output dimension of 256, text features through a fully connected layer with an output dimension of 128, and numerical features through a fully connected layer with an output dimension of 64. The category feature branch flattens the output of the embedding layer into a one-dimensional vector. Two fully connected layers are then used to perform nonlinear transformations with dimensions of 256 and 128, respectively. A Dropout layer is added in between to prevent overfitting. Finally, the Sigmoid activation function is used to output prediction values within the [0,1] range, indicating the user's preference for the book [21].

4 Experimental Results and Analysis

The model training uses a binary cross-entropy loss function and Adam optimizer, with a learning rate set to 0.001. To prevent overfitting, an early stopping strategy and corresponding model checkpoint callback are used. The evaluation metrics include accuracy and AUC values to comprehensively assess the model's recommendation effectiveness. The training data is divided into a training set and a test set at a ratio of 8:2 to ensure that the model has excellent generalization capabilities.

During the evaluation phase, the evaluate_model function is used to assess the performance of the trained model. This function accurately calculates metrics such as precision, recall, F1 score, and root mean square error (RMSE) for the model on both the training and testing datasets. These metrics reflect the model's recommendation effectiveness from different perspectives: precision and recall indicate the accuracy and completeness of the model's predictions for positive samples, the F1 score represents a balance between precision and recall, and RMSE measures the difference between predicted values and actual values.

From the experimental results, by selecting 100 different weight distributions, the RMSE values obtained on the training set and test set are shown in Fig. 1. A clear minimum RMSE value can be observed. RMSE represents the square root of the average of the squares of the differences between the predicted values and the actual values, and it measures the accuracy of the model's predictions. A low RMSE value indicates that the difference between the model's predicted values and the actual values is small, meaning the prediction results are closer to the actual values. Since the RMSE value is optimal, in the recommendation engine module, we selected a weighting distribution where image feature similarity accounts for 25%, text feature similarity accounts for 3%, numerical feature similarity accounts for 27%, and category feature similarity accounts for 45%. This weighting distribution enables more precise recommendations for users.

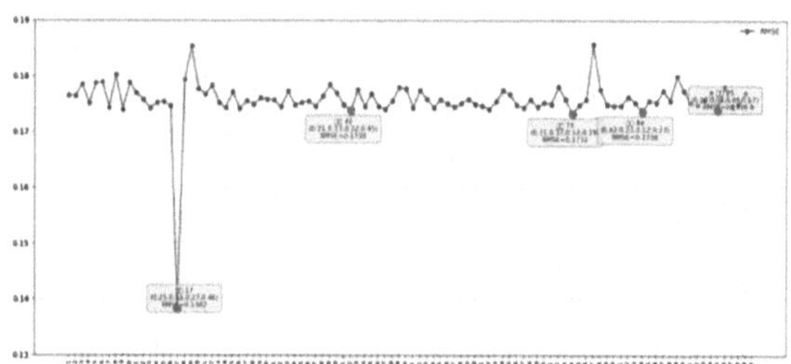

Fig. 1. Weighting Indicator Results Diagram

In 'recommendation_model.py', the visualization analysis of model performance comprehensively evaluates the effectiveness of the book recommendation system through a series of charts. First, the system trains on 100 different weight combinations, with each combination corresponding to a bar chart that displays the performance of four key metrics: Precision, Recall, F1 Score, and RMSE. These charts visually compare the stability and generalization capabilities of the model on the training set under different weighting schemes, aiding in the selection of the optimal parameter combination. Subsequently, the system retained the best-performing weight combination, retrained the model, and generated a bar chart representing the final model performance. The Precision, Recall, and F1 scores are close to ideal values, and the RMSE remains at a low level, indicating that the model has strong discriminative ability in the binary classification task of predicting whether a user will "recommend" a book.

Furthermore, the system recorded the metric changes of the best model at each epoch during training, plotting four smooth curves corresponding to the trends of Precision, Recall, F1, and RMSE as the epoch increases. These curves indicate that the model converges rapidly within the first 20 epochs, with F1 and Recall steadily increasing, Precision remaining at a high level, and RMSE gradually decreasing. This suggests that the training process is stable and does not exhibit significant overfitting or oscillation. Additionally, the system generates separate trend charts for each metric, making it easy to highlight the performance of a specific dimension in reports or presentations, such as the improvement process of F1 or the convergence path of RMSE. Overall, these charts not only validate the model's effectiveness but also provide reliable basis for subsequent parameter tuning or deployment (Fig. 2).

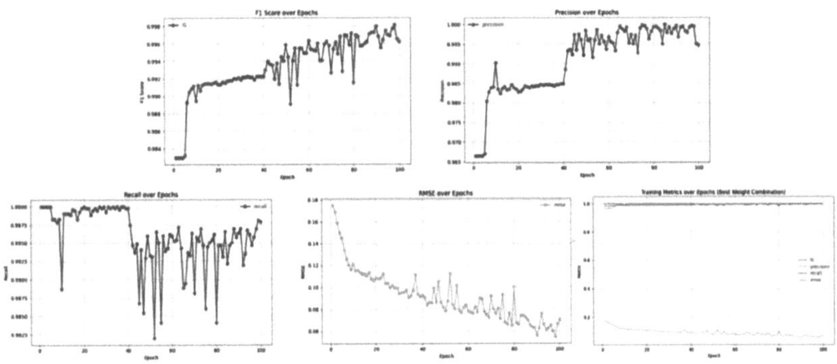

Fig. 2. Model Indicator Diagram

Compared with traditional recommendation systems, the innovation of this model lies in its deep integration of visual and semantic features of books. The cover image serves as an essential visual clue that reflects the style and theme of the book, while the text description provides a more direct indication of the content. This multimodal integration approach significantly improves the relevance and diversity of recommendations.

5 Summary and Outlook

This paper proposes a book recommendation model that integrates multi-modal features, combining four types of heterogeneous information: images, text, numerical data, and categories. The model achieves unified modeling and score prediction through a deep neural network. Experimental results on real-world datasets demonstrate that this model outperforms traditional collaborative filtering and single-modal recommendation methods in terms of recommendation performance and robustness.

The model's structure is well-designed, and its feature fusion strategy is efficient. Experimental results show that this method achieves high levels of accuracy and recall rate, demonstrating its promising application prospects for various recommendation scenarios such as book e-commerce and online reading platforms.

Future research can be expanded in the following areas:

1) Incorporate user behavior data (such as clicks, views, and bookmarks) to build user profiles and enable personalized modeling;
2) Apply attention mechanisms or graph neural networks to uncover higher-order relationships between features and enhance the model's expressive capabilities;
3) Integrate large language models (such as ChatGPT and BERT) to generate context-aware text features;
4) Explore cross-platform or cross-domain transfer learning mechanisms to improve the model's generalization capabilities.

In summary, this paper provides a feasible deep learning solution for multimodal recommendation systems, laying the theoretical and technical foundation for future personalized recommendation research and practical system deployment.

References

1. Afoudi, Y., Lazaar, M., Al, A.M.: Hybrid recommendation system combined content-based filtering and collaborative prediction using artificial neural network. Simul. Model. Pract. Theory **113**, 102375 (2021)
2. Widayanti, R., Chakim, M.H.R., Lukita, C., et al.: Improving recommender systems using hybrid techniques of collaborative filtering and content-based filtering. J. Appl. Data Sci. **4**(3), 289–302 (2023)
3. Chen, X., Yang, R., Xue, Y., et al.: A novel momentum prototypical neural network to cross-domain fault diagnosis for rotating machinery subject to cold-start. Neurocomputing **555**, 126656 (2023)
4. Mishra, A., Latorre, J.A., Pool, J., et al.: Accelerating sparse deep neural networks. arxiv preprint arxiv:2104.08378 (2021)
5. Shou, Y., Meng, T., Ai, W., et al.: Conversational emotion recognition studies based on graph convolutional neural networks and a dependent syntactic analysis. Neurocomputing **501**, 629–639 (2022)
6. Taye, M.M.: Theoretical understanding of convolutional neural network: concepts, architectures, applications, future directions. Computation **11**(3), 52 (2023)
7. Muduli, D., Dash, R., Majhi, B.: Automated diagnosis of breast cancer using multi-modal datasets: a deep convolution neural network based approach. Biomed. Signal Process. Control **71**, 102825 (2022)
8. Kong, Z., Zhang, M., Zhu, W., et al.: Multi-modal data Alzheimer's disease detection based on 3D convolution. Biomed. Signal Process. Control **75**, 103565 (2022)
9. Lu, S., Ding, Y., Liu, M., et al.: Multiscale feature extraction and fusion of image and text in VQA. Int. J. Comput. Intell. Syst. **16**(1), 54 (2023)
10. Bhuvaneswari, R., Krishna, C.A.: Deep learning for facial expression recognition using EfficientNetB0. In: 2023 Global Conference on Information Technologies and Communications (GCITC), pp. 1–6. IEEE (2023)

11. Naeem, M.Z., Rustam, F., Mehmood, A., et al.: Classification of movie reviews using term frequency-inverse document frequency and optimized machine learning algorithms. PeerJ Comput. Sci. **8**, e914 (2022)
12. Dey, R.K., Das, A.K.: Modified term frequency-inverse document frequency based deep hybrid framework for sentiment analysis. Multimedia Tools Appl. **82**(21), 32967–32990 (2023)
13. Al-Obaydy, W.N.I., Hashim, H.A., Najm, Y.A., et al.: Document classification using term frequency-inverse document frequency and K-means clustering. Indon. J. Electr. Eng. Comput. Sci. **27**(3), 1517–1524 (2022)
14. Ahmed, E., Letta, A.: Book recommendation using collaborative filtering algorithm. Appl. Comput. Intell. Soft Comput. **2023**(1), 1514801 (2023)
15. Zhang, Y., Liu, Z., Sang, C.: Unifying paragraph embeddings and neural collaborative filtering for hybrid recommendation. Appl. Soft Comput. **106**, 107345 (2021)
16. Koren, Y., Rendle, S., Bell, R.: Advances in collaborative filtering. In: Recommender Systems Handbook, pp. 91–142 (2021)
17. Na, L., Ming-xia, L., Hai-yang, Q., et al.: A hybrid user-based collaborative filtering algorithm with topic model. Appl. Intell. **51**(11), 7946–7959 (2021)
18. Ajaegbu, C.: An optimized item-based collaborative filtering algorithm. J. Ambient. Intell. Humaniz. Comput. **12**(12), 10629–10636 (2021)
19. Ahmadian, S., Joorabloo, N., Jalili, M., et al.: Alleviating data sparsity problem in time-aware recommender systems using a reliable rating profile enrichment approach. Expert Syst. Appl. **187**, 115849 (2022)
20. Yuan, H., Hernandez, A.A.: User cold start problem in recommendation systems: a systematic review. IEEE access **11**, 136958–136977 (2023)
21. Waoo, A.A., Soni, B.K.: Performance analysis of sigmoid and relu activation functions in deep neural network. In: Intelligent Systems: Proceedings of SCIS 2021, pp. 39–52. Springer, Singapore (2021)

Handwritten Signature Verification via Multimodal Fusion with Global-Local Features

Zhaosen Shi and Fagen Li(✉)

School of Computer Science and Engineering, University of Electronic Science and Technology of China, Chengdu 611731, Sichuan, People's Republic of China
fagenli@uestc.edu.cn

Abstract. The handwritten signature is a widely used and reliable biometric characteristic for personal verification. Traditionally, verification methods focus on a single modality, either dynamic sequences or static images, which limits their ability to capture complementary information. In this paper, we propose a multimodal Siamese Network that integrates both dynamic and static modalities. In addition to the original sequences and images, we incorporate pressure maps, local features, and global features to enhance representation. The global features are ranked and selected based on the silhouette index to improve discriminability. Three specialized subnetworks, 1D ResNet, 2D ResNet, and DNN, are employed to map the multimodal data into a hyperspace for joint forgery detection. Extensive experiments on two benchmark datasets, MSDS and SigComp2009, validate the effectiveness of our multimodal fusion strategy and feature selection approach.

Keywords: Handwritten signature · Multimodality · Global features · Siamese network

1 Introduction

The handwritten signature is an important biological trait widely used in legal, financial, and commercial fields. It is essential to authenticate identities based on the signature characteristics. With the increasing usage of electronic documents and devices, handwritten signatures are typically divided into two modalities: offline and online. Offline signatures refer to static images, which are scanned from papers or generated on tablets. Online signatures consist of dynamic sequences collected from sensors during the writing process, including coordinates, pressure, angles, and so on. In a typical process of handwritten signature verification (HSV), one or more genuine signatures are used to compare with a questioned signature [6]. Features of the genuine signatures are extracted and stored in a database. When performing the verification, the feature of questioned signature is obtained by a discriminative extractor and compared with that of the claimed writer, and then determining whether to accept this signature or

J. Yu and Y. Chen (Eds.): FCS 2025, CCIS 2805, pp. 265–280, 2027.
https://doi.org/10.1007/978-981-95-9514-3_18

not [6]. Moreover, a random string independent of the writer's name can be employed to achieve both identity authentication and privacy protection [32].

The main challenge in signature verification is mainly caused by the writer variability. On the one hand, there are intrinsic differences between genuine signatures from the same person. On the other hand, the imposter's imitation leads to the similarity between forgeries and genuine signatures. The forgeries are mainly divided into three types: (1) Random forgery, where an imposter has no prior knowledge of the targeted person and uses his or her own signature. (2) Simple forgery, where the imposter only knows the target's name. (3) Skilled forgery, where the imposter has access to samples of the target's signature and imitates the writing style. Verification of skilled forgeries is the most difficult [2,22]. Additionally, a replayed attack involves directly using the writer's genuine signatures [21], which has a higher level of threat.

Feature extraction plays an important part in signature verification and can be categorized into global and local features [20]. Global features model the signature as a holistic multidimensional vector, including information such as total time, number of strokes, average speed from online signatures and histogram, aspect ratio from offline signatures. Local features are manual time functions, including velocity, angle, and acceleration from online sequences and local orientation histograms from offline images. Moreover, the deep learning models, such as Convolutional Neural Networks (CNN) and Recurrent Neural Networks (RNN), are used to extract abstract features. The autoencoder is used to extract representative features [1], learning a mapping between signature space and latent space.

The verification can be either writer-dependent, where a model is built for each individual, or writer-independent, which a unique model is shared across all writers. According to the modality, the verification can be offline, online or a combination of them. Various mathematical methods have been applied to this task, including the Gaussian mixture module [26] and dynamic time warping [10, 19], symmetric positive definite matrices [34], and so on. Additionally, machine learning and deep learning techniques, such as top-rank learning [9], learning with rejection [9], and Siamese Network, also contribute to addressing this issue.

In this paper, we propose a multimodal Siamese Network for handwritten signature verification, with a particular focus on detecting skilled forgeries. Our contributions are summarized as follows:

- We extract diverse information from multimodal signatures, including pressure maps, local features, and global features. The fusion of these modalities enhances the effectiveness of signature verification.
- We perform a comprehensive analysis of multiple global features and select the most discriminative ones based on the silhouette index.
- We conduct extensive experiments to evaluate the proposed network architecture. Ablation studies highlight the necessity of multimodal fusion, and comparative results demonstrate the superior performance of our approach.

The remainder of this paper is organized as follows: Sect. 2 reviews related work on signature verification. Section 3 describes the materials and methods, including dataset, feature extraction, and network architecture. Experimental results are discussed in Sect. 4, and Sect. 5 concludes this paper.

2 Related Work

Traditional signature verifications rely on the mathematical methods. For offline signatures, the template matching directly compares the pixel similarity between a query and a template image. The feature-based classifier firstly extracts hand-crafted features such as histogram of oriented gradients, and then uses classifier such as the support vector machine and k-nearest neighbor to decide whether a signature is genuine or not. For online signatures, the Dynamic Time Warping (DTW) and hidden Markov models are commonly used methods, whereas DTW still achieves good performance in current studies [10,29]. Considering most cases that online sequences are in different length, the DTW finds the optimal alignment between time series and calculates their distance.

A number of studies focus on the deep learning techniques due to their remarkable performance in signature verification. These methods automatically perform feature extraction and achieve an end-to-end verification. (1) For offline signatures, Hafemann et al. [8] designed a 2D CNN feature extractor to effectively distinguish genuine signatures and skilled forgeries. This approach was later improved by Maruyama et al. [18] to address the few-shot problem. Wei et al. [25] introduced the inverse discriminative network to address the sparse information problem in offline images. Additionally, the transformer [15] also achieves effectiveness to recognize forgeries. (2) For online signatures, RNN and its variants are commonly used for feature extraction. Lai et al. [11] proposed a recurrent adaptation network to provide better generalization performance. Lai et al. [12,13] also employed 1D CNN to extract discriminative and holistic feature vectors from time-series data of online signatures. Zhang et al. [30] introduced a temporal-frequency synergistic model, integrating multi-scale temporalfrequency interactions and multi-domain representations to enhance the discrimination between genuine and forged handwriting. (3) The combination of both online and offline modalities has provided better performance than unimodal system [28,33]. Despite the common two modalities, Li et al. [16] proposed a verification framework that fuses the static features derived from signature images and dynamic features derived from pen stroke audio signals. Das et al. [4] explored the fusion of signature data and electroencephalogram (EEG) signals. Fierrez et al. [7] proposed a verification system by fusing local and global features, where the 100 global features are ordered according to scalar inter-writer separability. Martinez et al. [17] proposed to use another set of global features, where time, speed, direction, and geometry information are all considered.

Inspired by the excellent performance of Siamese network in the similarity comparison, researchers have applied it to the signature verification task. It consists of two identical subnetworks that share the same weights. The similarity between input pairs is learned by comparing their feature representations. Dey et al. [5] firstly proposed a convolutional Siamese network for offline signature verification. Xiao et al. [27] enhanced the convolutional Siamese network by employing a two-stage verification for both original and enhanced images. Tolosana et al. [23] proposed to employ Long Short-Term Memory (LSTM)-based and Gated Recurrent Unit (GRU)-based Siamese networks to address the online signature verification. As an improvement of Siamese network, the 2-channel network has been shown to achieve better performance [14]. Moreover, as another variant of Siamese network, the triplet network uses three inputs: positive, negative, and anchor, which leads to better performance but more complex input combinations. Xie et al. [28] employed it together with multimodal fusion to further improve signature verification performance.

3 Materials and Methods

3.1 Multimodal Datasets

We employ two datasets in our present study:

- **MSDS**[1] [31] It includes 402 writers, performing two writing sessions. In each session, every writer contributes to 10 genuine signatures and 10 skilled forgeries. The dynamic information such as x and y coordinates, pressure p, time stamps, and pen-up status, are recorded in text files. The corresponding static rendered images are separately saved.
- **SigComp2009**[2] [3] The training set and evaluation set are collected by two institutes, named NISDCC and NFI, respectively. The NISDCC set includes 12 authentic writers, each with 5 genuine signatures. Other 31 forging writers contribute 5 forgeries for each genuine signature. The NFI includes 100 authentic writers and 33 forging writers. Each writer gives 12 genuine signatures and each genuine one has 6 forgeries. The dynamic information such as x and y coordinates, pressure p, azimuth angle, and elevation angle are recorded in text files. The pen-up marks are implied by pressure information. The scanned offline signatures are separately saved.

The two datasets are selected for their diversity. Firstly, they respectively record Chinese signatures and Western signatures. For online modality, the MSDS dataset does not uniformly sample signatures over time and lacks information of pen-up process. In contrast, the SigComp2009 dataset uniformly samples signatures at 200 Hz (thus without timestamps) and includes pen-up information. For offline modality, the MSDS dataset generates electronic images one by one, while the SigComp2009 dataset scans multiple signatures written on A4 paper.

3.2 Preprocessing and Feature Extraction

Preprocessing of Offline Data. In the MSDS dataset, images are stored in RGBA format, where only Alpha channel contains information. Therefore, the

[1] MSDS dataset is available at https://github.com/HCIILAB/MSDS.

[2] SigComp2009 dataset is available at http://www.iapr-tc11.org/mediawiki/index.php/ICDAR_2009_Signature_Verification_Competition_(SigComp2009).

Table 1. Features extracted from time functions.

No.	Features	Description
1	$\dot{x}$	time derivative of x
2	$\dot{y}$	time derivative of y
3	$v = \sqrt{\dot{x}^2 + \dot{y}^2}$	path velocity magnitude
4	$\dot{v}$	time derivative of v
5	p	pressure
6	$\theta = \arctan(\dot{y}/\dot{x})$	path-tangent angle
7	$\cos\theta$	cosine of the path-tangent angle
8	$\sin\theta$	sine of the path-tangent angle
9	$\dot{\theta}$	time derivative of θ
10	$\triangle v = v \cdot \dot{\theta}$	velocity variation magnitude
11	$\sqrt{\dot{v}^2 + \triangle v}$	acceleration

data can be treated as single-channel. In the SigComp2009 dataset, the images are stored in greyscale, which is also single-channel. We invert the greyscale images such that the background pixels have the value of 0. To ensure the subsequent model focuses on the shape information, we uniformly remove blank space around the signatures. Then, we employ bilinear interpolation to resize images from MSDS to 64×192 and those from SigComp2009 to 128×384, both at a ratio of 1:3.

Preprocessing of Online Data. For online signatures, given coordinates x_i and y_i, we apply:

$$\begin{aligned} x_i &\leftarrow x_i - x_{\min}, \\ y_i &\leftarrow y_i - y_{\min}, \end{aligned} \tag{1}$$

such that minimum x and y values are zero. This is particularly necessary for the SigComp2009 dataset, where multiple signatures are written on the same sheet of paper.

Feature Extraction. In this part, we extract three types of features specially from online signatures. Firstly, we calculate 11 time functions [31] from raw physical signals of x and y coordinates, pressure p and time information, which are detailed in Table 1. It should be noted that the time interval $\triangle t$ in SigComp2009 dataset is always 1. To address variations in time duration among different writers, we standardize the sequences to a uniform length of 500 by using linear interpolation. Secondly, pressure maps are generated from online signatures. We initialize a zero matrix $\boldsymbol{M}$ of size $x_{\max} \times y_{\max}$ and assign the value p_i to the corresponding matrix element m_{y_i,x_i}. The pressure maps are resized to the same size as offline images. Finally, we extract global features [20] from online signatures. The global feature is described as a multi-dimensional vector, where each element represents one specific feature.

3.3 Models

We use the multimodal Siamese network for signature verification. The network consists of two subnetworks with identical parameters and shared weights. It maps two inputs into a common hyperspace, where genuine signatures from the same person are grouped together, while genuine and forged signatures are separated. The optimization problem is formulated as:

$$\min loss = (1 - y) \times D(f_1, f_2) + y \times \max(0, m - D(f_1, f_2)), \tag{2}$$

where $D(\cdot)$ calculates the Euclidean distance, label $y = 0$ if it is genuine pair and otherwise $y = 1$, f_1 and f_2 are feature vectors, and m is the ideal maximum margin.

In this work, each subnetwork consists of three feature extractors that map the inputs into 1D feature vectors. The three parts are 2D Residual Network (ResNet) $F_{\text{off}}(\cdot)$ for static images, 1D ResNet $F_{\text{on}}(\cdot)$ for dynamic sequences, and Deep Neural Networks (DNN) $F_{\text{glob}}(\cdot)$ for global features, respectively. The architecture of our multimodal Siamese networks is shown in Fig. 1.

Let the i-th multimodal signature be represented as $s_i = [st_i, dy_i]$. We extract the pressure map m_i, time feature tf_i, and global feature gf_i by using the three extractors:

$$\begin{aligned} f^i_{\text{off}} &= F_{\text{off}}(m_i \parallel st_i), \\ f^i_{\text{on}} &= F_{\text{on}}(tf_i), \\ f^i_{\text{glob}} &= F_{\text{glob}}(gf_i), \end{aligned} \tag{3}$$

where $||$ denotes concatenation along channels, and f^i_{off}, f^i_{on}, and f^i_{glob} are the 1D feature vectors. The pressure maps and offline images are both in the 2D form, so we perform an early fusion for them to get additional information. Considering that the network takes two multimodal signatures as input, three distances are calculated based on the respective feature vectors:

$$\begin{aligned} d_{\text{off}} &= D(f^i_{\text{off}}, f^j_{\text{off}}), \\ d_{\text{on}} &= D(f^i_{\text{on}}, f^j_{\text{on}}), \\ d_{\text{glob}} &= D(f^i_{\text{glob}}, f^j_{\text{glob}}), \end{aligned} \tag{4}$$

Our Siamese network employs a combined contrastive loss:

$$\begin{aligned} loss =& (1 - y) \times (d_{\text{off}} + d_{\text{on}} + d_{\text{glob}}) \\ &+ y \times [\max(0, m_{\text{off}} - d_{\text{off}}) + \max(0, m_{\text{on}} - d_{\text{on}}) + \max(0, m_{\text{glob}} - d_{\text{glob}})], \end{aligned} \tag{5}$$

where m_{off}, m_{on}, and m_{glob} are three margins. In the verification stage, we calculate a weighted hyperspace distance between a genuine reference signature and a questioned signature:

$$d = \lambda_{\text{on}} d_{\text{on}} + \lambda_{\text{off}} d_{\text{off}} + \lambda_{\text{glob}} d_{\text{glob}}, \tag{6}$$

where λ_{on}, λ_{off}, and λ_{glob} are hyperparameters that control the contribution of different modalities. The distance measures an overall semantic difference

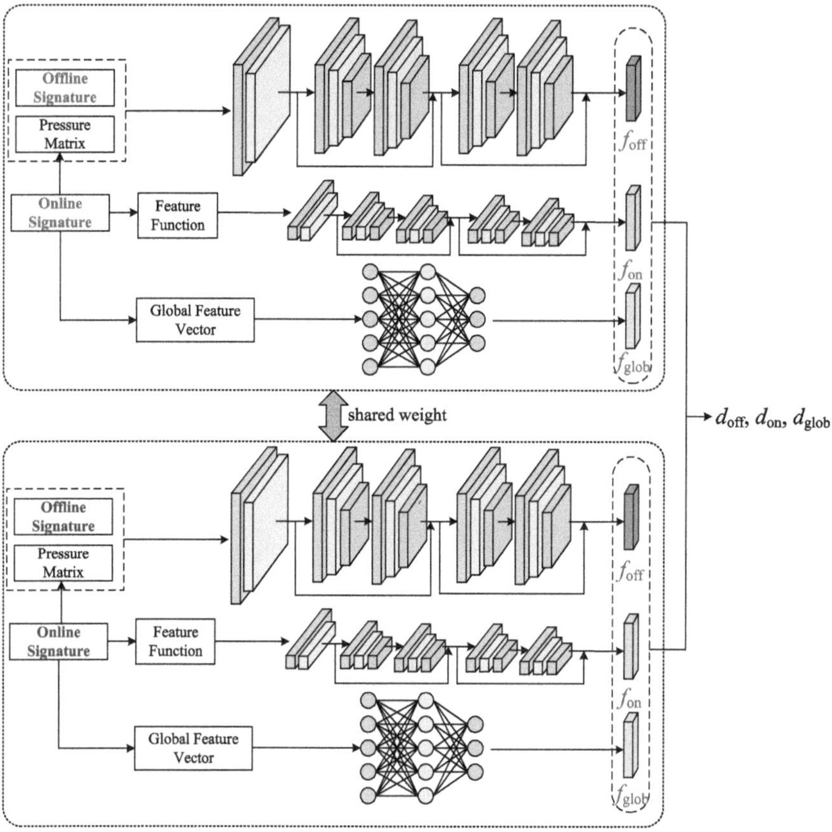

Fig. 1. Multimodal Siamese networks.

between the two signatures. Given a threshold t, if $d < t$, the questioned signature is classified as genuine, otherwise it is regarded as a forgery. Specifically, we empirically set $\lambda_{on} = 0.55$, $\lambda_{off} = 0.10$, and $\lambda_{glob} = 0.35$. We evaluate the model using Equal Error Rate (EER) and Accuracy (Acc). The EER is point where false rejection rate equals false acceptance rate, where a smaller EER indicates better system performance. It is calculated by varying the threshold and the threshold that yields EER is used to calculate Acc.

4 Experiments

4.1 Construction of Signature Pairs

For MSDS dataset, we use 402 writers in session 1. We divide them into 202 for constructing training set and 200 for testing set. In the training set, we use combinations of genuine signatures of each writer to construct $202 \times 10 \times 9 = 18180$ positive data pairs, and use combinations between genuine signatures and skilled forgeries of each writer to construct $202 \times 10 \times 10 = 20200$ negative data pairs. In the testing set, we only use the first genuine signature together with other genuine ones of each writer to construct $200 \times 9 = 1800$ positive pairs, and together with other forgeries to construct $200 \times 10 = 2000$ negative pairs, as shown in [31].

For SigComp2009 dataset, there are only unimodal data for a few signatures due to an error during recording. Therefore, we first exclude them and reserve the aligned online and offline signatures. For the 12 authentic writers in NISDCC training set, each with 5 genuine signatures and f_i ($145 \leq f_i \leq 155$) skilled forgeries. We get total $12 \times (5 \times 4) + \sum_{i=1}^{12}(5 \times f_i) = 240 + 9115 = 9355$ data pairs (240 for positive and 9115 for negative). For NFI evaluation set, there are 79 writers, 19 of whom have skilled forgeries. The number of genuine data g_i ranges from 6 to 12, and the number of forgeries f_i ranges from 26 to 36. We get total $\sum_{i=1}^{79} \times (g_i \times (g_i - 1)) + \sum_{i=1}^{19}(g_i \times f_i) = 9938 + 7242 = 17180$ data pairs (9938 for positive and 7242 for negative).

4.2 Global Features

Before starting the training, we select discriminative global features for the two datasets. Firstly, we define a set of candidate features, including: (1) signature total duration T_s, (2) total signature length L, (3) standard deviation of $y/(y_{\max}-y_{\min})$, (4) number of pen-ups, (5) number of sign changes of v_x and v_y (6) number of sign changes of p (7) number of local maxima in x, (8) number of local maxima in y, (9) number of local maxima in p. Besides, we compute the sum, average value (avg), standard deviation (std), and root mean square (rms) for the following features: x, y, p, v_x, v_y, v_p, v, a_x, a_y, a_p, a and j. The v, a, and j represent first, second, and third time derivative, respectively, for the path or pressure.

In all, we get $9 + 12 \times 4 = 57$ global features. For each writer in the training set, we use the Silhouette Index (SI) to measure the discrimination between genuine signatures and skilled forgeries. Given a specific global feature, the signatures of writer i are divided into two clusters: one for genuine signatures and the other for skilled forgeries. The silhouette value $s_i(j)$ for each signature j under this global feature is calculated as:

$$s_i(j) = \frac{b(j) - a(j)}{\max\{a(j), b(j)\}}, \tag{7}$$

where $a(j)$ is the average distance between signature j and all other signatures within the same cluster, and $b(j)$ is the average distance between j and all signatures in the other cluster. The silhouette index for writer i is then obtained by averaging $s_i(j)$ over all signatures:

$$s_i = \frac{1}{N_i} \sum_{j=1}^{N_i} s_i(j), \tag{8}$$

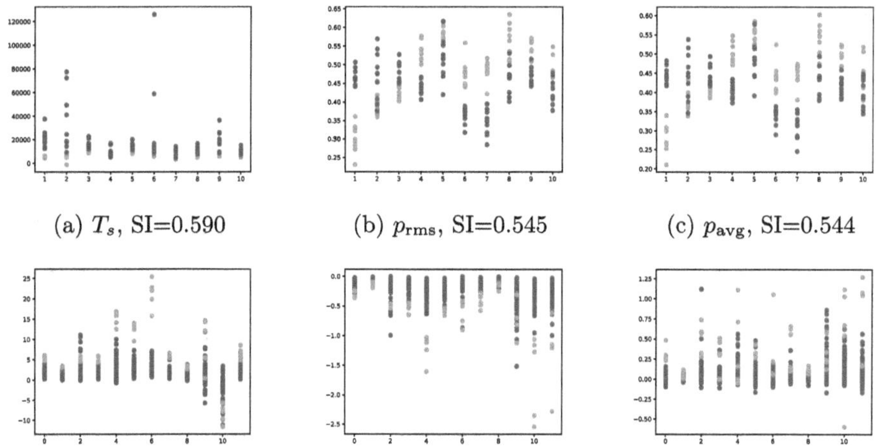

(a) T_s, SI=0.590 (b) p_{rms}, SI=0.545 (c) p_{avg}, SI=0.544

(d) $(v_x)_{\mathrm{avg}}$, SI=0.749 (e) $(a_p)_{\mathrm{avg}}$, SI=0.732 (f) $(v_p)_{\mathrm{avg}}$, SI=0.725

Fig. 2. Each vertical group represents a writer and each dot represents a signature. The red stands for genuine signatures and the blue stands for forgeries. (a)–(c) are for MSDS and (d)–(f) are for SigComp2009. (Color figure online)

where N_i is the number of signatures for writer i. Finally, the overall SI under this specific global feature is computed by averaging s_i over all writers. The value of SI ranges from -1 to 1, where a value closer to 1 indicates better discrimination between genuine signatures and forgeries. For MSDS dataset, the most discriminative features ranked from top to bottom are: T_s, p_{rms}, p_{avg}, y_{sum}, x_{sum}, v_{avg}, p_{sum}, v_{rms}, $(v_x)_{\mathrm{rms}}$, $(v_x)_{\mathrm{std}}$, $(v_p)_{\mathrm{rms}}$, and $(v_y)_{\mathrm{rms}}$. For SigComp2009 dataset, the most discriminative features ranked from top to bottom are: $(v_x)_{\mathrm{avg}}$, $(a_p)_{\mathrm{avg}}$, $(v_p)_{\mathrm{avg}}$, $(a_y)_{\mathrm{avg}}$, $(a_x)_{\mathrm{avg}}$, v_{avg}, j_{avg}, $(v_p)_{\mathrm{rms}}$, $(v_p)_{\mathrm{std}}$, a_{avg}, $(a_p)_{\mathrm{rms}}$, and $(a_p)_{\mathrm{std}}$. The top 3 global features of two datasets and the average SI index are shown in Fig. 2. It can be seen that some characteristics have shown evident differences between genuine and forged signatures.

4.3 Verification Results

The settings of three subnetworks are as follows: (1) The 2D ResNet begins with a convolutional layer that increases the input channels from 2 to 32, followed by three residual layers. The channels of residual blocks are 32, 64, 128, and strides are 1,1,2. The output is processed through adaptive average pooling and followed by a fully connected layer that outputs a 128-length vector. (2) The 1D ResNet begins with a convolutional layer that increases the input channels from 11 to 32, and residual blocks are similar with its 2D version. A 128-length vector is also output through adaptive average pooling and a fully connected layer. (3) The DNN has two hidden layers with 128 units and output a vector of length 64. The three networks all use the ReLU activation function and batch

Table 2. EER (%) of our methods and other works.

Dataset	Modality	Method	EER
MSDS	offline	SigNet [5]	34.48
	online	Bi-LSTM+Siamese [23]	6.25
	online	Sig2Vec [13]	6.76
	online	TA-RNNs [24]	6.99
	multimodal	ours	**5.98**
SigComp2009	offline	SigNet [5]	28.30
	online	Bi-LSTM+Siamese [23]	25.24
	multimodal	ours	**8.01**

normalization. The models were trained with the Adam optimizer for 50 epochs. The learning rate is 0.001 and with decay rate 0.95 every epoch. The batch size is 64, and m_{off}, m_{on}, and m_{glob} are all 2. Additionally, we use weighted sampler for SigComp2009 training set considering its imbalance positive and negative samples.

The comparison results are shown in Table 2. The MSDS dataset employs additional Sig2Vec and TA-RNNs because they are used in its original dataset verification [31]. It can be seen that our multimodal network outperforms the unimodal methods. Our multimodal model reduces the relative EER of best MSDS by 4.32% ((6.25-5.98)/6.25×100%); for SigComp2009, the relative EER reduction is 68.26% ((25.24-8.01)/25.24×100%).

We also perform ablation experiments to show the effectiveness of multimodal fusion. We consider cases that: (1) three subnetworks are separately employed, (2) two subnetworks of the three are used, (3) all subnetworks are used, where the three distances share the equal importance (**triple-**). Specifically, we exclude the feature map input for $F_{\text{off}}(\cdot)$ when $F_{\text{on}}(\cdot)$ is not used. The results are shown in Table 3. The Receiver Operating Characteristic (ROC) curves are shown in Fig. 3 and Fig. 4. It can be seen that the combination of two modalities and multiple kinds features contributes to the most effective verification. The online features are the most discriminative, followed by the global features. Their combination leads to a better performance. In contrast, the offline modality shows moderate performance, but when combined with other modalities, it leads to an improved result. Besides, an appropriate weight assignment for these three kinds of distances is also necessary.

To show the effectiveness of the selected global features, we compare them with other combinations: the usage of sum, rms, avg, and std of 12 features are considered, respectively. The results are shown in Table 4. It can be seen that a best performance is achieved with the guidance of SI.

Additionally, we implement experiments by using different subnetworks. The dynamic sequences are fed into 1D CNN, 1D ResNet, LSTM, Bi-LSTM, respectively. The static images are fed into 2D CNN or 2D ResNet. When varying the

Table 3. EER(%) and Acc(%) of different subnetwork combinations.

Dataset	Subnetwork	EER	Acc
MSDS	on	9.16	90.92
	off	31.22	62.82
	glob	10.02	89.82
	on+off	9.79	90.24
	on+glob	7.43	92.76
	off+glob	13.98	85.92
	triple-	7.29	92.71
	triple	**5.98**	**94.18**
SigComp2009	on	11.40	88.57
	off	24.31	75.89
	glob	16.15	83.89
	on+off	16.14	83.91
	on+glob	10.97	89.09
	off+glob	19.18	80.90
	triple-	12.70	87.27
	triple	**8.01**	**91.96**

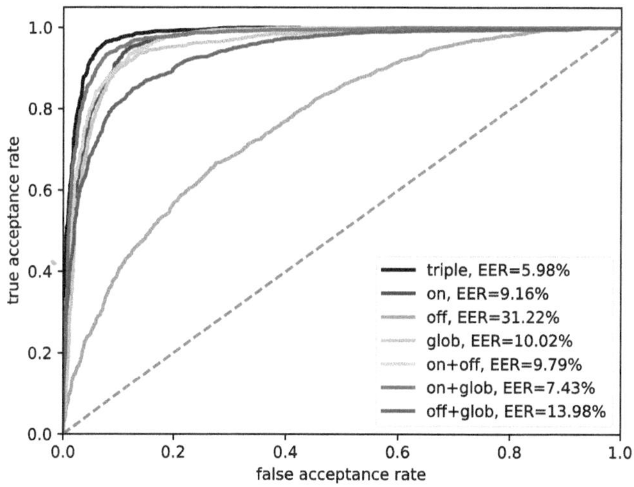

Fig. 3. ROC curves under different combinations on MSDS.

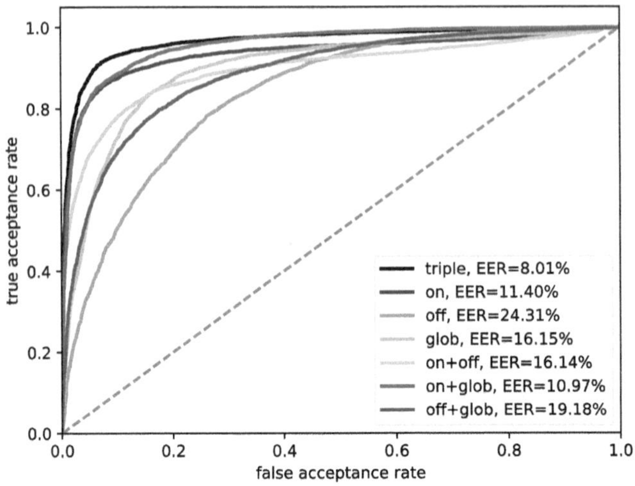

Fig. 4. ROC curves under different combinations on SigComp2009.

Table 4. EER(%)/Acc(%) of different global features.

Dataset	Global features	Unimodality	Multimodality
MSDS	sum	20.14/80.02	7.06/93.04
	rms	24.69/75.25	7.03/92.89
	avg	22.95/77.12	7.54/92.47
	std	23.75/76.38	7.90/92.18
	top 12 SI	**10.02/89.82**	**5.98/94.18**
SigComp2009	sum	21.32/78.96	10.74/89.13
	rms	16.93/82.99	11.76/88.12
	avg	19.55/80.39	11.15/88.78
	std	16.72/83.20	10.20/89.75
	top 12 SI	**16.15/83.89**	**8.01/91.96**

subnetwork of one modality, subnetwork of the other modality is maintained with original ResNet. The results are shown in Table 5. It can be seen that: for the offline modality, ResNet basically outperforms CNN, and for the online modality, ResNet also outperforms other networks, which means the ResNet extracts more discriminative features.

Table 5. EER(%)/Acc(%) of different subnetworks.

Dataset	Online subnetwork	Offline subnetwork	
		CNN-2D	ResNet-2D
MSDS	CNN-1D	13.18/86.64	11.91/88.12
	ResNet-1D	6.25/93.77	**5.98/94.18**
	LSTM	8.09/91.99	7.65/92.44
	Bi-LSTM	11.76/88.16	9.27/90.80
SigComp2009	CNN-1D	15.62/84.38	12.98/87.02
	ResNet-1D	8.82/91.18	**8.01/91.96**
	LSTM	12.28/87.74	13.17/86.86
	Bi-LSTM	13.33/86.80	11.89/88.13

5 Conclusions

In this work, we proposed a multimodal Siamese network for handwritten signature verification that integrates both offline and online modalities. By incorporating pressure maps, manual time features, and global features, the network is enhanced to effectively discriminate between genuine signatures and forgeries. The silhouette index was employed to select the most discriminative global features, further improving verification accuracy. Extensive experiments were conducted on two quite different bimodal datasets, MSDS and SigComp2009, to evaluate the effectiveness of the proposed method. The results demonstrate that our multimodal network outperforms unimodal systems as well as other approaches. The experiments also highlight the importance of the designed subnetworks.

In the future, we plan to explore triplet network and additional feature extractors. We will also investigate obtaining global features from offline images for further improvements. It should be noted that, although multimodal signature data were leveraged to strengthen authentication, other biometric modalities, such as keystroke dynamics or facial recognition, were not employed. We will also combine more modalities to broaden the applicability and robustness of the system for security applications.

Acknowledgements. This work is supported by the Sichuan Science and Technology Program (grant no. 2025YFHZ0320).

Disclosure of Interests. The authors have no competing interests to declare that are relevant to the content of this article.

References

1. Ahrabian, K., BabaAli, B.: Usage of autoencoders and Siamese networks for online handwritten signature verification. Neural Comput. Appl. **31**(12), 9321–9334 (2018). https://doi.org/10.1007/s00521-018-3844-z

2. Bhavani, S.D., Bharathi, R.K.: A multi-dimensional review on handwritten signature verification: strengths and gaps. Multimedia Tools Appl. (2023)
3. Blankers, V.L., van den Heuvel, C.E., Franke, K.Y., Vuurpijl, L.G.: Icdar 2009 signature verification competition. In: 2009 10th International Conference on Document Analysis and Recognition, pp. 1403–1407. IEEE (2009)
4. Das Chakladar, D., Kumar, P., Roy, P.P., Dogra, D.P., Scheme, E., Chang, V.: A multimodal-siamese neural network (mSNN) for person verification using signatures and EEG. Inf. Fusion **71**, 17–27 (2021). https://doi.org/10.1016/j.inffus.2021.01.004
5. Dey, S., Dutta, A., Toledo, J.I., Ghosh, S.K., Llados, J., Pal, U.: Signet: convolutional siamese network for writer independent offline signature verification (2017)
6. Diaz, M., Ferrer, M.A., Impedovo, D., Malik, M.I., Pirlo, G., Plamondon, R.: A perspective analysis of handwritten signature technology. ACM Comput. Surv. **51**(6) (2019). https://doi.org/10.1145/3274658
7. Fierrez-Aguilar, J., Nanni, L., Lopez-Peñalba, J., Ortega-Garcia, J., Maltoni, D.: An on-line signature verification system based on fusion of local and global information. In: Audio- and Video-Based Biometric Person Authentication, pp. 523–532. Springer, Heidelberg (2005)
8. Hafemann, L.G., Sabourin, R., Oliveira, L.S.: Learning features for offline handwritten signature verification using deep convolutional neural networks. Pattern Recogn. **70**, 163–176 (2017). https://doi.org/10.1016/j.patcog.2017.05.012
9. Ji, X., Suehiro, D., Uchida, S.: Paired contrastive feature for highly reliable offline signature verification. Pattern Recogn. **144**, 109816 (2023). https://doi.org/10.1016/j.patcog.2023.109816
10. Jiang, J., Lai, S., Jin, L., Zhu, Y.: DsDTW: local representation learning with deep soft-DTW for dynamic signature verification. IEEE Trans. Inf. Forensics Secur. **17**, 2198–2212 (2022). https://doi.org/10.1109/TIFS.2022.3180219
11. Lai, S., Jin, L.: Recurrent adaptation networks for online signature verification. IEEE Trans. Inf. Forensics Secur. **14**(6), 1624–1637 (2019). https://doi.org/10.1109/TIFS.2018.2883152
12. Lai, S., Jin, L., Lin, L., Zhu, Y., Mao, H.: Synsig2vec: learning representations from synthetic dynamic signatures for real-world verification (2019)
13. Lai, S., Jin, L., Zhu, Y., Li, Z., Lin, L.: SynSig2Vec: forgery-free learning of dynamic signature representations by sigma lognormal-based synthesis and 1D CNN. IEEE Trans. Pattern Anal. Mach. Intell. **44**(10), 6472–6485 (2022). https://doi.org/10.1109/TPAMI.2021.3087619
14. Li, C., Lin, F., Wang, Z., Yu, G., Yuan, L., Wang, H.: Deephsv: user-independent offline signature verification using two-channel CNN. In: 2019 International Conference on Document Analysis and Recognition (ICDAR), pp. 166–171 (2019). https://doi.org/10.1109/ICDAR.2019.00035
15. Li, H., Wei, P., Ma, Z., Li, C., Zheng, N.: TransOSV: offline signature verification with transformers. Pattern Recogn. **145**, 109882 (2024). https://doi.org/10.1016/j.patcog.2023.109882
16. Li, Q., Wang, Z., Jin, L., Yadikar, N., Ubul, K.: Mmhsv: a multimodal handwritten signature verification fusing dynamic and static feature. In: ICASSP 2024 - 2024 IEEE International Conference on Acoustics, Speech and Signal Processing (ICASSP), pp. 4730–4734 (2024). https://doi.org/10.1109/ICASSP48485.2024.10447071
17. Martinez-Diaz, M., Fierrez, J., Krish, R.P., Galbally, J.: Mobile signature verification: feature robustness and performance comparison. IET Biometrics **3**(4), 267–277 (2014)

18. Maruyama, T.M., Oliveira, L.S., Britto, A.S., Sabourin, R.: Intrapersonal parameter optimization for offline handwritten signature augmentation. IEEE Trans. Inf. Forensics Secur. **16**, 1335–1350 (2021). https://doi.org/10.1109/TIFS.2020.3033442
19. Matsuo, S., et al.: Deep attentive time warping. Pattern Recogn. **136**, 109201 (2023). https://doi.org/10.1016/j.patcog.2022.109201
20. Plamondon, R., Pirlo, G., Impedovo, D.: Online Signature Verification, pp. 917–947. Springer, Heidelberg (2014)
21. Shi, Z., Li, F., Hao, D., Sun, Q.: Handwritten signature verification via multimodal consistency learning. IEEE Trans. Inf. Forensics Secur. **20**, 3995–4007 (2025). https://doi.org/10.1109/TIFS.2025.3557674
22. Soelistio, E.A., Hananto Kusumo, R.E., Martan, Z.V., Irwansyah, E.: A review of signature recognition using machine learning. In: 2021 1st International Conference on Computer Science and Artificial Intelligence (ICCSAI), vol. 1, pp. 219–223 (2021)
23. Tolosana, R., Vera-Rodriguez, R., Fierrez, J., Ortega-Garcia, J.: Exploring recurrent neural networks for on-line handwritten signature biometrics. IEEE Access **6**, 5128–5138 (2018). https://doi.org/10.1109/ACCESS.2018.2793966
24. Tolosana, R., Vera-Rodriguez, R., Fierrez, J., Ortega-Garcia, J.: Biotouchpass2: touchscreen password biometrics using time-aligned recurrent neural networks. IEEE Trans. Inf. Forensics Secur. **15**, 2616–2628 (2020). https://doi.org/10.1109/TIFS.2020.2973832
25. Wei, P., Li, H., Hu, P.: Inverse discriminative networks for handwritten signature verification. In: 2019 IEEE/CVF Conference on Computer Vision and Pattern Recognition (CVPR), pp. 5757–5765 (2019). https://doi.org/10.1109/CVPR.2019.00591
26. Xia, X., Chen, Z., Luan, F., Song, X.: Signature alignment based on GMM for online signature verification. Pattern Recogn. **65**, 188–196 (2017). https://doi.org/10.1016/j.patcog.2016.12.019
27. Xiao, W., Wu, D.: An improved siamese network model for handwritten signature verification. In: 2021 IEEE International Conference on Networking, Sensing and Control (ICNSC), vol. 1, pp. 1–6 (2021). https://doi.org/10.1109/ICNSC52481.2021.9702190
28. Xie, L., Wu, Z., Zhang, X., Li, Y.: Synchronous spatio-temporal signature verification via fusion triplet supervised network. Eng. Appl. Artif. Intell. **123**, 106378 (2023). https://doi.org/10.1016/j.engappai.2023.106378
29. Yu, H., Shi, P.: A novel deep ensemble framework for online signature verification using temporal and spatial representation. In: Wang, D., Yung, M., Liu, Z., Chen, X. (eds.) Information and Communications Security. LNCS, pp. 534–549. Springer, Singapore (2023). https://doi.org/10.1007/978-981-99-7356-9_32
30. Zhang, P., Ding, K., Jin, L.: Capturing more: Learning multi-domain representations for robust online handwriting verification. arXiv preprint arXiv:2508.01427 (2025)
31. Zhang, P., Jiang, J., Liu, Y., Jin, L.: MSDS: a large-scale Chinese signature and token digit string dataset for handwriting verification. In: NeurIPS (2022)
32. Zhang, P., Liu, Y., Lai, S., Li, H., Jin, L.: Privacy-preserving biometric verification with handwritten random digit string. IEEE Trans. Pattern Anal. Mach. Intell. **47**(4), 3049–3066 (2025). https://doi.org/10.1109/TPAMI.2025.3529022

33. Zhou, Y., Zheng, J., Hu, H., Wang, Y.: Handwritten signature verification method based on improved combined features. Appl. Sci. Basel **11**(13) (2021)
34. Zois, E.N., Tsourounis, D., Kalivas, D.: Similarity distance learning on SPD manifold for writer independent offline signature verification. IEEE Trans. Inf. Forensics Secur. **19**, 1342–1356 (2024). https://doi.org/10.1109/TIFS.2023.3333681

A Quantum-Secure Certificateless Aggregate Signature Scheme for Emergency Communication

Zixuan Yan, Kexin Tian, and Shuanggen Liu(✉)

School of Cyberspace Security, Xi'an University of Posts and Telecommunications, Xi'an, Shaanxi, China
liushuanggen201@xupt.edu.cn

Abstract. Distributed ad-hoc networks must be used for communication in some emergency situations (like earthquakes and conflicts) since centralized communication hubs are susceptible to harm. Traditional PKI is unsuccessful since it relies on CAs. Examples of emergency communication information that needs to be kept confidential for a long time include rescue instructions and strategic deployments. Conventional aggregate signature schemes that rely on massive prime factorization and discrete logarithm problems fail to prevent quantum attacks. At disaster sites, resources are scarce, and multiple rescue nodes must send data at once, creating a high demand for communication. Therefore, the three primary criteria for emergency communication are decentralization, resistance to quantum attacks, and efficient aggregation with low bandwidth. This solution, which is based on the NTRU algorithm, combines aggregate signature technology and certificateless architecture to address the single point of failure of CAs in traditional PKI in disaster scenarios. It also withstands the threat of quantum computing to signature security and achieves efficiency and practicality in the emergency communication environment with limited resources.

Keywords: Anti-quantum attack · Decentralization · Efficient aggregation · Emergency communication · Efficiency

1 Introduction

1.1 Research Background

The existing signature schemes have exposed many limitations in emergency communication. The schemes based on PKI rely on complex certificate management processes. However, in the dynamic network environment of emergency communication, the communication infrastructure may suffer severe damage, causing the certificate server to fail properly, which makes certificate management extremely difficult and generates excessive communication and computational overhead, making it difficult to meet the efficiency requirements of

J. Yu and Y. Chen (Eds.): FCS 2025, CCIS 2805, pp. 281–292, 2027.
https://doi.org/10.1007/978-981-95-9514-3_19

emergency communication. Literature [1] proposed Identity-Based Cryptography (IBC) systems. Such schemes simplify the key management process by directly using the user's identity as the public key. However, the IBC-based schemes have the problem of key escrow. Once the PKG is compromised, the security of the entire system will be severely compromised, which is unacceptable under the security requirements of emergency communication.

The certificateless aggregate signature scheme(CLAS) combines the advantages of certificateless cryptography and aggregate signatures. On the one hand, it not only gets rid of the tedious process of certificate management but also avoids the security risks of key escrow. On the other hand, by compressing multiple signatures into one, it significantly reduces the demand for communication bandwidth and verification costs, which aligns with the dynamic network characteristics of frequent node joining and leaving in emergency communications. However, the schemes in [2] and [3] are based on classical number theory and will not suitable for the quantum era [4]. Meanwhile, the schemes proposed in references [5,6], and [7] are based on bilinear pairings, which have high computational complexity in operations, imposing an excessive computational burden on resource-constrained emergency communication devices and making them unsuitable for practical emergency communication scenarios.

As quantum computing technology continues to advance toward maturity, ensuring the post-quantum security of emergency communication has become an urgent matter. Lattices-based cryptographic algorithms have been proven to have good quantum resistance [8]. Both references [9] and [10] introduce lattices-based aggregate signature schemes. Therefore, combining lattices-based algorithms with certificateless aggregate signatures [11], such schemes can not only effectively resist quantum attacks and ensure the long-term security of emergency data but also adapt to the dynamic and resource-constrained characteristics of emergency networks through lightweight design and aggregation capabilities. Thus, in-depth research on the application of certificateless quantum aggregate signature schemes in emergency communication has significant practical significance for improving the security and response capabilities of emergency response systems.

1.2 Contributions

The main contributions of this paper are as follows:

1. Innovative scheme design: Propose a quantum-secure CLAS scheme, which integrates the lattice-based quantum-resistant cryptographic algorithm with the certificateless aggregate signature technology. Utilizing the quantum-resistant feature of lattice-based cryptography to ensure the security of emergency communication data under quantum computing threats and leveraging the advantages of certificateless aggregate signature, which reduces certificate management costs and lowers communication costs, to meet the characteristics of dynamic networks and resource-constrained environments of emergency communication.

2. Security Enhancement: In the design of the scheme, various security threats are fully taken into account to ensure that the scheme can still provide reliable security protection even when facing common security risks such as quantum attacks, malicious node forgery of signatures, and key leakage. This provides a solid foundation for the authenticity, integrity, and non-repudiation of information in emergency communication.
3. Efficiency Optimization: The key operations, such as signature generation, verification, and aggregation, have been optimized to reduce computational complexity and communication overhead. Through theoretical analysis and experimental verification, compared with existing related schemes, this paper's solution can significantly enhance the signature processing efficiency on resource-constrained emergency communication devices, reduce communication bandwidth usage, and better meet the strict requirements of emergency communication for real-time and efficiency.

1.3 Structure

The structure of the rest of this paper is as follows. Section 2 will cover foundational concepts such as aggregate signatures. Section 3 will elaborate on the concrete implementation of our CLAS scheme. In Sect. 4, a thorough security evaluation and performance analysis of the scheme will be presented. Finally, Sect. 5 will summarize the paper, along with discussions on its strengths, limitations, and potential avenues for future research.

2 Preliminary Preparatory Work

2.1 Certificateless Aggregate Signature

Certificateless aggregation signature is a new cryptographic scheme that integrates certificateless cryptography with aggregation signature technology. Its core principle is to eliminate the problems of certificate management and key custody while aggregating the signatures of multiple users into a short signature, achieving efficient verification.

The private key is made up of two components: the part of the private key generated by the Key Generation Center (KGC), such as $d_i = (d_{i,1}, d_{i,2})$, and the secret value chosen by the user, such as $s_i = (s_{i,1}, s_{i,2})$. The complete private key is $sk_i = (d_i, s_i)$, which neither the KGC nor the user can generate independently, avoiding the key custody problem in the IBC System. The public key is generated by the user based on the private key, such as $pk_i = H_0(ID_i) + \alpha_i \cdot (s_{i,1} + s_{i,2} \cdot h)$, without the need for certificate binding the identity and the public key, solving the certificate management cost problem of the PKI.

Aggregation signatures achieve the compression of n users' signatures σ_1,σ_2, ..., σ_n for different messages into a single aggregated signature σ_{agg} through a "many-to-one" mapping relationship. The verifier can confirm the validity of all signatures with just one verification, significantly reducing communication bandwidth and storage costs, and being particularly suitable for multi-node communication scenarios. The process of certificateless aggregation signatures (CLAS) revolves around four core aspects: Key Generation, Signature Generation, Signature Aggregation, and Verification. By combining the non-delegated characteristics of certificateless cryptography with the efficiency of aggregation signatures, it is adapted to the requirements of dynamic network environments.

2.2 Lattices

A lattice is a collection of points formed by n linearly independent vectors $v_1, v_2, \cdots, v_n$ within the linear space R^n, which can be expressed as $L = \{t_1 \cdot v_1 + t_2 \cdot v_2 + \cdots + t_n \cdot v_n \mid t_i \in \mathbb{Z}\}$. For polynomials $f = \sum_{i=0}^{n-1} f_i x^i$ and $g = \sum_{i=0}^{n-1} g_i x^i$ in R_q, they can both be represented by their coefficients $[f_0, f_1, \cdots, f_{n-1}]$. There are two mathematical operations for f and g, namely $f + g$ and $f \cdot g$. In addition,f,g and both operation result belonging to R_q. The following is an introduction to the NTRU lattice used in this paper.

1. Define an n-dimensional anti-cyclic matrix $A_n(f)$ for $f = \sum_{i=0}^{N-1} f_i x^i$ (which can be denoted as $A(f)$ with the subscript omitted) [12]:

$$A_n(f) = \begin{pmatrix} f \\ x \cdot f \\ \vdots \\ x^{N-1} \cdot f \end{pmatrix} = \begin{pmatrix} f_0 & f_1 & \cdots & f_{N-1} \\ -f_{N-1} & f_0 & \cdots & f_{N-2} \\ \vdots & \vdots & \vdots & \vdots \\ -f_1 & -f_2 & \cdots & f_0 \end{pmatrix} \tag{1}$$

2. Let q be a prime number satisfying $q > 5$, and n be an integer that is a power of two, $f, g \in R_q$. Let $h = g \cdot f^{-1} \mod q$. Then the NTRU lattice is defined as follows [12]:

$$\mathcal{L}_{h,q} = \{(u, v) \in R^2 \mid u + h \cdot v \equiv 0 \mod q\} \tag{2}$$

$L_{h,q}$ constitutes a full-rank lattice in $\mathbb{Z}^{2n}$, generated by the rows of the matrix:

$$\mathcal{A}_{h,q} = \begin{pmatrix} -A(h) & I_n \\ qI_n & O_n \end{pmatrix} \tag{3}$$

In the above matrix, I_n is the $n \times n$ identity matrix and O_n is the $n \times n$ zero matrix. When h is uniformly distributed in R_q, the Gram-Schmidt orthogonalization process of $\mathcal{A}_{h,q}$ is complex. In view of the above problem, literature [13] proposed a type of basis:

$$B_{f,g} = \begin{pmatrix} A(g) & -A(f) \\ A(G) & -A(F) \end{pmatrix} \tag{4}$$

where $F, G \in R_q$ and satisfy $f \cdot G - g \cdot F = q$.

3. For an integer n, a prime q, and $\sigma_f = 1.17\sqrt{q/(2n)}$ (satisfying $\|\tilde{B}_{f,g}\| \leq 1.17\sqrt{q}$), there exists an algorithm $TrapGen(q, n)$. This algorithm generates a polynomial $h = g \cdot f^{-1} \mod q$ and a short basis $B_{f,g}$ for $\mathcal{L}_{h,q}$ [12].

2.3 Gaussian Distributions

1. **Gaussian Distribution.** The Gaussian distribution over $\mathbb{R}^n$ is as follows [14]:

$$\forall x \in \mathbb{R}^n, \rho_{s,c}(x) = \exp\left(\frac{-\pi\|x-c\|^2}{s^2}\right) \tag{5}$$

where $s \in \mathbb{R}^+$ denotes the standard deviation, and $c \in \mathbb{R}^n$ is the center of this Gaussian distribution.

2. **Discrete Gaussian Distribution.** The Gaussian distribution over a lattice Λ is denoted as $\rho_{s,c}(\Lambda) = \sum_{x\in\Lambda} \rho_{s,c}(x)$. The discrete Gaussian distribution over Λ is defined as follows [14]:

$$D_{\Lambda,s,c} = \frac{\rho_{s,c}(x)}{\rho_{s,c}(\Lambda)} \tag{6}$$

When $c = 0$, $D_{\Lambda,s}$ is used to denote $D_{\Lambda,s,0}$.

3. **Sampling Algorithm.** Consider a prime number q, a short basis $B_{f,g}$ of the lattice $\Lambda_{h,q}$, and a polynomial $t \in R_q$ [12]. If $s \geq \|B_{f,g}\|\omega(\sqrt{\log(2n)})$ and $0 < \epsilon < 1$, then for $u = (t, 0) \in \mathbb{Z}_q^{2n}$, the following conclusions hold [12,14]:
(1)

$$\Pr\left[x \leftarrow D_{\Lambda_{h,q},s,u} : \|x - u\| > s \cdot \sqrt{2n}\right] \leq \frac{1+\epsilon}{1-\epsilon} \cdot 2^{-2n}$$

(2) A probabilistic polynomial-time algorithm $SampleGau(B_{f,g}, s, u)$ exists, which produces $(s_1, s_2) \in R_q^2$ that meet the conditions $s_1 + h \cdot s_2 = t$ and $\|(s_1, s_2)\| \leq s \cdot \sqrt{2n}$.

3 Our Scheme

3.1 System Model

The system model is presented in Fig. 1. The emergency communication system consists of five core entities:

- Key Generation Center (KGC): Initializes system parameters, generates partial private keys for users, and is deployed in the emergency command center to ensure the security of key distribution.
- Rescue Node (RN): Includes unmanned aerial vehicles, portable terminals, etc., generates signatures, and participates in information exchange. The private key consists of two parts: the partial private key issued by KGC and the secret value unique to itself.
- Aggregation Verification Node (AVN): Such as temporary base stations or command vehicles, responsible for aggregating signatures from multiple nodes and verifying them, supporting batch processing to improve efficiency.
- Trust Authority (TA): Monitors node behavior, handles accountability for malicious nodes, and prevents DoS attacks.
- Application Server (AS): AS collects information through secure connections, conducts an analysis of the data, and provides services such as traffic information, and emergency assistance information to nodes (Table 1).

Table 1. Symbol and explanation.

Symbol	Explanation
AS	Application server
TA	Trust authority
KGC	Key generation center
RN	Rescue node
AVN	Aggregation verification node
N_i	Single rescue node
ID_i	Identity of rescue node
msk	KGC's master private key
$param$	System parameter
$d_i = (d_{i,1}, d_{i,2})$	Partial private key of N_i
$s_i = (s_{i,1}, s_{i,2})$	Secret value of N_i
$ski = (d_i, s_i)$	Private key of N_i
pk_i	Public key of N_i
σ_i	Single signature
σ_{agg}	Aggregate signature
H_0, H_1, H_2	Hash function
t_i	Current timestamp

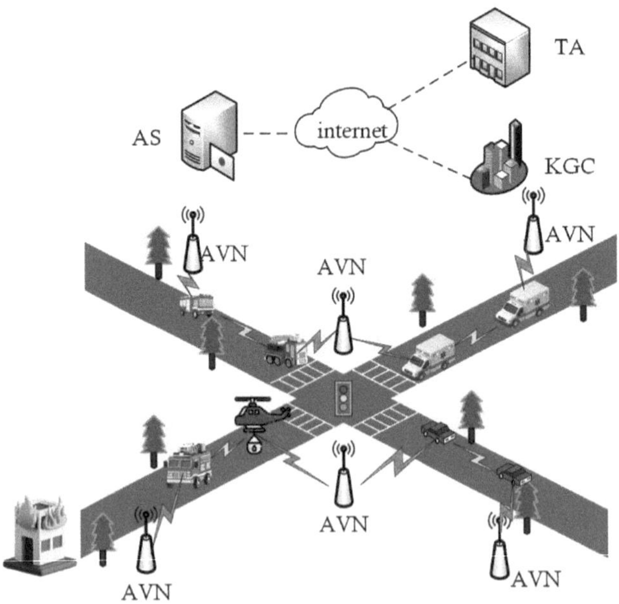

Fig. 1. System Model.

3.2 Scheme Proposal

Our scheme is divided into 9 parts, as comprehensively delineated in Ref. [15]. The flow of our scheme is as follows:

Setup. The emergency command center's KGC pre-generates parameters for quick networking after a disaster. First,1^n are entered as security parameters. The KGC chooses a prime number q and two positive integers s and σ. We use the $TrapGen(q,n)$ algorithm to compute the public key polynomial $h = g \cdot f^{-1} \mod q$ in order to construct h and create a short basis $B_{f,g}$ (Eq. (4)).

Subsequently, three hash functions are selected by the KGC. $H_0 : \{0,1\}^* \rightarrow \{v \in \mathbb{Z}_q^n\}$ maps the identities of rescue nodes; $H_1 : \mathbb{Z}_q^n \times \{0,1\}^* \rightarrow D_H : \{v \in \mathbb{Z}_q^n \mid 0 \leq \|v\| \leq \kappa, \kappa \ll q\}$. generates key parameters to adapt to the dynamic access of nodes;$H_2 : \{0,1\}^* \rightarrow \{v \in \mathbb{Z}_q^n\}$ processes messages and timestamps to prevent replay attacks. The KGC sets the $msk = B_{f,g}$ and stores it securely, defines the system parameters $params = (h, H_0, H_1, H_2, q, n)$, and outputs $params$ as public parameters.

Extract-Partial-Private-Key. Assign a portion of the private key to the newly added rescue node. First, provide the node identity ID_i. The KGC calculates $H_0(ID_i)$ and generates a polynomial pair $(d_{i,1}, d_{i,2})$ through $SampleGau(B_{f,g}, s, H_0(ID_i))$, satisfying $d_{i,1} + h \cdot d_{i,2} = H_0(ID_i) \mod q$ and $\|(d_{i,1}, d_{i,2})\| \leq s \cdot \sqrt{2n}$. The portion of the private key $d_i = (d_{i,1}, d_{i,2})$ is then sent to N_i via an encrypted channel.

Set-Secret-Value. The rescue node autonomously generates a secret value to avoid the KGC key being held in custody. N_i randomly selects polynomials $s_{i,1}, s_{i,2} \in D_{\mathbb{Z}^n,s}$, satisfying $\| (s_{i,1}, s_{i,2}) \| \leq s \cdot \sqrt{2n}$, and sets the $s_i = (s_{i,1}, s_{i,2})$.

Set-Private-Key. N_i integrates the partial private key with the secret value to generate the complete private key $ski = (d_i, s_i) = (d_{i,1}, d_{i,2}, s_{i,1}, s_{i,2})$.

Set-Public-Key. Node broadcasts the public key for other nodes to verify the signature without the need for a certificate. Input the complete private key sk_i of N_i, calculate $\alpha_i = H_1(ID_i, h, H_0(ID_i))$ and $P_i = s_{i,1} + s_{i,2} \cdot h \mod q$, and generate the public key $pk_i = H_0(ID_i) + \alpha_i \cdot P_i \mod q$.

CL-Sign. The rescue node generates a signature for the critical information and adds a timestamp to prevent replay. Given ID_i, message m_i, and private key sk_i. Randomly select $y_{i,2} \in D_{\mathbb{Z}^n,\sigma}$, calculate $y_{i,1} = y_{i,2} \cdot h$; integrate the timestamp t_i to calculate $c_i = H_2(m_i, y_{i,2}, t_i)$; calculate the signature components $z_{i,1} = (d_{i,1} + \alpha_i \cdot s_{i,1}) \cdot c_i + y_{i,1}$ and $z_{i,2} = (d_{i,2} + \alpha_i \cdot s_{i,2}) \cdot c_i - y_{i,2}$. The signature $\sigma_i = (c_i, z_i)$, where $z_i = z_{i,1} + z_{i,2} \cdot h$. N_i sends the message $msg_i = \{ID_i, m_i, \sigma_i, t_i\}$ to AVN.

CL-Verify. AVN verifies the legality of each message and filters out forged data. Firstly, it checks whether t_i is within the valid period. A validity period of 30 s can be set to adapt to emergency network delays. If within the valid period, the next step will be carried out. Subsequently, it verifies whether $z_i = pk_i \cdot c_i \mod q$ is valid.

CL-AggSign. The aggregation node compresses the signatures of multiple rescue nodes into one. Given valid signatures $\sigma_1,\ldots,\sigma_n$, the aggregation node calculates the aggregated signature $\sigma_{\text{agg}} = \sum_{i=1}^{n} z_i \mod q$.

CL-AggVerify. The verifier verifies the integrity of the batch messages and quickly confirms the validity of the data from multiple nodes. Verify $\sigma_{\text{agg}} = \sum_{i=1}^{n}(pk_i \cdot c_i) \mod q$, check $\|\sigma_{\text{agg}}\| \leq \beta_{\text{ver}}$. We assume $\beta_{\text{ver}} \geq N(s \cdot \sqrt{n}+1)(2s \cdot \sqrt{n}+n \cdot q \cdot \kappa)$. If all the above conditions are met, accept the aggregated signature; otherwise, reject the signature.

4 Performance Evaluation

4.1 Security Comparison

We selected three papers to compare with our solution, namely Ref. [16] and Ref. [17] based on ECC and Ref. [18] based on NTRU. Our proposed scheme, along with the NTRU-based Scheme [18], demonstrates quantum-resistant capabilities as they rely on lattice-based hard problems, rendering them immune to quantum algorithmic threats. In contrast, the ECC-based schemes, namely Scheme [16] and Scheme [17], which rely on traditional number-theoretic assumptions, are vulnerable to security breaches under quantum computational methods, such as Shor's algorithm. As illustrated in Table 2, our scheme and Scheme [18] exhibit superior security advantages compared to other schemes.

Table 2. Security comparison.

Security Properties	Ref. [16]	Ref. [17]	Ref. [18]	Our Scheme
Quantum-resistant	×	×	✓	✓
Unforgeability	✓	✓	✓	✓
MITM-resistant	✓	✓	✓	✓
Replay-resistant	✓	✓	✓	✓
Privacy preservation	✓	✓	✓	✓
Key escrow resistance	✓	✓	✓	✓

4.2 Computation Cost

We used the MIRACL library to measure the running time of the relevant operations in elliptic curve cryptography (ECC), and utilized the NFLlib lattice cryptography library to calculate the time consumption of operations related to lattices. All the codes were implemented in C++, and the algorithm running tests were completed on a laptop equipped with a Intel Xeon processor, 8GB of memory, and running 64-bit Ubuntu operating system. We employed an elliptic curve cryptography system with a security level of 80 bits: on a non-singular elliptic curve determined by the equation $E : y^2 = x^3 + ax + b \mod p$, where $a, b \in \mathbb{Z}_p^*$, a group G was defined. The order of this group is q, where q is a 160-bit prime number, and the generator of the group is a point P on the elliptic curve. In the lattice-based implementation scheme, we selected the $n = 257$ and set q to a 10-bit value.

We mainly focused on the two crucial stages, CL-AggSign and CL-AggVerify, which have a significant impact on the computational cost. By comparing these two stages, we can effectively analyze our computational cost. We considered a node quantity of $N = 100$, The relevant comparison are shown in Fig. 2, Table 3 and Table 4. In the aggregation signature stage, the signature times based on ECC for [16] and [17] are only 2.4 ms and 2.1 ms, respectively. The signature time of the scheme in [18] is same to that of our scheme, approximately 1.3 ms. Therefore, the latter two schemes are slightly superior to the first two schemes in this stage. However, in the aggregation verification stage, the verification times for [16] and [17] are 235.9 ms and 358.6 ms, respectively, which are hundreds of times longer than those of the latter two schemes. The scheme [18] takes 11.9ms. Meanwhile, our scheme takes 9.9 ms, which is less than scheme [18]. Scheme [18] and our scheme both employ the NTRU signature and verification method, so the two schemes have relatively little difference.

In conclusion, in terms of cost calculation, the two schemes based on NTRU are superior to the two schemes based on ECC. In the comparison of the two schemes based on NTRU, our scheme is slightly better overall than [18].

Table 3. The running time of various cryptographic operations.

Symbol	Explanation	Time (μs)
T_{pm}	Point multiplication	1153.06
T_{pa}	Point addition	10.58
T_{mpa}	Matrix modulo addition	13.19
T_{mpm}	Matrix modulo multiplication	86.52
T_{mpm}	Hash operation	31.66

Table 4. Comparison of computational costs.

Scheme	Agg. sign	Agg. verify
ref. [16]	$(2N+1)T_{pm}$	$(2N+1)T_{pm} + (3N-1)T_{pa}$
ref. [17]	$2(N-1)T_{pa}$	$3NT_h + (3N+1)T_{pa} + 3NT_{pm}$
ref. [18]	$(N-1)T_{mpa}$	$(N+1)T_{mpm}$
Ours	$(N-1)T_{mpa}$	$(N-1)T_{mpa} + NT_{mpm}$

4.3 Communication Overhead

We compared the communication overhead of our scheme with those in Literature [16,17], and [18]. First, we unified the security level of all these schemes at 80 bits. We set G_1 to 128 bytes and G to 40 bytes. Table 5 shows the individual signature lengths and aggregate signature lengths for the four schemes. In the table, we define $\sigma = 12s\kappa n$ and $nlog(2\sigma) = 321$ bytes according to Literature [19].

In Reference [16], the total transmission overhead is $4|G| + |Z_q^*| + |t_i| = 4 \times 40 + 20 + 4 = 184$ bytes. In Reference [17], the transmission overhead is $3|G| + 3|Z_q^*| + |t_i| = 3 \times 40 + 3 \times 20 + 4 = 184$ bytes. In Reference [18], the total transmission overhead reaches 1348 bytes. In contrast, the total transmission overhead of our proposed scheme is $20 \times 3 + 4 + 321 \times 2 = 684$ bytes. Figure 3 presents a comparison of the transmission overheads among the four schemes. The key factor we focus on is the signature length, which directly affects communication efficiency. From the comparison, it can be concluded that lattice-based schemes exhibit higher communication overhead compared to the two traditional signature schemes. However, compared with Reference [18], which is also based on the NTRU algorithm, our communication overhead is approximately 50% of that in Reference [18]. Therefore, in this aspect, our scheme demonstrates greater superiority over other lattice-based schemes.

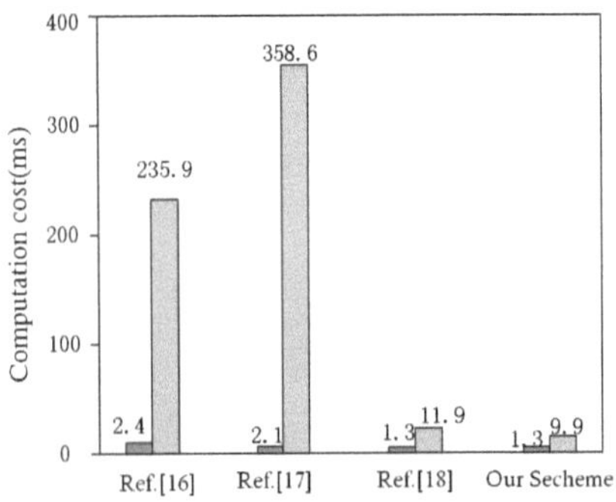

Fig. 2. Computational cost of aggregate signature and verification.

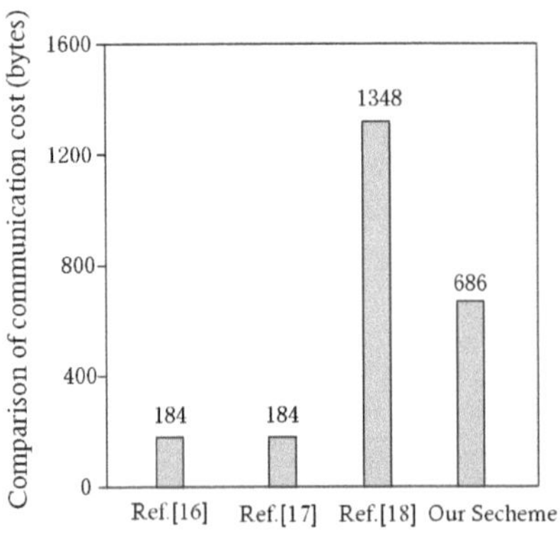

Fig. 3. Comparison of communication overhead.

Table 5. Signature length (Bytes).

Scheme	Single signature	Aggregate signature
ref. [16]	$\lvert G\rvert + \lvert Z_q^*\rvert$	$N\lvert G\rvert + \lvert Z_q^*\rvert$
ref. [17]	$\lvert G\rvert + \lvert Z_q^*\rvert$	$N\lvert G\rvert + \lvert Z_q^*\rvert$
ref. [18]	$4n \cdot log(2\sigma)$	$4n \cdot log(2\sigma)$
Ours	$2n \cdot log(2\sigma)$	$2n \cdot log(2\sigma)$

5 Conclusion

In this study, we propose an efficient and secure quantum-safe CLAS scheme for the field of emergency communications, which successfully avoids the complexities caused by certificate management and key escrow issues. Our scheme can individually sign various messages from different emergency nodes (rescue terminals, UAVs, sensors, etc.) and then aggregate these signatures into a single signature. This process notably streamlines verification time, reduces computation costs, and lowers bandwidth requirements. On the premise that lattice-based algorithms have been proven to be quantum-resistant, the proposed scheme boasts favorable quantum resistance, integrity, and unforgeability. Performance analysis results indicate that compared with other lattice-based schemes, our scheme also has a significant advantage in terms of cost calculation and communication expenses. Therefore, our proposed scheme is suitable for application in emergency communication field.

Lattice-based schemes do not have advantages over traditional cryptographic systems in terms of public and private key lengths and signature sizes, which results in higher storage and transmission overhead for lattice-based schemes. Therefore, in the future, we will devote more efforts to reducing communication costs and enhancing the adaptability of the schemes to dynamic and changing emergency network conditions.

References

1. Shamir, A.: Identity-based cryptosystems and signature schemes. In: Blakley, G.R., Chaum, D. (eds.) CRYPTO 1984. LNCS, vol. 196, pp. 47–53. Springer, Heidelberg (1985). https://doi.org/10.1007/3-540-39568-7_5
2. Kumar, P., Kumari, S., Sharma, V., et al.: Secure CLS and CL-AS schemes designed for VANETs. J. Supercomput. **75**(6), 3076–3098 (2019)
3. Jiang, Y., Deng, L., Ning, B.: Identity-based partially blind signature scheme: cryptanalysis and construction. IEEE Access **9**, 78017–78024 (2021)
4. Shor, P.W.: Polynomial-time algorithms for prime factorization and discrete logarithms on a quantum computer. SIAM Rev. **41**(2), 303–332 (1999)
5. Qian Mei, H., Xiong, J.C., et al.: Efficient certificateless aggregate signature with conditional privacy preservation in IoV. IEEE Syst. J. **15**(1), 245–256 (2020)
6. Zhiyan, X., He, D., Kumar, N., et al.: Efficient certificateless aggregate signature scheme for performing secure routing in VANETs. Secur. Commun. Netw. **2020**(1), 5276813 (2020)

7. Liang, Y., Liu, Y.: Analysis and improvement of an efficient certificateless aggregate signature with conditional privacy preservation in VANETs. IEEE Syst. J. **17**(1), 664–672 (2022)
8. Nejatollahi, H., Dutt, N., Ray, S., et al.: Post-quantum lattice-based cryptography implementations: a survey. ACM Comput. Surv. (CSUR) **51**(6), 1–41 (2019)
9. El Bansarkhani, R., Sturm, J.: An efficient lattice-based multisignature scheme with applications to bitcoins. In: Foresti, S., Persiano, G. (eds.) CANS 2016. LNCS, vol. 10052, pp. 140–155. Springer, Cham (2016). https://doi.org/10.1007/978-3-319-48965-0_9
10. Doröz, Y., Hoffstein, J., Silverman, J.H.: MMSAT: a scheme for multimessage multiuser signature aggregation. Cryptology ePrint Archive (2020)
11. Xie, J., Yupu, H., Gao, J., et al.: Certificateless sequential aggregate signature scheme on NTRU lattice. Chin. J. Electron. **28**(2), 294–300 (2019)
12. Hoffstein, J., Howgrave-Graham, N., Pipher, J., Silverman, J.H., Whyte, W.: NTRUSIGN: digital signatures using the NTRU lattice. In: Joye, M. (ed.) CT-RSA 2003. LNCS, vol. 2612, pp. 122–140. Springer, Cham (2003). https://doi.org/10.1007/3-540-36563-x_9
13. Ducas, L., Lyubashevsky, V., Prest, T.: Efficient identity-based encryption over NTRU lattices. In: Sarkar, P., Iwata, T. (eds.) ASIACRYPT 2014. LNCS, vol. 8874, pp. 22–41. Springer, Heidelberg (2014). https://doi.org/10.1007/978-3-662-45608-8_2
14. Gentry, C., Peikert, C., Vaikuntanathan, V.: Trapdoors for hard lattices and new cryptographic constructions. In: Proceedings of the Fortieth Annual ACM Symposium on Theory of Computing, pp. 197–206 (2008)
15. Huang, Y., Guangxia, X., Song, X., et al.: A quantum-secure certificateless aggregate signature protocol for vehicular ad hoc networks. Veh. Commun. **47**, 100775 (2024)
16. Thumbur, G., Rao, G.S., Vasudeva Reddy, P., et al.: Efficient and secure certificateless aggregate signature-based authentication scheme for vehicular ad hoc networks. IEEE Internet Things J. **8**(3), 1908–1920 (2021)
17. Zheng, H., Luo, M., Zhang, Y., et al.: A security-enhanced pairing-free certificateless aggregate signature for vehicular ad-hoc networks. IEEE Syst. J. **17**(3), 3822–3833 (2023)
18. Ming, X., Li, C.: An NTRU-based certificateless aggregate signature scheme for underwater acoustic communication. IEEE Internet Things J. **11**(6), 10031–10039 (2024)
19. Chen, C., Danba, O., Hoffstein, J., et al.: Algorithm specifications and supporting documentation. Brown University and Onboard Security Company, Wilmington USA (2019)

Cross-Sample Inconsistency-Based Video Forgery Detection Using Graph Convolutional Networks

Min Wu[1], Canghong Shi[1(✉)], and Yanli Chen[2(✉)]

[1] Xihua University, Chengdu 610039, Sichuan, China
canghongshi@163.com
[2] Guizhou Normal University, Guiyang 550018, China
yanli_027@163.com

Abstract. With the advancement of multimedia technology and the advent of the short video era, deepfake videos have become widespread in people's daily lives. This technology leverages artificial intelligence (AI) to generate or tamper with video content, disseminating it within the public sphere to cause detrimental effects. This paper proposes an efficient deepfake video detection method. Our approach capitalizes on the inconsistencies present in facial regions of deepfake video samples. These inconsistencies serve as key features for effective deepfake detection. The algorithm employs graph neural networks to establish cross-sample correlations, thereby enabling the detection of deepfake videos. Through experiments, we demonstrate the effectiveness and robustness of this method in video deepfake detection.

Keywords: Deepfake detection · Graph Convolutional Networks · Cross-Sample Inconsistency

1 Introduction

With the development of the internet and the widespread use of various video apps such as Douyin, Kuaishou, and Xigua Video [1], short videos have had an increasingly significant impact on people's lives. They have not only transformed how users consume information but also reshaped the way individuals express creativity, share opinions, and even conduct business. As short videos have grown rapidly, more and more people have gained access to video editing software, leading to a series of security and privacy issues [2].

In particular, the ease of access to AI-powered editing tools [3–6] has lowered the threshold for generating manipulated content, including face-swapped or deepfake videos. This democratization of video editing, while empowering for some, has also exacerbated issues related to copyright infringement, misinformation, and identity theft [7].

J. Yu and Y. Chen (Eds.): FCS 2025, CCIS 2805, pp. 293–306, 2027.
https://doi.org/10.1007/978-981-95-9514-3_20

Several high-profile video infringement incidents have recently drawn widespread public attention. Among them, a "video reposting software" developed by a post-90 s programmer has garnered particular attention. This software features functions such as watermark removal, mirror image modification, and face swapping, while supporting one-click cross-platform publishing to platforms like Douyin, Kuaishou, and Bilibili. This software has fueled users' practices of stealing original videos and artificially inflating follower counts. Such actions not only undermine the creative achievements of original content creators but also exacerbate the black market economy surrounding influencer marketing—monetizing traffic through unethical means. Given the detrimental societal impact, an urgent need exists for detection methods to prevent such phenomena.

In this paper, we describe a new method for detecting anomalies generated in video deepfake detection, specifically the insertion of a cropped facial portion from one frame into another frame within the video. Our method detects inconsistencies generated in video forgery while considering that focusing solely on inconsistencies between adjacent frames in the detection process may overlook inconsistencies across samples. Therefore, in addition to considering the consistency between adjacent frames, we propose a video forgery detection algorithm based on graph convolutional neural networks, which we name the TransGraph Forgery Detector (TGFD).

The graph convolutional network framework offers significant advantages over traditional methods for deepfake detection. By dynamically constructing cross-sample association graphs, the system achieves modeling of long-range dependencies between non-contiguous frames, effectively capturing the spatio-temporal inconsistencies characteristic of face-swapping operations; The message passing mechanism of graph convolutions amplifies subtle forgery traces that are difficult to detect within a single frame in the topological space, forming detectable strong signals; its non-Euclidean space modeling characteristics break through the limitations of temporal models, accurately identifying abnormal association patterns across time and space; the multi-layer topological filtering structure maintains sensitivity to forgery features while significantly enhancing robustness against post-processing interference; The node attention mechanism provides interpretable decision paths, enabling traceable identification of critical evidence frames. Compared to traditional methods relying on single-frame analysis or continuous temporal modeling, this framework achieves distributed perception of forgery traces and cross-sample collaborative verification through graph structure learning, opening new avenues for detecting highly sophisticated deepfake content.

2 Previous Work

With the advent of the short video era, the barriers to content creation and editing have been significantly lowered. An increasing number of individuals can now effortlessly utilize various video manipulation tools, which has led to a surge in incidents related to privacy violations, copyright infringement, and misinformation dissemination. Among these, deepfake video tampering has emerged as a particularly serious concern due to its high realism and potential for misuse. Deepfake forgeries are typically generated using convolutional neural networks (CNNs), which adopt an encoder-decoder architecture to map facial features from a source identity onto a target face, producing manipulated video frames with realistic expressions and movements [8]. The generation pipeline is illustrated in Fig. 1, where the model learns to encode facial attributes and reconstruct altered versions that closely resemble the original content while embedding forged elements.

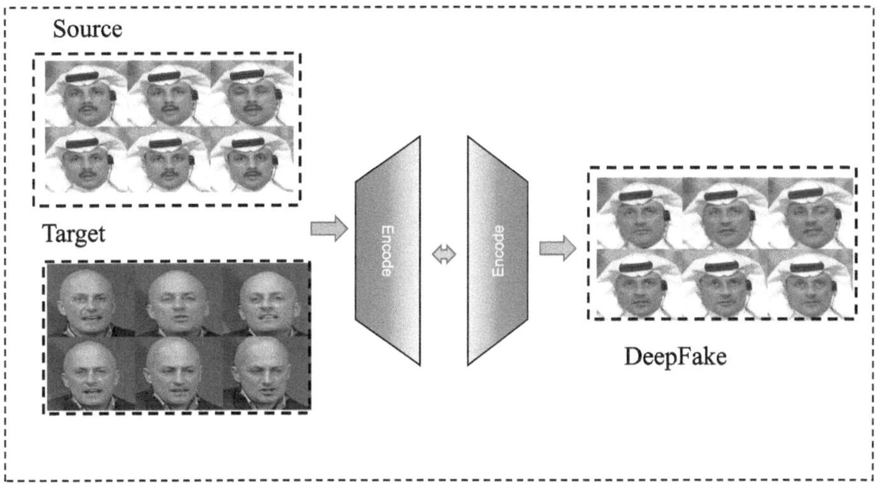

Fig. 1. The creation of video forgery

Most researchers using deep learning to create deepfake videos typically use RGB images as input for the network, then extract features through a feature extractor to classify the video. For example, researchers found that deepfake videos exhibit inconsistencies in media attributes, leading to the proposal of MesoNet [9], which uses two CNNs as feature extractors to detect inconsistencies in deepfake videos. Chollet et al. [10] proposed Xception as a benchmark for subsequent researchers. Nguyen et al. [11] discovered that facial pose changes can be used to distinguish between fake videos, leading to the proposal of capsule networks. PRRNet [12] achieves fake detection by constructing regional and relational features between fake frames.

Researchers found that using only RGB images as input could only extract superficial forgery features, neglecting deeper forgery features. They also discovered that forged images produce forgery artifacts in both the time domain and frequency domain. With the widespread development of attention mechanisms, researchers have further improved detection methods for deepfake video forgery.

F³Net [13] combines two branches, FAD and LFS, with DCT transformation and attention mechanisms to extract features, thereby improving detection accuracy. Zhao et al. [14] proposed the MADD model for forgery detection by combining features extracted using attention mechanisms with those obtained from shallow CNN models. Frank et al. [15] also used DCT transformation combined with deep learning for forgery detection. Lu et al. [16] proposed a spatiotemporal model featuring an innovative long-range attention mechanism. By integrating global semantic information to guide the network's focus on critical local regions, it effectively captures subtle spatial artifacts and temporal inconsistencies in deepfake videos, significantly enhancing detection accuracy.

More recently, transformer-based [17] approaches and cross-domain generalization frameworks have been explored to handle increasingly sophisticated forgery patterns. Although researchers have achieved significant results in deepfake video forgery detection classification using deep learning, existing methods are still limited to continuous frames or isolated samples during modeling, failing to effectively capture cross-sample correlation characteristics between non-continuous frames. This limitation results in insufficient detection capabilities for forgery patterns such as face transplantation that rely on cross-frame inconsistencies.

3 Method

3.1 Graph Convolutional Networks

Graph Convolutional Networks (GCNs) [18] provide an efficient framework for modeling graph-structured data by extending convolution to non-Euclidean data domains. Given a graph structure $\mathcal{G} = (\dot{\mathcal{V}}, \mathcal{E})$, Among the nodes $v_i \in \mathcal{V}$, carrying the feature vector $\mathbf{x}_i$, GCN realizes feature propagation through a neighborhood aggregation mechanism. The formula is as follows:

$$H^{(l+1)} = \sigma\left(\tilde{D}^{-\frac{1}{2}}\,\tilde{A}\,\tilde{D}^{-\frac{1}{2}}H^{(l)}W^{(l)}\right), \tag{1}$$

where $\tilde{A} = A + I_N$ is the adjacency matrix with added self-loops, $\tilde{D}$ is the diagonal degree matrix with entries $\tilde{D}_{ii} = \sum_j \tilde{A}_{ij}$, and $\tilde{D}^{-\frac{1}{2}}$ denotes the symmetric normalized inverse square root of $\tilde{D}$. Here, $H^{(l)}$ is the node feature matrix at the l-th layer, $W^{(l)}$ is the trainable weight matrix of the l-th layer, and $\sigma(\cdot)$ denotes the activation function.

After L layers of propagation, the node represents the fusion of L neighborhood information:

$$z_i = \sum_{k=0}^{L} \alpha_k\, h_i^{(k)}, \tag{2}$$

where $h_i^{(k)}$ denotes the representation of node i after the k-th layer of propagation, α_k is the learnable or predefined weight coefficient used to balance the contributions of different propagation depths, and L is the total number of GCN layers. The final representation z_i is thus obtained as a weighted fusion of multi-layer neighborhood information, integrating both shallow and deep structural features.

This work innovatively maps video frames to graph nodes and uses a dynamically constructed cross-frame correlation topology, denoted as ε, to enable the Graph Convolutional Network (GCN) to capture discontinuous forgery traces (such as face transplant operations) that are difficult to model using traditional time series models. Here, ε represents the set of edges connecting nodes (frames) based on their feature similarities or temporal-correlation patterns. This architecture significantly enhances the propagation and collaborative detection capabilities of weak forgery signals.

3.2 TGFD Model

To overcome the limitations of existing models, which only consider isolated samples during detection, this paper proposes a detection framework that combines graph convolutional neural networks with deep learning algorithms, based on existing mainstream detection models, as shown in Fig. 2. The proposed model initiates with a preprocessing stage, wherein the input video dataset is first decomposed into individual frames. Each frame undergoes a dedicated processing pipeline to normalize resolution, eliminate noise, and enhance salient features that may aid in forgery detection. Following this, the model focuses on identifying regions within the frames that exhibit prominent visual inconsistencies often indicative of manipulation artifacts introduced during the forgery process.

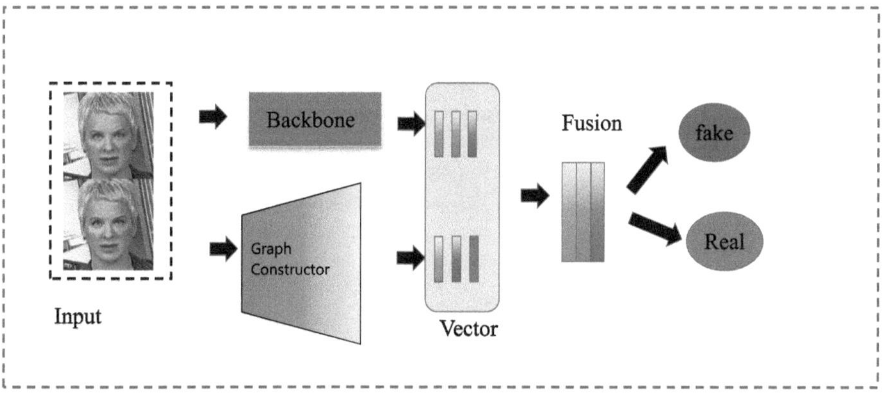

Fig. 2. The Structural Framework of TGFD

These inconsistency regions are then leveraged to guide the construction of a learning framework. A VGG19 model pre-trained on ImageNet is employed to extract low-level visual features from the identified regions, capturing local texture and structural anomalies. Recognizing that forgeries in videos may generate inconsistencies not only between temporally adjacent frames but also across non-consecutive ones, the model accounts for long-range temporal dependencies as well.

To effectively capture such cross-frame and cross-sample inconsistencies something traditional CNN architectures struggle with a graph-based representation is constructed. In this graph, nodes represent frames or frame patches associated with suspicious regions, and edges are defined based on the degree of inconsistency between them. This graph structure enables the model to learn relational information among disparate regions within and across frames.

Subsequently, deep feature extraction is carried out on this graph structure, allowing the model to capture high-level semantic inconsistencies. By integrating both the shallow features obtained via the CNN and the deep relational features from the graph, the model is able to more comprehensively characterize forgery artifacts. This multi-level feature fusion strategy contributes to enhanced generalization and significantly improves the accuracy of deepfake video detection. The specific process is shown in Algorithm 1.

Algorithm 1. TGFD: TransGraph Forgery Detection Algorithm

1: $\mathcal{V}$: Input video
2: k: Neighbor size for graph construction (default: 10)
3: τ: Cosine similarity threshold (default: 0.7)
4: L: Number of GCN layers (default: 6)
5: *predictions*: Forgery detection results per frame

Step 1: Preprocessing Decompose $\mathcal{V}$ into individual frames $\mathcal{F} = \{f_1, f_2, \ldots, f_N\}$ each frame $f_i \in \mathcal{F}$ Normalize resolution and preprocess f_i Detect and crop facial region r_i from f_i

Step 2: Feature Extraction Load pre-trained VGG19 backbone each region r_i Extract feature vector $\mathbf{x}_i = \text{VGG19}(r_i)$ L_2-normalize $\mathbf{x}_i$

Step 3: Graph Construction Initialize empty graph $\mathcal{G} = (\mathcal{V}, \mathcal{E})$ $\mathcal{V} \leftarrow \{\mathbf{x}_1, \mathbf{x}_2, \ldots, \mathbf{x}_N\}$ Nodes are frame features each node $v_i \in \mathcal{V}$ each node $v_j \in \mathcal{V} \setminus \{v_i\}$ $s_{ij} = \frac{\mathbf{x}_i \cdot \mathbf{x}_j}{\|\mathbf{x}_i\|\|\mathbf{x}_j\|}$ Compute cosine similarity $s_{ij} > \tau$ Add edge (v_i, v_j) to $\mathcal{E}$ Connect similar nodes Prune edges to maintain maximum k connections per node

Step 4: Graph Convolution $\mathbf{H}^{(0)} \leftarrow [\mathbf{x}_1, \mathbf{x}_2, \ldots, \mathbf{x}_N]^T$ Initial node features $l = 0$ to $L - 1$ $\mathbf{H}^{(l+1)} = \sigma\left(\tilde{\mathbf{D}}^{-\frac{1}{2}}\tilde{\mathbf{A}}\tilde{\mathbf{D}}^{-\frac{1}{2}}\mathbf{H}^{(l)}\mathbf{W}^{(l)}\right)$ Apply dropout to $\mathbf{H}^{(l+1)}$ with rate $p = 0.2$ $\mathbf{Z} = \sum_{l=0}^{L} \alpha_l \mathbf{H}^{(l)}$ Multi-layer feature fusion

Step 5: Classification & Loss each node v_i $\mathbf{p}_i = \text{Softmax}(\text{MLP}(\mathbf{z}_i))$ $\mathcal{L} = -\frac{1}{N}\sum_{i=1}^{N}[y_i \log(p_i) + (1 - y_i)\log(1 - p_i)]$ Cross-entropy loss Compute loss only on seed nodes (original batch samples) Apply gradient clipping ($\|\mathbf{g}\|_2 \leq 1.0$)

Return $\{\mathbf{p}_i\}$ for all frames $f_i \in \mathcal{F}$

3.3 Loss Function

To optimize the training process of our proposed framework, we adopt a standard cross-entropy loss function tailored for binary classification tasks. Specifically, the loss function is formulated as:

$$\mathcal{L}_{CE} = -\frac{1}{N}\sum_{i=1}^{N}\sum_{c=1}^{2} y_{i,c}\log(p_{i,c}), \tag{3}$$

where N denotes the number of seed nodes within each mini-batch, and $C = 2$ represents the two classes in our binary classification setting. Here, $y_{i,c}$ is the ground truth label (one-hot encoded), and $p_{i,c}$ is the predicted probability for class c of sample i.

4 Experiments

4.1 Datasets and Implementation Details

Datasets. The experiment uses the widely recognized FaceForensics++ (FF++) [19] benchmark dataset, which contains more than 1,000 original videos and tampered videos generated by four mainstream deepfake methods (Deepfakes, Face2Face, FaceSwap, NeuralTextures). All samples undergo multi-level compression processing to simulate real network transmission environments, providing a standardized testing platform for evaluating model robustness.

Implementation Details. All experiments were conducted on a workstation equipped with an NVIDIA GeForce RTX 4060 GPU (8GB VRAM) and 2TB system RAM, utilizing PyTorch 2.0 with CUDA 11.8 acceleration alongside PyTorch Geometric 2.3 for graph operations and OpenCV 4.7 for image processing. Feature extraction employed a VGG19 backbone pretrained on ImageNet, generating 512-dimensional feature vectors from 224 × 224 RGB images with region-specific features obtained through masked spatial average pooling. Graph construction created undirected k-NN graphs ($k = 10$) using cosine similarity thresholding ($\tau = 0.7$) with block-wise processing (2,000 nodes/block) and L^2-normalized features to ensure memory efficiency. It is worth noting that although our graph construction uses cosine similarity to connect nodes (frames), which superficially measures consistency between samples, this design aims to reveal inconsistencies embodied by forged features. The core principle is this: when edges in the graph represent high feature similarity, genuine frames naturally form tightly connected clusters. Conversely, while forged frames may exhibit high visual similarity to genuine frames, they often originate from different sources or processing workflows. This results in anomalous positions within the graph topology—either forming weak, unstable connections to the main clusters of genuine samples, or creating deviant subgraphs that break standard relational structures. Consequently, the training of Graph Convolutional Networks

(GCNs) focuses on identifying these structural anomalies and associative contradictions rather than direct feature differences. By learning to propagate information through similarity-based graph structures, the model effectively flags nodes whose connectivity patterns deviate from mainstream norms. This transforms consistency graphs into powerful tools for detecting inconsistencies. Network details are shown in Table 1. Training implemented a 6-layer LightGCN architecture with 128-dimensional hidden states optimized by AdamW ($\eta = 0.001$, $\lambda = 10^{-4}$) using ReduceLROnPlateau scheduling, with critical stability measures including gradient clipping ($\|\mathbf{g}\|_2 \leq 1.0$), dropout ($p = 0.2$), and full-precision arithmetic throughout. Batch processing was strategically configured with 16 images per batch during feature extraction and 32 nodes per batch during graph training, complemented by active VRAM monitoring to prevent resource exhaustion during large-scale graph operations.

Additionally, in our experiments, the video is decomposed into independent frames, with nodes constructed by uniformly sampling one frame every ten frames. This interval sampling strategy balances computational efficiency with temporal coverage: it reduces redundancy between highly similar consecutive frames while preserving sufficient temporal dynamics to capture potential forgery traces. Each sampled frame undergoes face region detection and feature extraction before being processed as a graph node. This ensures the constructed graph structure preserves both local details and long-range inconsistencies across frames.

Table 1. Implementation details of our experimental setup.

Category	Details
Feature Extraction	512-dim features from 224 $\times$ 224 RGB images, masked spatial average pooling
Graph Construction	Undirected k-NN (k = 10), cosine similarity ($\tau = 0.7$), blockwise (2000 nodes/block), L2-normalization
Design Principle	Genuine frames $\rightarrow$ tight clusters; forged frames $\rightarrow$ anomalous connections/deviant subgraphs
GCN Architecture	6-layer LightGCN, 128-dim hidden states
Optimization	AdamW ($\eta = 0.001$, $\lambda = 10^{-4}$), ReduceLROnPlateau
Stability Measures	Gradient clipping ($\|g\|_2 \leq 1.0$), dropout ($p = 0.2$), full-precision arithmetic
Batch Processing	16 images/batch (feature extraction), 32 nodes/batch (graph training), active VRAM monitoring

4.2 Comparative Experiment

To further validate the effectiveness of our proposed method, we conducted a comparative experiment against several state-of-the-art face forgery detection methods on the FF++(LQ) [19] dataset. As shown in Table 3, our method (TGFD) consistently outperforms previous approaches in both accuracy and AUC across all manipulation types. Compared with EfficientNet-B4, which already performs strongly, our method achieves slightly higher AUC on DF, F2F, and FS, while significantly outperforming it on NT with an AUC improvement of 3.56%. In terms of accuracy, our model achieves 96.98%, 93.22%, 95.01%, and 89.34% on DF, F2F, FS, and NT respectively, each surpassing EfficientNet-B4's results, with the largest gain again on FS by 1.56%. These improvements demonstrate that our approach not only excels in extracting single-frame forgery features but also benefits from the use of dynamic graph convolutions to model cross-frame associations, enabling a more comprehensive capture of forgery patterns and significantly enhancing detection performance.

4.3 Choosing Different Backbones

While modern architectures like ResNet or EfficientNet offer advantages in many tasks, we selected VGG19 for its structural simplicity and proven effectiveness in capturing low-to-mid level texture features [20], which are crucial for identifying forgery artifacts. More complex architectures may introduce unnecessary parameters and risk overfitting on smaller forensic datasets. As shown in Table 2, we compared VGG19 against several modern backbones. We conclude that VGG19 achieves the best performance on this task, hence selecting it as the backbone network for this paper.

Table 2. Comparison of Different Backbone Networks in Terms of Accuracy(%) on DF, F2F, FS, and NT.

Backbone	DF	F2F	FS	NT
VGG19	**99.44**	**94.58**	**96.55**	**90.99**
ResNet-50	99.40	94.20	96.10	90.50
EfficientNet-B0	99.42	94.55	96.50	90.80

4.4 Graph Construction Hyperparameters

The choices of the neighborhood size $k = 10$ and the cosine similarity threshold $\tau = 0.7$ were determined through empirical analysis on a validation set. The parameter k controls the density of the graph and the receptive field of each node. A small k may lead to isolated subgraphs that hinder information propagation, while a very large k can introduce noisy edges that blur the distinction between consistent and inconsistent samples. Similarly, τensures that only sufficiently similar nodes are connected, filtering out weak correlations. As illustrated in Fig. 3, we evaluated a range of values for k and τ. The combination of k = 10 and $\tau = 0.7$ was found to be a robust setting that delivered stable and high performance across all manipulation types in the FF++ dataset, justifying our selection.

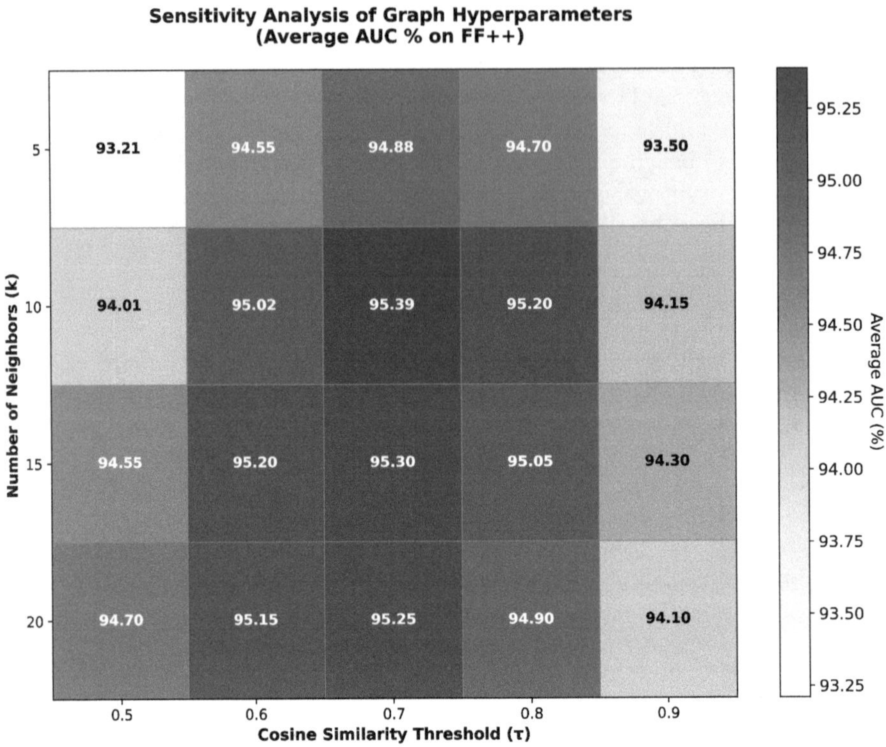

Fig. 3. Graph Construction Hyperparameters

4.5 Ablation Experiment

To verify the effectiveness and necessity of the proposed TGFD (TransGraph Forgery Detector) module, we conducted an ablation study by removing the TGFD component from the model and comparing the performance across four benchmark datasets: DF (Deepfakes), F2F (Face2Face), FS (FaceSwap), and NT (NeuralTextures). The experimental results are summarized in Table 4.

Without the TGFD module, the model achieves an accuracy (ACC) of 96.45% and AUC of 99.11% on the DF dataset. After incorporating the TGFD module, the accuracy improves to 96.98% and AUC increases to 99.44%, indicating a noticeable enhancement in distinguishing deepfake content. On the F2F dataset, the performance gain is even more pronounced: the accuracy improves from 90.12% to 93.22%, and AUC rises from 94.35% to 94.58%, demonstrating

Table 3. Comparison with previous face forgery detectors in binary detection accuracy on LQ (Low Quality) FF++.

Methods	DF		F2F		FS		NT	
	ACC (%)	AUC (%)	ACC (%)	AUC (%)	ACC (%)	AUC (%)	ACC (%)	AUC (%)
MesoNet [9]	88.27	–	56.30	–	61.19	–	40.67	–
MesoInception [9]	78.56	85.59	80.87	84.20	70.39	78.48	65.61	71.45
EfficientNet-B4 [21]	96.79	99.40	92.34	94.53	93.45	93.45	88.32	87.43
Xception [10]	94.32	98.17	86.18	93.06	92.11	96.32	75.64	83.61
F^3Net [13]	90.59	99.17	90.85	93.14	91.65	94.36	87.85	90.93
SPSL [22]	93.45	98.89	86.02	94.32	93.21	93.45	76.43	80.52
PRRNET [12]	95.63	97.89	91.32	94.46	94.93	93.22	80.01	84.32
TGFD (Ours)	**96.98**	**99.44**	**93.22**	**94.58**	**95.01**	**96.55**	**89.01**	**90.99**

TGFD's superior capability in capturing the subtle manipulations characteristic of reenactment-based forgery.

Similarly, for the FS dataset, the model without TGFD obtains 92.36% accuracy and 94.87% AUC, while the inclusion of TGFD leads to a significant performance boost to 95.01% accuracy and 96.55% AUC. The most substantial improvement is observed on the NT dataset, which is considered one of the most challenging due to its high-quality synthesis. The accuracy increases from 85.12% to 89.01%, and the AUC from 88.30% to 90.99%, highlighting TGFD's effectiveness in capturing temporal inconsistencies and cross-frame manipulation artifacts that traditional frame-based models may overlook.

These results collectively demonstrate that the TGFD module plays a crucial role in enhancing detection performance across multiple types of forgery. By leveraging temporal and relational information between frames, TGFD significantly improves both the discriminative capacity and generalization ability of the model. Therefore, the ablation study provides strong empirical evidence supporting the inclusion of TGFD as a core component of our detection framework.

Table 4. Ablation study on the TGFD module.

TGFD	DF		F2F		FS		NT	
	ACC	AUC	ACC	AUC	ACC	AUC	ACC	AUC
	96.45	99.11	90.12	94.35	92.36	94.87	85.12	88.30
✓	**96.98**	**99.44**	**93.22**	**94.58**	**95.01**	**96.55**	**89.01**	**90.99**

4.6 Robustness Evaluation

To further assess the robustness of our proposed method, we evaluate TGFD under several perturbation scenarios that simulate real-world degradations, including Gaussian blur, JPEG compression, spatial masking, and resolution downscaling. As shown in Table 5, our method consistently outperforms state-of-the-art baselines across all perturbation types. These results indicate that TGFD effectively captures underlying forgery patterns even when superficial visual clues are degraded, demonstrating strong resilience to common image-level distortions and post-processing operations.

Table 5. Robustness evaluation under various perturbation settings on FF++(LQ).

Perturbation Type	TGFD (Ours)	EfficientNet-B4 [21]	Xception [10]
Gaussian Blur ($\sigma = 1.0$)	**92.35**	88.43	85.12
JPEG Quality = 50	**91.74**	87.01	83.44
Random Masking (20%)	**90.62**	85.23	82.67

5 Discussion

Although TGFD demonstrates encouraging performance in the FF++(LQ) benchmark and effectively captures cross-sample inconsistencies through its graph convolution structure, there are still several directions worth exploring in the future. One potential direction is to optimize the dynamic graph construction process by incorporating richer semantic or temporal cues, thereby enhancing the model's ability to capture complex inter-frame relationships. Additionally, integrating complementary forgery clues (such as frequency-based artifacts or subtle physiological signals) may further enhance the model's robustness in detecting increasingly complex manipulation behaviors. Expanding the evaluation scope to emerging and diverse deepfake datasets (especially those designed to evade existing detection methods) is crucial for assessing the model's generalization capabilities. Methods such as few-shot learning or self-supervised learning can be explored to improve performance under limited supervision. Additionally, reducing the computational cost of graph generation and inference through model simplification or lightweight GCN architectures will facilitate real-time or edge deployment.

Acknowledgments. This work was supported by the National Key Research and Development Program of China (Grant No. 2023YFC3304504), the 2024 Xizang Autonomous Region Central Guided Local Science and Technology Development Fund Project (Grant No. XZ202401YD0015), and the Basic Research Program for Natural Science of Guizhou Province (Grant No. QIANKHEHEJICHU-ZD[2025]-043).

References

1. Wang, Q., Zhang, W., Wang, H.: Privacy concerns toward short-form video platforms: scale development and validation. Front. Psychol. **13**, 954964 (2022)
2. Wu, J.: The disruptive nature of short video-based social media platforms in China: the case of TikTok. In: Government Response to Disruptive Innovation: Perspectives and Examinations, pp. 252–274. IGI Global Scientific Publishing (2023)
3. Deepfakesweb. Deepfakesweb (2021). Accessed August 2021
4. FaceApp. Faceapp (2021). Accessed August 2021
5. Deepfakes. Deepfakes (2021). Accessed August 2021
6. Kowalski, M.: Faceswap (2021). Accessed August 2021
7. Becker, C., Laycock, R.: Embracing deepfakes and ai-generated images in neuroscience research. Eur. J. Neurosci. **58**(3), 2657–2661 (2023)
8. Li, L., et al.: Face X-ray for more general face forgery detection. In: Proceedings of the IEEE/CVF Conference on Computer Vision and Pattern Recognition (CVPR) (2020)
9. Afchar, D., Nozick, V., Yamagishi, J., Echizen, I.: MesoNet: a compact facial video forgery detection network. In: 2018 IEEE International Workshop on Information Forensics and Security (WIFS), pp. 1–7. IEEE (2018)
10. Chollet, F.: Xception: deep learning with depthwise separable convolutions. In: Proceedings of the IEEE Conference on Computer Vision and Pattern Recognition, pp. 1251–1258 (2017)
11. Nguyen, H.H., Yamagishi, J., Echizen, I.: Capsule-forensics networks for deepfake detection. In: Rathgeb, C., Tolosana, R., Vera-Rodriguez, R., Busch, C. (eds.) Handbook of Digital Face Manipulation and Detection. Advances in Computer Vision and Pattern Recognition, pp. 275–301. Springer, Cham (2022). https://doi.org/10.1007/978-3-030-87664-7_13
12. Shang, Z., Xie, H., Zha, Z., Lingyun, Yu., Li, Y., Zhang, Y.: PRRNet: pixel-region relation network for face forgery detection. Pattern Recogn. **116**, 107950 (2021)
13. Qian, Y., Yin, G., Sheng, L., Chen, Z., Shao, J.: Thinking in frequency: face forgery detection by mining frequency-aware clues. In: Vedaldi, A., Bischof, H., Brox, T., Frahm, J.-M. (eds.) ECCV 2020. LNCS, vol. 12357, pp. 86–103. Springer, Cham (2020). https://doi.org/10.1007/978-3-030-58610-2_6
14. Zhao, H., Zhou, W., Chen, D., Wei, T., Zhang, W., Yu, N.: Multi-attentional deepfake detection. In: Proceedings of the IEEE/CVF Conference on Computer Vision and Pattern Recognition, pp. 2185–2194 (2021)
15. Frank, J., Eisenhofer, T., Schönherr, L., Fischer, A., Kolossa, D., Holz, T.: Leveraging frequency analysis for deep fake image recognition. In: International Conference on Machine Learning, pp. 3247–3258. PMLR (2020)
16. Wei, L., et al.: Detection of deepfake videos using long-distance attention. IEEE Trans. Neural Netw. Learn. Syst. **35**(7), 9366–9379 (2023)
17. Gong, L.Y., Li, X.J., Chong, P.H.J.: Swin-fake: a consistency learning transformer-based deepfake video detector. Electronics **13**(15), 3045 (2024)
18. Kipf, T.N.: Semi-supervised classification with graph convolutional networks. arXiv preprint arXiv:1609.02907 (2016)
19. Rossler, A., Cozzolino, D., Verdoliva, L., Riess, C., Thies, J., Nießner, M.: FaceForensics++: learning to detect manipulated facial images. In: Proceedings of the IEEE/CVF International Conference on Computer Vision, pp. 1–11 (2019)
20. Simonyan, K., Zisserman, A.: Very deep convolutional networks for large-scale image recognition. arXiv preprint arXiv:1409.1556 (2014)

21. Tan, M., Le, Q.: EfficientNet: rethinking model scaling for convolutional neural networks. In: International Conference on Machine Learning, pp. 6105–6114. PMLR (2019)
22. Liu, H., et al.: Spatial-phase shallow learning: rethinking face forgery detection in frequency domain. In: Proceedings of the IEEE/CVF Conference on Computer Vision and Pattern Recognition, pp. 772–781 (2021)

Watermarking Quantum Neural Networks Based on Sample Grouped and Paired Training

Limengnan Zhou[1], Chaoyue Huang[2], and Hanzhou Wu[2,3](✉)

[1] School of Electronic and Information Engineering, University of Electronic Science and Technology of China, Zhongshan Institute, Zhongshan 528400, China

[2] School of Communication and Information Engineering, Shanghai University, Shanghai 200444, China

h.wu.phd@ieee.org

[3] School of Big Data and Computer Science, Guizhou Normal University, Guiyang 550025, China

Abstract. Quantum neural networks (QNNs) leverage quantum computing to create powerful and efficient artificial intelligence models capable of solving complex problems significantly faster than traditional computers. With the fast development of quantum hardware technology, such as superconducting qubits, trapped ions, and integrated photonics, quantum computers may become reality, accelerating the applications of QNNs. However, preparing quantum circuits and optimizing parameters for QNNs require quantum hardware support, expertise, and high-quality data. How to protect intellectual property (IP) of QNNs becomes an urgent problem to be solved in the era of quantum computing. We make the first attempt towards IP protection of QNNs by watermarking. To this purpose, we collect classical clean samples and trigger ones, each of which is generated by adding a perturbation to a clean sample, associated with a label different from the ground-truth one. The host QNN, consisting of quantum encoding, quantum state transformation, and quantum measurement, is then trained from scratch with the clean samples and trigger ones, resulting in a watermarked QNN model. During training, we introduce sample grouped and paired training to ensure that the performance on the downstream task can be maintained while achieving good performance for watermark extraction. When disputes arise, by collecting a mini-set of trigger samples, the hidden watermark can be extracted by analyzing the prediction results of the target model corresponding to the trigger samples, without accessing the internal details of the target QNN model, thereby verifying the ownership of the model. Experiments have verified the superiority and applicability of this work.

Keywords: Quantum model watermarking · quantum neural networks · intellectual property protection · quantum computing · security

J. Yu and Y. Chen (Eds.): FCS 2025, CCIS 2805, pp. 307–318, 2027.
https://doi.org/10.1007/978-981-95-9514-3_21

1 Introduction

Quantum computing (QC) represents a profound paradigm shift from classical computation by leveraging quantum-mechanical phenomena such as superposition, entanglement, and interference. Recent years have witnessed significant progress in quantum hardware, with superconducting qubits (e.g., IBM's 1,121-qubit Condor processor[1]), trapped ions (e.g., IonQ's 35-qubit systems[2]), and photonic platforms (e.g., Xanadu's Borealis [1]) achieving milestones in scalability and error mitigation. Despite these advances, challenges such as decoherence, high error rates, and the demand for fault-tolerant quantum error correction persist, limiting near-term applications. Concurrently, quantum machine learning and quantum-enhanced artificial intelligence (AI) algorithms have emerged as promising interdisciplinary fields. Hybrid quantum-classical models, e.g., variational quantum eigensolvers, and quantum neural networks (QNNs) demonstrate potential for exponential speedup in optimization and big data analysis. However, the current noisy intermediate-scale quantum (NISQ) era restricts practical deployments, necessitating further research into error-resilient algorithms and hardware-software co-design. The synergy between QC and AI continues to evolve, driven by theoretical breakthroughs and experimental validations.

Despite the aforementioned challenges in quantum computing, rapid advancements in quantum hardware and error mitigation techniques are steadily bridging the gap towards practical quantum computers. Once these technical hurdles are addressed, the realization of large-scale, fault-tolerant QC appears increasingly plausible. This progress underscores the critical importance of quantum AI (QAI) represented by QNNs that are quantum-enhanced algorithms providing exponential speedups for optimization, learning, and simulations, being capable of solving complex problems. Therefore, it can be foreseen that QNN models will become popular and quite important in the era of QC.

However, similar to neural network (NN) models built upon classical circuits, creating ansatz architectures and determining near-optimal parameters for QNN models require quantum hardware support, expertise, and high-quality data. As a result, the question of how to protect the intellectual property (IP) of QNNs becomes an urgent problem to be solved in the process of QAI industrialization. Fortunately, digital watermarking [2], as a kind of information hiding technology, enables us to insert a watermark into a neural model covertly by adjusting the network without reducing utilization, which is typically referred to as *model watermarking*. When disputes arise, the ownership of the target model can be reliably identified by extracting the watermark from the model, providing a very promising solution for IP protection of NN models. Therefore, it is straightforward to take into account model watermarking for protecting QNN models.

Many model watermarking methods have been reported in past years. These methods can be divided into three categories that include white-box watermarking, black-box watermarking, and box-free watermarking. White-box watermark-

[1] https://en.wikipedia.org/wiki/IBM_Condor.

[2] https://ionq.com/blog/how-we-achieved-our-2024-performance-target-of-aq-35.

ing requires the full or partial access to the internal details of the target model during watermark extraction. It is typically realized by modulating network parameters [3,4] or structures [5]. Black-box watermarking, on the other hand, does not require the access to the internals of the target model during extraction, but needs to interact with the target model. It is typically realized by injecting the watermark into a neural model through training the model with clean samples and trigger ones. These trigger samples are generated by two ways, i.e., using those samples not related to the downstream task [6], and using specifically crafted samples related to the downstream task [7–10]. The watermark is extracted by analyzing the consistency between the assigned labels and the prediction results when inputting a set of trigger samples, for classification models. Box-free watermarking is a special case of black-box watermarking [2,11,12]. It extracts the watermark from the output of the target model without interacting with the model directly, which facilitates copyright protection and source tracking of AI-generated content (AIGC). This paper investigates black-box watermarking since black-box is more common than white-box.

Back to QC, a quantum is the smallest discrete unit of a physical property in quantum mechanics, which can be used for various computing tasks. The qubit is an elementary concept in quantum computing, which is the quantum version of a classical bit, capable of being in a superposition of '0' and '1' simultaneously, expressed as $|\psi\rangle = \alpha|0\rangle + \beta|1\rangle$, where $\alpha^2 + \beta^2 = 1$. When measuring a qubit in the computational basis, it will collapse either to the state $|0\rangle$ or $|1\rangle$ with the probability α^2 and β^2. Quantum gates manipulate qubits-quantum systems. A quantum gate is a fundamental building block of quantum circuits, analogous to classical logic gates in conventional computing. Quantum gates perform reversible operations, transforming qubit states through unitary transformations, enabling phenomena like superposition, entanglement, and interface.

A QNN contains three modules, i.e., *quantum encoding*, *quantum state transformation*, and *quantum measurement*. Quantum encoding, e.g., maps classical input data into quantum states by amplitude encoding or angle encoding. Quantum state transformation, as the "quantum hidden layer" composed of trainable quantum gates, allows non-linear transformations and representations of the input quantum states. For quantum measurement, it extracts classical outputs by measuring qubits. During model training, a classical optimizer such as stochastic gradient descent (SGD) or Adam can be appropriately adjusted to optimize the tunable parameters within the quantum gates.

The proposed method marks a QNN by training the QNN from scratch with the to-be-quantum-encoded classical clean samples and trigger ones. The trigger sample is constructed by adding a perturbation (trigger) to a clean sample associated with a label different from the ground-truth. To reduce the computational complexity and maintain the performance of the QNN on its original task, we experimentally use a classical NN to reduce the dimension of the input sample and propose a sample grouped and paired strategy for training the QNN. The NN can be considered as an internal module of quantum encoding, thereby requiring the NN to be trained together with other modules. When disputes arise, by collecting a mini-set of trigger samples, the watermark can be extracted by analyzing the prediction results of the target QNN model corresponding to these trigger samples, without accessing the internal details of the target QNN model, thereby verifying the ownership of the model. Experiments show that the performance of the watermarked QNN model on the downstream task can be maintained while achieving satisfactory performance on verification.

The rest structure is organized as follows. We introduce the proposed method in Sect. 2, followed by experimental results and analysis in Sect. 3. Finally, we conclude this paper and provide discussion in Sect. 4.

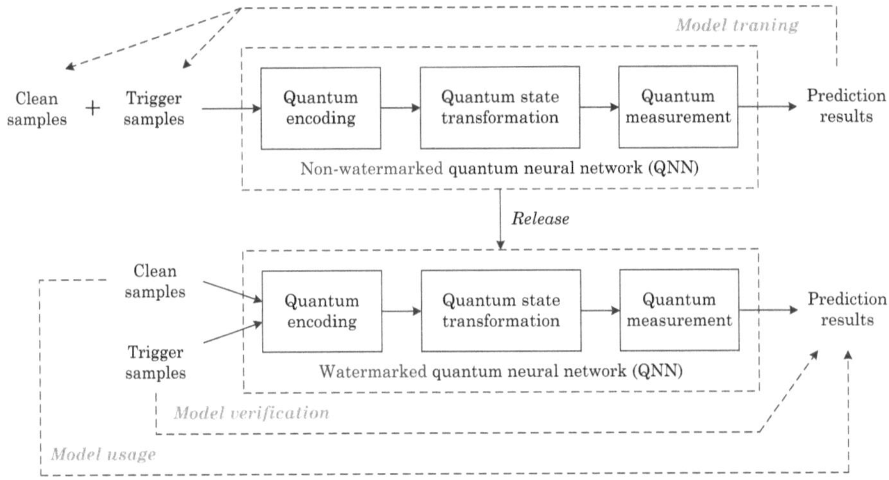

Fig. 1. General framework for the proposed method.

2 Proposed Method

Figure 1 presents the general framework including three stages, i.e., model training, model usage, and model verification. The purpose of model training is to produce the watermarked model by training the model from scratch using clean samples and trigger samples. After model training, the watermarked model is put into use, which corresponds to model usage. By analyzing the degree of consistency between the labels and the prediction results in correspondence to a set of trigger samples, the ownership of the target QNN model can be verified, which corresponds to model verification. In the following, we provide more details.

2.1 Model Training

Without the loss of generalization, we limit the original task of the benign QNN $\mathcal{Q}$ to be marked as image classification. This indicates that the non-marked QNN model can be generated by training with a clean image dataset $\mathcal{D} = \{(\mathbf{x}_i, y_i) | 1 \leq i \leq |\mathcal{D}|\}$. Here, $\mathbf{x}_i \in \mathbb{R}^{n \times m \times d}$ is a clean image sample, and $y_i \in \{0, 1, ..., c-1\}$ represents the ground-truth label of $\mathbf{x}_i$. It is noted that $\mathbf{x}_i$ is a classical sample, which means that $\mathbf{x}_i$ will be encoded before entering quantum hidden layer.

We mark $\mathcal{Q}$ by backdooring. To achieve this goal, a certain number of trigger samples $\mathcal{T} = \{(\mathbf{x}'_i, y'_i) \mid 1 \leq i \leq |\mathcal{T}|\}$ should be constructed in advance. There are various ways to construct the trigger samples. We do not limit the construction of trigger samples. However, in our experiments, to verify the applicability of the proposed method, we construct trigger samples by simply adding a perturbation to clean samples. More details can be found in the experimental section.

Assuming that we have already constructed $\mathcal{D}$ and $\mathcal{T}$, $\mathcal{Q}$ will be trained from scratch with $\mathcal{D}$ and $\mathcal{T}$. During model training, our purpose is to find the near-optimal parameters so that the loss function used for fitting the clean dataset and trigger dataset can be minimal, i.e., the marked model $\mathcal{Q}_{\text{marked}}$ is given by

$$\mathcal{Q}_{\text{marked}} = \arg\min_{\mathcal{Q}^*} \mathbb{E}_{(\mathbf{x},y)\sim\mathcal{D}}||\mathcal{Q}(\mathbf{x}) - y|| + \mathbb{E}_{(\mathbf{x}',y')\sim\mathcal{T}}||\mathcal{Q}(\mathbf{x}') - y'||. \tag{1}$$

It is required that $\mathcal{Q}_{\text{marked}}$ not only performs well on its original task, but also exhibits satisfactory performance on watermark verification. To balance the two requirements, during model training, we adjust the order of the input samples.

Mathematically, let $\mathcal{S}_{\text{DT}} = \{(\mathbf{x}_1, y_1), (\mathbf{x}_2, y_2), ..., (\mathbf{x}_{|\mathcal{S}_{\text{DT}}|}, y_{|\mathcal{S}_{\text{DT}}|})\}$ denote the entire sample sequence to be orderly fed into the QNN during model training. For any $1 \leq i \leq |\mathcal{S}_{\text{DT}}|$, we have $y_i \in [0, c-1]$, where c is the total classes of the samples. $\mathcal{S}_{\text{DT}}$ can be orderly divided into *disjoint* groups (subsets), i.e.,

$$\mathcal{S}_{\text{DT}} = \mathcal{S}_{\text{D}_1} \cup \mathcal{S}_{\text{T}_1} \cup \mathcal{S}_{\text{D}_2} \cup \mathcal{S}_{\text{T}_2} \cup ... \cup \mathcal{S}_{\text{D}_q} \cup \mathcal{S}_{\text{T}_q}, \tag{2}$$

where $|\mathcal{S}_{\text{D}_1}| = |\mathcal{S}_{\text{D}_2}| = ... = |\mathcal{S}_{\text{D}_q}| = n_{\text{D}}$ and $|\mathcal{S}_{\text{T}_1}| = |\mathcal{S}_{\text{T}_2}| = ... = |\mathcal{S}_{\text{T}_q}| = n_{\text{T}}$.

Obviously, $q(n_{\text{D}} + n_{\text{T}}) = |\mathcal{S}_{\text{DT}}|$. Each $\mathcal{S}_{\text{D}_i}$, $i \in [1, q]$, contains n_{D}/c samples with label $j \in [0, c-1]$. Therefore, $n_{\text{D}} \bmod c = 0$. On the other hand, $n_{\text{T}} \bmod 2 = 0$, implying that the samples in each $\mathcal{S}_{\text{T}_i}$, $i \in [1, q]$, are paired. It means that the samples in $\mathcal{S}_{\text{T}_i}$ can be divided into a total of $n_{\text{T}}/2$ disjoint pairs, each of which consists of a clean sample and its trigger version. For better understanding, the samples in $\mathcal{S}_{\text{D}_i}$, e.g., can be orderly expressed as

$$\{(\mathbf{a}_1, 0), (\mathbf{a}_2, 1), ..., (\mathbf{a}_c, c-1), ..., (\mathbf{a}_{n_{\text{D}}-c+1}, 0), (\mathbf{a}_{n_{\text{D}}-c+2}, 1), ..., (\mathbf{a}_{n_{\text{D}}}, c-1)\}, \tag{3}$$

where the first term $\mathbf{a}_j$ for j-th sample point represents the sample data, and the second term represents the corresponding label. Similarly, the samples in $\mathcal{S}_{\text{T}_i}$, e.g., can be orderly expressed as

$$\{(\mathbf{b}_1, l_1), (\mathbf{b}'_1, t), (\mathbf{b}_2, l_2), (\mathbf{b}'_2, t), ..., (\mathbf{b}_{n_{\text{T}}/2}, l_{n_{\text{T}}/2}), (\mathbf{b}'_{n_{\text{T}}/2}, t)\}, \tag{4}$$

where $\mathbf{b}'$ is the triggered one of the clean sample $\mathbf{b}$, and $t \in [0, c-1]$ is the new label assigned to all trigger samples, satisfying $l_j \neq t$ for any $j \in [1, n_{\text{T}}/2]$. It is possible that one may arrange the sequence by other ways. However, our experiments show that the above order already ensures satisfactory balance between watermarking and task functionality.

2.2 Model Usage and Verification

For model usage, it is required that the generalization of $\mathcal{Q}$ after watermarking should be well maintained, i.e., for any to-be-tested clean sample $\mathbf{x}$,

$$\Pr\{\mathcal{Q}(\mathbf{x}) \neq \mathcal{Q}_{\text{marked}}(\mathbf{x})\} < \epsilon_0 \in [0,1], \tag{5}$$

where ϵ_0 is a small threshold approaching zero. Meanwhile, it is expected that $\mathcal{Q}$ and $\mathcal{Q}_{\text{marked}}$ exhibit significantly different prediction performance on the trigger set, i.e., for any to-be-tested trigger sample $\mathbf{x}'$,

$$\Pr\{\mathcal{Q}(\mathbf{x}') = \mathcal{Q}_{\text{marked}}(\mathbf{x}')\} < \epsilon_1 \in [0,1], \tag{6}$$

where ϵ_1 is a small threshold approaching zero. From a statistical detection point of view, in order to identify the ownership of the target QNN model, we aim to maximize the difference of prediction accuracy between $\mathcal{Q}$ and $\mathcal{Q}_{\text{marked}}$ over the to-be-tested trigger set $\mathcal{R} = \{(\mathbf{x}_i'', y_i'') \mid 1 \leq i \leq |\mathcal{R}|\}$, i.e.,

$$\left| \frac{1}{|\mathcal{R}|} \sum_{i=1}^{|\mathcal{R}|} \delta(\mathcal{Q}_{\text{marked}}(\mathbf{x}_i''), y_i'') - \frac{1}{|\mathcal{R}|} \sum_{i=1}^{|\mathcal{R}|} \delta(\mathcal{Q}(\mathbf{x}_i''), y_i'') \right| > \Delta \in [0,1], \tag{7}$$

where $\delta(x, y) = 1$ if $x = y$ otherwise $\delta(x, y) = 0$, and Δ is a *large* threshold.

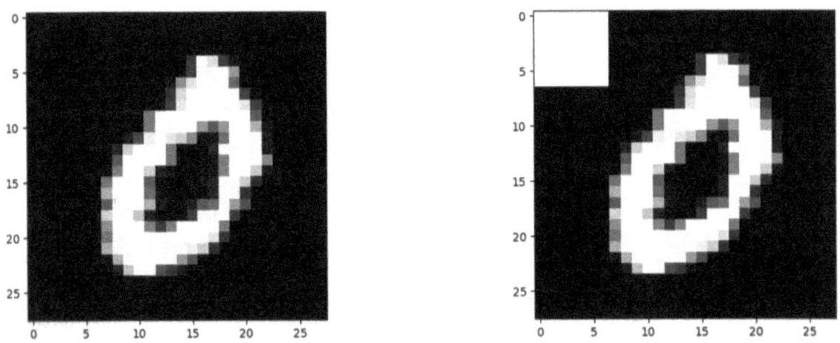

Fig. 2. An example of the trigger sample: clean sample (left) and trigger sample (right).

3 Experimental Results and Analysis

3.1 Setup

We conduct experiments on MNIST, a widely used benchmark in deep learning. MNIST consists of 70,000 grayscale images of handwritten digits $(0, 1, 2, ..., 9)$, each sized 28×28 $(= 784)$ pixels. Due to the constraints in computing resources and quantum computing performance, we only select a certain number of samples for simulation. Moreover, in experiments, we only consider binary classification since QNN cannot perform very well on ordinary classification between ten digits, e.g., we have tested the performance on classifying digits '0' and '1'. We argue that reducing the number of classes to be predicted does not affect the feasibility of the proposed method since the prediction performance relies heavily on model architecture, which were not deeply studied in this paper. Our main purpose is to introduce a general framework for QNN watermarking.

Given a clean sample, its trigger version is generated by adding a visual white block to the clean sample. Figure 2 provides an example of the trigger sample, in which the size of the white block is 7×7 and the value of each pixel within the white block is set to 255. Although the design of this trigger is simple, it is enough to demonstrate the effectiveness of the proposed method. In fact, the proposed method is not limited to this trigger design, indicating that alternative approaches for trigger sample generation can also be devised.

As mentioned above, we investigate binary classification in this paper. Along this line, for each experiment, we use 100×2 samples for model training. The size of the test set used for model usage is set to 100×2, and that for model verification is set to 100×1. We did not use the validation set since the size of the training set is not large-scale. Taking classification between digits '0' and '1' for explanation, the training set for training a clean model consists of 100 samples for each category. To train a marked model, we use 200 samples as well, among which 10 samples are trigger samples and the rest are clean ones. For model usage, the test set consists of 100 clean samples for each category. For model verification, the test set contains 100 trigger samples. The trigger is added to the clean sample with label '0' so that the label of trigger sample is '1'.

In addition, the batch size is set to 4. The number of epochs is set to 4. We use binary cross entropy for loss determination, and Adam for loss optimization. The learning rate is set to 0.01. We use a signal GPU for acceleration.

3.2 Model Architecture

There are no special requirements for the architectural design of QNN. In this paper, we use hybrid quantum-classical model. A hybrid quantum-classical neural network typically consists of a classical neural network coupled with a parameterized quantum circuit in a co-processing framework. The classical network handles feature extraction, dimensionality reduction, or pre/post-processing, while the quantum component performs specialized computations such as quantum feature embedding, kernel estimation, or variational optimization.

Along this line, we design the QNN as follows. We first apply a fully-connected neural network (FNN) to reduce the 784-d input image into a 16-d real-valued vector, which corresponds to quantum encoding in Fig. 1. Then, the 16-d real-valued vector is fed into the parameterized quantum circuit to learn the quantum feature embedding, which corresponds to quantum state transformation. Finally, the quantum feature embedding is processed by another classical network (which accepts a value as input and produces a value as output) to refine predictions. This process corresponds to quantum measurement. Although the architecture designed in this paper is simple and not a purely quantum network architecture, it demonstrates the effectiveness of the proposed method. We use *Qiskit*[3], which is an open-source, Python-based, high-performance software stack for quantum computing, originally developed by IBM, for simulation.

```
def create_qnn(n_qubits, reps):
    input_params = ParameterVector('input', length=n_qubits)
    learn_params = ParameterVector('weight', length=reps * n_qubits)
    qc = QuantumCircuit(n_qubits)
    for i in range(n_qubits):
        qc.ry(input_params[i], i)
    for rep in range(reps):
        for i in range(n_qubits - 1):
            qc.cx(i, i + 1)
        for i in range(n_qubits):
            qc.rz(learn_params[rep * n_qubits + i], i)
    return qc, input_params, learn_params


class QNNModel(nn.Module):
    def __init__(self, n_qubits, reps):
        super(QNNModel, self).__init__()
        qc, input_params, learn_params = create_qnn(n_qubits, reps)
        self.qnn = EstimatorQNN(circuit=qc, input_params=input_params,
weight_params=learn_params, input_gradients=True)
        self.quantum_layer = TorchConnector(self.qnn)
        self.post_process = nn.Linear(1, 1)
        self.classical = nn.Sequential(nn.Linear(784, 16), nn.ReLU())

    def forward(self, x):
        x = self.classical(x)
        x = self.quantum_layer(x)
        x = self.post_process(x)
        x = torch.sigmoid(x)
        return x
```

Fig. 3. Source code for the model architecture (with Python, PyTorch, and Qiskit).

Figure 3 shows the Python code for building the QNN model. As shown in Fig. 3, the total number of trainable parameters for quantum state transformation is `reps`$*$`n_qubits`. In our experiments, we set `reps` $= 2$ and `n_qubits` $= 16$. Instead of applying amplitude encoding or angle encoding to the output of the classical neural network, we here directly fed the classical output into the hidden quantum layer since the classical network and the hidden quantum layer will be trained together so that the classical output could adapt to the hidden input.

It can also be seen that quantum state transformation uses CX (Controlled-X) Gate, RY (Rotation-Y) Gate and RZ (Rotation-Z) Gate. The CX gate is a two-qubit entangling gate that flips the target qubit (applies an X gate) if and only if the control qubit is in state $|1\rangle$. It is fundamental for creating quantum correlations (entanglement) and forms the basis of many quantum algorithms.

[3] https://www.ibm.com/quantum/qiskit.

The RY gate performs a single-qubit rotation around the Y-axis of the Bloch sphere by angle θ. It is widely used for preparing superpositions and variational quantum circuits. The RZ gate rotates a qubit around the Z-axis by θ, introducing a phase difference between the $|0\rangle$ and $|1\rangle$ states without changing their probabilities, which is critical for quantum phase manipulation.

Table 1. Performance evaluation of model watermarking on binary classification.

Category 1	Category 2	#{trigger set}	Acc. (clean set)	Acc. (trigger set)
0	1	0	86%	23%
0	1	10	83.5%	97%
2	3	0	81.5%	37%
2	3	10	79%	98%
4	5	0	85%	37%
4	5	10	83.5%	82%
6	7	0	77%	32%
6	7	10	75%	79%
8	9	0	93%	40%
8	9	10	90.5%	80%

Table 2. Performance evaluation after increasing the size of trigger set.

Category 1	Category 2	#{trigger set}	Acc. (clean set)	Acc. (trigger set)
4	5	10	83.5%	82%
4	5	20	80.5%	100%
6	7	10	75%	79%
6	7	20	72.5%	100%
8	9	10	90.5%	80%
8	9	20	89%	100%

3.3 Results

We are ready to report our experimental results. Table 1 shows the performance on model watermarking, in which different digit pairs are evaluated. When the size of the trigger set is zero, it means that we are concerned about the performance of the non-watermarked model on the original classification task. We set the size of the trigger set to 10 by default to mark the QNN. It is noted that, for those watermarked models, they are trained with the proposed strategy, while the non-watermarked ones are trained with random sample permutation. In addition, we generate the trigger samples by adding the trigger pattern to clean samples belonging to Category 1. We set $n_{\mathrm{T}} = 2$ and $|\mathcal{S}_{\mathrm{DT}}| = 200$ by default.

As shown in Table 1, different parameter settings generally result in different performance. Overall, given the baseline, after watermarking, the performance on the original task tends to decline, which is reasonable since more tasks require more representation abilities of the model, thereby reducing the generalization ability on a specific task. However, the performance degradation on the original task can be kept within a relatively low level. Moreover, the difference of the classification accuracy on the trigger set between the non-watermarked model and the watermarked model is significant for all five cases, indicating that the proposed method enables us to reliably embed and extract the watermark. This shows that the proposed method is effective for IP protection of QNN models.

It can be observed from Table 1 that some digit pairs give relatively smaller classification accuracy values on the trigger set, which should be affected by the size of the trigger set. To confirm this, we increase the size of the trigger set. As shown in Table 2, the trigger samples can be perfectly classified, at the expense of slight performance decline on the original task. Therefore, in applications, it is suggested to properly adjust the ratio between the size of the clean samples and the size of the trigger samples so that a good performance trade-off between the original task and the watermarking task can be achieved.

We have also evaluated the performance using the proposed training strategy and random ordering strategy (i.e., the training samples are randomly ordered). Simulation results indicate that the proposed method gives better watermarking performance and has better stability than random ordering in terms of watermark extraction, i.e., random orders give significant different performance, e.g., some random orders cause watermark extraction to fail completely in our experiments, while this can be largely avoided with the proposed method.

4 Conclusion and Discussion

In the era of rapid advancements in QC and AI, protecting intellectual property and ensuring model integrity have become paramount. Watermarking quantum neural network models is a critical step to safeguard innovation, prevent unauthorized use, and maintain accountability in AI deployments. In this paper, we present an effective watermarking framework for QNN models. The main idea is to use trigger samples to train the clean model so that the resulting watermarked model enables us to verify its ownership by analyzing the prediction results given the input samples. In order to balance the performance of the model on the original task and the watermarking task, we introduce sample grouped and paired training strategy for updating the trainable quantum parameters. Experimental results verify its feasibility and effectiveness.

The rapid progress in QC suggests that practical, large-scale quantum machines will likely become a reality in the foreseeable future. As these systems approach computational supremacy, they will inevitably power the next generation of artificial intelligence, enabling breakthroughs in optimization, cryptography, and machine learning. However, this transformative potential also introduces unprecedented security risks. Quantum-driven AI models, with their enhanced capabilities, could be exploited for malicious intents. It is now the time to develop safeguards, before quantum AI becomes mainstream, ensuring a secure and trustworthy foundation for the future of intelligent systems. We hope this preliminary exploration could motivate researchers focused on information hiding to pay more attention on IP protection of quantum AI models.

Acknowledgement. This work was partly supported by the Basic Research Program for Natural Science of Guizhou Province under Grant Number QIANKEHEJICHU-ZD[2025]-043, and also partly supported by the Guangdong Basic and Applied Basic Research Foundation under Grant Number 2023A1515010815. It was also supported by the 2024 Xizang Autonomous Region Central Guided Local Science and Technology Development Fund Project under Grant Number XZ202401YD0015.

References

1. Madsen, L., Laudenbach, F., Falamarzi, M., et al.: Quantum computational advantage with a programmable photonic processor. Nature **606**, 75–81 (2022)
2. Wu, H., Liu, G., Yao, Y., Zhang, X.: Watermarking neural networks with watermarked images. IEEE Trans. Circuits Syst. Video Technol. **31**(7), 2591–2601 (2021)
3. Uchida, Y., Nagai, Y., Sakazawa, S., Satoh, S.: Embedding watermarks into deep neural networks. In: Proceedings of the ACM International Conference on Multimedia Retrieval, pp. 269–277 (2017)
4. Wang, J., Wu, H., Zhang, X., Yao, Y.: Watermarking in deep neural networks via error back-propagation. In: Proceedings of the IS&T Electronic Imaging, Media Watermarking, Security and Forensics, pp. 22-1–22-9 (2020)
5. Zhao, X., Wu, H., Zhang, X.: Structural watermarking to deep neural networks via network channel pruning. In: Proceedings of the IEEE International Workshop on Information Forensics and Security, pp. 1–6 (2021)
6. Adi, Y., Baum, C., Cisse, M., Pinkas, B., Keshet, J.: Turning your weakness into a strength: watermarking deep neural networks by backdooring. In: Proceedings of the 27th USENIX Conference on Security Symposium, pp. 1615–1631 (2018)
7. Zhao, X., Wu, H., Zhang, X.: Watermarking graph neural networks by random graphs. In: Proceedings of the IEEE International Symposium on Digital Forensics and Security, pp. 1–6 (2021)
8. Zhao, X., Wu, H., Zhang, X.: Transferable watermarking to self-supervised pre-trained graph encoders by trigger embeddings. In: Proceedings of the IEEE International Workshop on Information Forensics and Security, pp. 1–6 (2024)
9. Wang, Y., Wu, H.: Protecting the intellectual property of speaker recognition model by black-box watermarking in the frequency domain. Symmetry **14**(3), 619 (2022)
10. Liu, Y., Wu, H., Zhang, X.: Robust and imperceptible black-box DNN watermarking based on Fourier perturbation analysis and frequency sensitivity clustering. IEEE Trans. Dependable Secure Comput. **21**(6), 5766–5780 (2024)

11. Zhang, L., Liu, Y., Zhang, X., Wu, H.: Suppressing high-frequency artifacts for generative model watermarking by anti-aliasing, In: Proceedings of the ACM Workshop on Information Hiding and Multimedia Security, pp. 223–234 (2024)
12. Zhang, L., et al.: Generative model watermarking based on human visual system. In: Proceedings of the 19th International Forum of Digital Multimedia Communication, pp. 136–149 (2022)

SPHINCS+ Based Certification Scheme for Federated Learning Client Models

Xiao Hu, Yuxin Hu, Rui Wu, and Shuanggen Liu(✉)

School of Cyberspace Security, Xi'an University of Posts and Telecommunications, Xi'an 710121, China
{huxiao,huyuxin,wurui}@stu.xupt.edu.cn, liushuanggen201@xupt.edu.cn

Abstract. Although federated learning as distributed machine learning effectively solves the data privacy problem and the data silo problem, the client model update still faces security threats such as malicious tampering and poisoning attacks. Additionally, existing RSA/ECDSA-based authentication schemes are problematic in resisting quantum computing attacks, and post-quantum security privacy protection techniques are extremely needed. This paper proposes a federated learning client model update authentication scheme based on SPHINCS+ to achieve authentication of client model updates and ensure the secure aggregation of global models. The scheme uses hash signatures to ensure quantum-resistant capabilities. Meanwhile, based on the binding of WOTS+ key chains and Hypertree paths, the scheme associates multiple updates from the same client, which can effectively resist persistent poisoning attacks. Ultimately, experimental evaluation and performance analysis show that the scheme achieves efficient key management while safeguarding post-quantum security and defending against persistent poisoning attacks, and provides a new paradigm with enhanced deployability for federated learning privacy protection for post-quantum security.

Keywords: federated learning · SPHINCS+ · secure aggregation · post-quantum cryptography

1 Introduction

With the rapid expansion of intelligent models in the era of large-scale models, federated learning has become a key technology for resolving the conflict between data privacy and collaboration. As a distributed machine learning paradigm [1], federated learning ensures that raw data always remains on the client side, with only encrypted model parameters being uploaded. This way, not only is the leakage of sensitive information prevented, but the problem of data silos [2] is also solved. Its feature of "data available but not visible" not only meets the demands of highly regulated industries but also provides a secure path for the large-scale application of AI in fields such as finance, healthcare, and IoT. However, during model training, the client model update process still faces security threats

J. Yu and Y. Chen (Eds.): FCS 2025, CCIS 2805, pp. 319–335, 2027.
https://doi.org/10.1007/978-981-95-9514-3_22

such as malicious tampering and poisoning attacks [3]. Traditional public-key cryptography (such as ECDSA) cannot resist quantum attacks, and federated learning models typically require long-term deployment. Besides, traditional signature schemes lack the capability to continuously track malicious clients and cannot resist persistent poisoning attacks. This issue means that once an attacker exploits this feature to launch an attack, the server will be unable to identify client models that have been maliciously tampered with. That is, incorrect client models will still be aggregated into the global model, causing serious damage to model training [4].

With the continuous development of quantum technology, existing encryption technologies are facing unprecedented security threats. Authentication schemes based on RSA/ECDSA rely on traditional number theory problems. However, Shor's algorithm (Peter Shor, 1994) shows that quantum computers are capable of efficiently solving these problems in polynomial time, rendering traditional public key encryption schemes completely ineffective [5]. Therefore, in order to ensure that federated learning architectures keep being secure in the future, research and adoption of quantum-resistant encryption schemes are imperative.

Currently, NIST has launched standardization work on post-quantum cryptography, and SPHINCS+ has been included by NIST as one of four accepted PQC standards as well as one of only three recommended schemes for digital signatures [6]. SPHINCS+ is a stateless, hash-based signature scheme that protects data from quantum computing attacks. This paper proposes a federated learning client model update authentication scheme based on SPHINCS+ that aims to utilize post-quantum security to provide a new authentication mechanism for federated learning client model updates. This solution utilizes hash-based post-quantum signature technology to ensure high security even under the threat of quantum computing. While it is based on WOTS+ keychains and Hypertree path binding to associate multiple updates from the same client, thereby realizing resistance to persistent poisoning attacks.

1.1 Related Work

Federated Learning Client Model Update and Privacy Protection. To ensure the secure aggregation of global models in federated learning, an authentication mechanism has been added to model updates in the federated learning client with the aim of ensuring the security of model updates. In [7], McMahan et al. first used RSA-2048 signatures to verify client identities in the FedAvg framework, but this method is not resistant to quantum attacks and lacks continuous tracking capabilities. literature [8] designed a one-time password (OTP) and model hash binding scheme (CCS'20), but it relies on a centralized token distribution server, which is prone to single points of failure. literature [9] uses Intel SGX to create trusted containers, whose memory encryption mechanism can defend against conventional memory sniffing attacks. However, subsequent research found that this solution has a risk of measurement channel leakage. Malicious attackers can recover 98.7% of the gradient data through CacheBank attacks, and its security depends on specific hardware, which greatly increases deployment costs. The Krum algorithm proposed in [10] detects abnormal updates by counting characteristics on the server side, but its effectiveness depends on the attack ratio threshold. When attackers use low-and-slow attacks, the detection accuracy drops to 31.7%.

Post-Quantum Cryptography and Hash Signatures. With the expansion of post-quantum cryptography, signature schemes based on hash functions have attracted widespread attention due to their security and efficiency. As an emerging distributed machine learning technique, federated learning has emerged as a promising solution for privacy protection in federated learning. literature [11] provides a detailed introduction to the SPHINCS+ architecture, including the WOTS+, Hypertree, and FORS components, and analyzes its application in post-quantum security. literature [12] proposed Smart Digital Signatures (SDS), a compact and efficient hash-based signature scheme for distributed ledger technology (DLT) to replace ECDSA and resist quantum computing attacks. Moreover, in terms of research on domestic post-quantum signature algorithms, literature [13] uses the SM3 hash function to improve the XMSS algorithm, which makes it more suitable for domestic applications. literature [14] systematically reviews hash-based post-quantum signature schemes, analyzing their security, optimization methods, and standardization progress.

To sum up, both federated learning and post-quantum signature schemes are currently undergoing rapid development. However, existing schemes still face challenges such as quantum resistanceThis paper applies post-quantum signature algorithms to the authentication of client model updates in federated learning. Moreover, it uses WOTS+ keychains and Hypertree path binding to associate multiple updates from the same client, thereby enhancing the overall security and privacy protection capabilities of the federated learning framework.

1.2 Paper Contributions

The main contributions of this paper are threefold: First, this paper proposes a federated learning client model update authentication scheme based on SPHINCS+, which signs client model updates to ensure the secure aggregation of global models, while also providing post-quantum security to resist quantum computing attacks.

Second, utilizing the algorithm structure of SPHINCS+ and the characteristics of federated learning client model updates, based on WOTS+ key chains and Hypertree path binding, multiple updates from the same client are associated, which achieves defense against persistent poisoning attacks.

Third, the performance of SPHINCS+ with different parameter sets in federated learning is tested through systematic experiments. It is also compared with existing schemes, and it is found that this paper has significant advantages in key management.

1.3 Paper Structure

The remainder of this paper is organized as follows. Section 2 introduces knowledge about federated learning frameworks, poisoning attacks, SPHINCS+, and other topics. Section 3 proposes a federated learning client model update authentication scheme based on SPHINCS+ and provides a detailed introduction to the scheme process. Section 4 presents a preliminary implementation of the proposed scheme and performs a performance analysis, comparing it with existing schemes to verify its practicality. Finally, Sect. 5 summarizes the research results of this paper and discusses possible directions for future improvements.

2 Preliminary Knowledge

2.1 Federated Learning Framework and Poisoning Attacks

Federation Learning

Theorem 1. *Federated learning is a privacy-preserving distributed machine learning paradigm whose core characteristics can be formally defined as a triplet:*

$$FL = (C, S, A) \tag{1}$$

where:

- $C = \{c_1, \ldots, c_N\}$*represents the set of clients participating in training.*
- S *is the central server responsible for model aggregation.*
- $A : R^d \times R^d \rightarrow R^d$ *is the aggregation function (such as FedAvg).*

Federated learning allows clients to keep raw data locally while training models. The client only needs to upload the trained model parameters to the server, thus avoiding the leakage of sensitive information. The central server aggregates the parameters transmitted by each client and uses the aggregated parameters to update the global model. After each round of communication, the server distributes the updated global model to the clients participating in the training for a new round of training updates until the model converges. Federated learning is widely used in the large-scale implementation of AI in fields such as finance, healthcare, and IoT (Fig. 1).

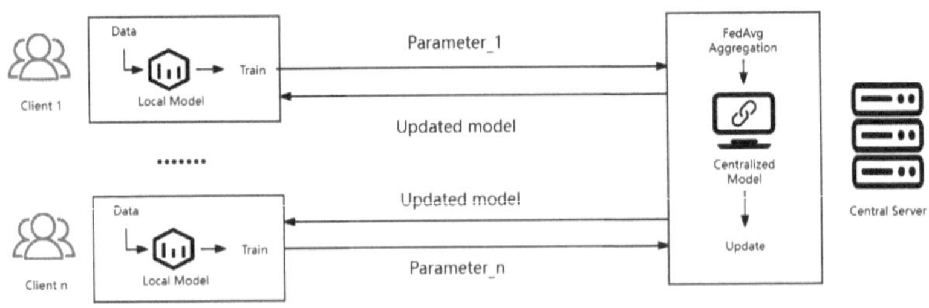

Fig. 1. Schematic diagram of the core process of federated learning.

Poisoning Attack. In terms of federated learning, a poisoning attack refers to an attacker providing carefully constructed model updates to the central server, thus the central server's global model participating in the global aggregation with the model updates provided by the malicious attacker during aggregation, which causes the updated global model to exhibit unexpected behavior on the test data.

Theorem 2. *(Persistent Poisoning Attack (PPA)) means that a malicious client gradually (Δ_t) destroys the performance of the model through low-amplitude, high-frequency gradient perturbations over multiple iterations of federated learning, satisfying.*

$$\lim_{T\to\infty}\sum_{t=1}^{T}\Delta_t = \omega^* - \omega^{bengin} \quad , ||\Delta_t|| \leq \in /\sqrt{T} \tag{2}$$

where:

- *t denotes a single-round gradient perturbation*
- *ω denotes the convergence state of the model when the attack is successful (usually the malicious model desired by the attacker)*
- *ω^{bengin} denotes the ideal global model parameters that converge normally in federated learning without attack.*
- *T denotes the number of iterations in federated learning.*

2.2 SPHINCS+

SPHINCS+ (Stateless Hash-based Digital Signature) is a stateless signature scheme based on hash functions. SPHINCS+ introduces a few-signature scheme (Forest of Random Subsets, FORS) based on Hypertree trees and one-time signature schemes (OTS). Among them, FORS acts as a message compression layer, decomposing high-dimensional hashes into low-dimensional subsets. WOTS+ acts as a signature layer, signing the reduced-dimensional subsets. Hypertree organizes large-scale key organization and tracing through the organization of hypertree structures. The following introduces the main components of SHPINCS+ and its process:

FORS. The FORS signature scheme is based on hash functions, and the signature key can be used to generate multiple signatures. Its security depends on the collision resistance of the hash function. FORS is a message compression layer with k independent subtrees mainly used for hash compression. Where k is the number of signature indexes of FORS signature scheme and t is the value space of each index. The FORS signature is mathematically expressed as follows:

$$M \to digest = H(M) \tag{3}$$

$$idx_i = split\_digest[i] mod t(t = 2^a) \tag{4}$$

$$\sigma_{ORS} = \cup_{i-1}^{k}(s_k, auth_p, ath_i) \tag{5}$$

WOTS+. The less signature scheme WOTS+ is based on the hash function, which can sign a message once and control the balance of signature size and computational efficiency by Winternitz parameter w. The WOTS+ algorithm consists of three parts: key generation algorithm, signature generation algorithm, and signature verification algorithm, and the detailed explanation of the WOTS+ algorithm can be studied in the literature [15].

Hypertree Structure. Hypertree is a supertree structure consisting of multiple layers of Merkle trees, Through path proofs of the hypertree, signature verification can be performed using root hashes. The structure of Merkle tree is as follows (Fig. 2):

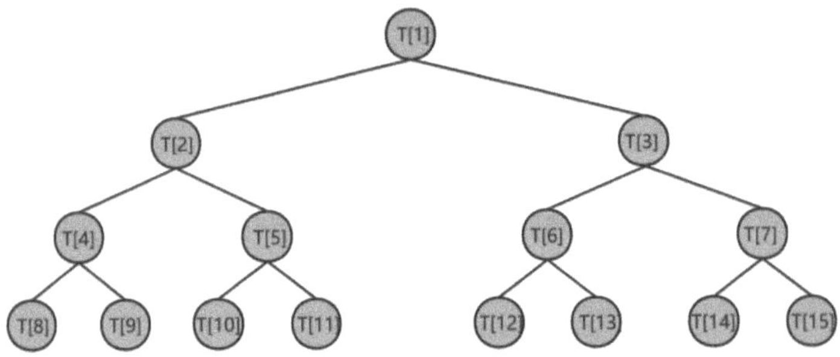

Fig. 2. Merkle tree structure with leaf tree of 8.

In the above figure, the height of the Merkle tree is h = 3, and there are 8 leaves. Each leaf node corresponds to a WOTS+ example. Two adjacent child nodes are linked to the parent node through a hash function, and this process is repeated iteratively until the root node is reached. A combination of multi-layered Merkle trees forms a Hypertree structure, which is structured as follows (Fig. 3):

SPHINCS+ Process. First the message hash is compressed by FORS and the root of k independent subtrees is used as the public key of FORS. After FORS signs the message, then WOTS+ signs the public key of FORS (i.e., the root node of the FORS tree), and finally generates the authentication path that generates the signature through the Hypertree structure. A detailed explanation of the SPHINCS+ algorithm can be found in literature [6], and its core formula can be abstracted as follows:

$$\sigma_{SPHINCS+} = (FORS_{Sig} \oplus WOTS + \oplus Hypertree_{Auth}) \tag{6}$$

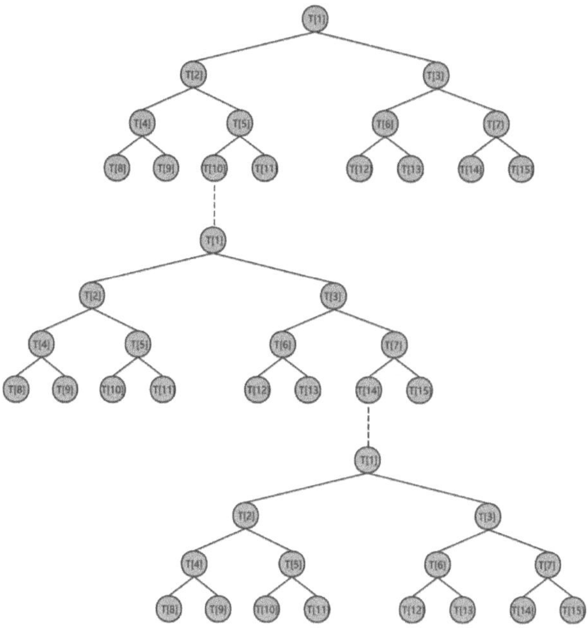

Fig. 3. "Structure of 3-layer Hypertree with height 9."

2.3 Security Foundations Against Quantum Attacks

With the emergence of quantum computers, in the specific context of cryptography, their supercomputational power enables them to efficiently crack large integer prime factorization and discrete logarithm problems in polynomial time. The SPHINCS+ signature algorithm used in this paper's scheme, on the other hand, does not rely on any number-theoretic difficulties, and its security relies exclusively on the properties of the hash function: unidirectionality, collision resistance, and second-phase attack resistance. While quantum computers can speed up computation on certain problems, there are no effective quantum algorithms for the properties of hash functions. Currently, the most effective algorithm for hash functions is Grover, but the algorithm's acceleration of the hash search is capped at the square level, unable to achieve an exponential breakthrough, and therefore unable to break the SPHINCS+ signature in polynomial time (Fig. 4).

3 Authentication Scheme for Federated Learning Client Model Update Based on SPHINCS+

In the application scenario of privacy protection in federated learning, the client model update authentication scheme based on SPHINCS+ can continuously

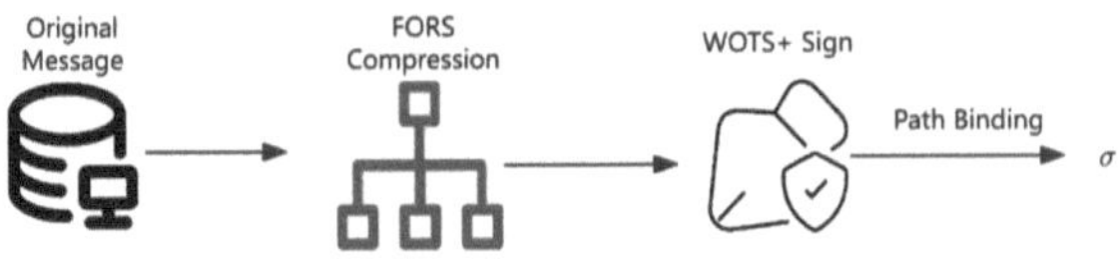

Fig. 4. Sketch of SPHINCS+ process.

track multiple rounds of training for the same client, which is highly compatible with the characteristics of multiple rounds of iteration in federated learning. Below, we will provide a detailed introduction to the federated learning client model update authentication scheme designed based on SPHINCS+ below. Table 1 provides an introduction to the parameters and functions used in the scheme.

Table 1. Introduction to Parameters and Functions used by the scheme

Introduction to Parameters and Functions:
m: Initial message n: Output length of the hash function h/d: Total tree height/layer ClientID: Client unique identifier used to bind the key to the client PRF: Pseudorandom function used for generating pseudo-random numbers $WOTS_{pre\_pk}$:WOTS + public key used for last round of signatures WOTS_KeyGen: Derive the Round key pair from the key chain extract(): Information extraction function

3.1 Initialization Phase

The initialization phase of the scheme consists of two phases: building the federated learning system model and key generation. These two phases are described separately in the following.

Modeling the Federated Learning System

1. Determine the clients C to participate in the federated learning training
   The client ($C = \{c_1, \ldots, c_N\}$) is responsible for executing local training, uploading the parameters θ_i of the training-trained model and signing it with SPHINCS+.
2. Determine the center server S
   The center server is responsible for collecting the parameters uploaded by each client and verifying the SPHINCS+ signature. Only clients that pass the verification have their model parameters aggregated by the central server, which distributes the updated global model to each client after aggregation.

The central server needs to initialize the global model as the base model for federated communication.

Key Generation

1. Client Generation of Key Pairs The client generates a key pair (PK_i, SK_i) using the SPHINCS+ algorithm.
2. The central server stores the public key The client submits the public key (PK_i) to the central server for storage.
3. Private key local storage The client's private key (SK_i) is stored in the local device (Fig. 5).

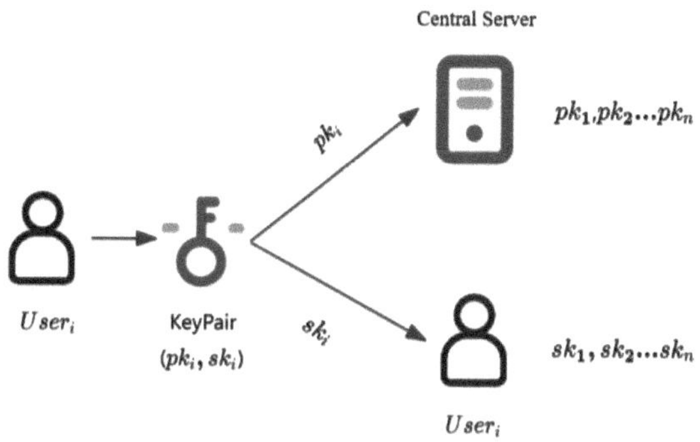

Fig. 5. Flowchart of key generation.

Algorithm 1:Key Generation:$(pk_i, sk_i, WOTS_{chain\_seed})\leftarrow$ KeyGen(n,d,ClientID)

Input: n,d, ClientID
Output: $(pk_i, sk_i, WOTS_{chain\_seed})$
global_seed = PRF(n,ClientID)
sk_seed = PRF(global_seed)
pk_seed = PRF(global_seed)
$WOTS_{chain\_seed}$ = PRF(sk_seed)
root = Gengerate_root(sk_seed, pk_seed,d,ClientID)
pk_i = pk_seed ||root||ClientID
sk_i= sk_seed || pk_i || $WOTS_{chain\_seed}$
Return $(pk_i, sk_i, WOTS_{chain\_seed})$

3.2 Data Training and Signature Phase

1. Each client participating in federated learning training (C = $\{c_1, \ldots, c_N\}$) receives the initialized global model from the central server and trains the data locally to derive the trained model θ_i.

2. Each client signs the trained model parameters using the Sphincs+ signature algorithm to obtain theσ_i, and uploads the model parameters together with the signatures to the central server, i.e., the central server is subjected to the set of model parameters $\{\theta_1, \ldots, \theta_N\}$ and the set of signatures$\{\sigma_1, \ldots, \sigma_N\}$.

Algorithm 2:Signing:$\sigma \leftarrow$Sign(m,sk_i,n,h,d,ClientID,round, $WOTS_{pre\_pk}$)

Input:m,sk_i,n,h,d,ClientID,round, $WOTS_{pre\_pk}$
Output: σ
sk_seed,pk_seed,$WOTS_{chain\_seed}$,root= extract(sk_i)
$WOTS_{sk}, WOTS_{pk}$ = WOTS_KeyGen,($WOTS_{chain\_seed}$,round)
If round>1 and not WOTS_Verify_Chain($WOTS_{chain_seed}, WOTS_{pre_pk}, WOTS_{pk}$):
  Return 0
msg_digest = H(m‖ ClientID ‖ round)
fors_sig,fors_pk = fors.sign(sk_seed, pk_seed, msg_digest,ClientID)
wots_sig = wots.sign($WOTS_{sk}$,fors_pk ‖ ClientID ‖ round)
tree_idx = compute_tree_idx(h,d,ClinetID)
leaf_idx = compute_leaf_idx(h,d,ClientID)
auth_path = generate_auth_path(leaf_idx,tree_idx,h,d,sk_seed,pk_seed)
σ= (Client‖rand‖fors_sig‖fors_pk‖wots_sig‖$WOTS_{pk}$‖auth_path‖$WOTS_{pre.pk}$)
Return σ

(See Fig. 6).

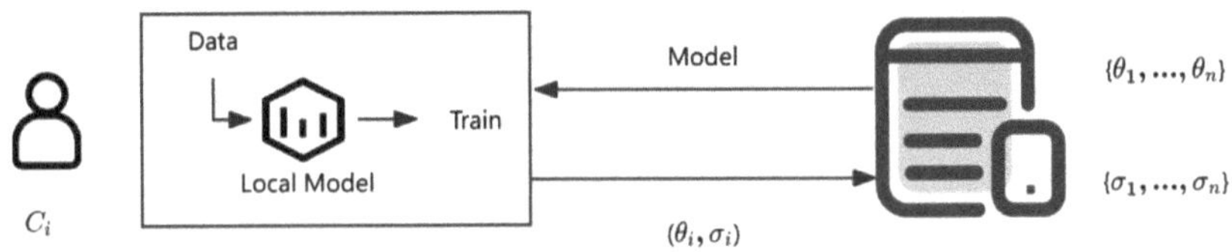

Fig. 6. Flowchart of data training and signature phase.

3.3 Signature Verification and Parameter Aggregation Phase

(1) The central server receives the model parameter set $\{\theta_1, \ldots, \theta_N\}$ and the signature set $\{\sigma_1, \ldots, \sigma_N\}$ from the client, and the central server calls the SPHINCS+ signature verification algorithm to verify the signatures. Only the model parameters that pass the validation will participate in the next step of parameter aggregation, we assume that all parameters pass the validation, then the parameters that participate in the aggregation will be $\{\theta_1, \ldots, \theta_N\}$.
(2) The central server uses the Federated Average Algorithm (FedAvg) to aggregate the set of validated model parameters to obtain the new model parameters θ_*, and the global model is updated using θ_*. The next round of fed-

erated learning communication is then performed until the model converges (i.e., the model achieves the expected performance) (Fig. 7).

Algorithm 3:Verification:0/1 $\leftarrow$ Verify(σ,pk_i,m,h,d)

Input: σ,pk_i,m,h,d
Output: 0/1
pk_seed ,expected_root = extract(pk_i)
sig_ClientID,sig_round = extract(σ)
If sig_ClientID != extract(pk_i):
  Return 0
msg_digest = H(m || sig_ClientID || sig_round)
if not FORS_Verify(fors_sig, fors_pk, msg_digest, pk_seed, sig_ClientID):
  Return 0
wots_msg = fors_pk || sig_ClientID || sig_round
if not WOTS_Verify(wots_sig, WOTS_pk, wots_msg):
  Return 0
if sig_round and not WOTS_Verify_Chain(extract(pk_i), $WOTS_{pre.pk}$, $WOTS_{pk}$:
  Return 0
tree_idx = compute_tree_idx(sig_ClientID, d,h)
leaf_idx = compute_leaf_idx(sig_round, d,h)
computed_root = Verify_Auth_Path(auth_path,$WOTS_{pk}$, tree_idx, leaf_idx, d, pk_seed)
if computed_root != expected_root:
  Return 0
Return 1

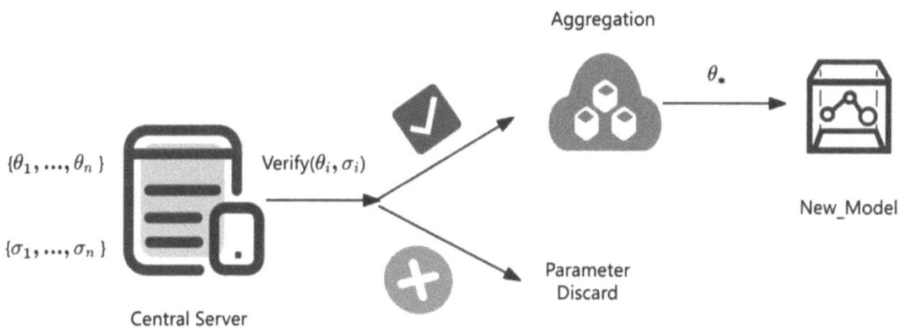

Fig. 7. Flowchart of signature verification and aggregation phase.

4 Performance Analysis

In this section, the scheme of this paper is experimentally evaluated, and its performance is analyzed and compared with the schemes in the related literature. First, we analyze it in terms of key generation time, key length, and key signing and verification time; Then the comparison is made in terms of safety and functionality. The performance of the implementation is provided in Table 2, which was measured on a test machine running Linux Ubuntu 22.04, equipped with an NVIDIA RTX 3080 Ti GPU. Table 3 gives the parameter configuration information for the experiment.

Table 2. Initial testing of program performance

Parameter Set	PubKeySize	KeyGenTime(s)	SignTime(s)	SigSize(Byte)	VerifyTime(s)
sha2_128f	32	0.005187	0.12108	17088	0.007247
sha2_128s	32	0.253504	1.471815	7856	0.001742
sha2_192f	48	0.007693	0.200775	35664	0.011389
sha2_192s	48	0.496450	3.018012	16224	0.002696
sha2_256f	64	0.028690	0.558922	49856	0.015818
sha2_256s	64	0.445695	4.286768	29792	0.005482
shake_128f	32	0.008690	0.204294	17088	0.011811
shake_128s	32	0.552542	3.578244	7856	0.002773
shake_192f	48	0.012756	0.319707	35664	0.018170
shake_192s	48	0.808280	5.034817	16224	0.005138
shake_256f	64	0.033517	0.662727	49856	0.019483
shake_256s	64	0.535460	4.786419	29792	0.007028

Table 3. Configuration of experimental parameters

Modelr Set	Task	Clients	Epochs	Batch Size	LR
MLP	MedMNIST	10	10	32	1e-5

4.1 Federal Performance

Figure 8 shows the variation of the client loss function and accuracy during federation learning, and Fig. 9 shows the model accuracy after completing the entire federation communication. The federation performance of this program is as expected.

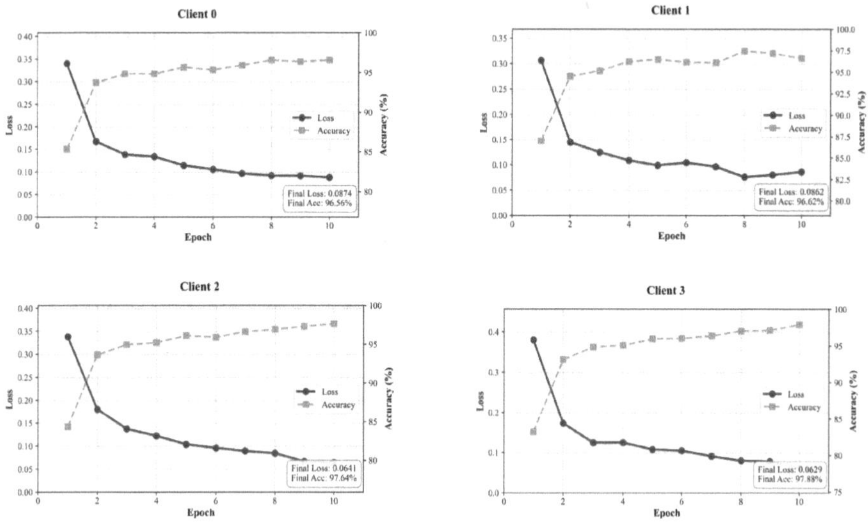

Fig. 8. Client-side model loss function and test correctness.

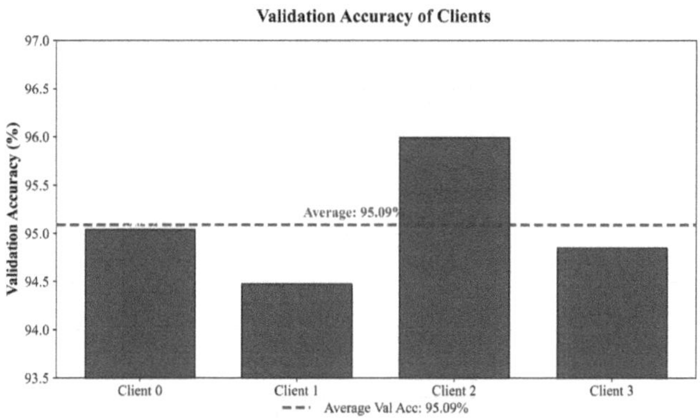

Fig. 9. Accuracy of the model on the evaluation set.

4.2 Efficiency Analysis

Considering that sphincs+ has multiple parameter sets and different parameter sets perform differently in the experiments, we will use sha_128f as an example for comparison (all parameter sets used in the experiments are anti-quantum).

(1) Key generation time
The key generation time in this paper is 0.005 s, much lower than that of XMSS (12.09 s) in literature [16], which indicates that the Hypertree structure is more efficient than the XMSS scheme in generating large-scale keys.

(2) Key size
The size of the public key in this paper is 32 bytes, which is much lower than the 640 bytes of XMSS, 2688 bytes of Dillithium5, and 1760 bytes of Falcon-1024 in the literature [16]. This will greatly reduce the communication overhead and key storage overhead in the key exchange phase.

(3) Signature and verification time
The signature time of this paper's scheme is 0.0052 s, and the verification time is 0.0072 s, which is higher than the signature and verification time of the lattice-based scheme in literature [17]. This is because the scheme in this paper needs to perform a large number of hash iteration operations, increasing the time complexity, but in exchange for higher security.
Taken together, this paper's scheme is slightly less efficient in the signature verification phase, but has obvious advantages in key generation time and key size (Fig. 10).

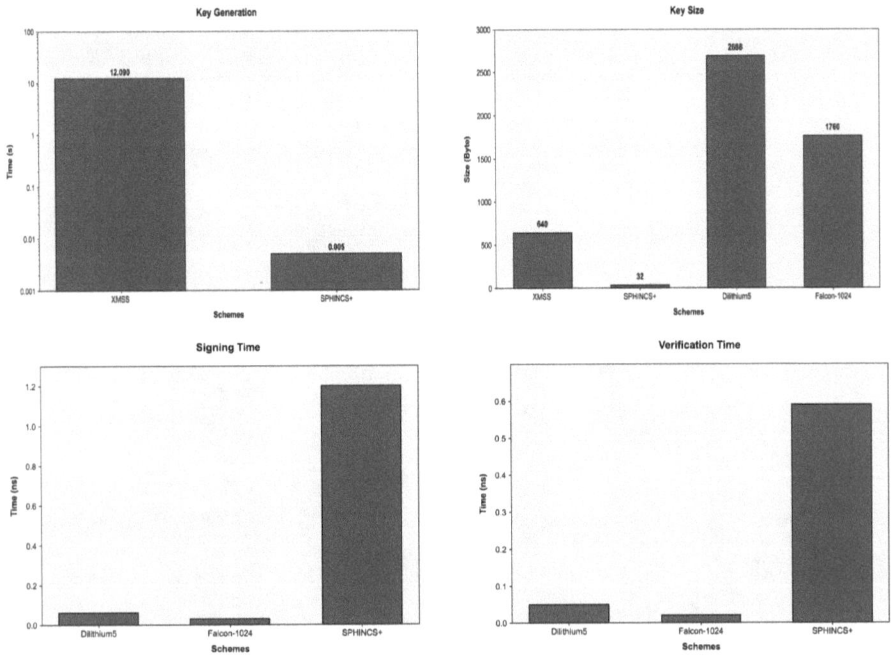

Fig. 10. Efficiency Comparison.

4.3 Safety and Functionality Comparison

(1) Quantum offensive resistance
The scheme in this paper is based on hash function construction. It can resist quantum attacks, similar to literature [16] (Gerki signatures combined with XMSS) and literature [17], but superior to literature [18] (which is not quantum-resistant).

(2) Resistant to sustained poisoning attacks
In this paper, the scheme can correlate multiple updates from the same client through WOTS+key chain and Hypertree path binding, and effectively defends against low-amplitude and high-frequency gradient perturbation attacks. In contrast, the Krum algorithm of the literature [10] reduced the detection accuracy to 31.7% under a low-amplitude sustained poisoning attack. The Gerky scheme in literature [17] is also not resistant to sustained poisoning attacks.
(3) Key State Management
The stateless design of the Sphincs+signature scheme in this paper eliminates the need for state storage and synchronization and reduces security risks due to state loss, tampering, or desynchronization. In contrast, XMSS in the literature [16] is required to perform state storage and synchronization, which will be a burden for resource-constrained devices.
(4) Federal Learning Applicability
This paper's scheme is suitable for federated learning scenarios (e.g., healthcare and finance) that require long-term deployment. Although the signature volume of sphincs+ is large, the model itself that needs to be transmitted by federated learning is much larger than the signature volume, which has a limited impact on FL communication. In addition, FL usually has a long local training time, and FL model updates do not need to be real-time, the latency impact of sphincs+ signatures is largely negligible.

In summary, the scheme in this paper is more efficient in terms of key generation time and has a smaller key length; in addition, its stateless nature avoids complex state storage and synchronization. The solution ensures resistance to quantum attacks while excelling in resistance to sustained poisoning attacks. Although there is a slight compromise in the signature time and verification time, the scheme in this paper is still suitable for federated learning scenarios that require high privacy protection requirements in the long run.

5 Summarize

In this paper, we propose a federated learning client model update authentication scheme based on SPHINCS+ to authenticate client model updates and secure global model aggregation. The scheme utilizes hash signatures to ensure resistance to quantum computation. Meanwhile, based on the binding of the WOTS+ key chain to the Hypertree path, the scheme associates multiple updates from the same client, which can effectively counter persistent poisoning attacks.

We experimentally analyze the security and performance of our scheme. The results show that our scheme is resistant to quantum attacks and persistent poisoning attacks, and has obvious advantages in key length and key generation time.

Although the scheme proposed in this paper has achieved some improvements in privacy protection, there are still some issues that need further research. For example, how to further optimize the signature and verification process to reduce the computational overhead and improve the applicability of the scheme for high-frequency signature federation learning. In addition, future research could explore how to incorporate other privacy-preserving techniques, such as Multiparty Secure Computing (MPC) or Lattice-based Post-Quantum Encryption Algorithm (KYBER), further to improve the security and privacy of federated learning frameworks.

References

1. McMahan, B., Moore, E., Ramage, D., Hampson, S., y Arcas, B.A.: Communication-efficient learning of deep networks from decentralized data. In: Artificial Intelligence and Statistics, pp. 1273–1282. PMLR (2017)
2. Kairouz, P., et al.: Advances and open problems in federated learning. Found. Trends® Mach. Learn. **14**(1–2), 1–210 (2021)
3. Baruch, G., Baruch, M., Goldberg, Y.: A little is enough: circumventing defenses for distributed learning. Adv. Neural Inf. Process. Syst. **32** (2019)
4. Shejwalkar, V., Houmansadr, A.: Manipulating the byzantine: optimizing model poisoning attacks and defenses for federated learning. In: NDSS (2021)
5. Steane, A.: Quantum computing. Rep. Prog. Phys. **61**(2), 117 (1998)
6. Hülsing, A., Kudinov, M.: Recovering the tight security proof of sphincs. In: Agrawal, S., Lin, D. (eds.) Advances in Cryptology – ASIACRYPT 2022. ASIACRYPT 2022. LNCS, vol. 13794, pp. 3–33. Springer, Cham (2022). https://doi.org/10.1007/978-3-031-22972-5_1
7. McMahan, B., Moore, E., Ramage, D., Hampson, S., y Arcas, B.A.: Communication-efficient learning of deep networks from decentralized data. In: Artificial Intelligence and Statistics, pp. 1273–1282. PMLR (2017)
8. Liu, Y., Kang, Y., Xing, C., Chen, T., Yang, Q.: A secure federated transfer learning framework. IEEE Intell. Syst. **35**(4), 70–82 (2020)
9. Mo, F., Haddadi, H., Katevas, K., Marin, E., Perino, D., Kourtellis, N.: PPFL: privacy-preserving federated learning with trusted execution environments. In: Proceedings of the 19th Annual International Conference on Mobile Systems, Applications, and Services, pp. 94–108 (2021)
10. Fang, M., Cao, X., Jia, J., Gong, N.: Local model poisoning attacks to {Byzantine-Robust} federated learning. In: 29th USENIX Security Symposium (USENIX Security 20), pp. 1605–1622 (2020)
11. Bernstein, D.J., Hülsing, A., Kölbl, S., Niederhagen, R., Rijneveld, J., Schwabe, P.: The sphincs+ signature framework. In: Proceedings of the 2019 ACM SIGSAC Conference on Computer and Communications Security, pp. 2129–2146 (2019)
12. Shahid, F., Khan, A.: Smart digital signatures (sds): a post-quantum digital signature scheme for distributed ledgers. Futur. Gener. Comput. Syst. **111**, 241–253 (2020)
13. Sun, S., et al.: Xmss-sm3 and mt-xmss-sm3: instantiating extended merkle signature schemes with sm3. Cryptology ePrint Archive (2022)
14. Fathalla, E., Azab, M.: Beyond classical cryptography: a systematic review of post-quantum hash-based signature schemes, security, and optimizations. IEEE Access (2024)

15. Buchmann, J., Dahmen, E., Ereth, S., Hülsing, A., Rückert, M.: On the security of the winternitz one-time signature scheme. Int. J. Appl. Cryptogr. **3**(1), 84–96 (2013)
16. Zhou, Z., Tian, Y., Xiong, J., Peng, C., Li, J., Yang, N.: Blockchain and signcryption enabled asynchronous federated learning framework in fog computing. Digit. Commun. Netw. (2024)
17. Li, P., Chen, T., Liu, J.: Enhancing quantum security over federated learning via post-quantum cryptography. In: 2024 IEEE 6th International Conference on Trust, Privacy and Security in Intelligent Systems, and Applications (TPS-ISA), pp. 499–505. IEEE (2024)
18. Mo, F., Haddadi, H., Katevas, K., Marin, E., Perino, D., Kourtellis, N.: PPFL: privacy-preserving federated learning with trusted execution environments. In: Proceedings of the 19th Annual International Conference on Mobile Systems, Applications, and Services, pp. 94–108 (2021)

Research on Integrated Adversarial Attack and Defense Method for Object Detection Models

Chengbang Yang, Xu Zhang, Kun Chen, and Anzhi Wang(✉)

School of Big Data and Computer Science, Guizhou Normal University, Guiyang 550025, China
cvml16102@163.com

Abstract. Deep learning-based object detection models have been widely applied in autonomous driving, industrial quality inspection, and other fields, with security issues triggered by adversarial perturbations receiving significant attention. However, current adversarial attack and defense methods for object detection models predominantly focus on single attack or defense strategies. This limitation hinders their ability to counter diverse unknown attacks, resulting in poor attack adaptability and insufficient defensive generalization capabilities of the models. To address these challenges, this paper proposes an integrated adversarial attack and defense method for object detection models. By integrating multiple mainstream attack methods to generate diversified adversarial examples, and combined with adversarial training for defense. Experiments conducted on both general object detection and specific object detection tasks demonstrate that the integrated attacks significantly diminish model accuracy, while the integrated defense mechanism promptly restores model performance. This approach effectively enhances model security in complex adversarial scenarios.

Keywords: Object Detection · Adversarial Perturbations · Security · Adversarial Attack · Adversarial Defense

1 Introduction

With the rapid development of artificial intelligence technology, deep learning-based object detection models have achieved remarkable progress and are extensively applied in fields such as autonomous driving [1], medical diagnosis [2], and industrial quality inspection [3]. However, object detection models are susceptible to malicious attacks from adversarial samples, which seriously reduces the detection performance of the model and poses serious security risks in practical applications. For example, in the domain of industrial building quality inspection, concrete bubble detection serves as a critical application of object detection in building engineering quality assessment. Through precise identification, detection, and quantitative analysis of concrete bubbles, it provides essential basis for evaluating the durability, strength, frost resistance, and

C. Yang, X. Zhang and K. Chen—Contributing authors.

J. Yu and Y. Chen (Eds.): FCS 2025, CCIS 2805, pp. 336–350, 2027.
https://doi.org/10.1007/978-981-95-9514-3_23

impermeability of concrete structures. The accuracy and reliability of its detection results directly affect the safety acceptance of the project. Once the detection model fails under adversarial disturbances, it is likely to lead to missed or false detections, thereby causing risks to concrete quality and safety risks.

However, most of the adversarial attack and defense methods of existing object detection models [4–6] focus on a single adversarial attack or defense strategies, resulting in problems such as poor adaptability to adversarial attacks and insufficient generalization ability of defense. The single adversarial attack methods can only generate specific adversarial samples, which is difficult to reflect the different adversarial attacks that the model faces in real complex scenarios; Meanwhile, a single adversarial defense can only defend against specific attack patterns, making it difficult for the model to effectively respond to diverse unknown attacks. Furthermore, existing ensemble attack methods [7, 8] primarily focus on model ensembles, treating proxy models as fixed, immutable units, and lack adaptability.

To address the aforementioned issues, this paper proposes an integrated adversarial attack and defense method (IAAD) for object detection models. In terms of adversarial attacks, this article integrates multiple mainstream adversarial attack algorithms (e.g., gradient -based and optimization-based methods) to generate diverse and more transferable adversarial examples, reflecting the vulnerability of the model to different adversarial attacks in realistic and complex scenarios. In terms of adversarial defense, building a defense mechanism that combines adversarial training with integrated strategies enhances the model's ability to defend against various known and unknown adversarial attacks, further improving the model's robustness. The experimental results demonstrate that the proposed integrated adversarial attack method can significantly reduce the accuracy of the model, while the integrated adversarial defense strategy can effectively restore the performance of the model, proving that this method can solve the problems of poor attack adaptability and insufficient defense generalization performance of singular adversarial attack and defense approaches, and improve the security and robustness of the model in real-world complex scenarios.

Overall, this article has three main contributions:

(1) To solve the problems of poor adaptability to adversarial attacks and insufficient generalization ability in existing single object detection adversarial attack/defense methods and integrated attack approaches, this paper proposes an integrated adversarial attack and defense method, which achieves the reliability and robustness of the object detection model through a collaborative adversarial attack and defense integration strategy.
(2) In order to compensate for the shortcomings of spatial attacks (pixel perturbations) in global information attacks, this paper proposes a frequency-domain based adversarial attack method called FFGSM based on Fast Gradient Sign Method (FGSM).
(3) The experimental results on general object detection task and specific object detection task (concrete bubble detection) show that the integrated adversarial attack and defense method proposed in this paper can effectively enhance the reliability and robustness of the model.

2 Related Work

As one of the core tasks of computer vision, object detection aims to localize and recognize objects of interest within images or videos. It has been widely deployed in fields including autonomous driving, medical diagnosis, and industrial quality inspection. Object detection models exhibit vulnerability to adversarial perturbations, which can lead to a series of security issues. Therefore, exploring the security of object detection models holds substantial research merit and promoting practical applications is of great significance. To enhance the security of object detection models, researchers have conducted security research on object detection models from two aspects: adversarial attacks and adversarial defenses.

In adversarial attacks, object detection adversarial attack methods [9] are mainly divided into optimization iterative attack methods [10, 11], generator generated attack methods [12, 13], query based attack methods [14], and transferability based attack methods [15, 16]. Xie [10] et al. proposed the first algorithm DAG that applies adversarial attacks to object detection and semantic segmentation tasks. By optimizing the loss function of multiple objects (such as pixels or candidate boxes) to generate adversarial samples, the model generates erroneous predictions in detection and segmentation tasks. This method validates the vulnerability of object detection models in complex tasks. Suryanto [11] et al. proposed an innovative physical camouflage attack method called DTN, which can learn and simulate the physical imaging effect of object texture changes, generate physically feasible adversarial textures, and enable object objects to evade detection by object detectors in the physical world, ensuring that the generated adversarial samples remain aggressive in real-world scenarios. The above attack methods based on optimization iteration has a simple way of generating adversarial samples, but they suffer from slow training speed and low efficiency. Considering the limitations of other attack methods in generating large-scale adversarial samples, Generative Adversarial Networks (GAN) [17] can quickly generate a large number of adversarial samples through forward propagation. Therefore, Hu [12] et al. proposed a method based on GAN, which generates adversarial patches that are highly similar to real images by sampling the image manifolds learned by BigGAN and StyleGAN, achieving the transition of adversarial attacks from obvious perturbations to natural concealment. Hu [13] et al. proposed an adversarial texture that can be covered on clothing of any shape, and any local area has adversarial properties, effectively solving the problem of missing adversarial patch segments in multiple perspectives and achieving effective attacks on pedestrian detectors in the physical world from multiple perspectives. Liang [14] et al. proposed the PRFA framework, which generates adversarial perturbations through random search and focuses on attacking the predicted boxes and confidence scores of the object detector, significantly improving the attack efficiency of object detection in black box scenarios. However, most attack methods based on optimization iteration are implemented in simple scenarios and are difficult to reflect the situation in complex scenarios. Therefore, Huang [15] et al. proposed a self-integration strategy that combines data augmentation, model perturbation, and random pruning of adversarial patches to improve the transferability of adversarial patches on black box models.

In adversarial defense, object detection and adversarial defense methods are primarily classified into adversarial training [18, 19], denoising [20, 21], and adversarial noise detection methods [22]. Goodfellow [18] et al. proposed a simple and efficient adversarial sample generation method, FGSM, which calculates the gradient of the loss function relative to the input, adds perturbations along the gradient direction, generates adversarial samples, and uses the generated adversarial samples for adversarial training to improve the performance of the model. Using the idea of adversarial training, Chen [19] et al. proposed a class aware robust adversarial training method called CWAT to address the issue of class imbalance in object detection. By globally optimizing and normalizing the loss of various types of objects, the method enhances the robustness of the model to multi class objects. Liu [20] et al. proposed the SAC framework, which combines adversarial patch detection and image completion techniques to defend object detectors from adversarial patch attacks, achieving efficient adversarial patch defense against object detectors. Yin [22] et al. proposed a method for detecting adversarial attacks using the relationships between multiple objects in an image. By analyzing the intrinsic correlations between objects and checking the intrinsic consistency of input data, adversarial samples can be identified.

However, most of the above-mentioned object detection models only study a single adversarial attack and defense strategy, lacking a reflection of the diverse attack patterns that the model is subjected to in real scenarios, making it difficult to characterize the true vulnerability of the model. Furthermore, a single adversarial defense strategy leads to insufficient generalization ability of the model when facing combined or unknown attacks. However, most object detection methods [23, 24] focus on improving model accuracy and have not Insufficient attention the impact of adversarial attacks on detection reliability. To bridge this gap, this article proposes an integrated adversarial attack and defense method for object detection models, which achieves the reliability and robustness of the object detection model through a collaborative adversarial attack and defense integration strategy.

3 Methodology

3.1 Frequency Domain Based Adversarial Attack Method

Existing adversarial attack methods predominantly generate adversarial samples through spatial-domain pixel perturbations, thereby disrupting the original image feature distribution. However, spatial pixel perturbation only perturbs local information (e.g., edges and textures) of the image and does not reflect the global perturbation information of the image. Given that frequency domain information can better reflect the global image characteristics (such as color composition and semantics). To explore adversarial samples with larger distribution differences, this paper proposes a frequency-domain based adversarial attack method called FFGSM based on Fast Gradient Sign Method (FGSM) [18]. Its structure is shown in Fig. 1.

Firstly, the input image is processed into two branches. One branch, like FGSM attack, generates corresponding gradient information through the object detection model and symbolizes it as adversarial noise; The other branch applies a Gaussian filter (high-pass/low-pass) after performing the Fourier transform in Eq. (1) to separate high-frequency information (edges, textures, details) from low-frequency information (overall contours, background). The Gaussian high-pass filter is expressed as Eq. (2), and

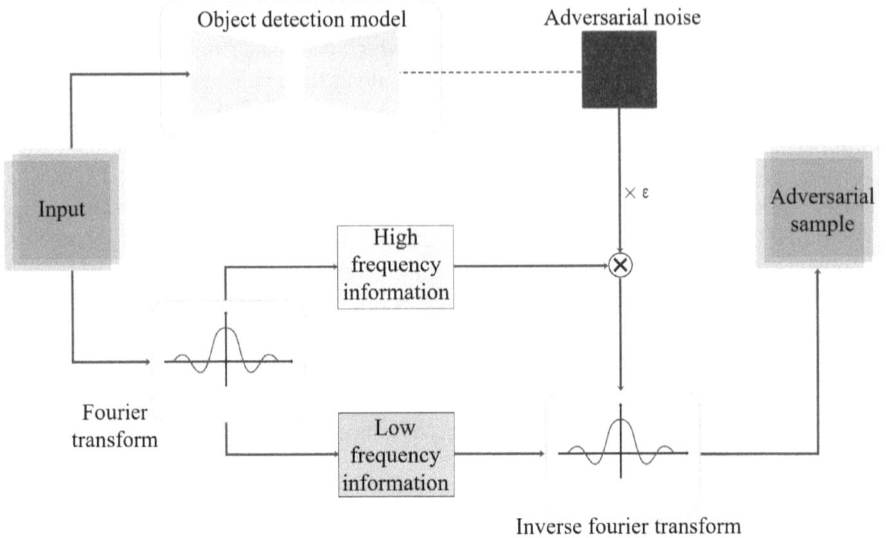

Fig. 1. Frequency domain based adversarial attack method

high-frequency information is obtained after processing via Eq. (3). Subsequently, high-frequency information is element-wise multiplied with adversarial noise and injected into multiplicative noise to enhance feature differences. Finally, the fused high-frequency and low-frequency information is transformed by inverse Fourier transform to restore the spatial image and obtain the final adversarial samples. Here, the perturbation strength ε is set to 0.2.

$$F(u, v) = \sum_{x=0}^{M-1} \sum_{y=0}^{N-1} f(x, y) \bullet e^{-2\pi i(\frac{ux}{M} + \frac{uy}{N})} \tag{1}$$

$$H(u, v) = 1 - e^{-\frac{D(u,v)^2}{2D_0^2}} \tag{2}$$

$$F_{high}(u, v) = F(u, v) \bullet H(u, v) \tag{3}$$

Among them, F (u, v) is the representation of image f (x, y) in the frequency domain, M and N represent the rows and columns of the image, respectively, and u and v represent the time-frequency coordinates. D(u, v) is the distance from the frequency point (u, v) to the center of the frequency domain, and D0 denotes the cutoff frequency, which is set to 30.

This method focuses on high-frequency information of the image (corresponding to details such as object edges and textures), and changes its distribution by adding noise. Due to the high dependence of object detection (especially small objects) on detailed features, such perturbations can effectively reduce the model's localization and classification performance, thereby expanding the distribution difference between the original samples and adversarial samples.

3.2 Integrated Adversarial Attack and Defense Strategy

Currently, adversarial attack and defense methods for object detection models mostly focus on a single adversarial attack or defense strategy, resulting in problems such as poor adaptability to adversarial attacks and insufficient generalization ability of defense. Meanwhile, existing ensemble attack approaches primarily concentrate on model integration, treating proxy models as fixed, immutable units, and thus lack adaptability. Inspired by literature [25], this paper proposes an integrated adversarial attack and defense method (IAAD) for object detection models. This method integrates multiple mainstream attacks and introduces a dynamic scheduling mechanism, significantly improving the efficiency of adversarial sample generation and attack success rate.

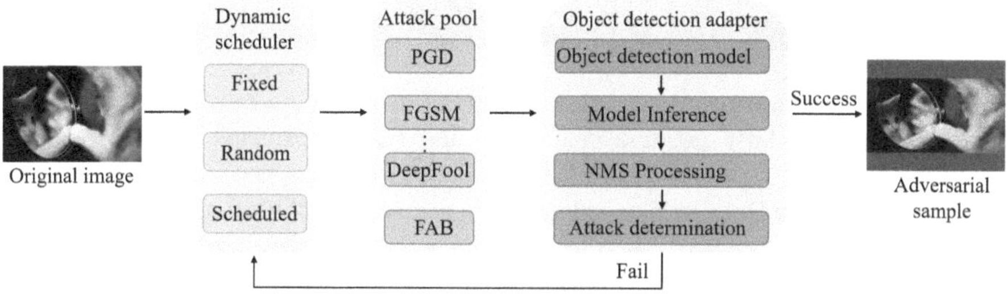

Fig. 2. Integrated Adversarial Attack and Defense Structure Diagram

The IAAD method mainly includes three core components: adversarial attack pool, dynamic scheduler, and object detection adapter. Its structure is illustrated in Fig. 2. The adversarial attack pool is used to integrate multiple pixel space attack methods (such as Projected Gradient Descent (PGD) [26], Adversarial Projected Gradient Descent (APDG) [27], Fast Gradient Sign Method (FGSM) [18], Iterative FGSM (IFGSM) [28], Momentum IFGSM (MIFGSM) [29], Nesterov IFGSM (NIFGSM) [30], FFGSM, Fast Adaptive Boosting Attack (FAB) [31], DeepFool [32], etc.). The dynamic scheduler supports three attack sequence strategies: fixed order, random order, and planned scheduling to schedule multiple attack methods in the adversarial attack pool. Planned scheduling employs a Dynamic Scheduling Matrix (DSM) and the Hungarian algorithm to dynamically adjust attack sequences, optimizing attack outcomes. The DSM quantifies the weights for selecting attack methods at different stages and learns scheduling strategies through dynamic updates. The Hungarian algorithm generates the optimal attack sequence from the DSM matrix. Through this collaboration, the DSM matrix learns synergistic relationships among attack methods, while the Hungarian algorithm translates these learnings into executable optimal sequences, ultimately achieving more efficient attack. The object detection adapter uses Non Maximum Suppression (NMS) to perform confidence threshold filtering based on the model inference results to determine whether the attack is successful. IAAD coordinates multiple adversarial attack methods through a dynamic scheduler to achieve efficient attacks on the model.

4 Experiments

4.1 Experimental Setup

Datasets and Evaluation Metrics. This article uses two datasets, MS COCO [33] and ZS-CBD [34]. The former is used for testing general object detection tasks, while the latter is used for testing specific object detection tasks (concrete bubble detection). The average precision (AP) [35] commonly used in object detection tasks is used to evaluate the performance of the model, which can reflect the impact of adversarial attacks on the model. The calculation formula is shown in Eq. (4).

$$\begin{cases} AP = \int_0^1 P(R)dR \\ P = \frac{TP}{TP+FP} \\ R = \frac{TP}{TP+FN} \end{cases} \tag{4}$$

where P denotes precision, R represents recall, TP is the count of true positive detections, FP indicates false positive detections, and FN corresponds to false negative detections.

Implementation Details. This article uses the concrete bubble detection model DB-YOLO proposed in reference [34] as the baseline and conducts experiments on NVIDIA GeForce RTX 4070Ti with MS COCO and ZS-CBD datasets. The initial learning rate is 0.01, the input image size is 640 × 640, the batch size is 8, and the training epochs are 800 epochs.

To comprehensively evaluate the effectiveness and transferability of the proposed method in this article, we conducted experiments not only on general object detection, but also on specific object detection tasks (concrete bubble detection). The implementation details of the relevant experiments are as follows.

4.2 Adversarial Attack and Defense Experiment for General Object Detection

Validity Verification of Adversarial Attacks. To validate the effectiveness of the proposed frequency-domain adversarial attack method (FFGSM) and the integrated adversarial attack method (IAAD), we trained models on the MS COCO-1K dataset and compared the attack effectiveness of five mainstream adversarial attack methods (PGD [26], APDG [27], FGSM [18], MIFGSM [29], FAB [31]) against our proposed methods (FFGSM, IAAD), as shown in Table 1. Here, "No attack" denotes the performance without any attack, and lower metrics indicate more effective attacks.

As shown in Table 1, various adversarial attack methods can affect models to varying degrees, among which gradient attacks (FGSM, MIFGSM) have the most significant impact on model performance. This is because their perturbation patterns are highly intense and can directly cover object features. In contrast, the iterative attack PDG exhibits the weakest effectiveness. This is due to the fixed step size employed by PGD, which limits its efficacy. The adoption of an adaptive step size in APDG further enhances the attack's effectiveness. Compared to iterative attacks, the proposed frequency-domain attack FFGSM achieves higher success rates by introducing gradient perturbations across global image information, demonstrating the critical importance of frequency-domain

Table 1. Performance comparison of general object detection models attacked by different adversarial attack methods

Attack Type	P	R	AP_{50}	$AP_{50:95}$
No attack	0.656	0.518	0.575	0.383
PGD	0.421	0.170	0.202	0.109
APDG	0.167	0.146	0.082	0.040
FGSM	**0.007**	**0.030**	**0.006**	**0.002**
MIFGSM	0.085	0.099	0.038	0.017
FAB	0.252	0.160	0.133	0.070
FFGSM(Ours)	0.389	0.125	0.149	0.071
IAAD(Ours)	0.050	0.280	0.075	0.031

features for detection. Furthermore, compared to single adversarial attack methods (e.g., PGD, APDG, MIFGSM, FAB, FFGSM), the proposed integrated adversarial attack method IAAD rapidly degrades model prediction performance, demonstrating the vulnerability of models under integrated adversarial attacks. This stems from IAAD's ability to generate diverse adversarial samples that cover a broader range of vulnerabilities by integrating multiple attack techniques. For instance, IAAD reduces model accuracy from 65.6% to 5%, further demonstrating the destructive impact of the proposed integrated adversarial attack method on model performance.

Validity Verification of Adversarial Defenses. To evaluate the model's defense performance against various adversarial attack methods, this paper employs adversarial training [36] as the baseline defense. The experimental results are shown in Table 2, where ↑/↓ represents the improvement/decrease in defense ability compared to adversarial attacks, and the higher the indicator, the more effective the defense. Through experimental testing, it has been proven that the accuracy of the model after defense is not as high as the original model (No attack), because the defense capability of the model is achieved at the expense of a certain level of accuracy. The IAAD method is slightly inferior to other defense methods (such as PGD and FGSM) in terms of accuracy P index, but it outperforms other methods in AP, verifying that the integrated defense strategy proposed in this paper is slightly better than a single defense strategy.

The visualization results of adversarial defense are shown in Fig. 3. After defense, the model's attention to the object is significantly improved, but it is still affected by noise. The reason is that defense training may overly focus on adversarial noise, sacrificing the model's generalization ability to normal samples. Therefore, in the future, dynamic defense strategies can be considered to enhance the adaptability of the model.

Table 2. Performance comparison of general object detection models after adversarial defense

Attack Type	P	R	AP_{50}	$AP_{50:95}$
No attack	0.656	0.518	0.575	0.383
PGD	0.460(0.039 ↑)	0.233(0.063 ↑)	0.281(0.079 ↑)	0.131(0.022 ↑)
APDG	0.231(0.064 ↑)	0.284(**0.138** ↑)	0.229(0.147 ↑)	0.124(0.084 ↑)
FGSM	0.580(**0.573** ↑)	0.113(0.083 ↑)	0.189(0.183 ↑)	0.093(0.091 ↑)
MIFGSM	0.473(0.388 ↑)	0.203(0.104 ↑)	0.243(0.205 ↑)	0.124(0.107 ↑)
FAB	0.131(0.121 ↓)	0.410(0.250 ↑)	0.232(0.099 ↑)	0.116(0.046 ↑)
FFGSM(Ours)	0.485(0.096 ↑)	0.166(0.041 ↑)	0.214(0.065 ↑)	0.094(0.023 ↑)
IAAD(Ours)	0.270(0.220 ↑)	0.309(0.029 ↑)	0.294(**0.219** ↑)	0.160(**0.129** ↑)

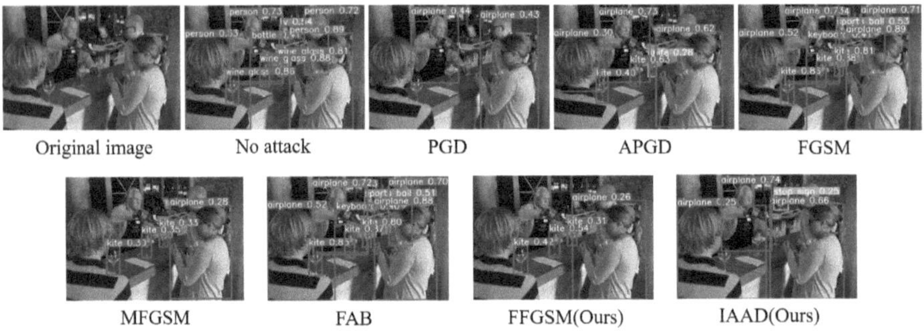

Fig. 3. The detection performance of general object detection models after defense using different adversarial defense methods

4.3 Adversarial Attack and Defense Experiment for Specific Object Detection

In order to verify the generality of the integrated adversarial attack and defense method proposed in this paper, experiments were conducted on specific object detection (concrete bubble detection) tasks. The experimental details are as follows.

Validity Verification of Adversarial Attacks. To validate the effectiveness of the FFGSM method and the IAAD method, this paper trained the model on the ZS-CBD dataset and compared the attack effects of eight mainstream adversarial attack methods on the model, including PGD [26], APDG [27], FGSM [18], IFGSM [28], MIFGSM [29], NIFGSM [30], FAB [31], DeepFool [32], as shown in Table 3.

Table 3. Performance comparison of specific object detection models attacked by different adversarial attack methods

Attack Type	P	R	AP_{50}	$AP_{50:95}$
No attack	0.794	0.705	0.711	0.503
PGD	0.538	0.374	0.316	0.144
APDG	0.528	0.386	0.323	0.114
FGSM	0.353	0.251	0.194	0.073
NIFGSM	0.336	0.261	0.179	0.087
MIFGSM	0.308	0.238	0.148	0.037
IFGSM	0.536	0.380	0.319	0.112
FAB	0.628	0.439	0.408	0.171
DeepFool	0.675	0.450	0.483	0.278
FFGSM(Ours)	0.425	0.247	0.209	0.084
IAAD(Ours)	**0.119**	**0.045**	**0.026**	**0.010**

Among single adversarial attack methods, gradient attacks (FGSM, NIFGSM, MIFGSM) have a significant impact on model performance. Among them, MIFGSM, which introduces momentum technology, has the most stable attack direction and can effectively avoid gradient oscillations, resulting in the best performance; Iterative attacks (PGD, NIFGSM) did not perform optimally in the task because the fixed step size used by PGD limited its effectiveness, while NIFGSM iteratively added noise to make it suitable for dense prediction in object detection, further indicating that the attack method needs to be adapted to the task scenario; The frequency domain attack FFGSM proposed in this article reduces the model accuracy to 42.5% and significantly lowers the model performance, verifying the importance of frequency domain information for detection. Meanwhile, compared to single adversarial attack methods (such as PGD and FGSM), the proposed integrated adversarial attack method IAAD can rapidly reduce the predictive performance of the model, verifying its vulnerability. This is because IAAD generates diverse adversarial samples through integrated diversification attacks, covering a wider range of vulnerabilities. For example, IAAD reduced the accuracy of the model from 79.4% to 11.9%, while MIFGSM only reduced the model performance to 30.8%, further demonstrating the destructive effect of the proposed ensemble adversarial attack method on model performance.

Validity Verification of Adversarial Defenses. To evaluate the model's defense performance against various adversarial attack methods, this paper also conducted quantitative experimental comparisons of eight mainstream adversarial attack methods.

The experimental results are shown in Table 4. Through experimental testing, it has been proven that the integrated adversarial defense method is inferior to other defense methods (such as PGD) in terms of defense performance, but to some extent superior to certain single adversarial defense methods (such as FGSM). The reason for its suboptimal defense performance is that the defense generalization performance of the IAAD method sacrifices a certain degree of accuracy. In addition, compared to other defense methods, the IAAD method can effectively restore detection performance (by 59.1%), significantly reducing the success rate of attacks. Among them, the visualization results of adversarial defense are shown in Fig. 4.

Table 4. Performance comparison of specific object detection models after adversarial defense

Attack Type	P	R	AP_{50}	$AP_{50:95}$
No attack	0.794	0.705	0.711	0.503
PGD	0.749(0.221 ↑)	0.616(0.242 ↑)	0.617(0.301 ↑)	0.345(0.201 ↑)
APDG	0.754(0.226 ↑)	0.645(0.259 ↑)	0.658(0.335 ↑)	0.375(0.261 ↑)
FGSM	0.688(0.335 ↑)	0.592(0.341 ↑)	0.589(0.395 ↑)	0.323(0.250 ↑)
NIFGSM	0.718(0.382 ↑)	0.585(0.324 ↑)	0.586(0.407 ↑)	0.313(0.226 ↑)
MIFGSM	0.723(0.415 ↑)	0.580(0.342 ↑)	0.593(0.445 ↑)	0.317(0.280 ↑)
IFGSM	0.763(0.227 ↑)	0.659(0.279 ↑)	0.672(0.353 ↑)	0.394(0.282 ↑)
FAB	0.734(0.106 ↑)	0.671(0.232 ↑)	0.664(0.256 ↑)	0.383(0.212 ↑)
DeepFool	0.706(0.031 ↑)	0.632(0.182 ↑)	0.641(0.158 ↑)	0.357(0.079 ↑)
FFGSM(Ours)	0.693(0.268 ↑)	0.479(0.232 ↑)	0.478(0.269 ↑)	0.249(0.165 ↑)
IAAD(Ours)	0.710(**0.591 ↑**)	0.512(**0.467 ↑**)	0.536(**0.510 ↑**)	0.302(**0.292 ↑**)

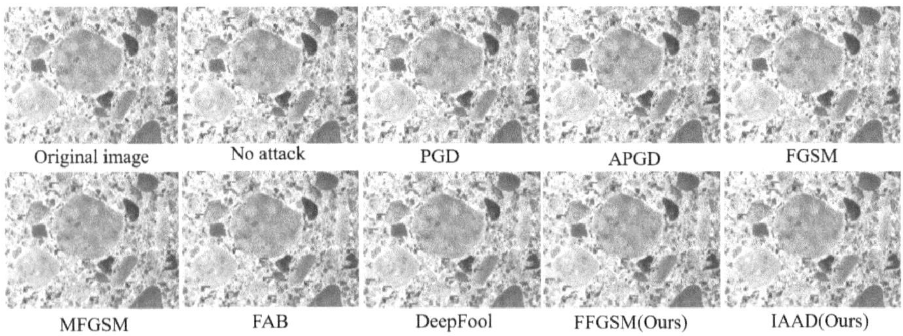

Fig. 4. The detection performance of specific object detection models after defense using different adversarial defense methods

4.4 Ablation Study

To validate the defense capability of integrated adversarial defense method against unknown adversarial attacks, this paper compares the performance of single adversarial defense methods (such as PGD and FGSM) with IAAD adversarial defense method when subjected to unknown noise FFGSM on a specific object detection task. The experimental results are shown in Table 5. The defense effectiveness of the integrated adversarial defense method outperforms that of single adversarial defense methods, demonstrating that the integrated approach exhibits greater generalization against unknown noise.

Table 5. Comparison of generalization performance of integrated adversarial defense methods for unknown noise

Attack Type	P	R	AP_{50}	$AP_{50:95}$
PGD	0.474	**0.246**	0.240	0.010
FGSM	0.366	0.227	0.184	0.071
IAAD(Ours)	**0.556**	0.217	**0.271**	**0.119**

To validate the attack effectiveness of three attack sequences (such as fixed order, random order, and scheduled order) in the dynamic scheduler, this paper conducts ablation experiments on a specific object detection task. As shown in Table 6, the fixed strategy exhibits the weakest degradation of model performance, while the random strategy causes the most significant damage. The scheduled strategy demonstrates the second-best attack effectiveness, yet it achieves the highest P-metric attack performance.

Table 6. Performance comparison of three attack sequences in dynamic schedulers.

Attack Sequences	P	R	AP_{50}	$AP_{50:95}$
fixed order	0.134	0.066	0.038	0.015
random order	0.119	**0.045**	**0.026**	**0.010**
scheduled order	**0.107**	0.052	0.030	0.011

5 Conclusion

This article focuses on the adversarial vulnerability of object detection models, and proposes an integrated adversarial attack and defense method to address the single adversarial attack or defense method problem faced by general object detection models and specific object detection models (concrete bubble detection) in practical engineering applications. The aim is to improve the reliability and security of the model in complex real-world scenarios. The experimental results validated the effectiveness of the proposed method. Integrated adversarial attacks significantly reduce the accuracy of model detection and fully expose its vulnerability in complex scenarios; The integrated defense strategy effectively restores the performance of the model, proving that this method can solve the problems of poor attack adaptability and insufficient defense generalization ability that exist in single adversarial methods. At the same time, in order to comprehensively evaluate the effectiveness and transferability of the proposed method, we validated it on both general object detection and specific object detection tasks (concrete bubble detection), ultimately improving the safety and robustness of the detection model in real complex engineering scenarios, providing technical support for the reliable application of deep learning models in construction quality monitoring.

Acknowledgments. This study was partially supported by the National Natural Science Foundation of China under Grant (62162013), Natural Science Foundation of Guizhou Province Science and Technology Department (QianKeHeJiChu MS-[2025]-249), Diting Quality Inspection-Pioneering Automated Detection of Concrete Air Bubbles, Mingjing Smart Screening-The Pioneer in Precision Intelligent Polyp Detection.

Disclosure of Interests. The authors declare no competing interests relevant to the content of this article.

References

1. Cao, X., Wang, P., Zhang, Z., et al.: RAFDet: a novel camera-radar fusion framework for robust 3d object detection in autonomous driving. In: ICASSP 2025–2025 IEEE International Conference on Acoustics, Speech and Signal Processing (ICASSP). IEEE, pp. 1–5 (2025)
2. Mantri, R., Khan, R.A.H., Mane, D.: Leveraging transfer learning for object detection in medical imaging. In: 2025 International Conference on Emerging Smart Computing and Informatics (ESCI). IEEE, pp. 1–5 (2025)
3. Shirazi, M., Schmitz, M., Janssen, S., et al.: Verifying the applicability of synthetic image generation for object detection in industrial quality inspection. In: 2021 20th IEEE International Conference on Machine Learning and Applications (ICMLA). IEEE, pp. 1365–1372 (2021)
4. Wang, T., Wang, Z., Wang, C., et al.: Can't slow me down: learning robust and hardware-adaptive object detectors against latency attacks for edge devices. In: Proceedings of the Computer Vision and Pattern Recognition Conference, pp. 19230–19240 (2025)
5. Lin, Z., Zhao, Y., Chen, K., et al.: I don't know you, but I can catch you: real-time defense against diverse adversarial patches for object detectors. In: Proceedings of the 2024 on ACM SIGSAC Conference on Computer and Communications Security, pp. 3823–3837 (2024)
6. Pathak, S., Shrestha, S., AlMahmoud, A.: Model agnostic defense against adversarial patch attacks on object detection in unmanned aerial vehicles. In: 2024 IEEE/RSJ International Conference on Intelligent Robots and Systems (IROS). IEEE, pp. 2586–2593 (2024)
7. Xiong, Y., Lin, J., Zhang, M., et al.: Stochastic variance reduced ensemble adversarial attack for boosting the adversarial transferability. In: Proceedings of the IEEE/CVF Conference on Computer Vision and Pattern Recognition, pp. 14983–14992 (2022)
8. Tang, B., Wang, Z., Bin, Y., et al.: Ensemble diversity facilitates adversarial transferability. In: Proceedings of the IEEE/CVF Conference on Computer Vision and Pattern Recognition, pp. 24377–24386 (2024)
9. Xinxin, W., Jing, C., Kun, H., et al.: A review of adversarial attacks and defenses for object detection. J. Commun. **44**(11), 260–277 (2023)
10. Xie, C.H., Wang, J.Y., Zhang, Z.S., et al.: Adversarial examples for semantic segmentation and object detection. In: Proceedings of IEEE International Conference on Computer Vision (ICCV). Piscataway: IEEE Press, pp. 1378–1387 (2017)

11. Suryanto, N., Kim, Y., Kang, H., et al.: DTA: physical camouflage attacks using differentiable transformation network. In: Proceedings of IEEE/CVF Conference on Computer Vision and Pattern Recognition (CVPR). Piscataway: IEEE Press, pp. 15284–15293 (2022)
12. Hu, Y.C.T, Chen, J.C., Kung, B.H., et al.: Naturalistic physical adversarial patch for object detectors. In: Proceedings of IEEE/CVF International Conference on Computer Vision (ICCV). Piscataway: IEEE Press, pp. 7828–7837 (2022)
13. Hu, Z.H., Huang, S.Y., Zhu, X.P., et al.: Adversarial texture for fooling person detectors in the physical world. In: Proceedings of IEEE/CVF Conference on Computer Vision and Pattern Recognition (CVPR). Piscataway: IEEE Press, pp. 13297–13306 (2022)
14. Liang, S.Y., Wu, B.Y., Fan, Y.B., et al.: Parallel rectangle flip attack: a query-based black-box attack against object detection. In: Proceedings of IEEE/CVF International Conference on Computer Vision (ICCV). Piscataway: IEEE Press, pp. 7677–7687 (2022)
15. Huang, H., Chen, Z.Y., Chen, H.R., et al.: T-SEA: transfer-based self-ensemble attack on object detection. In: Proceedings of IEEE/CVF Conference on Computer Vision and Pattern Recognition (CVPR). Piscataway: IEEE Press, pp. 20514–20523 (2023)
16. Cai, Z.K., Rane, S., Brito, A.E., et al.: Zero-query transfer attacks on context-aware object detectors. In: Proceedings of IEEE/CVF Conference on Computer Vision and Pattern Recognition (CVPR). Piscataway: IEEE Press, pp. 15004–15014 (2022)
17. Goodfellow, I.J., Pouget-Abadie, J., Mirza, M., et al.: Generative adversarial networks. Commun. ACM **63**(11), 139–144 (2020)
18. Goodfellow, I.J., Shlens, J., Szegedy, C.: Explaining and harnessing adversarial examples, 2014. arXiv Preprint, arXiv: 1412.6572
19. Chen, P.C., Kung, B.H., Chen, J.C.: Class-aware robust adversarial training for object detection. In: Proceedings of IEEE/CVF Conference on Computer Vision and Pattern Recognition (CVPR). Piscataway: IEEE Press, pp. 10415–10424 (2021)
20. Liu, J., Levine, A., Lau, C.P., et al.: Segment and complete: defending object detectors against adversarial patch attacks with robust patch detection. In: Proceedings of IEEE/CVF Conference on Computer Vision and Pattern Recognition (CVPR). Piscataway: IEEE Press, pp. 14953–14962 (2022)
21. Li, H.F., Li, G.B., Yu, Y.Z.: ROSA: robust salient object detection against adversarial attacks. IEEE Trans. Cybern. **50**(11), 4835–4847 (2020)
22. Yin, M.J., Li, S.S., Cai, Z.K., et al.: Exploiting multi-object relationships for detecting adversarial attacks in complex scenes. In: Proceedings of IEEE/CVF International Conference on Computer Vision (ICCV). Piscataway: IEEE Press, pp. 7838–7847 (2022)
23. Inoue, R., Tsuchiya, M., Yasui, Y.: Decoupled PROB: decoupled query initialization tasks and objectness-class learning for open world object detection. In: 2025 IEEE/CVF Winter Conference on Applications of Computer Vision (WACV). IEEE, pp. 8207–8216 (2025)
24. Wang, H., Li, Q., Sun, H., et al.: VaBUS: edge-cloud real-time video analytics via background understanding and subtraction. IEEE J. Sel. Areas Commun. **41**(1), 90–106 (2022)
25. Sun, J., Yao, W., Jiang, T., et al.: ${A^{3} D} $: a platform of searching for robust neural architectures and efficient adversarial attacks. IEEE Trans. Pattern Anal. Mach. Intell. (2025)
26. Madry, A., Makelov, A., Schmidt, L., et al.: Towards Deep Learning Models Resistant to Adversarial Attacks, 2017. https://doi.org/10.48550/arXiv.1706.06083
27. Cinà, A.E., Rony, J., Pintor, M., et al.: Attackbench: evaluating gradient-based attacks for adversarial examples. In: Proceedings of the AAAI Conference on Artificial Intelligence, vol. 39, no. 3, pp. 2600–2608 (2025)
28. Kurakin, A., Goodfellow, I.J., Bengio, S.: Adversarial Examples in the Physical World. Artificial Intelligence Safety and Security. Chapman and Hall/CRC, Boca Raton, pp. 99–112 (2018)

29. Zhou, R.: Study of optimiser-based enhancement of adversarial attacks on neural networks. In: 2024 International Conference on Interactive Intelligent Systems and Techniques (IIST). IEEE, pp. 740–747, 2024
30. Wang, Y., Wu, Y., Wu, S., et al.: Boosting the transferability of adversarial attacks with frequency-aware perturbation. IEEE Trans. Inf. Forensics Secur. (2024)
31. Costa, J.C., Roxo, T., Proença, H., et al.: How deep learning sees the world: a survey on adversarial attacks & defenses. IEEE Access (2024)
32. Abroshan, M., Moosavi-Dezfooli, S.M.: SuperDeepFool: a new fast and accurate minimal adversarial attack. Adv. Neural. Inf. Process. Syst. **37**, 98537–98562 (2024)
33. Lin, T.Y., Maire, M., Belongie, S., et al.: Microsoft COCO: Common Objects in Context. In: Fleet, D., Pajdla, T., Schiele, B., Tuytelaars, T. (eds.) Computer Vision – ECCV 2014. ECCV 2014. LNCS, vol. 8693, pp. 740–755. Springer, Cham (2014). https://doi.org/10.1007/978-3-319-10602-1_48
34. Chen, K., Wang, A.: BD-YOLO: high-precision lightweight concrete bubble detector based on YOLOv7. In: Lin, Z., et al. (eds.) Pattern Recognition and Computer Vision. PRCV 2024. LNCS, vol. 15033, pp. 379–393. Springer, Singapore (2024). https://doi.org/10.1007/978-981-97-8502-5_27
35. Padilla, R., Netto, S.L., Da Silva, E.A.B.: A survey on performance metrics for object-detection algorithms. In: 2020 International Conference on Systems, Signals and Image Processing (IWSSIP). IEEE, pp. 237–242 (2020)
36. Zhang, H., Wang, J.: Towards adversarially robust object detection[C]//Proceedings of the IEEE/CVF International Conference on Computer Vision, pp. 421–430 (2019)

Privacy Protection

Conditional Privacy-Preserving Authentication Scheme for VANETs Using Anonymous Credentials

Wangjunchen He, Qiang Li, Bing Liu(✉), and Ling Xiong

School of Computer and Software Engineering, Xihua University, Chengdu, China
liubing@mail.xhu.edu.cn

Abstract. With the rapid advancement of intelligent transportation systems, vehicle communication security has emerged as a critical issue requiring urgent attention. Existing communication schemes often fall short in ensuring privacy protection and identity authentication, leaving them vulnerable to privacy breaches, identity theft, and other security risks that could lead to traffic chaos and safety incidents. To address these challenges, this paper proposes a novel vehicle communication security scheme based on zero-knowledge proof and anonymous credentials. By leveraging randomized anonymous credentials and zero-knowledge proof technology, the scheme ensures identity authentication, privacy protection, non-forgery, traceability, and unlinkability in vehicular communications. Formal and informal analyses demonstrate the scheme's security and reliability, while comparisons of computational and communication overhead highlight its efficiency advantages. Experimental results show that the proposed scheme effectively balances security and efficiency, making it highly suitable for practical vehicular network applications.

Keywords: VANETs · Privacy-Preserving · Authentication · Anonymous credential · Unlinkability

1 Introduction

As unmanned vehicles gradually become a main mode of urban transportation, the key technologies of intelligent vehicles and Intelligent Transportation Systems (ITS) are playing an increasingly significant role in the intelligent transportation system. These intelligent vehicles, equipped with high-performance computing devices, are of great importance in achieving efficient and safe traffic management. However, in information systems serving a large user base, the security risks caused by identity privacy disclosure are also increasing significantly. For example, the PlayStation network experienced external intrusion incidents [9], where 77 million user account information were leaked, serving as

This work was supported by the National Natural Science Foundation of China (Grant No. 12501736).

J. Yu and Y. Chen (Eds.): FCS 2025, CCIS 2805, pp. 353–369, 2027.
https://doi.org/10.1007/978-981-95-9514-3_24

a wake-up call to user privacy. In a networked environment, the consequences of identity privacy breaches may be even more severe: attackers could potentially use leaked information to release false traffic data, induce traffic chaos, and even cause fatal road accidents [11]. Therefore, how to effectively protect user privacy has become a key issue that needs to be addressed.

Traditional identity authentication mechanisms often rely on centralized identity verification systems. However, these schemes frequently expose user identity information during the authentication process, thereby increasing the risks of privacy leakage and other security threats [17,18]. To address these challenges, authentication technology based on anonymous credentials has emerged. The core idea of this technology is to ensure user identity anonymity, thereby avoiding direct leakage of sensitive information during the certification process. Nevertheless, complete anonymity has practical drawbacks: excessive anonymity may allow malicious vehicles to exploit the mechanism to disseminate false information, potentially causing traffic accidents. Consequently, it is of paramount importance to design a conditional anonymous authentication scheme that can guarantee anonymity while enabling necessary traceability. Accordingly, a trusted third party should be able to reveal the identity of malicious actors.

To balance anonymity and traceability, this research adopts a combined technology of anonymous credentials with the ElGamal algorithm scheme [13,14]. This approach aims to protect user privacy while allowing for the conditional revelation of malicious vehicle identities. Furthermore, during session interactions between vehicles, it is essential to ensure unlinkability between each authentication session. Even if a user participates in multiple authentication sessions, the credential information generated in different sessions should not be associable, thereby maximizing the protection of user identity privacy. However, while achieving unlinkability and privacy protection, traditional anonymous authentication schemes often encounter issues of low authentication efficiency and high computational overhead. These problems limit their promotion in practical application scenarios such as the Internet of Vehicles (IoV).

To this end, this paper proposes a conditional privacy-preserving authentication scheme based on anonymous credentials. The main contributions of this work are summarized as follows:

- **Conditional anonymity with traceability:** By integrating anonymous credentials and the ElGamal algorithm, the scheme enables identity authentication with conditional anonymity. This ensures that the real identity of a user remains hidden during authentication, while still allowing a trusted third party to trace and reveal the identity of malicious vehicles when necessary.
- **Un-linkability via zero-knowledge proofs:** The use of zero-knowledge proof (ZKP) technology ensures that vehicles cannot be linked across multiple authentication sessions. This significantly enhances user privacy by preventing adversaries from correlating different interactions of the same vehicle.
- **Improved efficiency and robustness:** The scheme is designed to reduce computational overhead and improve authentication efficiency, making it more practical for real-world vehicular environments. It also provides

enhanced security features, contributing to the overall robustness of the Internet of Vehicles (IoV) system.

The rest of the paper is organized as follows: Sect. 2 reviews related work on authentication schemes in VANETs, highlighting limitations in privacy, unlinkability, and efficiency. Section 3 presents the system model and security requirements, defining the roles of TA, RSUs, and vehicles, and outlining the adversary model. Section 4 details the proposed authentication protocol, including initialization, registration, authentication, and tracing phases using anonymous credentials and zero-knowledge proofs. Section 5 provides formal and informal security analyses, proving the scheme's robustness under the random oracle model. Section 6 evaluates performance in terms of computational and communication overhead, showing superior efficiency compared to existing schemes. Section 7 concludes the paper by summarizing contributions and suggesting future research directions.

2 Related Work

With the rapid development of Vehicular Ad-hoc Networks (VANETs), privacy-preserving authentication has attracted extensive attention. Early identity-based methods (e.g., Sun et al. [18], Zhang et al. [24]) provided authentication and traceability but suffered from high computational cost and limited unlinkability. Group-signature approaches (e.g., Jiang et al. [12]) improved anonymity and traceability but introduced complex group management and communication overhead.

More recently, blockchain-assisted schemes have emerged. For example, Lin et al. [15] and Zhou et al. [25] leveraged blockchain for traceability, but their performance is constrained by the latency of consensus mechanisms. Bao et al. [2] proposed a user-controlled data linkability framework, yet it still lacks efficient unlinkability. In 2024–2025, lightweight blockchain-based protocols (e.g., Chen et al. [4], Wang et al. [19]) further improved scalability and reduced communication cost, but they do not integrate zero-knowledge proofs for unlinkability.

Compared to these works, our scheme combines anonymous credentials, PUF-based security, and zero-knowledge proofs, achieving conditional anonymity, unlinkability, and efficiency simultaneously.

3 System Architecture and Security Model

3.1 System Architecture

The system model of this research scheme is shown in Fig. 1, and there are three main entities in the system model, namely, the trusted authority, the roadside unit, and the vehicle.

Trusted authorities (TA) is a fully trusted entity with strong storage and computing capabilities. It generates system parameters and issues anonymous credentials to registered users, ensuring secure and controllable access. It also

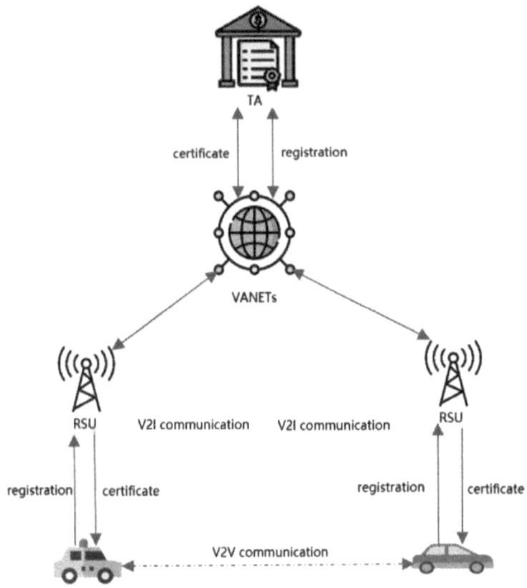

Fig. 1. System model diagram.

enables traceability of malicious vehicles to maintain system reliability and protect user privacy [15,24].

Road-Side Units (RSUs) are fixed infrastructure deployed at roadsides, intersections, and parking areas. They facilitate communication between vehicles and the network using wireless technologies like DSRC or C-V2X. RSUs support real-time traffic data collection, enhance communication coverage, and help optimize traffic flow and safety [20,23].

Vehicles, equipped with onboard units (OBUs), are key network components responsible for sensing, data transmission, and communication with RSUs and other vehicles. They collect environmental data, receive traffic updates, and support cooperative driving and intelligent navigation. In the IoV, vehicles act as both data sources and relays, enabling wide-area connectivity and contributing to intelligent traffic management [22].

3.2 Adversary Model

Assume that A is a probabilistic polynomial-time adversary. The main purpose of is to forge information about a car. A may have the following attack capabilities.

A has the ability to intercept, modify, and replay messages transmitted by vehicles.

A may be a legitimate vehicle in VANET that wants to impersonate another legitimate vehicle.

A may be a legitimate RSU in a VANET that wants to impersonate another legitimate vehicle.

3.3 Security Requirements

Traceable authentication and authorization scheme satisfies the following security requirements: identity privacy protection, unforgeability, unlinkability, and traceability.

Identity privacy protection: the user's real identity should be protected during the authentication process. The intercepted information will not reveal any valuable information of the sender to the adversary.

Unforgeability: It ensures that the sender's message was sent by a legitimate entity and the receiver can verify the message, while ensuring that the risk of data forgery is negligible.

Unlinkability: unable to connect two transmissions from the same vehicle, in order to prevent the attacker seized on public channel of information, to analyze and monitor the motion of the vehicle path.

Traceability: In order to prevent vehicles from randomly sending false information, a third-party trusted agency can disclose malicious vehicle information.

The ability to resist various attacks to ensure the security of the anonymous authentication system. Ensure that the anonymous authentication scheme in vehicular AD hoc networks can resist various attacks, including physical attacks, impersonation attacks, and replay attacks.

4 Program Description

4.1 Initialization Phase

Given a security parameter λ as input, the Trusted Authority Center $TA_i(1 \leq i \leq n)$ generates all system parameters according to the following steps, and the description of other symbols of the system is shown in Table 1.

All trusted authorities cooperate to select an elliptic curve additive cyclic group of order q. g_1 is the generator of the group $\mathbb{G}_1$, g_2 is the generator of the group $\mathbb{G}_2$, and a bilinear pair $e : \mathbb{G}_1 \times \mathbb{G}_2 \rightarrow \mathbb{G}_T$ is selected.

TA_i selects a hash function H: $\{0,1\}^* \rightarrow \mathbb{Z}_q^*$. The trust authority selects $x \in \mathbb{Z}_q^*$ as its private key and computes the public key $X = g_1^x$. Then, TA_i publishes the system parameters $\{G_1, G_2, G_T, g_1, g_2, q, H, X\}$ to the RSUs in its region.

4.2 Registration Phase

The registration process involves vehicles providing authentic identity information such as ID cards and phone numbers to the trusted authority. After verification by the identity registration center, a certificate is issued to the user. The specific steps are as follows:

Step 1: The user sends the vehicle's authentic identity information ID_{vi} and personal information PI (ID card, phone number) to the trusted authority TA_i via a secure channel.

Table 1. Identification Description

Symbols	Definition
x	Trusted authority's private key
X	The public key of the trusted authority
ID_{vi}	The vehicle's unique identification
PI	The user's real identity information
g_1	The generator element of $\mathbb{G}_1$
g_2	The generator of $\mathbb{G}_2$
C	The challenge value generated by the vehicle
PUF_{vi}	Physical unclonable functions
Re, Si	Random integers
$r, a_1, a_2, a_3, a_4, b_1$	Random integers
H	Hash function
σ_{vi}	Anonymous credentials of the vehicle
ΔT	Time threshold
m_i	A message sent by the vehicle
π	Zero-knowledge proof
C_1, C_2	ElGamal ciphertext
$\oplus$	XOR operation
T_1	The time at which the vehicle sends the message
T_2	The time the vehicle receives the message

Step 2: Upon receiving the user's authentic identity, the trusted authority first verifies it by checking the uploaded ID card or the verification code sent to the user's phone number. Upon successful verification, the trusted authority computes a certificate $\sigma_{vi} = g_1^{\frac{1}{x+ID_{vi}}}$ for the user and sends it securely.

Step 3: After receiving the certificate from the trusted authority, the user verifies whether the following equation holds. If it does, the user generates a random number C, computes $Re = H(PUF_{vi}(C))$, and $Si = Re \oplus \sigma_{vi}$. The user then stores C, Si, and ID_{vi} locally; otherwise, the data is rejected by the verifier.

$$e(\sigma_{vi}, Xg_2^{ID_{vi}}) = e(g_1, g_2) \tag{1}$$

4.3 Information Sharing and Authentication Phase

When vehicles need to share traffic information, the recipient must verify whether the message sent by the sender is generated with a legitimate certificate. If not, the information is rejected by the verifier; otherwise, the information is shared.

The sender performs the following operations:

Step 1: Using the locally stored random number C, compute $Re = H\big(PUF_{vi}$ $(C)\big)$, then calculate the certificate $\sigma_{vi} = Re \oplus Si$. Next, select a random number $r \in \mathbb{Z}_q^*$ to blind the certificate: $\hat{\sigma}_{vi} = \sigma_{vi}^r$. The user also needs to prove: (1) the blinded certificate is generated from a legitimate certificate,(2) the embedded attribute ID_{vi} in the certificate is valid and genuine, without revealing the hidden certificate information and the real-time information ID_{vi}. To achieve this, a zero-knowledge proof is constructed for verification.

$$\pi' \in \text{ZKP}\{(ID_{vi}, r) : A = g_1^r \hat{\sigma}_{vi}^{-ID_{vi}}\} \tag{2}$$

Step 2: Compute the tuple $(C_1, C_2) = \left(g_1^r, ID_{vi}X^r\right)$. (C_1, C_2) is used for tracing malicious vehicle identities and must prove that the real identity is bound to a legitimate certificate, rather than being fabricated. This means proving that ID_{vi} in C_2 is the ID_{vi} in the certificate σ_{vi}. This relationship can be further expressed as $\hat{\sigma}_{vi}^{C_2 X^{-r}} = A^{-1} g_1^r$. Let $B = -C_2 X^{-r}$, which is equivalent to:

$$A = g_1^r \hat{\sigma}_{vi}^B, \quad -B^{-1} C_2 = X^r \tag{3}$$

Combining the equations from (2), all proof statements are assembled into the final zero-knowledge proof π:

$$\pi \in \text{ZKP}\{(ID_{vi}, r) : A = g_1^r \hat{\sigma}_{vi}^{-ID_{vi}} \wedge A = g_1^r \hat{\sigma}_{vi}^B \wedge -B^{-1} C_2 = X^r \wedge C_1 = g_1^r\} \tag{4}$$

Intuitive Description of Zero-Knowledge Proof. To enhance clarity, we provide a high-level description of the zero-knowledge proof procedure. Intuitively, the prover (vehicle) constructs four sub-proofs (π_1–π_4) to convince the verifier that:

1. it possesses a valid credential bound to a legitimate identity;
2. the blinded credential is correctly derived;
3. the embedded identity matches the credential;
4. the randomization factor is fresh.

These proofs are assembled into a non-interactive ZKP using the Fiat–Shamir heuristic. Algorithm 1 provides a pseudocode description of the interaction steps.

Algorithm 1 Zero-Knowledge Proof Interaction

1: Vehicle selects random $r \in \mathbb{Z}_q^*$ and blinds credential.
2: Vehicle constructs sub-proofs $\pi_1 \ldots \pi_4$.
3: Prover sends $(C_1, C_2, \hat{\sigma}_{vi}, \pi)$ to verifier.
4: Verifier checks time freshness $|T_2 - T_1| < \Delta T$.
5: Verifier validates π using public parameters.
6: If all checks pass, message m_i is accepted.

Finally, the sender transmits the data package $\left((C_1, C_2), m_i, T_1, \hat{\sigma}_{vi}, \pi\right)$ to the recipient.

The recipient performs the following operations:

Step 1: Verify whether the inequality $|T_2 - T_1| < \Delta T$ holds, where T_2 is the current timestamp and ΔT is a threshold. If the inequality does not hold, the message is rejected.

Step 2: Verify that the message m_i sent by the sender is generated with a legitimate certificate, i.e., validate the proof π. The proof π can be divided into four sub-proofs:

$$\pi_1 \in \text{ZKP}\{(ID_{vi}, r) : A = g_1^r \hat{\sigma}_{vi}^{-ID_{vi}}\}$$

$$\pi_2 \in \text{ZKP}\{(r) : A = g_1^{\bar{a}_1} \hat{\sigma}_{vi}^{-\bar{b}_1}\}$$

$$\pi_3 \in \text{ZKP}\{(r) : -B^{-1}C_2 = X^r\}$$

$$\pi_4 \in \text{ZKP}\{(r) : c_1 = g_1^r\}$$

- **Proof Process of π_1:**
  - The prover executes the following steps:
    * Select two random numbers $a_1, b_1 \in \mathbb{Z}_q^*$.
    * Select the challenge value $C_1 = H(\bar{y} \parallel m_i)$, where $\bar{y} = g_1^{a_1} \hat{\sigma}_{vi}^{-b_1}$, and m_i is the shared information.
    * Compute $y_1 = g_1^r \hat{\sigma}_{vi}^{-ID_{vi}}$, $\bar{a}_1 = a_1 + C_1 r$, $\bar{b}_1 = b_1 + C_1 ID_{vi}$.
    * Send $(y_1, C_1, \bar{a}_1, \bar{b}_1, m_i)$ to the verifier.
  - The verifier calculates: $y_1 = y_1^{-C_1} g_1^{\overline{a_1}} \hat{\sigma}_{vi}^{\overline{-b_1}} = g_1^{-C_1 r} \hat{\sigma}_{vi}^{-C_1 ID_{vi}} g_1^{a_1 + C_1 r} \hat{\sigma}_{vi}^{-b_1 - C_1 ID_{vi}} = g_1^{a_1} \hat{\sigma}_{vi}^{-b_1}$.
    Check whether $C_1 = H(\bar{y} \parallel m_i)$.
- **Proof Process of π_2:**
  - The prover executes the following steps:
    * Select a random number $a_2 \in \mathbb{Z}_q^*$.
    * Select the challenge value $C_2 = H(\overline{y_2})$, where $\overline{y_2} = g_1^{a_2}$.
    * Compute $\overline{a}_2 = a_2 + C_2 r$, $y_2 = y_1 \hat{\sigma}_{vi}^{-B}$.
    * Send $(y_2, C_2, \overline{a}_2)$ to the verifier.
  - The verifier calculates: $\overline{y_2} = g_1^{\overline{a_2}} y_2^{-C_2} = g_1^{a_2 + C_2 r} g_1^{-C_2 r} = g_1^{a_2}$.
    Check whether $C_2 = H(\overline{y_2})$.
- **Proof Process of π_3:**
  - The prover executes the following steps:
    * Select a random number $a_3 \in \mathbb{Z}_q^*$.
    * Select the challenge value $C_3 = H(\overline{y_3})$, where $\overline{y_3} = X^{a_3}$.
    * Compute $\overline{a}_3 = a_3 + C_3 r$, $y_3 = -B^{-1} C_3$.
    * Send $(y_3, C_3, \overline{a}_3)$ to the verifier.
  - The verifier calculates: $\overline{y_3} = X^{\overline{a_3}} y_3^{-C_3} = X^{a_3 + C_3 r} X^{-C_3 r} = X^{a_3}$
    Check whether $C_3 = H(\overline{y_3})$.
- **Proof Process of π_4:**
  - The prover executes the following steps:
    * Select a random number $a_4 \in \mathbb{Z}_q^*$.
    * Select the challenge value $C_4 = H(\overline{y_4})$, where $\overline{y_4} = g_1^{a_4}$.
    * Compute $\overline{a_4} = a_4 + C_4 r$, $y_4 = C_1$.
    * Send to the verifier $(y_4, C_4, \overline{a_4})$.
  - The verifier calculates: $\overline{y_4} = g_1^{\overline{a_4}} y_4^{-C_4} = g_1^{a_4 + C_4 r} g_1^{-C_4 r} = g_1^{a_4}$
    Check whether $C_4 = H(\overline{y_4})$.

If all verifications are successful, the message m_i is deemed authentic, and the information is accepted. Otherwise, the authentication fails.

4.4 Tracing Phase

If a vehicle sends malicious information, its real identity will be exposed by TA_i. The process is as follows:

Step 1: When the RSU detects malicious behavior of a vehicle, it sends the data packet (C_1, C_2) from the vehicle to the trusted authority.

Step 2: The trusted authority, upon receiving it, computes $\frac{C_1}{C_2^x} = ID_{vi}$. Here, x is the private key of the trusted authority. Then it searches the vehicle identity information database $\{ID_{vi}, PI\}$ to reveal the user's real identity.

5 Safety Analysis

5.1 Formal Analysis

Theorem 4.1. The zero-knowledge proof π satisfies completeness, reliability, and honest zero-knowledge.

Proof: The completeness can be directly verified through the above formula process. In the zero-knowledge proof π, the secret information that needs to be preserved is r. To prove its reliability, the extractor gets $(y_1, C_1, \overline{a_1}, \overline{b_1}, m_i)$ and $(y_1', C_1', \overline{a_1}', \overline{b_1}', m_i')$ in two sessions. By the formula $r = (C_1 - C_1')/(\overline{a_1} - \overline{a_1}')$, the secret information r is obtained. To prove its zero-knowledge property, assume there exists a simulator S, which generates parameters $(r', a_1', b_1' \in \mathbb{Z}_q^*)$, and computes the session as $(y_i', C_i', \overline{a_i}', \overline{b_i}', m_i')$. This is an acceptable conversation. It is indistinguishable from the parameters produced by the honest prover.

Theorem 4.2. If π is reliable, the scheme ensures the reliability of credential transmission.

Proof: The system publishes parameters $\{G_1, G_2, G_T, g_1, g_2, q, H, X\}$. The user selects a blinding factor $r \in \mathbb{Z}_q^*$, randomizes the credential, generates $\hat{\sigma}_{vi} = \sigma_{vi}^r$, and computes $A = g_1^r \hat{\sigma}_{vi}^{-ID_{vi}}$. Generate:

$$\pi' \in ZKP\{(ID_{vi}, r) : A = g_1^r \hat{\sigma}_{vi}^{-ID_{vi}}\}$$

The zero-knowledge property of ZKP ensures that no information about the credential σ_{vi} is revealed. Moreover, using the blinding factor to randomize the credential ensures its unlinkability, and attackers will not obtain any valuable information.

Theorem 4.3. If π is reliable, the scheme achieves attribute privacy.

Proof: The system publishes parameters $\{G_1, G_2, G_T, g_1, g_2, q, H, X\}$. The user selects $r \in \mathbb{Z}_q^*$, retains identity privacy information ID_{vi}, generates $y_1 = g_1^r \hat{\sigma}_{vi}^{-ID_{vi}}$, and let $A = y_1$. Generate:

$$\pi_1 \in ZKP\{(ID_{vi}, r) : A = g_1^r \hat{\sigma}_{vi}^{-ID_{vi}}\}$$

The zero-knowledge property of ZKP ensures that no information about identity privacy ID_{vi} is revealed. At the same time, the randomized credential $\hat{\sigma}_{vi} = \sigma_{vi}^r = g_1^{\frac{1}{x+ID_{vi}}}$ is uniformly random and does not leak any information.

Theorem 4.4. If π is reliable, the scheme achieves unlinkability.

Proof: The system publishes parameters $\{G_1, G_2, G_T, g_1, g_2, q, H, X\}$. The user selects $r \in \mathbb{Z}_q^*$, retains identity privacy information ID_{vi}, generates $(C_1, C_2) = (g_1^r, ID_{vi}X^r)$, $\hat{\sigma}_{vi} = \sigma_{vi}^r$. And generates zero-knowledge proof π:

$$\pi_1 \in ZKP\{(ID_{vi}, r) : A = g_1^r \hat{\sigma}_{vi}^{-ID_{vi}}\}$$
$$\pi_2 \in ZKP\{(r) : A = g_1^r \hat{\sigma}_{vi}^B\}$$
$$\pi_3 \in ZKP\{(r) : -B^{-1}C_2 = X^r\}$$
$$\pi_4 \in ZKP\{(r) : C_1 = g_1^r\}$$

In the data packet $(C_1, C_2, m_i, T_1, \hat{\sigma}_{vi}, \pi)$, the zero-knowledge property of ZKP ensures that no information about identity privacy ID_{vi} is revealed. At the same time, the randomized credential $\hat{\sigma}_{vi} = \sigma_{vi}^r = g_1^{\frac{1}{x+ID_{vi}}}$ is uniformly random and does not leak any information, thus ensuring the unlinkability of the messages sent by the vehicles.

Theorem 4.5. Assume A is an adversary attacking within polynomial time t, then the advantage of A in forging anonymous credentials within polynomial time is negligible.

Proof: The challenger C announces the system parameters to the adversary$A$: $\{G_1, G_2, G_T, g_1, g_2, q, H, X\}$. The proof consists of a series of games $G_i (1 \leq i \leq 5)$ to simulate the attack initiated by A, and $W_i (1 \leq i \leq 5)$ indicates A successfully guessing the random bit r in G_i.

G$_0$: This game simulates the original attack. According to the definition of semantic security [23], we can obtain: $Adv_P(A) = |2Pr[W_0] - 1|$.

G$_1$: This game simulates a passive attack. Adversary A eavesdrops on the public channel messages $(y_i, C_i, \overline{a_i}, \overline{b_i}, m_i)$. However, due to the existence of DDH, A's winning probability does not increase. Therefore, we get: $Pr[W_1] = Pr[W_0]$.

G$_2$: G_2 is the same as G_1, with only one difference. C runs a simulator S identical to Theorem 4.1, and replaces all zero-knowledge proofs $(\pi_1, \pi_2, \pi_3, \pi_4)$. The simulation is indistinguishable from the real one, and does not use real secret parameters. This setup is indistinguishable to A. If A can distinguish G_2 and G_1 with non-negligible advantage, then A can break the zero-knowledge proof. Therefore, the adversary's winning probability in G_2 satisfies the following equation: $|Pr[W_2] - Pr[W_1]| \leq \epsilon_{ZK}$.

G$_3$: G_3 is the same as G_2, with only one difference. Random numbers (r_1', r_2') replace $g_1^r \hat{\sigma}_{vi}^{-ID_{vi}}$. From the perspective of A, these two settings are indistinguishable. Specifically, C generates $y_1 = g_1^{r_1'} \hat{\sigma}_{vi}^{r_2'}$, $\overline{a_1} = a_1 + C_1 r_1'$, $\overline{b_1} = b_1 + C_1 r_2'$, and adversary computes $y_1 = y_1^{-C_1} g_1^{\overline{a_1}} \hat{\sigma}_{vi}^{-b_1}$ to get $g_1^{r_1} \sigma_{vi}^{-b_1}$. If A can distinguish G_3 and G_2 with non-negligible advantage, then A can break DDH. Therefore, the adversary's winning probability in G_2 satisfies the following equation: $|Pr[W_3] - Pr[W_2]| \leq \epsilon_{DDH}$.

5.2 Informal Analysis

Here, we provide a detailed analysis of the scheme to demonstrate its security features.

Identity Authentication

Vehicles use randomized anonymous credentials and zero-knowledge proofs to authenticate themselves. Receivers verify credential validity through bilinear pairings, ensuring only legitimate vehicles can participate.

Identity Privacy Protection

Vehicles communicate using anonymous credentials, allowing message verification without revealing real identities. Zero-knowledge proofs ensure ownership without exposing private information [21].

Unforgeability

When vehicles register, they send their unique identity identifier ID_{vi} to the trusted authority, which generates an anonymous credential $\sigma_{vi}^r = g_1^{\frac{1}{x+ID_{vi}}}$. Each time a vehicle sends information, it sends a randomized anonymous credential $\hat{\sigma}_{vi} = \sigma_{vi}^r$. If the credential is legitimate and valid, it must contain the vehicle's unique identity identifier ID_{vi}. Because the vehicle will perform zero-knowledge proofs for its unique identity identifier, if the vehicle's credential does not contain the identity identifier ID_{vi} or the identity identifier is incorrect, the verification will fail. Moreover, the vehicle also sends $(C_1, C_2) = (g_1^r, ID_{vi} X^r)$, which also contains the vehicle's identity identifier. If the vehicle fakes ID_{vi}, the trusted authority computes $\frac{C_1}{C_2^x} = ID_{vi}$ to obtain the vehicle's false identity information, thus locating the malicious vehicle. Therefore, our scheme achieves unforgeability [1,5].

Traceability

In our scheme, at least one TA can obtain the real identity to locate the malicious vehicle from (C_1, C_2). By calculating $\frac{C_1}{C_2^x} = ID_{vi}$ to obtain the vehicle's real identity information, and uses the vehicle identity information database $\{ID_{vi}, PI\}$, to get the user's identity information. Due to the difficulty problem of discrete logarithms, without the private key of the trusted institution, no polynomial-time adversary can decrypt the real identity of the malicious vehicle from (C_1, C_2). Therefore, our scheme can track malicious vehicles. [12,25]

Unlinkability

In this scheme, unlinkability is achieved by the randomized anonymous credential $\hat{\sigma}_{vi} = \sigma^{r}_{vi}$. When vehicles send information to different vehicles, the anonymous credentials generated in each communication are different. Moreover, when vehicles perform verification, the non-interactive zero-knowledge proof formulas $y_1 = g_1^{a_1}\hat{\sigma}_{vi}^{-b_1}, \overline{y_2} = g_1^{a_2}, \overline{y_3} = X^{a_3}, \overline{y_4} = g_1^{a_4}$ are also different. Because the random numbers $(a_1, b_1, a_2, a_3, a_4)$ generated each time are different. If an attacker intercepts multiple messages transmitted by vehicles $(C_1, C_2), m_i, T_1, \hat{\sigma}_{vi}, \pi)$, he cannot link this information together, thus each vehicle-generated data packet is new and different, and it is impossible to infer the user's real identity and intentions from the known data. Therefore, the scheme can achieve unlinkability.

Resistance to Impersonation Attacks

To impersonate a vehicle, an attacker must forge a valid anonymous credential, which requires the vehicle's unique identity ID and the trusted authority's private key. Since credentials are generated securely and verified through zero-knowledge proofs without revealing ID, no polynomial-time adversary can produce a valid credential. Therefore, the scheme is secure against impersonation attacks under simulation.

Resistance to Replay Attacks

During the authentication phase, the data packet sent by the vehicle includes the current timestamp T_1. The receiving vehicle, upon receiving the data, obtains the current timestamp T_2 and verifies whether the equation $|T_2 - T_1| < \Delta T$ holds, where ΔT is the maximum communication delay allowed by both parties. If the equation holds, further verification steps are performed. Therefore, the scheme can resist replay attacks.

Resistance to Physical Attacks

In this paper, a physical attack is defined as an attacker attempting to obtain the anonymous credential stored in the vehicle for their own benefit. In our scheme, the vehicle does not directly store the anonymous credential σ_{vi}, but instead calculates $Re = H(PUF_{vi}(C))$, and then computes $\sigma_{vi} = Re \oplus Si$ to obtain the credential. Due to the physical unclonable function's physical characteristics, even if an attacker obtains the vehicle's PUF, they cannot compute $Re = H(PUF_{vi}(C))$. Because when anyone tries to alter the behavior of the device, the PUF will become invalid. Therefore, the scheme can resist physical attacks.

Discussion on PUF Practicality. While PUFs provide strong tamper resistance, their deployment in real-world vehicular environments poses challenges. Environmental variations such as temperature and voltage may affect stability, and large-scale deployment raises cost and integration concerns. Recent studies [10,19] suggest employing error-correction codes and fuzzy extractors to enhance reliability. In practice, PUFs can be embedded into on-board modules as secure key generators, offering an additional hardware layer of security in VANETs.

6 Security Features Comparison

Table 2 shows the security feature comparison of this scheme with Feng et al. [7], Bao et al. [2], and Liu et al. [16].

Table 2. Security Features Comparison

Scheme	Auth.	Privacy	Unforge.	Trace.	Unlink.	Replay	Physical
Feng et al. [7]	✓	✓	–	✓	–	✓	–
Bao et al. [2]	✓	✓	–	✓	✓	✓	×
Liu et al. [16]	✓	✓	✓	✓	×	×	×
Zhou et al. [25]	✓	✓	✓	✓	✓	✓	–
Proposed	✓	✓	✓	✓	✓	✓	✓

Note: ✓/× respectively indicate whether the scheme can or cannot provide the corresponding security requirement, - indicates that the scheme does not have the corresponding security requirement.

7 Performance Evaluation

This section compares the performance of the proposed scheme with existing related schemes Feng et al. [7], Bao et al. [2], and Liu et al. [16] in terms of computational cost and communication cost. Since the initialization phase is not frequently used, this experiment focuses on the performance analysis of the registration and authentication phases.

The experiment was conducted using Python on a computer with the following configuration: CPU Intel(R) Core(TM) i7-8750H CPU@2.20 GHz, 16 GB RAM, running Windows 10 Professional. The experiment code was based on JDK1.8 and JPBC encryption library.

7.1 Computational Overhead

To evaluate the computational overhead in the process of credential generation and verification, the key operations involved include: exponentiation (T_e), multiplication (T_m), modular inversion (T_{mi}), point addition (T_a), homomorphic encryption multiplication (T_{mm}), homomorphic encryption exponentiation (T_{me}), bilinear pairing (T_{pt}), attribute encryption (T_{ae}), and attribute decryption

(T_{ad}). The time consumed by each operation is listed in Table 3. The experimental results are compared with the schemes of Feng et al. [7], Bao et al. [2], and Liu et al. [16]. In the vehicle information transmission process, the anonymous credential of the vehicle needs to be randomized first, then a zero-knowledge proof is constructed, and the relevant data is sent to the receiving vehicle for verification. To evaluate the performance of the scheme, this study mainly compares the computational overhead of the user credential generation and verification phases. Table 4 shows the comparison of this scheme with other schemes in terms of computational overhead. To simplify the evaluation process, this study ignores the computational time of hash operations and addition operations.

According to the experimental results, the computational overhead for generating an anonymous credential in our scheme is 0.000053 ms, while the computational overhead for the receiving vehicle to verify the credential is 0.000031 ms. In Feng et al.'s scheme, the computational overhead for generating an anonymous credential is $T_{ae} + 2T_m = 0.025586$ ms, and the computational overhead for the receiving vehicle to verify is $T_{ad} + 2T_m + T_{mi} = 0.027909$ ms. In Bao et al.'s scheme requires a computational overhead of $7T_m + 17T_a + 3T_{pt} = 0.031551$ ms for generating an anonymous credentials, and $T_{ad} + 2T_m + T_{mi} = 0.031551$ ms for vehicle verification. In Rui et al.'s scheme requires a computational overhead of $3E_p + 8T_{me} + 4T_{me} = 29.069808$ ms for generating anonymous credentials, and $3T_{me} + 3T_{me} = 12.416943$ ms for vehicle verification. Considering the computational overhead, our scheme outperforms other three schemes in terms of computational time for both in user credential generation and server verification. Our scheme has significant advantages in computational performance, especially in vehicular network environments where communication efficiency is essential.

Table 3. Time Overhead for Various Operations

Operation Type	Time Overhead (ms)
T_e	0.000003
T_m	0.000001
T_{mi}	0.000007
T_a	0.000532
T_{mm}	0.808408
T_{me}	3.330573
T_{pt}	0.007500
T_{ae}	0.025584
T_{ad}	0.027900

7.2 Communication Overhead

In this section, we compare the communication overhead of the proposed scheme with those of Feng et al. [7], Bao et al. [2], and Liu et al. [16]. The following

Table 4. Total Computational Overhead

Scheme	Create Credential	Verify Credential
Feng et al.	$T_{ae} + 2T_m$	$T_{ad} + 2T_m + T_{mi}$
Bao et al.	$7T_m + 17T_a + 3T_{pt}$	$7T_m + 17T_a + 3T_{pt}$
Liu et al.	$3T_{mm} + 8T_{me}$	$3T_{mm} + 3T_{me}$
Our Scheme	$11T_e + 6T_m + 2T_{mi}$	$9T_e + 4T_m$

notations are used: $\tilde{Z}_q$(96 bits),$\tilde{G_1}$ (192 bits), $\tilde{G_2}$ and $\tilde{G_T}$ (256 bits each), $\tilde{T}$ (8 bits), $\tilde{M}$ (160 bits), and $|ABE|$ (4096 bits). The proposed scheme requires 2568 bits per message, while Feng et al.'s scheme requires 5128 bits, Bao et al.'s scheme requires 3747 bits, and Liu et al.'s scheme requires 2984 bits. As shown in Table 5, the proposed scheme achieves the lowest communication overhead among the compared schemes. This reduction in data transmission enhances overall communication efficiency, making the scheme more suitable for resource-constrained vehicular environments (Fig. 2).

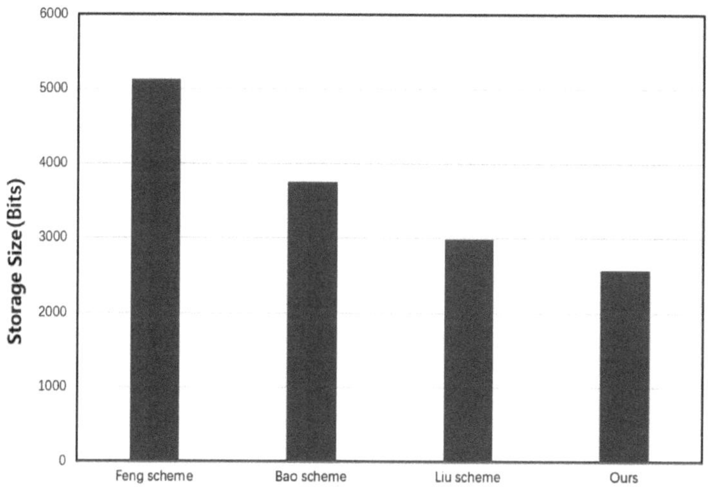

Fig. 2. Communication Overhead of Various Schemes.

Parameter Justification: The parameter settings follow the recommendations of elliptic curve cryptography. Specifically, $\mathbb{Z}_q = 96$ bits ensures hardness of the discrete logarithm problem, $G_1 = 192$ bits corresponds to a ~96-bit security level, and G_2, $G_T = 256$ bits provide pairing security comparable to AES-128. These settings balance efficiency and security for vehicular environments, aligning with NIST guidelines.

Table 5. Communication Overhead Comparison

Scheme	Communication Cost	Length (bits)
Feng et al.	$\tilde{G}_1 + \tilde{Z}_q + \lvert ABE\rvert + \tilde{T} + \tilde{M}$	5128
Bao et al.	$2\tilde{G}_1 + 9\tilde{Z}_q + 9\tilde{G}_T + \tilde{T} + \tilde{M}$	3747
Liu et al.	$8\tilde{G}_1 + 4\tilde{G}_2 + \tilde{G}_T + \tilde{T} + \tilde{M}$	2984
Proposed Scheme	$8\tilde{G}_1 + 9\tilde{Z}_q + \tilde{T} + \tilde{M}$	2568

8 Conclusion

This paper presents a conditional privacy protection authentication scheme based on anonymous credentials. By introducing anonymous credentials and ElGamal algorithms [6], the scheme achieves conditional anonymity in identity authentication, ensuring user anonymity while allowing the revelation of malicious vehicle identities through a trusted third party when necessary. Furthermore, zero-knowledge proof technology is incorporated to ensure unlinkability of vehicles in multiple interactions [3], thereby enhancing the system's privacy protection capabilities.

The security of the scheme is proven under the random oracle model, and it is shown to offer more security features compared to several related schemes. Through experimental analysis, it is demonstrated that the scheme exhibits better performance in terms of computational overhead and communication costs, demonstrating strong practical applicability and robustness [8,10].

References

1. Ali, I., Gervais, M., Ahene, E., et al.: A blockchain-based certificateless public key signature scheme for vehicle-to-infrastructure communication in vanets. J. Syst. Architect. **99**, 101636 (2019)
2. Bao, Z., He, D., Wang, H., et al.: Bap: a blockchain-assisted privacy-preserving authentication protocol with user-controlled data linkability for vanets. IEEE Trans. Intell. Veh. **9**(2), 4206–4220 (2023)
3. Blum, M., Feldman, P., Micali, S.: Non-interactive zero-knowledge and its applications. In: Proceedings of the 20th ACM Symposium on Theory of Computing, pp. 103–112 (1988)
4. Chen, X., Li, Y., Zhang, T.: Lightweight blockchain-assisted authentication for vanets. IEEE Internet Things J. (2024)
5. Cui, J., Zhang, J., Zhong, H., et al.: Spacf: a secure privacy-preserving authentication scheme for vanet with cuckoo filter. IEEE Trans. Veh. Technol. **66**(11), 10283–10295 (2017)
6. ElGamal, T.: A public key cryptosystem and a signature scheme based on discrete logarithms. IEEE Trans. Inf. Theory **31**(4), 469–472 (1985)
7. Feng, Q., He, D., Zeadally, S., et al.: Bpas: blockchain-assisted privacy-preserving authentication system for vehicular ad hoc networks. IEEE Trans. Industr. Inf. **16**(6), 4146–4155 (2019)

8. Fiat, A., Shamir, A.: How to prove yourself: practical solutions to identification and signature problems. In: Odlyzko, A.M. (eds.) Advances in Cryptology — CRYPTO' 86. CRYPTO 1986. LNCS, vol. 263, pp. 186–194. Springer, Berlin, Heidelberg (1987). https://doi.org/10.1007/3-540-47721-7_12
9. He, D., Zeadally, S., Xu, B., et al.: An efficient identity-based conditional privacy-preserving authentication scheme for vehicular ad hoc networks. IEEE Trans. Inf. Forensics Secur. **10**(12), 2681–2691 (2015)
10. Herder, C., Yu, M.D., Koushanfar, F., et al.: Physical unclonable functions and applications: a tutorial. Proc. IEEE **102**(8), 1126–1141 (2014)
11. Jan, S.A., Amin, N.U., Othman, M., et al.: A survey on privacy-preserving authentication schemes in vanets: attacks, challenges and open issues. IEEE Access **9**, 153701–153726 (2021)
12. Jiang, Y., Ge, S., Shen, X.: Aaas: an anonymous authentication scheme based on group signature in vanets. IEEE Access **8**, 98986–98998 (2020)
13. Johnson, D., Menezes, A., Vanstone, S.: The elliptic curve digital signature algorithm (ecdsa). Int. J. Inf. Secur. **1**, 36–63 (2001)
14. Koblitz, N., Menezes, A., Vanstone, S.: The state of elliptic curve cryptography. Des. Codes Crypt. **19**, 173–193 (2000)
15. Lin, C.H., He, D., Huang, X., et al.: Bcppa: a blockchain-based conditional privacy-preserving authentication protocol for vehicular ad hoc networks. IEEE Trans. Intell. Transp. Syst. **22**(12), 7408–7420 (2020)
16. Liu, J., Li, X., Jiang, Q.: Bua: a blockchain-based unlinkable authentication in vanets. In: IEEE ICC, pp. 1–6 (2020)
17. Raya, M., Hubaux, J.P.: Securing vehicular ad hoc networks. J. Comput. Secur. **15**(1), 39–68 (2007)
18. Sun, J., Zhang, C., Zhang, Y., et al.: An identity-based security system for user privacy in vehicular ad hoc networks. IEEE Trans. Parallel Distrib. Syst. **21**(9), 1227–1239 (2010)
19. Wang, L., Zhao, H., Liu, Q.: Puf-enhanced conditional privacy authentication in vehicular networks. IEEE Trans. Veh. Technol. (2025)
20. Wang, S., Yao, N.: Liap: a local identity-based anonymous message authentication protocol in vanets. Comput. Commun. **112**, 154–164 (2017)
21. Yang, Y.T., Chou, L.D., Tseng, C.W., et al.: Blockchain-based traffic event validation and trust verification for vanets. IEEE Access **7**, 30868–30877 (2019)
22. Yao, Y., Chang, X., Mišić, J., et al.: Bla: blockchain-assisted lightweight anonymous authentication for distributed vehicular fog services. IEEE Internet Things J. **6**(2), 3775–3784 (2019)
23. Zhang, C., Lin, X., Lu, R.: Raise: an efficient rsu-aided message authentication scheme in vehicular communication networks. In: IEEE ICC, pp. 1451–1457. (2008)
24. Zhang, L., Wu, Q., Domingo-Ferrer, J., et al.: Distributed aggregate privacy-preserving authentication in vanets. IEEE Trans. Intell. Transp. Syst. **18**(3), 516–526 (2016)
25. Zhou, X., He, D., Khan, M.K., et al.: An efficient blockchain-based conditional privacy-preserving authentication protocol for vanets. IEEE Trans. Veh. Technol. **72**(1), 81–92 (2022)

T-DAC: Time-Driven Access-Controlled Cryptography for Streaming Data

Qiang Tong(✉), Zhicheng Li, Ze Huang, and Jian Xu

School of Software, Northeastern University, Shenyang, China
tongq@swc.neu.edu.cn, 2410594@stu.neu.edu.cn , xuj@mail.neu.edu.cn

Abstract. The rapid growth of real-time streaming data in fields such as IoT, healthcare, and intelligent transportation poses significant challenges for data privacy and access control. Current encryption and authorization frameworks often depend on centralized architectures, exhibit high latency, and cannot satisfy the security requirements of resource-limited devices. We introduce time-driven access-controlled (T-DAC) cryptography for streaming data, a framework designed for secure, decentralized sharing of streaming data. T-DAC incorporates a time-indexed symmetric encryption scheme with differential subkeys, a collaborative access authorization protocol utilizing obfuscated token generation, and a stateless decryption approach that removes the need to store historical keys. The design ensures user-centric data sovereignty, low computational overhead, and robust resistance to collusion and chosen plaintext attacks. We formally define the system model, threat assumptions, and protocol algorithms, and offer rigorous security proofs under semi-honest adversary models. Experimental evaluations on Windows and Raspberry Pi platforms show that T-DAC achieves microsecond-level processing and significantly outperforms traditional schemes like Paillier, HEAC, and ASHE in efficiency and adaptability. Under a 64-bit security setting, T-DAC achieves encryption and decryption times of 0.34 µs and 0.32 µs on Windows 10, and 1.9 µs on Raspberry Pi 4B. Even in its 128-bit configuration, T-DAC remains efficient, incurring only a 6.67% time overhead compared to previous schemes. These findings indicate that T-DAC is well-suited for real-time applications demanding high-throughput and secure data access across heterogeneous platforms.

Keywords: Streaming data encryption · Access control · Symmetric cryptography · Privacy-preserving

1 Introduction

The convergence of Internet of Things (IoT), cloud computing, and 5G technologies has led to the explosive growth of real-time stream data [1–3], widely distributed across domains such as industrial control systems, urban sensing networks, remote healthcare, and intelligent transportation [4–6]. Unlike traditional

J. Yu and Y. Chen (Eds.): FCS 2025, CCIS 2805, pp. 370–384, 2027.
https://doi.org/10.1007/978-981-95-9514-3_25

static data, stream data is characterized by high volume, strong temporal continuity, and heterogeneity. Over 83% of edge devices are constrained by limited computational power (typically below 1 TOPS) and storage capacity (averaging less than 128GB), pushing the computation and storage workloads to centralized cloud platforms. While this offloading alleviates resource bottlenecks, it has resulted in a structural shift in data control. Third-party service providers now dominate data lifecycle management, which severely weakens user autonomy over data usage, visibility, and disposal, and significantly increases the risks of data leakage.

Recent reports highlight an alarming rise in global data breaches [7,8]. From 2022 to 2024, confirmed data privacy incidents surged from 5,212 to 10,626, more than doubling in two years. Industrial sectors are disproportionately affected. Ransomware in manufacturing has become one of the most severe threats. IBM's Data Breach Reports from 2022 to 2024 show that the healthcare sector has suffered the highest cost per breach for consecutive years, with nearly 40% of leak cases involving multi-platform data spanning across cloud and edge. The protection of stream data, which contains sensitive signals related to physical infrastructure, human health, and national security, has thus become a problem of both privacy preservation and public interest.

1.1 Motivation

Despite the growing demand for secure stream data analytics, existing systems face several critical limitations [9,10]:

Cloud-Centric Security Risks. Current stream processing pipelines heavily rely on cloud-based data custodianship. Users must trust third-party platforms for secure storage and policy enforcement. However, 65% of critical data breaches in recent years were caused by insider misuse or certificate authority compromise, and user data often deviates from its declared access intent even under secure transport.

Lack of User-Centric Access Control. Although regulations such as the Personal Information Protection Law explicitly grant users control over their data, many service providers enforce unilateral policies that force users into privacy-for-service tradeoffs. Around 80% of platforms adopt privacy-hostile default configurations, and nearly 30% of mobile applications engage in covert data harvesting. On average, each app collects 7.5 types of user data beyond necessity.

Centralized Authorization Bottlenecks. Centralized access control infrastructures are vulnerable to single points of failure. In 2024, a major identity management outage caused over 1,200 enterprises to suffer a 7-hour service disruption. Moreover, metadata such as access time and IP addresses can enable statistical inference attacks, which may infer user activity attributes with up to 76.8% accuracy.

These issues indicate the need for a new model that ensures data privacy, user sovereignty, and resilience in the face of adversarial compromise and platform failure, especially for large-scale, continuous stream data environments.

1.2 Contribution

To address the above challenges, we propose T-DAC, a cryptographic framework that provides secure and fine-grained access control over encrypted stream data in an end-to-end and decentralized manner. Our contributions are summarized as follows:

1. **Time-indexed encryption with differential keys.** We design a novel stream encryption mechanism based on evolving subkeys derived from a compact master key. Each data chunk is encrypted with a time-differential key and embedded with subkey transitions to enable later verification or decryption.
2. **Decentralized token-based access authorization.** We introduce a distributed token aggregation protocol in which a group of Access Policy Controllers (APCs) collaboratively produce an obfuscated transformation token using local subkey transitions and canceling noise. This ensures that no single party can reconstruct or forge access capability.
3. **Stateless decryption with zero key exposure.** Our decryption algorithm allows Data Consumers to recover plaintext using only the ciphertext and a global token. It does not require storing or reconstructing any historical keys, which achieves full statelessness and minimal trust assumptions.

1.3 Related Work

The exponential growth of real-time data streams driven by the convergence of IoT, cloud computing, and 5G technologies [11–13] has led to an urgent demand for privacy-preserving access control schemes. Although numerous institutions and scholars have proposed privacy-preserving mechanisms to mitigate data exposure risks, many of these solutions still face critical limitations in real-world deployments. Most existing methods [14–16] struggle to handle the temporal correlation and continuous nature of streaming data. Attackers can infer sensitive patterns by analyzing structural relationships in streaming flows. As the data accumulates across stages, the risk of privacy leakage becomes compounded and long-term. In response, many works have focused on developing stream-specific encryption and access control techniques to meet privacy protection needs.

Several efforts have applied dynamic noise injection to mitigate privacy inference risks. For example, Ce et al. [17] proposed a differential privacy-based end-to-end streaming encryption mechanism that adds calibrated noise to encrypted data, aiming to improve privacy without degrading utility. However, their approach assumed that the original data are non-sensitive, limiting its applicability in sensitive scenarios. Dan et al. [18] introduced the PrivStream framework to enhance edge-side privacy resistance by integrating training-stage noise fusion. Nevertheless, their design did not account for dynamic stream features and real-time responsiveness, restricting practical deployment.

Guo et al. [19] proposed TABE-DAC, a solution that compresses and encrypts data segments during transmission while optimizing decryption and access performance. Despite improved efficiency, the scheme depends on trusted third-party services, leaving residual risks. Cui et al. [20] developed a domain-specific secure medical access control model combining attribute-based encryption and policy matching, but its complexity hinders deployment across heterogeneous systems.

Xin et al. [21] proposed a hybrid enforcement mechanism combining function-based access and blockchain-enhanced control to address the issues of centralized risk and insufficient privacy guarantees. Their solution enables decentralized access under heterogeneous environments. However, performance evaluation revealed scalability limitations when tested in realistic deployment settings. Zhang et al. [22] proposed a dynamic revocation-aware encryption tree model, but it suffered from high network latency and communication failures during revocation events.

Other works [23–25] emphasized hierarchical control and adaptive noise regulation. However, these solutions often required centralized key orchestration, posing single points of failure and limiting scalability. Most recently, Yin et al. [26] examined access privilege delegation in distributed environments but relied on centralized authority for key distribution, failing to realize full decentralization.

2 Preliminaries

2.1 Additive Homomorphic Symmetric Encryption

Assumes the data stream is a sequence of events $\{e_1, e_2, \ldots, e_i, e_{i+1}, \ldots\}$, where each event $e_i = \{t_i, m_i\}$ contains a timestamp t_i and a message m_i. A pseudorandom function (PRF) f_k keyed by k is used to generate a pseudorandom keystream value based on t_i. The encryption process is defined as:

$$\mathrm{Enc}(k, t_{i-1}, e_i) = (t_i, t_{i-1}, m_i + k_i - k_{i-1} \mod M) \tag{1}$$

where

$$k_i = f_k(t_i), \quad k_{i-1} = f_k(t_{i-1}) \tag{2}$$

To decrypt the aggregated sum over a time window $[t_i, t_j]$, only $k_i = f_k(t_i)$ and $k_j = f_k(t_j)$ are required:

$$\sum_{i=0}^{n} m_i = \mathrm{Dec}_{k_i,k_j}\left(\sum_{i=0}^{n} c_i\right) = \sum_{i=0}^{n} c_i - k_j + k_i \mod M \tag{3}$$

2.2 Random Graph

The Erdős–Rényi random graph model [27], introduced by Paul Erdő s and Alfréd Rényi in 1959, defines two canonical random graph constructions: $G(n,p)$ and $G(n,M)$. In $G(n,p)$, each of the $\binom{n}{2}$ possible edges is included independently with probability p; in $G(n,M)$, exactly M edges are randomly chosen among n nodes. This work adopts an optimized $G(n,p)$ model, where the edge probability p is selected to maintain stable probabilistic connectivity. When $n \to \infty$ and $p \geq (\ln n)/n$, the graph becomes connected with high probability and provides the following advantages:

(1) **Inherent Resilience:** Random connectivity reduces vulnerability to node exposure attacks and improves system robustness;
(2) **Degree Controllability:** The average degree $k = p(n-1)$ is tunable, enabling balanced communication loads across nodes;

Combining the generated communication topology with the encryption mechanism ensures secure transmission while minimizing overall communication overhead.

2.3 Secure Multi-party Computation

Traditional computation models require participants to submit private data to a centralized entity, posing privacy risks. Secure Multi-Party Computation (SMPC) enables multiple parties to jointly compute a function while keeping their inputs private. The protocol leverages cryptographic primitives such as homomorphic encryption and zero-knowledge proofs to ensure confidentiality during computation. n parties $P_1, P_2, \ldots, P_n$ hold private inputs $T_1, T_2, \ldots, T_n$ and collaboratively compute the aggregation function $f(T_1, \ldots, T_n) = \sum_{i=1}^{n} T_i$ via the following steps:

1. All parties jointly generate random values $\{r_i\}_{i=1}^{n}$ such that $\sum_{i=1}^{n} r_i = 0$;
2. Each party computes its masked value $V_i = T_i + r_i$ and sends it to the designated aggregator C via a secure channel;
3. The aggregator computes:

$$V = \sum_{i=1}^{n} V_i = \sum_{i=1}^{n} (T_i + r_i) = \sum_{i=1}^{n} T_i \tag{4}$$

This protocol ensures the correctness of the final result while preserving the privacy of each participant's input.

3 System Model

3.1 System Entities

As illustrated in Fig. 1, we propose a three-entity system model for secure, time-indexed data sharing with access control. The system architecture uses Kafka, a high-throughput distributed publish-subscribe messaging system. The system involves the following roles: the **Data Owner (DO)**, who encrypts sequential data using time-derived differential keys; the **Access Policy Controllers (APCs)**, a distributed set of key holders that jointly authorize access by contributing masked subkey transitions; and the **Data Consumer (DC)**, who decrypts the data using only a compact transformation token aggregated from the APCs.

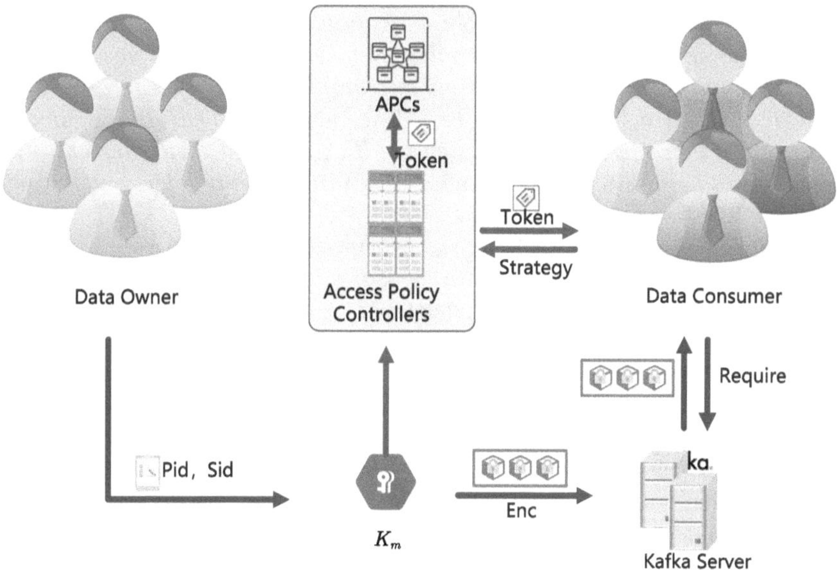

Fig. 1. System Model.

1. Data Owner (DO): The Data Owner is responsible for generating the root master key and encrypting the time-indexed data stream. The DO segments the stream into fixed intervals and derives subkeys k_i via $\text{GenKey}(K_m, t_i)$, then computes the differential key $K_i = k_i - k_{i-1}$ (with a virtual k_0 for the first chunk). Each plaintext m_i is encrypted as $c_i = \text{Enc}_{K_i}(m_i)$, and encoded as $e_i = c_i + (-k_i + k_{i-1}) \bmod M$. The ciphertexts $\{e_i\}$ are uploaded to the Data Server, while the DO retains control of the master key K_m.
2. Multi-Access Policy Controllers (APC): A group of n controllers jointly enforce access policies. Each APC p holds its own subkey path $\{k_i^{(p)}\}$ and computes a local token $\tau_p = -k_i^{(p)} + k_{i-1}^{(p)} \bmod M$, along with a canceling noise term k_p derived from randomized pairwise values $\{k_{p,q}\}$. These are summed to produce a global token $\mathsf{Token}_{\mathsf{MPC}} = \sum_{p=1}^{n}(\tau_p + k_p) \bmod M = -k_i + k_{i-1} \bmod M$. The use of canceling noise ensures unlinkability and distributed control, preventing any single APC from authorizing access alone.
3. Data Consumer (DC): The Data Consumer is an authorized party that receives encrypted chunks $\{e_i\}$ and the aggregated token $\mathsf{Token}_{\mathsf{MPC}}$. It performs decryption by computing $K_i = -\mathsf{Token}_{\mathsf{MPC}} \bmod M$, recovering $c_i = e_i - \mathsf{Token}_{\mathsf{MPC}} \bmod M$, and finally obtaining the plaintext as $m_i = \text{Dec}_{K_i}(c_i)$. The DC does not require access to any subkeys or master secrets, ensuring stateless and privacy-preserving access.

Definition 1. *The* T-DAC *is a tuple of six algorithms defined as follows:*

GenMainkey$(\mathsf{Pid}, \mathsf{Sid}) \rightarrow K_m$. *The main key generation algorithm is run by the data owner. It takes as input a policy identifier* Pid *and a stream identifier* Sid, *and outputs the master key* K_m.

GenKey(K_m, t_i) $\rightarrow k_i$. *The subkey derivation algorithm is run by either the data owner or an access policy controller. It takes as input the master key* K_m *and a timestamp* t_i*, and outputs a time-synchronized subkey* k_i*.*

Enc($K_m, \{t_i\}, \{m_i\}, interval$) $\rightarrow \{e_i\}$. *The encryption algorithm is run by the data owner. It takes as input the master key* K_m*, a sequence of timestamps* $\{t_i\}$*, the corresponding messages* $\{m_i\}$*, and a time interval parameter, and outputs a sequence of encoded ciphertexts* $\{e_i\}$*.*

GenCancelGraph(n, p, b) $\rightarrow \{k_{p,q}\}$. *The canceling graph generation algorithm is jointly run by all access policy controllers. It takes as input the number of controllers* n*, the edge probability parameter* $p = 2^{-b}$*, and outputs pairwise canceling values* $\{k_{p,q}\}$ *based on a randomized graph.*

TransformToken($\{\tau_p\}, \{k_{p,q}\}$) $\rightarrow$ Token. *The transformation token generation algorithm is jointly executed by the access policy controllers. Each controller computes a local token* τ_p *and a canceling term* k_p*, and the algorithm outputs the global transformation token* Token.

Dec(e_i, Token) $\rightarrow m_i$. *The decryption algorithm is run by the data consumer. It takes as input an encoded ciphertext* e_i *and the transformation token* Token*, and outputs the decrypted plaintext* m_i *without requiring knowledge of any subkeys.*

3.2 Threat Model

We assume that the DO (Data Owner), DC (Data Consumer), and APCs (Access Policy Controllers) are fully trusted and execute the protocol as specified. All communication between these entities is protected by secure channels to ensure confidentiality and integrity. Other participants, including external observers or corrupted components, are modeled as *semi-honest* but *curious* adversaries. They follow the protocol but attempt to infer sensitive data from observable information. The adversary $\mathcal{A}$ is defined as a probabilistic polynomial-time (PPT) algorithm with the following capabilities:

Definition 2. *Let* $\tau = \{\mathsf{DO}, \mathsf{DC}, \mathsf{APC}_1, \ldots, \mathsf{APC}_n\}$ *be the set of trusted protocol-compliant entities. Let* ρ *denote the set of external or adversarial parties. An adversary* $\mathcal{A}$ *satisfies:*

- ***Partial Corruption.*** $\mathcal{A}$ *may statically corrupt up to* $t < n/2$ *of the APCs, denoted* $\mathsf{Corrupt}(\mathcal{A}) \subset \mathsf{APC}$*, and obtains their internal state (including* k_p*,* τ_p*).*
- ***Public View Access.*** $\mathcal{A}$ *can observe public ciphertexts* $\{e_i\}$ *and the global token* $\mathsf{Token}_{\mathsf{MPC}}$ *transmitted to the data consumer.*
- ***Key Hiding.*** $\mathcal{A}$ *has no access to the master key* K_m *or to the subkey paths* $\{k_i^{(p)}\}$ *of any honest* APC.
- ***PRF Security.*** *The random noise values* k_p *are generated via a secure pseudorandom function and are computationally indistinguishable from uniform random values under PPT constraints.*

- ***Token Collusion Bound.*** *$\mathcal{A}$ cannot forge a valid token unless it obtains correct $\tau_p + k_p$ contributions from a threshold of APCs. In our setting, the threshold is $n/2$, ensuring collusion resistance.*

4 T-DAC's Detailed Design

The construction of our scheme is detailed as follows:

GenMainkey$(Pid, Sid) \leftarrow K_m$. First, both the strategy identifier (Pid) and the stream identifier (Sid) are independently generated using a pseudo-random function. Each falls within the range $[0, 2^{n-1} - 1]$, where n denotes the desired key length in bits. Given Pid and Sid, the key derivation process computes a 128-bit intermediate secret, K_{128}, by applying a key generator function to the concatenation $Pid \| Sid$. Specifically, $K_{128} \leftarrow \text{KeyGenerator}(Pid \| Sid)$, where the KeyGenerator is a pseudorandom key derivation function (KDF) that securely expands the input pair into a 128-bit secret. Finally, the main key K_m is derived by hashing the concatenation of Pid, Sid, and K_{128} using the SHA-256 function:

$$K_m \leftarrow \text{SHA}_{256}(Pid \| Sid \| K_{128}), \tag{5}$$

The resulting hash is truncated to 16 bytes to produce a fixed-length symmetric key for downstream encryption.

GenKey$(K_m, t_i) \rightarrow k_i$. Given the main key K_m and timestamp t_i, the subkey generation algorithm first computes a nonce by concatenating the two values, i.e., $t_{\text{nonce}} \leftarrow t_i + K_m$. This is followed by an AES-based expansion to produce a 16-byte block, denoted as t_{AES}, such that $t_{\text{AES}}.\text{length} = 16$ bytes. This step increases the entropy space to 2^{128}. Next, the algorithm computes a SHA-256 hash of the nonce $t_{\text{hash}} \leftarrow \text{SHA}_{256}(t_{\text{nonce}})$. Finally, the subkey k_i is derived by applying a bitwise XOR to the left and right halves of the hash output:

$$k_i \leftarrow \text{Key\_XOR}(t_{\text{hash}}.R_{16},\ t_{\text{hash}}.L_{16}), \tag{6}$$

where $t_{\text{hash}}.R_{16}$ and $t_{\text{hash}}.L_{16}$ represent the rightmost and leftmost 16 bytes of the digest, respectively. This construction yields a time-indexed subkey k_i, which is suitable for streamwise encryption or token generation at time t_i.

Enc$(K_m, \{t_i\}, \{m_i\}, \text{interval}) \rightarrow \{e_i\}$. Given the main key K_m derived from GenMainkey(), the encryption process applies a time-synchronized subkey derivation scheme based on a pre-defined interval parameter interval. For each timestamp t_i, the corresponding subkey k_i is computed via GenKey(K_m, t_i). In the case of the first data block, where no previous subkey k_{i-1} exists, the algorithm sets a virtual reference time $t_0 = t_1 - \text{interval}$ to derive k_0. Once subkeys k_i and k_{i-1} are available, the scheme computes a differential key $K_i = k_i - k_{i-1}$. The original plaintext message m_i is then encrypted using a symmetric encryption function with K_i as the key as $c_i = \text{Enc}_{K_i}(m_i)$, where $\text{Enc}_{K_i}(\cdot)$ denotes a secure block cipher such as AES. To further enhance modularity and support token-based decryption, the scheme encodes c_i into a final ciphertext e_i via a key-difference-based transformation:

$$e_i = c_i + (-k_i + k_{i-1}) \bmod M, \tag{7}$$

where M is a public modulus that bounds the ciphertext domain. This encoding embeds the subkey transition into the ciphertext and supports efficient authorization.

GenCancelGraph$(n, p, b) \rightarrow \{k_{p,q}\}$. Given n access policy controllers (APCs), the randomized pairwise canceling values $\{k_{p,q}\}$ are generated based on the Erdős-R 'enyi random graph model $G(n,p)$, where $p = 2^{-b}$ determines the probability of each potential edge. To build the canceling structure, the algorithm proceeds as follows:

1. Divide the PRF output into b-bit segments to derive up to 2^b candidate random graphs.
2. For each segment, use it as seed to construct a random graph via $G(n,p)$.
3. Each edge (u,v) in a selected graph is assigned a random canceling value $k_{u,v}$.
4. The values are distributed such that node u adds $-k_{u,v}$ and node v adds $+k_{u,v}$.
5. For final canceling consistency, ensure $\sum_p k_p = 0$ by satisfying:

$$k_p = \sum_{p<q}(-k_{p,q}) + \sum_{p>q} k_{q,p}, \quad \sum_p k_p \equiv 0 \mod M. \tag{8}$$

This strategy allows each node to locally compute canceling noise with minimal overhead, while ensuring global cancellation and anonymized coordination across all participants.

TransformToken$_{\text{MPC}}(\{\tau_p\}, \{k_{p,q}\}) \rightarrow \mathsf{Token}_{\mathsf{MPC}}$. Let $k_i^{(p)}$ denote the subkey of participant p at time index i. Each participant p computes a local transformation token as $\tau_p = -k_i^{(p)} + k_{i-1}^{(p)} \mod M$, which captures the difference in subkeys between two adjacent time slots and reflects access control over a specific temporal window. The participant then derives a canceling term k_p from the pairwise values $\{k_{p,q}\}$ generated via a randomized graph-based mechanism. The final combined token is constructed as:

$$\mathsf{Token}_{\mathsf{MPC}} = \sum_{p \in P}(\tau_p + k_p) \mod M = -k_i + k_{i-1} \mod M, \tag{9}$$

where $k_p = \sum_{p>q} k_{p,q} - \sum_{p<q} k_{p,q}$, ensuring that $\sum_p k_p = 0$ holds modulo M.

Dec$(e_i, \mathsf{Token}_{\mathsf{MPC}}) \rightarrow m_i$. Given the encoded ciphertext e_i and a global transformation token $\mathsf{Token}_{\mathsf{MPC}}$, the decryption algorithm directly recovers the original plaintext message m_i without requiring access to intermediate subkeys. To decrypt, the algorithm first computes the differential encryption key K_i as the negation of the token $K_i \leftarrow -\mathsf{Token}_{\mathsf{MPC}} \mod M$. Next, it removes the embedded transformation from the ciphertext $c_i \leftarrow e_i - \mathsf{Token}_{\mathsf{MPC}} \mod M$. Finally, the plaintext message is recovered by applying a symmetric decryption algorithm, such as AES, with K_i as the decryption key $m_i \leftarrow \text{Dec}_{K_i}(c_i)$. This revised procedure eliminates the need for explicit subkey reconstruction and enables stateless decryption based solely on the received token.

5 Security Analysis

5.1 Resistance to Chosen Plaintext Attack

Theorem 1. *Under the assumption that the masking value k_p is drawn from a secure pseudorandom function (PRF), and the adversary cannot access the secret subkey τ_i, the obfuscated token $\tilde{\tau}_i = \tau_i + k_p \bmod M$ is indistinguishable from a uniform random value. Thus, the scheme satisfies IND-CPA security.*

Proof. We define the following CPA security game G_{CPA} between a challenger $\mathcal{C}$ and an adversary $\mathcal{A}$:

Initialization Phase. $\mathcal{C}$ selects n APCs and, for each one, independently samples a true token component τ_i and a random value $k_{p_i} \leftarrow \mathsf{PRF}(s_i)$.
Obfuscation Phase. The challenger computes the masked token $\tilde{\tau}_i = \tau_i + k_{p_i} \bmod M$ for each APC.
Challenge Phase. The adversary $\mathcal{A}$ selects an index $j \in [1, n]$ and receives $\tilde{\tau}_j$ from $\mathcal{C}$.
Guess Phase. $\mathcal{A}$ outputs a guess, attempting to determine whether $\tilde{\tau}_j$ corresponds to τ_j or a random value (i.e., distinguishes it from uniform).
Conclusion. Since k_{p_j} is pseudorandom and independent of τ_j, and the sum modulo M preserves uniformity over a hidden term, the masked value $\tilde{\tau}_j$ is computationally indistinguishable from random. Therefore:

$$\Pr[\mathcal{A} \text{ wins }] \leq \frac{1}{2} + \epsilon(\lambda) \tag{10}$$

where $\epsilon(\lambda)$ is a negligible function in the security parameter λ.

5.2 Collusion Resistance

Theorem 2. *Let $\mathcal{A}$ statically corrupt a subset of access policy controllers $\mathcal{V} \subset \{\mathsf{APC}_1, \ldots, \mathsf{APC}_n\}$ such that $|\mathcal{V}| < n/2$. Then, under the assumption that each k_p is independently pseudorandom and the canceling condition $\sum_p k_p \equiv 0 \bmod M$ holds, $\mathcal{A}$ cannot recover the remaining τ_q for $q \notin \mathcal{V}$ nor forge a valid transformation token* $\mathsf{Token}_{\mathsf{MPC}}$.

Proof. We define the collusion-resistance game G_{COL} between a challenger $\mathcal{C}$ and an adversary $\mathcal{A}$:

Setup Phase. $\mathcal{C}$ initializes n APCs and randomly selects a collusion set $\mathcal{V} \subset [n]$ of size $t < n/2$. For each APC p_i, it generates:

$$\text{Local Token: } \tau_i, \quad \text{Noise: } k_i \leftarrow \mathsf{PRF}(s_i) \tag{11}$$

ensuring $\sum_{i=1}^{n} k_i \equiv 0 \mod M$.

Observation Phase. $\mathcal{A}$ learns τ_i and k_i for all $i \in \mathcal{V}$, and observes the final transformation token:

$$\mathsf{Token}_{\mathsf{MPC}} = \sum_{i=1}^{n} (\tau_i + k_i) \bmod M \tag{12}$$

Goal of $\mathcal{A}$: Use the known shares to infer any τ_q for $q \notin \mathcal{V}$, or forge a valid token $\mathsf{Token}_{\mathsf{MPC}}'$ for an unauthorized time index.
Security Argument: At least $n - t > n/2$ values of $(\tau_i + k_i)$ remain hidden. Since each k_i is pseudorandom and τ_i is never exposed, the residual entropy prevents $\mathcal{A}$ from computing any missing τ_q. It is not difficult to find:

$$\Pr[\mathcal{A} \text{ wins}] \leq \text{negl}(\lambda) \tag{13}$$

6 Evaluation

The experimental environment is deployed on a Windows 10 system configured with 16 GB of RAM and an 11th Gen Intel(R) Core(TM) i7-11800 processor running at 2.30 GHz, or alternatively on a Raspberry Pi 4B configured with 32-bit architecture, 4 GB of RAM, and an ARM Cortex-A72 processor running at 1.5 GHz. The development is carried out using the IntelliJ IDEA integrated development environment (IDE), and the Java Development Kit (JDK) version 17 is used as the compilation tool.

6.1 Stream Data Processing

In the stream data processing phase, this section conducts experiments on the generation time of the master key, the generation time of the subkey, and the encryption time. The encryption data are randomly selected integers in the range of 0 to 10^{10}, covering almost all possible values encountered in stream data. The experiment is conducted on a Windows 10 platform using 64-bit keys. As shown in Fig. 2, the results indicate that the average encryption time for a single record is approximately 0.34μs, the average generation time of a master key per stream record is about 191μs, and the generation time of a subkey per data unit (1 s) is about 16μs.

Performance comparison experiments based on 64-bit keys were conducted on different computing platforms. On the Raspberry Pi 4B platform, the encryption time for a single stream of data is approximately 1.89 μs, the master key generation time is about 1012.4 μs, and the subkey generation time for data units of 1 s is around 544.7 μs. Compared to the Windows 10 platform, the Raspberry Pi 4B has a 400% increase in encryption and master key generation processing time, and a 3000% increase in subkey generation time. However, the total processing time for 1 s of data, which is the sum of the subkey generation time and encryption time, can still be controlled within 0.8 ms, fully meeting the real-time processing requirements, as shown in Fig. 3.

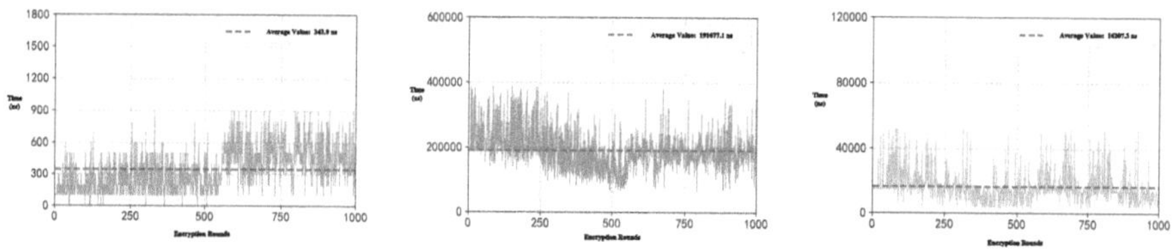

(a) Encryption Time (b) MasterKey Generation (c) Subey Generation

Fig. 2. Average Cost Time for Key Generation and Encryption.

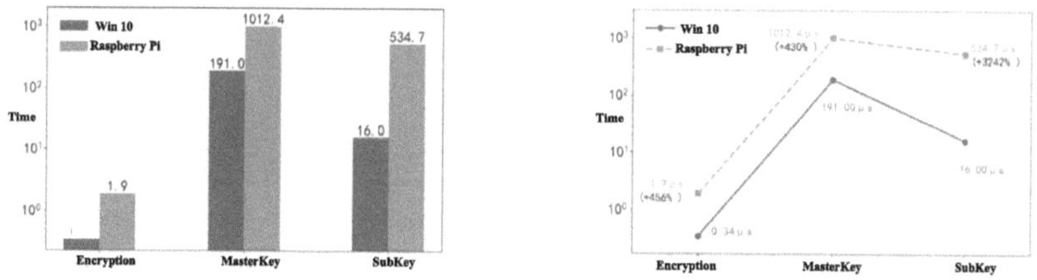

(a) Encryption Time (b) Encryption Efficiency

Fig. 3. Comparison of Encryption on Different Platforms.

6.2 Comparison of Encryption Methods

This paper presents the encryption schemes proposed in T-DAC and compares them with some existing encryption methods, including the ASHE scheme proposed by Papadimitriou et al. [28], the HEAC scheme proposed by Lukas Burkhalter et al. [29], as well as the classical Paillier encryption. Experimental results show that the decryption times for these methods vary, as shown in Table 1.

Table 1. Data setting and time consumption for different encryption methods

Method	Security Level	Raspberry Pi		Win 10	
		Enc	Dec	Enc	Dec
Paillier	80-bit	1.59 s	1.62 s	30 ms	15 ms
HEAC	80-bit	1.08 ms	1.08ms	5.1 μs	5.1 μs
ASHE	80-bit	0.3 ms	0.3 ms	1.5 μs	1.5 μs
T-DAC	64-bit	1.9 μs	1.9 μs	0.34 μs	0.32 μs
T-DAC	128-bit	0.28 ms	0.28 ms	0.84 μs	0.80 μs

The experimental results show that under the 80-bit security level, Paillier encryption on the resource-constrained Raspberry Pi 4B takes an average of

1.59 s for encryption and 1.62 s for decryption. HEAC demonstrates performance improvement over Paillier, with an average of 1.03 milliseconds on Raspberry Pi and 5.15 μs on Windows 10. ASHE further outperforms HEAC, achieving average encryption and decryption times of 0.3 ms on Raspberry Pi and 1.5 ms on Windows 10. Compared with ASHE, the 128-bit version of DESDC-ACM reaches 0.28 ms for both encryption and decryption on Raspberry Pi, showing a negligible performance difference—only about a 6.67% improvement in average time efficiency. On Windows 10, it achieves 0.84 microseconds for encryption and 0.80 microseconds for decryption, representing an average efficiency improvement of approximately 46.67%.

7 Conclusion

In this paper, we propose T-DAC, a novel cryptographic framework designed for secure, fine-grained, and decentralized access control over time-indexed streaming data. Through the integration of time-differential subkey encryption, collaborative token-based authorization, and stateless decryption, T-DAC addresses critical security challenges in real-time data environments, especially those involving resource-constrained devices. Experimental results demonstrate that T-DAC achieves microsecond-level encryption and decryption latency, significantly outperforming existing schemes such as Paillier, HEAC, and ASHE, while maintaining strong security guarantees. Our findings highlight the practicality of T-DAC in real-time stream processing scenarios.

Acknowledgement. This research was funded in part by the National Natural Science Foundation of China grant 62372096.

References

1. Zhong, Z., Yan, S., Li, Z., Tan, D., Yang, T., Cui, B.: Burstsketch: finding bursts in data streams. In: Proceedings of the 2021 International Conference on Management of Data, SIGMOD '21, pp. 2375–2383. New York, NY, USA (2021). Association for Computing Machinery
2. Nguyen, T.T., et al.: Detecting rumours with latency guarantees using massive streaming data. VLDB J. **32**(2), 369–387 (2022)
3. Patidar, N., et al.: Transparency in AI decision making: a survey of explainable AI methods and applications. Adv. Robot. Technol. **2**(1) (2024)
4. Sun, D., Xue, S., Huifeng, W., Jia, W.: A data stream cleaning system using edge intelligence for smart city industrial environments. IEEE Trans. Industr. Inf. **18**(2), 1165–1174 (2022)
5. Imtiaz, S., et al.:. Privacy Preserving Time-Series Forecasting of User Health Data Streams . In: 2020 IEEE International Conference on Big Data (Big Data), pp. 3428–3437, Los Alamitos, CA, USA. IEEE Computer Society (2020)

6. Jolfaei, A., Kant, K., Shafei, H.: Secure data streaming to untrusted road side units in intelligent transportation system. In: Proceedings 2019 18th IEEE International Conference on Trust, Security and Privacy in Computing and Communications/13th IEEE International Conference on Big Data Science and Engineering, IEEE Trustcom BigDataSE ISPA, pp. 793–798, United States (2019). Institute of Electrical and Electronics Engineers (IEEE). 18th IEEE International Conference on Trust, Security and Privacy in Computing and Communications, TrustCom 2019 ; Conference date: 05-08-2019 Through 08-08-2019
7. Hakim, A.R., Ramli, K., Gunawan, T.S., Windarta, S.: A novel digital forensic framework for data breach investigation. IEEE Access **11**, 42644–42659 (2023)
8. Sandeep, V., Sahoo, J., Bhambu, P.: Defending machine learning in agriculture against data breaches and misuse. In: 2024 15th International Conference on Computing Communication and Networking Technologies (ICCCNT), pp. 1–6 (2024)
9. Jang, J., Im, Y., Jung, J.H., Park: An effective handling of secure data stream in IoT. Appl. Soft Comput. **68**, 811–820 (2018)
10. Havet, A., et al.: Securestreams: A reactive middleware framework for secure data stream processing. In: Proceedings of the 11th ACM International Conference on Distributed and Event-Based Systems, pp. 124–133 (2017)
11. Hemalatha, S, Alzubaidi, L.H., Sundar, R., Priya, S., Gajbhiye, P., Sheela, M.S.: A development of 5g technology in cloud computing and its optimization technique. In: 2024 4th International Conference on Advance Computing and Innovative Technologies in Engineering (ICACITE), pp. 372–377. IEEE (2024)
12. Vu Khanh, Q., Hoai, N.V., Manh, L.D., Le, A.N., Jeon, G.: Wireless communication technologies for IoT in 5G: Vision, applications, and challenges. Wirel. Commun. Mob. Comput. **2022**(1), 3229294 (2022)
13. Rahimi, H., Zibaeenejad, A., Safavi, A.A.: A novel IoT architecture based on 5G-IoT and next generation technologies. In: 2018 IEEE 9th Annual Information Technology, Electronics and Mobile Communication Conference (IEMCON), pp. 81–88. IEEE (2018)
14. Guo, T., Sathe, S., Aberer, K.: Fast distributed correlation discovery over streaming time-series data. In: Proceedings of the 24th ACM International on Conference on Information and Knowledge Management, pp. 1161–1170 (2015)
15. Bermudez-Edo, M., Barnaghi, P., Moessner, K.: Analysing real world data streams with spatio-temporal correlations: Entropy vs. pearson correlation. Autom. Constr. **88**, 87–100 (2018)
16. Dao, M.-S., Zettsu, K., Pongpaichet, S., Jalali, L., Jain, R.: Exploring spatio-temporal-theme correlation between physical and social streaming data for event detection and pattern interpretation from heterogeneous sensors. In: 2015 IEEE International Conference on Big Data (Big Data), pp. 2690–2699. IEEE (2015)
17. Feng, C., Venkitasubramaniam, P.: Inferential separation for privacy: irrelevant statistics and quantization. IEEE Trans. Inf. Forensics Secur. **17**, 2241–2255 (2022)
18. Wang, D., Ren, J., Wang, Z., Zhang, Y., Shen, X.: Privstream: a privacy-preserving inference framework on IoT streaming data at the edge. Inf. Fusion **80**, 282–294 (2022)
19. Guo, L., Yang, X., Yau, W.-C.: Tabe-dac: Efficient traceable attribute-based encryption scheme with dynamic access control based on blockchain. IEEE Access **9**, 8479–8490 (2021)
20. Cui, J., Duan, L., Ni, W., Li, C.: Secure cross-domain medical data sharing based on distributed cloud and blockchain services. In: 2023 IEEE International Conference on Web Services (ICWS), pp. 698–700 (2023)

21. Wang, X., Jiang, P., Baker, T., Li, T., Zhu, L.: Enabling privacy and leakage resistance for dynamic blockchain-based access control systems. Comput. Stan. Interfaces **84** (2023). Publisher Copyright: 2022 Elsevier B.V
22. Zhang, Q., Chang, X., Zhong, H., Chengjie, G., Cui, J.: Revocable and efficient blockchain-based fine-grained access control against edos attacks in cloud storage. IEEE Trans. Comput. **73**(8), 2012–2024 (2024)
23. Russo Russo, G., Cardellini, V., Presti, F.L.: Hierarchical auto-scaling policies for data stream processing on heterogeneous resources. ACM Trans. Auton. Adapt. Syst. **18**(4), 1–44 (2023)
24. Cernuda, J., Ye, J., Kougkas, A., Sun, X.-H.: Hstream: a hierarchical data streaming engine for high-throughput scientific applications. In: Proceedings of the 53rd International Conference on Parallel Processing, pp. 231–240 (2024)
25. Ding, G., Wang, Y., Li, C., Sun, H., Li, C., Wang, L., Yin, H., Huang, T.: Hscfc: High-dimensional streaming data clustering algorithm based on feedback control system. Futur. Gener. Comput. Syst. **146**, 156–165 (2023)
26. Yin, H., Zhu, Y., Hua Deng, L.O., Qin, Z., Li, K.: Privacy-preservation enhanced and efficient attribute-based access control for smart health in cloud-assisted internet of things. IEEE Internet Things J. **12**(1), 894–903 (2025)
27. Shahraeini, M.: Modified erdős-rényi random graph model for generating synthetic power grids. IEEE Syst. J. **18**(1), 96–107 (2023)
28. Papadimitriou, A., et al.: Big data analytics over encrypted datasets with seabed. In: 12th USENIX Symposium on Operating Systems Design and Implementation (OSDI 16), pp. 587–602, Savannah, GA. USENIX Association (2016)
29. Burkhalter, L., Hithnawi, A., Viand, A., Shafagh, H., Ratnasamy, S.: TimeCrypt: encrypted data stream processing at scale with cryptographic access control. In: 17th USENIX Symposium on Networked Systems Design and Implementation (NSDI 20), pp. 835–850, Santa Clara, CA. USENIX Association (2020)

Privacy-Protected Classification of Sensitive Personal Information

Yancheng Dong(✉), Jian Yang, Song Jiang, and Pei Li

Civil Aviation Chengdu Electronic Technology Co., Ltd., Chengdu, China
{dongyancheng,yangjian,jiangsong,lipei}@caacetc.com

Abstract. Sensitive personal information, such as identity numbers, names, and health or financial records, must be classified securely and intelligently in the digital age. This paper addresses the challenges of large-scale data classification, particularly when performed by commercial cloud platforms, which may expose data to privacy breaches. We propose a novel method for classifying sensitive information based on homomorphic encryption, which protects data privacy while preserving classification accuracy and efficienc. Our method stands out by balancing the trade-off between privacy and computational feasibility, a crucial breakthrough in the field. We compare our method with existing ones, showing the advantages and disadvantages of different approaches, such as those based on differential privacy and secure multi-party computation. Our experiments on diverse domains demonstrates our method's effectiveness, achieving 90% accuracy in financial fraud detection and 82% in demographic data classification directly over encrypted data. Moreover, performance optimizations reduce computation time by 30% without compromising accuracy. This confirms the scheme's practical viability in cloud computing environments. While accuracy may trail some plaintext or hybrid baselines, the method is preferable for settings that demand end-to-end encrypted inference and model confidentiality, with no trusted third parties.

Keywords: Privacy preservation · Intelligent classification · Homomorphic encryption · Convolutional neural networks

1 Introduction

Handling and classifying sensitive personal information, such as identity numbers and financial data, poses a significant challenge to privacy protection in digital systems [1]. Intelligent classification, particularly through deep learning methodologies, has become a key tool in processing such data. However, the extensive computational and storage resources required for large-scale data classification have necessitated outsourcing data to cloud computing platforms. Yet, outsourcing to public commercial cloud platforms introduces new challenges and risks [2]. The potential exploitation and analysis of personal data by these platforms for commercial gains could lead to breaches in personal data privacy. Therefore,

J. Yu and Y. Chen (Eds.): FCS 2025, CCIS 2805, pp. 385–406, 2027.
https://doi.org/10.1007/978-981-95-9514-3_26

exploring and researching privacy protection methods in the intelligent classification of sensitive personal information is crucial.

While existing privacy protection schemes for intelligent classification, such as those based on differential privacy [3,4] and secure multi-party computation [5–7], have made significant strides, they often come with trade-offs like reduced data utility or limitations in computational efficiency. Recently, methods based on homomorphic encryption [8] have garnered attention for their ability to perform computations on encrypted data, maintaining data privacy. However, deep learning methods based on homomorphic encryption face technical challenges and limitations, such as low computational efficiency and difficulties in handling large-scale data.

This paper makes the following primary contributions:

- We propose a privacy-preserving convolutional neural network (CNN) classification method using the CKKS homomorphic encryption scheme that ensures both data privacy and model confidentiality against the cloud server.
- We adapt neural network components, such as the ReLU activation function, for encrypted computation and implement optimizations that reduce computation time by 30%.
- We conduct extensive experiments on diverse sensitive datasets, demonstrating a multi-objective balance among accuracy, computational cost, and two-sided confidentiality (data and model).

In the subsequent sections, we will explore the relevant background knowledge, system models, threat models, as well as our core and extended schemes. We will also provide a security analysis and performance evaluation of our approach, and conclude with a summary of our research findings and future research directions.

2 Problem Statement

2.1 System Model

The proposed system model, focusing on the intelligent classification of personal sensitive information over ciphertexts, includes three parties, the data provider (DP), the cloud server (CS), and the user (U), as illustrated in Fig. 1.

- Data provider (DP). A data provider is an individual or organization that provides sensitive personal information to be classified. It is responsible for encrypting the raw data into ciphertext and then uploading it to a cloud server for classification.
- Cloud server (CS). A cloud server is a service provider that offers computing and storage resources. It is responsible for receiving the encrypted data uploaded by the data provider and using a well-trained deep-learning model for classification. Encryption prevents the cloud server from accessing the plaintext data, thus ensuring its privacy. Especially, the deep learning model

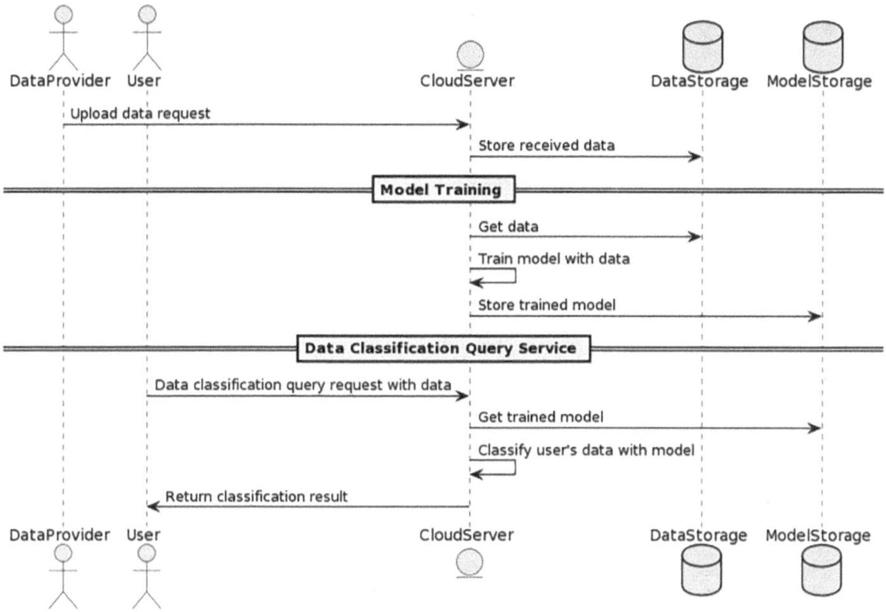

Fig. 1. System model

is the core component of the system used for data classification. It is a CNN-based model that can be trained and classified on encrypted data. The training and optimization of the model are done on the cloud server. The cloud server further contains two components, the data storage and the model storage.

- User (U). A user enters sensitive personal data to be classified and receives data classification results.

In this system, we operate under the premise that DP is both trustworthy and committed to the protection of data confidentiality and security. CS is considered a semi-honest entity. It adheres to the system model but may attempt to scrutinize or monitor transmitted and stored data due to potential commercial incentives. U is typically a benign client who submits sensitive data to be classified and receives the results. However, our threat model also considers the scenario where a U aims to illicitly obtain the proprietary CNN model through interactions with the CS. We assume that communication channels between all parties are secure. Consequently, our system model prioritizes the privacy of data and models, while also acknowledging and addressing the risks posed by external threats to privacy, integrity, and availability, including but not limited to safeguarding against side-channel attacks, model theft, and reverse engineering endeavors. This approach ensures a comprehensive and realistic security strategy for the entire system.

2.2 Design Goal

The system is required to complete the below design goals.

- Accuracy: High classification accuracy comparable to plaintext models. The performance must be sufficient to demonstrate the practical viability of the proposed method for sensitive tasks such as financial fraud detection and demographic analysis.
- Performance: Manageable computational overhead. The computational cost inherent to homomorphic encryption must be kept within a feasible range for high-performance computing environments, and the system must be amenable to performance optimizations that demonstrably reduce computation time without compromising accuracy.
- Data privacy: Robust confidentiality of sensitive personal information. Data must remain encrypted throughout its transmission and processing within the entire system.
- Model confidentiality: Strong protection of the model's intellectual property. Both model parameters and structure must be encrypted to prevent unauthorized access, theft, or reverse engineering.

3 Preliminaries

3.1 Taylor Series Approximation of the ReLU Function

In neural network processing under homomorphic encryption, a key challenge is managing the nonlinear nature of activation functions like Rectified Linear Unit (ReLU). The ReLU function, fundamental in neural networks and defined as $f(x) = \max(0, x)$, is known for its simplicity and efficiency. To adapt it for homomorphic encryption, which primarily supports addition and multiplication operations, we use a Taylor series approximation.

3.2 CKKS Homomorphic Encryption

Cheon *et al.* [9] introduced CKKS to perform homomorphic operations on encrypted approximate real and complex numbers, making it suitable for various applications involving encrypted data in the cloud. We give a brief description of CKKS below, and more intricate details are referred to [9]. The notations used in our description of the scheme are summarized in Table 1.

1. Parameter generation:
   - Choose n as a power of 2.
   - Select a distribution χ for secret key sampling.
   - Choose modulus chain: A decreasing sequence of moduli $q_0, q_1, \ldots, q_L$.
2. Key generation:
   - Secret key: Sample s from χ and set $sk = (1, s)$.
   - Public key: Sample a uniformly from R_{q_0} and e from a noise distribution. Compute $b = -(a \cdot s + e) \pmod{q_0}$. The public key is $pk = (a, b)$.

Table 1. Notations used and their meaning

Notation	Meaning
R	Polynomial ring $\mathbb{Z}[X]/(X^n+1)$
R_q	Residue ring R/qR, where q is a modulus
t	A small modulus, usually a prime close to 2^{32} or 2^{64}
χ	A noise distribution for secret key sampling

- Relinearization key: For $i = 1$ to L, sample two polynomials e_1, e_2 from a noise distribution and compute $u = s^2 + e_1$ and $v = -(a \cdot u + e_2) \pmod{q_i}$. The relinearization key for q_i is $rlk_i = (u, v)$.

3. Encryption:
   - For $m \in R_t$ and public key $pk = (a, b)$, sample s' from χ and two polynomials e_1, e_2 from a noise distribution. The ciphertext is $c = (c_0, c_1)$, where:

$$c_0 \equiv m + t\,e_1 + t\,s'\,b \pmod{q_0}, \tag{1}$$

$$c_1 \equiv t\,e_2 + t\,s'\,a \pmod{q_0}. \tag{2}$$

4. Decryption:
   - For a ciphertext $c = (c_0, c_1)$ and secret key $sk = (1, s)$, compute:

$$m' \equiv c_0 + c_1\,s \pmod{q_0}. \tag{3}$$

   The message m can be obtained from m' by rounding coefficients to the nearest multiple of q_0/t and then reducing modulo t.

5. Homomorphic operations:
   - Addition: Direct component-wise addition in R_q.
   - Multiplication: Multiply component-wise and use the relinearization key to reduce the size of the ciphertext.
   - Rescaling: Reduce the modulus of a ciphertext from q_i to q_{i+1} by dividing coefficients by q_i/q_{i+1} and rounding.

Note that in the context of the CKKS homomorphic encryption scheme (and many other lattice-based cryptographic schemes), the distribution χ is typically a noise distribution from which secret keys and error terms are sampled. The specific properties of χ ensure that the cryptographic scheme retains its security guarantees. Common choices for χ in lattice-based cryptography include:

- Binary distribution: Each coefficient is chosen independently from $\{0, 1\}$ or $\{-1, 0, 1\}$.
- Ternary distribution: Each coefficient is chosen independently from $\{-1, 0, 1\}$.
- Gaussian distribution: Coefficients are chosen according to a discrete Gaussian distribution centered around zero.

The choice of χ and its parameters often impacts both the security and efficiency of the cryptographic scheme.

4 Proposed Basic Scheme

4.1 Overview

This scheme involves several main stages: data preprocessing, encrypted convolution operations, encryption of model parameters, and the adaptation of loss functions and optimizers for an encrypted domain. Initially, raw data is cleansed, transformed, and augmented to prepare it for model training. Subsequently, the preprocessed data and convolution kernels are encrypted. This ensures that all subsequent operations are performed in the encrypted domain.

We perform convolution operations and adapt activation functions on the ciphertext, facilitating complex data analysis and model training. Concurrently, model parameters, including weights and biases, are encrypted prior to training, safeguarding the security of data and parameters throughout the training process. Ultimately, loss functions and optimizers are specifically chosen and adapted to suit the encrypted environment, ensuring the protection of data privacy during computational processes.

The core of this scheme is its ability to perform complex machine learning training directly on encrypted data, thus guaranteeing both data privacy and model confidentiality against the untrusted participants. While this approach introduces computational overhead compared to plaintext training, it provides a robust and secure solution for processing sensitive information.

4.2 Detailed Description

This comprehensive scheme incorporates crucial stages tailored for the intelligent classification of sensitive personal information, encompassing data preprocessing, encrypted convolution operations, encryption of model parameters, and the adaptation of loss functions and optimizers in an encrypted environment. Detailed formal descriptions of each phase are as follows:

1) Data preprocessing: In the data preprocessing stage, raw data—including personal identifiers, health records, financial information, and other sensitive details— undergoes cleansing, transformation, and augmentation. This process is crucial for preparing the data specifically for sensitive information classification, ensuring the integrity and relevance of the data are maintained.

- Data cleaning: For each data point x_i in the dataset, imputation is performed in cases of missing values using strategies such as mean, median, mode, or a specific value. Outliers are addressed through removal or correction.
- Data Transformation: The dataset undergoes standardization $x_i' = \frac{x_i - \mu}{\sigma}$ and normalization $x_i' = \frac{x_i - \min(X)}{\max(X) - \min(X)}$, where X represents the feature column, with μ and σ denoting the mean and standard deviation, respectively.
- Data augmentation: Each image I in the dataset is transformed through rotations by various angles (e.g., $\pm 15°$, $\pm 30°$) and flipping operations, both horizontally and vertically.

2) Encrypted convolution operations: Subsequently, both the preprocessed data and convolution kernels are encrypted. This step is pivotal in preserving the confidentiality of sensitive personal data during processing. Encrypted convolution operations enable the extraction of nuanced features from sensitive datasets, essential for accurate classification, while ensuring data privacy.

- Encryption of Data: Post-preprocessing, both the data $I_{i,j}$ and convolution kernels $K_{m,n}$ are encrypted using an encryption function Enc, yielding $E_{i,j} = \text{Enc}(I_{i,j})$ and $E^K_{m,n} = \text{Enc}(K_{m,n})$.
- Performing Encrypted Convolution: For each output location (p, q), convolution is computed as $C_{p,q} = \sum_{i=0}^{H-1} \sum_{j=0}^{W-1} E_{i,j} \cdot E^K_{m,n}$, where H and W represent the height and width of the convolution kernel, respectively.

3) Encryption of model parameters: Model parameters, including weights and biases, are encrypted before training. This encryption fortifies the security of both the data and the parameters during the training phase, an essential factor when dealing with personal sensitive information.

- Parameter Encryption: Prior to training, model parameters such as weights W and biases b are encrypted, resulting in $W_{\text{enc}} = \text{Encrypt}(W)$ and $b_{\text{enc}} = \text{Encrypt}(b)$.
- Utilization of Encrypted Parameters: Encrypted parameters and input data are used to compute the encrypted output $\text{Output}_{\text{enc}} = \text{Conv}(\text{Input}_{\text{enc}}, W_{\text{enc}}) + b_{\text{enc}}$ in forward propagation.

4) Adaptation of loss functions and optimizers: Loss functions and optimizers are meticulously selected and tailored to suit the encrypted environment, optimizing the classification process of sensitive personal data. This customization ensures that data privacy remains uncompromised during computational operations, an indispensable aspect of working with sensitive information.

- Loss Function Adaptation: Loss functions suitable for computation in the encrypted domain, such as Mean Squared Error (MSE), are chosen. The MSE is formalized as $L(\theta) = \frac{1}{N} \sum_{i=1}^{N} (y_i - f(x_i; \theta))^2$, where y_i represents the encrypted true labels, and $f(x_i; \theta)$ denotes the model predictions.
- Optimizer Adaptation: Gradient-based optimizers such as SGD are adapted for use, with the update rule in the encrypted domain expressed as $\theta_{t+1} = \theta_t - \alpha \nabla L(\theta_t)$.

The core advantage of this scheme lies in its dual capability to safeguard data security and privacy while facilitating the execution of complex machine learning algorithms on encrypted data. This approach is specifically beneficial for the intelligent classification of sensitive personal information, where data privacy is paramount. Despite its increased computational complexity and time consumption, this method offers a highly effective and secure solution for handling and classifying sensitive data.

4.3 Noise Control

In the domain of CKKS homomorphic encryption, particularly in its application to CNNs for secure and efficient computation, the concept of a noise budget assumes a paramount role. This subsection delineates the multifaceted process of noise budget determination and management, within the framework of CNNs.

1) Noise accumulation model: We begin by establishing a noise accumulation model in CKKS encryption, which can be expressed as:

$$\text{Noise}(x, n) = \sigma^2 \, n \, x. \tag{4}$$

where x is the encrypted data, n represents the number of operations, and σ^2 is the noise variance.

2) ModReduction technique: For the ModReduction technique, the noise reduction can be formalized as:

$$\text{ModReduction}(x) = x - \left\lfloor \frac{x}{q} \right\rfloor q. \tag{5}$$

where q is the modulus of the ciphertext, and $\lfloor \cdot \rfloor$ represents the floor function.

3) Parameter optimization: The parameter optimization is crucial for controlling the noise. This can be represented by adjusting the polynomial modulus N and scale Δ, which are crucial for controlling the noise growth:

$$\text{OptimizedNoise}(x) = \sigma^2 \, \frac{n}{N\Delta} \, x. \tag{6}$$

4) Activation function approximation: For activation function approximation, such as ReLU, we use a fourth-order Taylor series approximation at a point a, expressed as:

$$f(x) \approx f(a) + f'(a)(x-a) + \frac{f''(a)}{2!}(x-a)^2 + \frac{f'''(a)}{3!}(x-a)^3 + \frac{f^{(4)}(a)}{4!}(x-a)^4. \tag{7}$$

This approximation transforms ReLU's nonlinear characteristics into a polynomial form, facilitating computations on encrypted data while preserving privacy.

It is important to note that the selection of a fourth-order approximation is a strategic decision. This choice strikes a balance between computational efficiency and accuracy. Higher-order approximations increase accuracy but also computational complexity. This trade-off is critical in scenarios like encrypted data processing or real-time neural network computations, where both precision and speed are essential. Thus, the fourth-order approximation is selected for its ability to capture ReLU's key features with manageable complexity, balancing the need for accurate modeling against computational constraints.

In an illustrative application of the aforementioned noise control framework, consider a prototypical CNN comprising a singular convolution layer followed by a ReLU activation function, operating within a CKKS homomorphic encryption context. For this scenario, we set a noise variance σ^2 at 0.01. Assume the network undergoing a sequence of 5 operations, denoted as $n = 5$. In our optimized parameterization, the polynomial modulus N is set to 8192, and the scaling factor Δ is adjusted to 40. Correspondingly, the initial noise impact can be formulated as $0.01 \times 5 \times x$. By deploying the ModReduction technique with a modulus q of 786433, the noise influence is mitigated. Subsequently, we employ a fourth-order Taylor series approximation of the ReLU function, represented as $\text{ReLU}_{\text{approx}}(x) = a_0 + a_1x + a_2x^2 + a_3x^3 + a_4x^4$. The coefficients a_0, a_1, a_2, a_3, and a_4 are determined by fitting the ReLU function and its derivatives around $x = 0$. For instance, the coefficients might be selected as $a_0 = 0, a_1 = 0.5, a_2 = 0.25, a_3 = 0, a_4 = -0.05$, based on the behavior of the ReLU function near the origin. This example illustrates the application of our noise control framework to maintain computational accuracy in a CNN under CKKS homomorphic encryption, ensuring both functional integrity and computational accuracy.

Encrypting both inputs and model parameters does not preclude convergence. The noise-management pipeline above (bounded multiplicative depth, relinearization/rescaling, and modulus reduction), together with conservative learning-rate scheduling and gradient clipping, keeps the encrypted gradient signal within the valid precision window of CKKS across iterations. In practice, this preserves the descent direction of updates, so parameters move toward a stationary point as in plaintext training, while maintaining data and model confidentiality.

5 Evaluation

To validate the effectiveness and efficiency of our proposed homomorphic encryption-based method for the intelligent classification of sensitive personal information, we conducted extensive experiments. This section details the experimental setup, datasets used, evaluation metrics, comprehensive results with analysis, and comparisons with existing methods.

5.1 Experimental Setup

Our experiments were conducted on a high-performance computing platform to handle the computational demands of homomorphic encryption and deep learning operations. The detailed hardware and software configurations of our experimental environment are summarized in Table 2.

5.2 Datasets

We evaluated our method on several datasets containing sensitive personal information. Due to privacy concerns, actual sensitive data cannot be publicly used; thus, we utilized the following publicly available datasets, which serve as proxies for real-world sensitive data:

Table 2. Experimental Environment Configuration

Category	Details
Processor	Model: Intel Xeon E5-2698 v4 Core Count: 20 cores Thread Count: 40 threads Base Frequency: 2.20 GHz Maximum Frequency: 3.60 GHz
Memory	Capacity: 256 GB DDR4 Frequency: 2400 MHz
GPU	Model: NVIDIA Tesla V100 CUDA Cores: 5120 Memory: 32 GB HBM2
Storage	Type: SSD Capacity: 2 TB
Network	Bandwidth: 10 Gbps
Operating System	Version: Ubuntu 18.04 LTS
Software	Programming Language: Python 3.8 HE Library: Microsoft SEAL DL Framework: TensorFlow 2.4

1. **Adult Dataset (UCI Machine Learning Repository)** [10]: Contains demographic information used for income classification. Features include age, workclass, education, marital status, occupation, etc. The dataset comprises 48842 instances.
2. **MIMIC-III Clinical Database** [11]: Comprises de-identified health-related data from hospital patients. Features include vital signs, lab results, medications, diagnoses, etc. It includes data from over 40000 patients with 60000 hospital admissions.
3. **Credit Card Fraud Detection Dataset (Kaggle)** [12]: Includes transaction data for detecting fraudulent activities. Features encompass transaction amount, time, and anonymized features V1–V28. The dataset contains 284807 transactions with 492 fraud cases.
4. **Synthetic Personal Dataset (Generated using Synthea)** [13]: Artificially created dataset simulating personal health records. Features include patient demographics, medical histories, treatments, etc. For experiments, we generated 10000 synthetic patient records.

5.3 Model Architecture

We designed a CNN tailored for encrypted data processing. The model architecture included:

- Input Layer: Accepts encrypted feature vectors.
- Convolutional Layers: Two layers with encrypted convolution operations to extract features.
- Activation Function: Approximated ReLU using a fourth-order Taylor series, suitable for encrypted computations.
- Fully Connected Layers: Two layers for classification tasks.

For encryption adaptations, all weights and biases were encrypted. Activation functions and pooling operations were adapted to function in the encrypted domain. Figure 2 illustrates the model architecture.

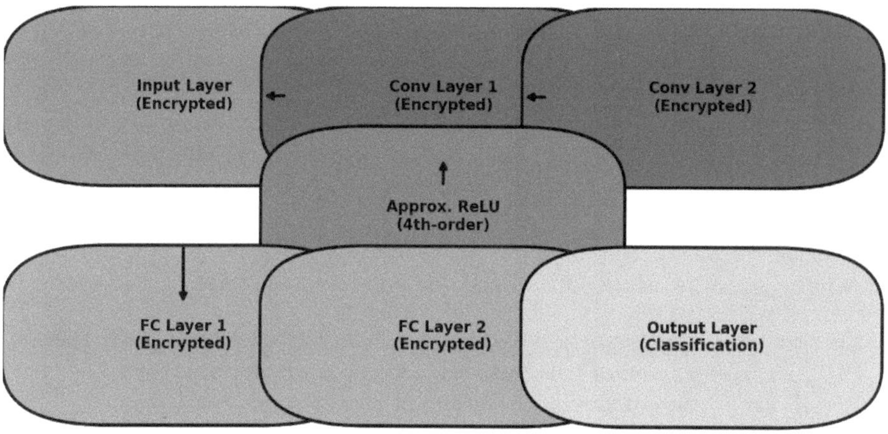

Fig. 2. Model Architecture Diagram

5.4 Data Preprocessing

During the data preprocessing phase, raw data—including personal identifiers, health records, financial information, and other sensitive details—undergoes a series of enhancements, such as cleansing, transformation, and augmentation. This process is essential for preparing the data specifically for the classification of sensitive information, ensuring that both integrity and relevance are maintained throughout. Data cleaning involves addressing missing values and outliers, while normalization is performed by scaling features to a standard range, thereby facilitating effective model training. Prior to inputting into the model, all data features are encrypted using the CKKS homomorphic encryption scheme, thereby ensuring data privacy throughout the computation process.

5.5 CKKS Parameter Selection

We selected CKKS parameters including polynomial degree $n = 16384$, scaling factor $\Delta = 2^{40}$, and a modulus chain supporting 30 levels. These parameters

Table 3. CKKS Parameter Selection

Parameter	Value
Polynomial Degree (n)	16384
Scaling Factor (Δ)	2^{40}
Ciphertext Modulus (q)	Modulus chain supporting 30 levels
Security Level	128-bit
Noise Distribution (χ)	Centered discrete Gaussian

ensure high security while facilitating efficient computations on encrypted data. The ciphertext modulus q was set as a modulus chain supporting 30 levels to accommodate the depth of the neural network. A scaling factor $\Delta = 2^{40}$ was used to maintain precision during computations. The security level was set to 128-bit, aligning with industry standards. The noise distribution χ was chosen as a centered discrete Gaussian. Table 3 lists the CKKS parameters used.

5.6 Evaluation Metrics

To comprehensively assess our method, we employed the following metrics:

- Accuracy (ACC): Proportion of correct predictions over total predictions.
- Precision: Proportion of true positives among predicted positives.
- Recall: Proportion of true positives among actual positives.
- F1-Score: Harmonic mean of precision and recall.
- Processing Time: Time taken for encryption, computation, and decryption.
- Communication Overhead: Data size transmitted between the client and server.

Taken together, these metrics provide a multi-objective evaluation. Under the required security constraints, they capture both model utility and computational efficiency, allowing us to highlight the overall balance achieved by our approach rather than focusing on a single performance number.

5.7 Experimental Results

1) Adult Dataset Results: The model achieved a training accuracy of 85% and a testing accuracy of 82% on the Adult dataset. Precision was 80%, recall was 78%, and the F1-score was 79%. The encryption time was 2.5 h, computation time was 5 h, and decryption time was 0.5 h. Communication overhead amounted to 1.2 GB. Figure 3 shows the Receiver Operating Characteristic (ROC) curve for this classification task. This curve visualizes the trade-off between the true positive rate and the false positive rate. Our model achieved an Area Under the Curve (AUC) of 0.82, indicating strong classification performance.

The model achieved comparable accuracy to plaintext models, demonstrating the practicality of homomorphic encryption in real-world applications. The slight

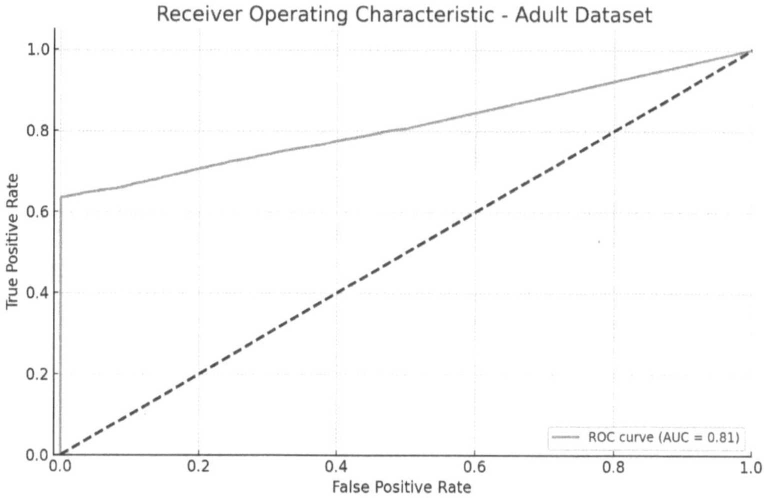

Fig. 3. ROC Curve for the Adult Dataset

decrease in performance is due to the approximation of nonlinear functions and noise accumulation in encrypted computations.

2) MIMIC-III Dataset Results: For the MIMIC-III dataset, the model achieved a training accuracy of 78% and a testing accuracy of 75%. Precision was 74%, recall was 72%, and the F1-score was 73%. Encryption time was 3 h, computation time was 6 h, and decryption time was 0.6 h. Communication overhead was 1.5 GB. Figure 4 presents the confusion matrix for the MIMIC-III dataset.

Despite the complexity and high dimensionality of medical data, the model demonstrated reasonable accuracy. The results indicate the method's suitability for sensitive healthcare applications, with acceptable trade-offs between security and performance.

3) Credit Card Fraud Detection Results: On the Credit Card Fraud Detection dataset, the model achieved a training accuracy of 92% and a testing accuracy of 90%. Precision was 88%, recall was 85%, and the F1-score was 86.5%. Encryption time was 1.8 h, computation time was 4.2 h, and decryption time was 0.4 h. Communication overhead was 900 MB. Figure 5 illustrates the precision-recall curve for fraud detection.

The high accuracy in fraud detection highlights the model's effectiveness in scenarios where the positive class (fraud cases) is significantly underrepresented. The encrypted computations did not substantially hinder performance, showcasing the method's robustness.

4) Synthetic Personal Dataset Results: Using the synthetic personal dataset, the model achieved a training accuracy of 82% and a testing accuracy of 80%. Precision was 79%, recall was 77%, and the F1-score was 78%. Encryption time was 2 h, computation time was 5 h, and decryption time was 0.5 h. Communication overhead was 1 GB. As further evidence of effective optimization

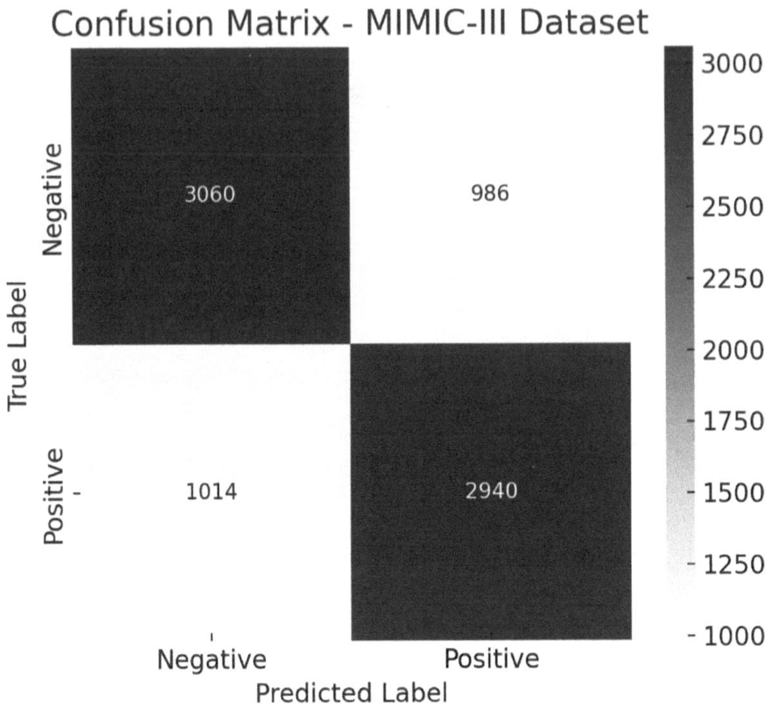

Fig. 4. Confusion Matrix for the MIMIC-III Dataset

under encryption, the training and validation accuracy curves in Fig. 6 increase smoothly and plateau, indicating stable convergence of gradient descent despite double encryption of inputs and parameters. The results on synthetic data confirm the model's generalizability and effectiveness across different types of sensitive personal information. The encrypted environment slightly impacts computation time but maintains acceptable accuracy levels.

Remark 1. In the Adult dataset, the model achieved a training accuracy of 85% and a testing accuracy of 82%. In contrast, the Credit Card Fraud Detection dataset exhibited a training accuracy of 92% and a testing accuracy of 90%. This discrepancy can be attributed to the differing characteristics of the datasets; for instance, the Adult dataset primarily comprises demographic information, while the Credit Card Fraud Detection dataset encompasses more complex transactional data patterns.

Remark 2. The encryption, computation, and decryption times varied across datasets (e.g., Adult dataset: encryption time 2.5 h; MIMIC-III dataset: 3 h). These differences are likely due to factors such as dataset size and complexity, which influence the computational load during processing.

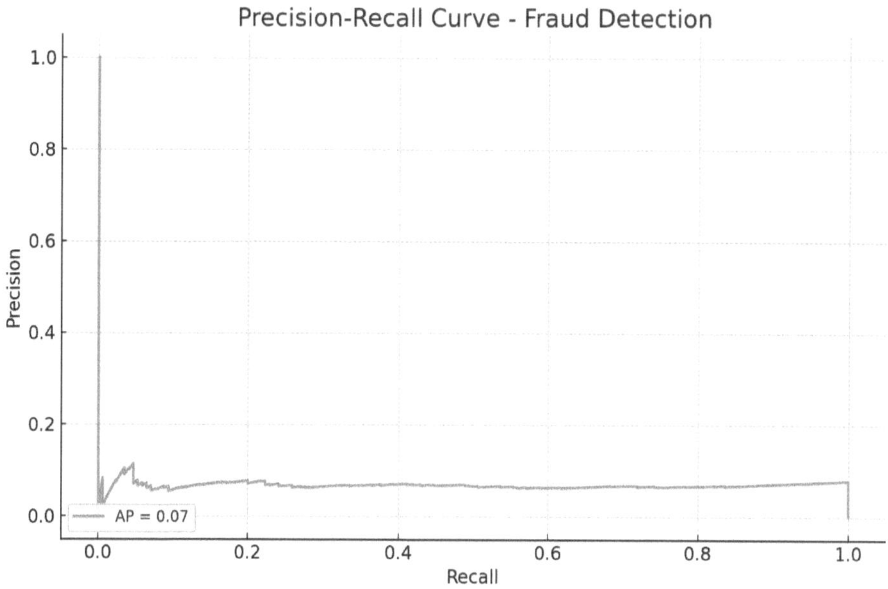

Fig. 5. Precision-Recall Curve for Fraud Detection

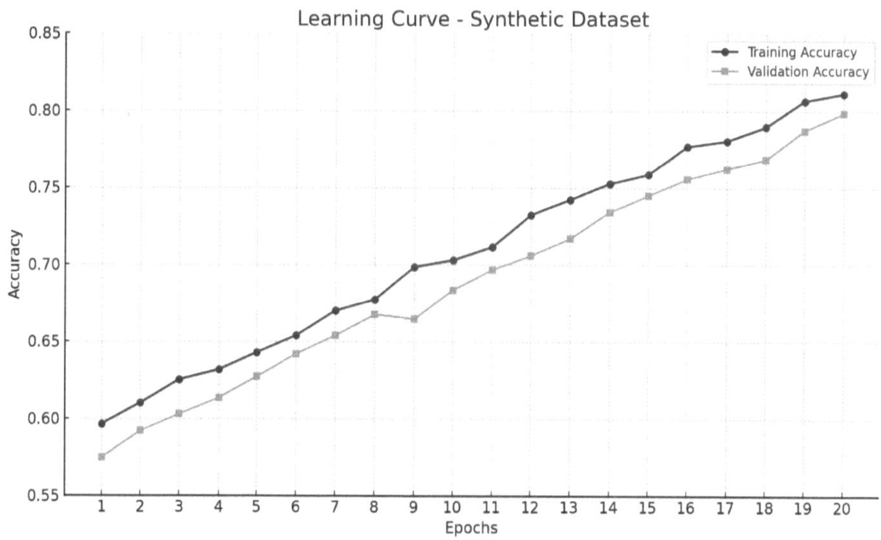

Fig. 6. Learning curve for the synthetic dataset

Remark 3. The communication overhead also varied between datasets (e.g., 1.2 GB for Adult vs. 900 MB for Credit Card Fraud). This variation can be attributed to the number of features and instances in each dataset, which affects the overall size of the data transmission.

5.8 Comparative Analysis

We compared our method with existing privacy-preserving classification approaches [14–16]. Table 4 provides a comparison with these methods.

Table 4. Comparison with existing privacy-preserving classification methods

Metric	CryptoNets [14]	DeepSecure [15]	Chameleon [16]	Proposed Method
Accuracy	98%	95%	97%	90%
Encryption Scheme[a]	LHE	SMPC	Hybrid Approaches	FHE
Computation Time	High	Very High	Moderate	Moderate
Security Level	128-bit	128-bit	128-bit	128-bit
Data Privacy	High	High	High	High
Model Confidentiality	Not Addressed	Not Addressed	Partial	High

[a] LHE: Leveled Homomorphic Encryption; SMPC: Secure Multi-Party Computation; FHE: Fully Homomorphic Encryption.

This analysis provides a comparative overview of four distinct schemes in the realm of homomorphic encryption-based CNNs, including our proposed scheme. The comparison is based on various parameters such as accuracy, training and testing times, security, and usability ratings. Taken together, these results illustrate a multi-objective comparison. Under the required security constraints, these metrics reflect both the utility of the learned model and the computational cost of privacy protection. This perspective highlights that our method achieves a balanced trade-off rather than optimizing a single metric, which is crucial for deployments where both data and model must remain confidential.

Gilad-Bachrach *et al.* [14] presented a novel approach, CryptoNets, to applying neural networks to encrypted data, achieving high throughput and accuracy. It set a precedent in secure data processing in neural networks, balancing computational efficiency with robust security protocols. Especially, CryptoNets achieves the accuracy of 99% on the MNIST dataset. Tran *et al.* [15] tackled the challenge of scalable and provably secure deep learning. It offered a blend of performance and security, making it suitable for applications requiring both accurate predictions and stringent data protection. Specifically, DeepSecure achieves the accuracy of 95% on the CIFAR-10 dataset. Mohassel [16] introduced a hybrid secure computation framework, Chameleon, tailored for machine learning applications. It excelled in providing a balance between computational efficiency and security, evidenced by its performance (the accuracy of 97%) on the MNIST dataset. Our approach leverages homomorphic encryption in convolutional neural networks, prioritizing data security while maintaining reasonable computational efficiency. It presents a novel methodology, especially potent in scenarios demanding the utmost data confidentiality. To provide a direct comparison against these established benchmarks, we evaluated our proposed scheme on the MNIST and CIFAR-10 datasets. On the MNIST dataset, our scheme achieves the accuracy of 95% and takes about 7 h for training, while on the CIFAR-10 dataset only 80% accuracy with about 10 h for training.

As a result, CryptoNets and Chameleon demonstrate high accuracy on the MNIST dataset, with DeepSecure also performing well on the CIFAR-10 dataset. Our scheme, while having lower accuracy on CIFAR-10, perform better on the simpler MNIST dataset. Besides, all schemes exhibit high security ratings, with our scheme offering the highest level of data protection.

Our method achieves competitive accuracy with moderate computation time compared to fully homomorphic encryption schemes, which often incur significant computational overhead. The CKKS scheme enables efficient encrypted computations, making it suitable for practical applications. Unlike some existing methods, our approach ensures both data privacy and model confidentiality, addressing potential threats of model inversion and theft.

Remark 4. While CryptoNets achieves high accuracy on simpler datasets like MNIST (99%), it may not perform as well on more complex datasets. Our method provides a balanced approach that prioritizes both data security and practical applicability in sensitive environments.

5.9 Security Analysis

1) Data confidentiality: In this scheme, data remains encrypted throughout transmission and processing, ensuring data confidentiality. This is mathematically substantiated by the security of the chosen homomorphic encryption scheme. For instance, the security of the CKKS scheme is demonstrated based on the complexity of ring learning problems. On the other hand, all computations, including convolution operations and activation function calculations, are conducted within the encrypted domain, further ensuring the confidentiality of the data.

2) Prevention of side-channel attacks: Judicious system design and parameter selection mitigate the likelihood of side-channel attacks. Randomization techniques can counter temporal side-channel attacks. Besides, physical layer protective measures, such as shielding techniques, reduce electromagnetic emissions, preventing electromagnetic side-channel attacks.

3) Prevention of model theft: The model parameters are encrypted, preventing model theft even if an attacker gains access to these parameters, as they cannot decrypt them to obtain the original parameters. The model structure can be safeguarded through encryption or obfuscation techniques to prevent theft of the model structure.

4) Prevention of model reverse engineering: Encrypting both the model parameters and structure effectively prevents reverse engineering attacks, making it impossible for attackers to extract useful information from the encrypted model.

5) noise control and security parameter selection: Through noise management techniques, security issues arising from noise accumulation can be prevented. Selecting appropriate security parameters, such as polynomial degree and scaling factor, ensures the security of the system.

6) CKKS security: The proof of security is based on the chosen homomorphic encryption scheme. The security of the CKKS scheme is predicated on the difficulty of ring learning problems.

7)Model complexity and time security: To further strengthen the security argument, we analyze the computational complexity required for an adversary to reconstruct the model from encrypted parameters. Under our chosen CKKS parameters with polynomial degree $n = 16384$, a 30-level modulus chain and a 128-bit security level, the underlying security reduces to solving an RLWE instance with $n = 16384$. Using the state-of-the-art LWE estimator, this instance corresponds to at least 2^{128} operations for the best known lattice attacks, which matches the NIST Level 1 post-quantum security standard. Therefore, the adversary's success probability within any feasible time window is negligible and the entire model remains computationally secure during training.

5.10 Performance Optimization

To mitigate the computational overhead introduced by homomorphic encryption, we adopted a set of well-established optimization techniques. First, we leveraged multi-core CPU and GPU resources to parallelize encryption and computation tasks, a strategy shown to significantly improve throughput in prior work on privacy-preserving neural network inference such as CryptoNets [14]. Second, we employed the batching capability of the CKKS scheme to pack multiple plaintext slots into a single ciphertext, thereby enabling simultaneous evaluation of multiple data points and reducing the total number of homomorphic multiplications [9]. Third, we approximated nonlinear activation functions with low-degree polynomials, which reduces the multiplicative depth and is a standard approach for encrypted neural network inference [14]. These optimizations collectively reduce both the number of expensive ciphertext-ciphertext multiplications and the number of key-switching operations, leading to a practical reduction of the overall computational complexity. As illustrated in Fig. 7, the combined effect of these optimizations results in approximately 30% shorter computation time compared with the unoptimized baseline, without sacrificing model accuracy.

5.11 Discussion

Our extensive experiments validate the practicality of the proposed method for privacy-preserving classification of sensitive personal information. While there is an inherent trade-off between security and computational efficiency, our method strikes a balance by maintaining high accuracy and acceptable performance levels. Moreover, the learning curves empirically confirm stable convergence of the encrypted training procedure, addressing concerns that double encryption might hinder iterative optimization.

Our method offers strong data and model confidentiality, competitive accuracy, and broad applicability across sensitive datasets. While it introduces higher

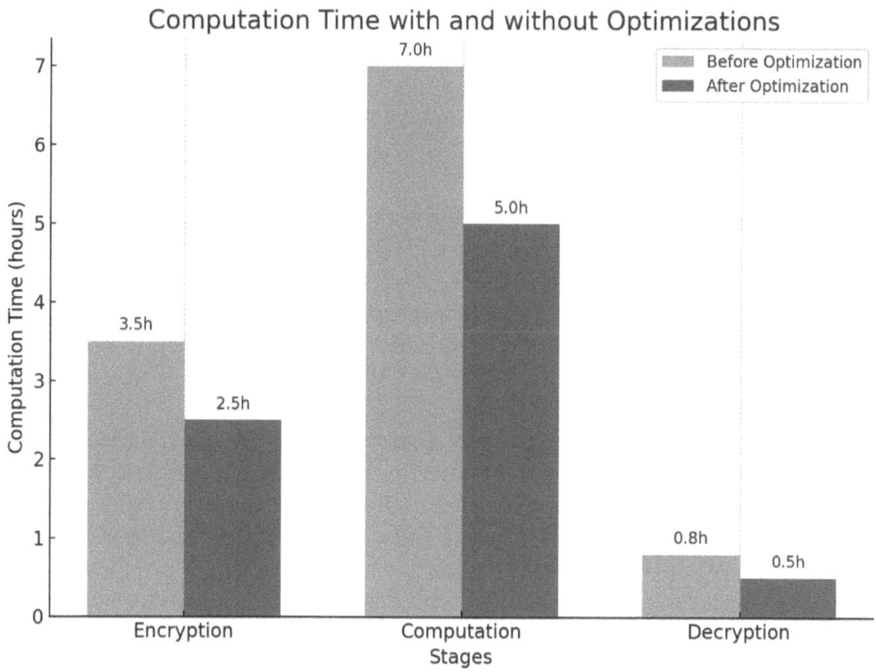

Fig. 7. Impact of optimizations on computation time

computational overhead for large-scale tasks, this trade-off is acceptable in settings where confidentiality and trust minimization are mandatory. In such regulated deployments, our method provides a well-balanced solution rather than aiming for universal superiority in raw accuracy.

Future work includes exploring more efficient homomorphic encryption schemes, implementing hardware accelerators specifically designed for encrypted computations, and extending the method to support more complex models like deep neural networks with additional layers.

6 Related Works

The burgeoning interest in safeguarding sensitive personal information while employing intelligent classification systems has led to a myriad of studies and advancements in this domain. This section reviews relevant literature, highlighting key contributions and methodologies.

Privacy-Preserving Techniques in Data Classification: Research directed towards developing privacy-preserving algorithms for data classification has been substantial. Kanagala *et al.* [17] explored cryptographic techniques to secure data during the classification process, while Fletcher *et al.* [18] focused on differential privacy methods. Similarly, Geng *et al.* [19] discussed federated learning with privacy preservation for IoT devices, and Wong and Liu [5] addressed secure multi-party computation for privacy-preserving data mining. Beyond classical MPC/DP baselines, cross-silo systems using multiparty homomorphic encryption have enabled privacy-preserving training and prediction for sequence models (RNNs) under passive-adversary settings [20]. For graph data, parallel-packed HE has further demonstrated practical encrypted GCN inference with notable speedups [21].

Machine Learning Approaches for Sensitive Data: The application of machine learning in handling sensitive data, especially in healthcare and finance, has seen significant interest. Alzubi *et al.* [22] presented a novel approach using deep learning for classifying medical records while maintaining patient confidentiality. Patel and Gupta [23] demonstrated the use of ensemble learning techniques in fraud detection. Additionally, Chen and Wu [24] highlighted privacy-aware machine learning algorithms for health data analysis.

Homomorphic Encryption in Data Classification: Recently, homomorphic encryption has emerged as a potent tool for secure data processing. Liu *et al.* [25] showcased how it can be applied to computations on encrypted data, thus preserving privacy during classification tasks. Li *et al.* [26] proposed a homomorphic encryption-based framework for privacy-preserving text classification on deep neural networks, enabling encrypted inference while maintaining model utility.

Synthetic Data Generation for Privacy Preservation: The generation of synthetic datasets, resembling real-world data while ensuring privacy, has been another area of focus. The Synthea framework [27] generates realistic but synthetic patient records for research purposes. Morgan and Patel [28] also contributed to the field with AI techniques for personal data anonymization.

Emerging Technologies in Health Data Management: The role of emerging technologies like blockchain in secure and transparent health data management has been explored by Brown and Green [29]. Evans and Sharma [30] discussed privacy-preserving analytics for cloud-based healthcare systems, further emphasizing the importance of privacy in healthcare data.

In summary, the field of privacy-protected intelligent classification of personal sensitive information is rapidly evolving, with research spanning from cryptographic methods to innovative machine learning algorithms and emerging technologies. As data privacy concerns continue to rise, these advancements play a crucial role in shaping the future of secure and efficient data classification.

7 Conclusion

In this paper, we proposed and evaluated a novel privacy-preserving classification method using a CNN over data encrypted with the CKKS homomorphic encryption scheme. The primary advantage and contribution of our work is a dual-protection framework that safeguards not only the privacy of user data but also the confidentiality of the proprietary model from theft and reverse engineering. Our comprehensive experiments on diverse datasets, including demographic, clinical, and financial data, empirically validated this approach. The results demonstrated that our method achieves high classification accuracy while operating entirely in the encrypted domain. Furthermore, performance optimizations were shown to significantly reduce computational overhead, confirming that our scheme is a secure, efficient, and practical solution for real-world sensitive data classification.

References

1. Mehnaz, S., Dibbo, S.V., Kabir, E., Li, N., Bertino, E.: Are your sensitive attributes private? novel model inversion attribute inference attacks on classification models. In: 31st USENIX Security Symposium, USENIX Association, pp. 4579–4596 (2022)
2. Alouffi, B., Hasnain, M., Alharbi, A., Alosaimi, W., Alyami, H., Ayaz, M.: A systematic literature review on cloud computing security: threats and mitigation strategies. IEEE Access **9**, 57792–57807 (2021)
3. Zhang, T., Zhu, Q.: Dynamic differential privacy for admm-based distributed classification learning. IEEE Trans. Inf. Forensics Secur. **12**(1), 172–187 (2016)
4. Ye, D., Shen, S., Zhu, T., Liu, B., Zhou, W.: One parameter defense–defending against data inference attacks via differential privacy. IEEE Trans. Inf. Forensics Secur. **17**, 1466–1480 (2022)
5. Wong, C., Liu, M.: Secure multi-party computation for privacy-preserving data mining. In: Proceedings of the International Conference on Data Mining, pp. 1145–1152 (2020)
6. Knott, B., Venkataraman, S., Hannun, A., Sengupta, S., Ibrahim, M., van der Maaten, L.: Crypten: secure multi-party computation meets machine learning. Adv. Neural. Inf. Process. Syst. **34**, 4961–4973 (2021)
7. Gao, C., Yu, J.: SecureRC: a system for privacy-preserving relation classification using secure multi-party computation. Comput. Secur. **128**, 103142 (2023)
8. Yuan, J., Liu, W., Shi, J., Li, Q.: Approximate homomorphic encryption based privacy-preserving machine learning: a survey. Artif. Intell. Rev. **58**(3), 82 (2025)
9. Cheon, J.H., Kim, A., Kim, M., Song, Y.: Homomorphic encryption for arithmetic of approximate numbers. In: 23rd International Conference on the Theory and Applications of Cryptology and Information Security, pp. 409–437 (2017)
10. Becker, B., Kohavi, R.: UCI machine learning repository: Adult dataset (1996). https://doi.org/10.24432/C5XW20
11. Johnson, A.E.W., et al.: MIMIC-III, a freely accessible critical care database. Sci. Data **3**, 160035 (2016)
12. Machine Learning Group - ULB, Credit card fraud detection dataset (2018). https://www.kaggle.com/datasets/mlg-ulb/creditcardfraud
13. Walonoski, J., et al.: Synthea: an approach, method, and software mechanism for generating synthetic patients and the synthetic electronic health care record. J. Am. Med. Inform. Assoc. **25**(3), 230–238 (2018)
14. Dowlin, N., et al.: Cryptonets: applying neural networks to encrypted data with high throughput and accuracy. In: Proceedings of the 33rd International Conference on Machine Learning, pp. 201–210 (2016)
15. Rouhani, B.D., Riazi, M.S., Koushanfar, F.: Deepsecure: scalable provably-secure deep learning. In: Proceedings of the 55th Annual Design Automation Conference, pp. 1–6 (2018)
16. Riazi, M.S., et al.: Chameleon: a hybrid secure computation framework for machine learning applications. In: Proceedings of the Asia Conference on Computer and Communications Security, pp. 707–721 (2018)

17. Kanagala, P., Jayaraman, R.: Effective encryption approach to improving the secure cloud framework through fuzzy-based encrypted cryptography. Soft Comput. 1–10 (2023)
18. Fletcher, S., Islam, M.Z.: Decision tree classification with differential privacy: a survey. ACM Comput. Surv. **52**(4), 1–33 (2019)
19. Geng, T., Liu, J., Huang, C.-T.: A privacy-preserving federated learning framework for IoT environment based on secure multi-party computation. in: IEEE Annual Congress on Artificial Intelligence of Things, pp. 117–122 (2024)
20. Sav, S., Diaa, A., Pyrgelis, A., Bossuat, J.-P., Hubaux, J.-P.: Privacy-Preserving federated recurrent neural networks. Proc. Priv. Enhancing Technol. **2023**(4), 500–521 (2023)
21. Ran, R., Xu, N., Liu, T., Wang, W., Quan, G., Wen, W.: Penguin: parallel-packed homomorphic encryption for fast graph convolutional network inference. In: Proceedings of the 37th International Conference on Neural Information Processing Systems, pp. 1–13 (2023)
22. Alzubi, J.A., Alzubi, O.A., Singh, A., Ramachandran, M.: Cloud-IIoT-based electronic health record privacy-preserving by CNN and blockchain-enabled federated learning. IEEE Trans. Industr. Inf. **19**(1), 1080–1087 (2022)
23. Patel, S., Gupta, R.: Ensemble learning for fraud detection in financial transactions. Int. J. Adv. Res. Sci. Commun. Technol. **5**(6), 403–409 (2020)
24. Chen, L., Wu, D.: Privacy-aware machine learning algorithms for health data analysis. Digit. Health **5**, 117–128 (2021)
25. Liu, Y., Luo, Y., Zhu, Y., Liu, Y., Li, X.: Secure multi-label data classification in cloud by additionally homomorphic encryption. Inf. Sci. **468**, 89–102 (2018)
26. Li, K., Huang, R., Yang, B.: Privacy-preserving text classification on deep neural network. Neural Process. Lett. **57**(2), 29 (2025)
27. Synthea Project Team: Synthea: synthetic data generation for healthcare research. J. Biomed. Inform. **27**(2), 103–110 (2022)
28. Morgan, T., Patel, V.: AI techniques for personal data anonymization. J. Artif. Intell. Res. **8**, 99–111 (2022)
29. Brown, A., Green, D.: Blockchain-based solutions for secure and transparent health data management. Healthc. Inf. Res. **27**(2), 103–110 (2021)
30. Evans, R., Sharma, P.: Privacy-preserving analytics for cloud-based healthcare systems. In: IEEE Symposium on Privacy and Security in Healthcare, pp. 657–662 (2020)

An Authenticated Certificateless Public Key Searchable Encryption Scheme

Bo Yang[1,2(✉)], Xiaoguang Liu[2,3,4], and Naixin Zhang[1,2]

[1] The School of Computing and Artificial Intelligence, Southwest Minzu University, Chengdu 610041, Sichuan, People's Republic of China
boyoung6@foxmail.com, zhangXXx1@foxmail.com

[2] Key Laboratory of National Civil Affairs Commission, Southwest Minzu University, Chengdu 610041, Sichuan, People's Republic of China
{liuxg,21700128,nxzhang}@swun.edu.cn

[3] Guangxi Key Laboratory of Cryptography and Information Security, Guilin University of Electronic Technology, Guilin 541004, Guangxi, People's Republic of China

[4] Guangxi Key Laboratory of Cryptography and Information Security, Guilin 541004, Guangxi, People's Republic of China

Abstract. With the rapid development of cloud computing, an increasing number of users are adopting cloud services for data storage and sharing. However, ensuring data security, privacy, and authenticity in cloud environments remains a critical challenge. Searchable encryption has emerged as a promising solution for secure and privacy-preserving data sharing in the cloud. We propose a novel searchable proxy re-encryption scheme that enables ciphertext search and secure multi-user data sharing. Our scheme leverages certificateless cryptography to eliminate the drawbacks of traditional public-key infrastructures, such as certificate management and key escrow problems. Additionally, the proposed scheme incorporates an authentication mechanism to ensure data integrity and authenticity during sharing. The security analysis demonstrates that our scheme achieves confidentiality, searchability, and authentication.

Keywords: Searchable Encryption · Certificateless Cryptography · Cloud Security · Authentication

1 Introduction

As a new cryptographic primitive, searchable encryption was proposed to achieve the retrieval function of ciphertext data. It mainly includes two categories: symmetric searchable encryption and public key searchable encryption. The former

Supported by the Sichuan Science and Technology Program (No. 2024YFHZ0013), the Graduate Innovative Key Program of Southwest Minzu University (No: YCZD2024005), the Fund of Guangxi Key Laboratory of Cryptography and Information Security (No. GCIS202121) and the Technological Innovation Team for Qinghai-Tibetan Plateau Research in Southwest Minzu University (No. 2024CXTD19).

J. Yu and Y. Chen (Eds.): FCS 2025, CCIS 2805, pp. 407–424, 2027.
https://doi.org/10.1007/978-981-95-9514-3_27

was proposed by Song et al. [1] for symmetric cryptosystems, while the latter was proposed by Boneh et al. [2] for public key crypto systems. In 2006, Byun et al. [3] first noticed significant flaws in keyword search schemes. Specifically, since the selection space of keywords is smaller than that of passwords, and users often use common keywords when searching for documents, this leads to the risk of keyword guessing attacks, thereby compromising keyword privacy. In 2016, Chen et al. [4] constructed a new search framework, namely dual-server public key encryption with keyword search, successfully addressing internal keyword guessing attacks initiated by malicious servers. In 2022, Ernura [5] proposed a new general construction method for searchable encryption schemes, whose core idea is consistent with that of Liu et al. [6], but this method only uses one smooth mapping hash function, while Liu et al. used two, making its construction more efficient. In 2025, Xu et al. [7] developed a searchable encryption scheme based on the national standard SM9 to protect the privacy of smart healthcare data. However, searchable encryption only allows data recipients to perform keyword searches and does not authorize other users to query and access ciphertexts. Therefore, how to authorize data users to search, decrypt, and view ciphertext data still requires further research.

Proxy re-encryption is designed to complete decryption authorization without exposing the private key of the data owner. The proxy server holding the re-encryption key can transform the ciphertext encrypted with the data owner's public key into a ciphertext that can be decrypted by the data user's private key, and during this process, the proxy server cannot obtain any data information. In 1988, Blaze et al. [8] first proposed a proxy re-encryption scheme, which enables a semi-trusted proxy to transform the ciphertext of the data sender so that the data receiver can decrypt it without revealing the plaintext message of the data sender. In 2005, Ateniese et al. [9] first provided a formal definition of proxy re-encryption and proposed a corresponding security model in their research. Their proposed scheme has anti-collusion characteristics and can ensure the security of the master key. Moreover, it is the first one-way proxy re-encryption scheme constructed using bilinear pairings. In 2013, Isshiki et al. [10] designed the first one-way multi-hop proxy re-encryption scheme in the standard model that achieves full chosen-ciphertext security and provided a specific construction. In 2015, Tang et al. [11] proposed a non-interactive one-way multi-hop proxy re-encryption scheme based on strong security assumptions in the multilinear setting. In 2020, Deng et al. [12] combined compact identity-based encryption technology with key randomization update technology to construct a new proxy re-encryption scheme, namely Hybrid Attribute-based Proxy Re-encryption (HAPRE). In 2025, Li et al. [13] proposed a cloud-chain collaborative attribute-based proxy re-encryption industrial Internet data sharing scheme, which allows the blockchain to handle the computational tasks of data re-encryption and ciphertext retrieval, thereby avoiding the problem of malicious proxies in traditional proxy re-encryption schemes due to the selection of semi-trusted clouds as proxies.

Therefore, by integrating the functional advantages of searchable encryption and proxy re-encryption, a searchable proxy re-encryption can be obtained, which combines the keyword search function with the decryption authorization function. In 2010, Shao et al. [14] first proposed the concept of a searchable proxy re-encryption scheme and designed a specific proxy re-encryption scheme, which enables data owners to grant both keyword search and decryption permissions to data users simultaneously. In the same year, Yua et al. [15] proposed the Re-dPEKS scheme, which differs from the former in that the algorithms for encrypting messages and keywords can be operated separately. In 2011, Chen et al. [16] proposed a proxy re-encryption scheme with private search, which enables data users and owners to efficiently query and access files stored in untrusted clouds while hiding query and data privacy from cloud providers. In 2012, Wang et al. [17] proposed a new scheme that can simultaneously achieve joint keyword search and selective delegation, i.e., multi-keyword search. In the same year, Fang et al. [18] first proposed a scheme with chosen-ciphertext security. In 2013, Yang et al. [19] proposed a new scheme supporting joint keyword search, in which the concepts of designated tester and proxy re-encryption function were defined, and the first designated tester proxy re-encryption joint keyword searchable encryption scheme was constructed. This scheme is highly efficient and can resist keyword attacks. In 2016, Chen et al. [20] proposed a new concept that integrates the proxy re-encryption function with keyword search and the threshold cryptosystem function to achieve time-controllable authorization. In the same year, Yang et al. [21] proposed the first time-related searchable encryption scheme, which has designated tester and proxy re-encryption functions with timing, and can revoke the authorized person's rights after a certain period. In 2017, Yang et al. [22] proposed a semantic keyword search scheme, which not only supports exact keyword search but also supports synonym keyword search. In 2019, Manzoor et al. [23] proposed a new blockchain-based scheme. In 2021, Yu et al. [24] proposed a lattice-based time-controllable multi-user connected keyword searchable proxy re-encryption scheme, which can resist quantum attacks. In 2025, S. Fugkeaw and J. Deevijit [25] proposed the SESPRE scheme, which combines searchable proxy re-encryption with access control, and introduces inverted indexes and B+ trees to improve the search efficiency, which guarantees secure querying and improves efficiency.

Motivated by the need for secure and efficient data sharing in cloud environments, we propose a certificateless searchable proxy re-encryption scheme, which integrates searchable encryption with proxy re-encryption to enable privacy-preserving multi-user data sharing. Our scheme eliminates key escrow and certificate management issues while ensuring data authenticity through an embedded authentication mechanism. The primary contributions of this study are as follows:

- We propose a novel certificateless searchable proxy re-encryption scheme, which combines the advantages of searchable encryption and proxy re-encryption. This allows data owners to delegate search capabilities to authorized users without compromising data confidentiality or introducing certificate management overhead.

- We design a signature mechanism to ensure the authenticity of data sources. By integrating identity-based authentication into the re-encryption process, our scheme prevents unauthorized access while maintaining ciphertext retrieval.
- The security analysis under the certificateless cryptography framework, proving that our scheme achieves indistinguishability against chosen-keyword attacks (IND-CKA), indistinguishability under the adaptive chosen ciphertext attacks (IND-CCA), and strong existential unforgeability under chosen message attacks (SUF-CMA).

2 Preliminaries

2.1 Bilinear Pairing

We assume that $\mathbb{G}_1$ and $\mathbb{G}_2$ are cyclic addition and multiplicative groups mediated by p, respectively. The bilinear pairing is a map $\hat{e}:\mathbb{G}_1\times\mathbb{G}_1\rightarrow\mathbb{G}_2$, and it has the following properties:

1. *Bilinearity*: $\forall P,Q\in\mathbb{G}_1$ and $\forall a,b\in\mathbb{Z}_q^*$, there is $\hat{e}\left(aP,bQ\right)=\hat{e}(P,Q)^{ab}$.
2. *Non-degeneracy*: $\exists P,Q\in\mathbb{G}_1$ such that $\hat{e}\left(P,Q\right)\neq 1\in\mathbb{G}_2$.
3. *Computability*: $\forall P,Q\in\mathbb{G}$, there is an effective algorithm to compute $\hat{e}\left(P,Q\right)$.

2.2 Bilinear Diffie-Hellman (BDH) Problem

BDH: Let $\hat{e}:\mathbb{G}_1\times\mathbb{G}_1\rightarrow\mathbb{G}_2$ be a bilinear pairing. Suppose that $P,aP,bP,cP\in\mathbb{G}_1$ are known points and $a,b,c\in_R\mathbb{Z}_q^*$ are unknown numbers, and then, computing the value of $\hat{e}(P,P)^{abc}\in\mathbb{G}_2$ is difficult.

2.3 Computational Diffie-Hellman (CDH) Problem

CDH: Let $\mathbb{G}_1$ be a cyclic additive group of prime order q with generator P. Given $(P,aP,bP)\in\mathbb{G}_1^3$ where $a,b\in_R\mathbb{Z}_q^*$ are unknown, the CDH problem is to compute $abP\in\mathbb{G}_1$.

2.4 System Model

The system model is primarily composed of four entities: the key generation center (KGC), the cloud server, the data owner, and the data receiver. The system model can be defined by the following ten probabilistic polynomial-time algorithms:

1. *Setup*: This algorithm takes a security parameter λ and outputs a system parameter $prms$ and a master secret key x.

2. *PartialKeyGen*: This algorithm takes system parameters $prms$, master key x, and user identity ID_i, and outputs partial private key U_i of the user.
3. *FullKeyGen*: This algorithm takes system parameters $prms$ and the user partial private key U_i, and outputs the full private key sk_i and the user public key pk_i.
4. *Encrypt*: This algorithm takes system parameters $prms$, the message m and the data owner's public key pk_1, and outputs the ciphertext CT.
5. *Sign*: This algorithm takes system parameters $prms$, the message m and the data owner's full private key sk_1, and outputs the signature σ.
6. *TrapDoorGen*: This algorithm takes system parameters $prms$, the keyword w' and the data receiver's full private key sk_2, and outputs the trapdoor CT.
7. *ReKeyGen*: This algorithm takes system parameters $prms$, the data owner's full privaye key sk_1 and the data receiver's public key pk_2, and outputs the re-encryption key RK.
8. *ReEncrypt*: This algorithm takes system parameters $prms$, the ciphertext CT, the trapdoor TD and the re-encryption key RK, and outputs the re-encryption ciphertext CT'.
9. *Decryption*: This algorithm takes the re-encryption ciphertext CT', and the data receiver's full private key sk_2, and outputs the message m'.
10. *Verify*: This algorithm takes system parameters $prms$, the message m', the signature σ and the data owner's public key pk_1, and outputs the verification result 0 or 1.

3 The Proposed Scheme

In this section, we elaborate on our network implementation and configuration details.

3.1 *Setup*(λ)

Given a security parameter λ, KGC selects a bilinear mapping $\hat{e} : \mathbb{G}_1 \times \mathbb{G}_1 \rightarrow \mathbb{G}_2$, three secure hash functions $H_1 : \{0,1\}^* \rightarrow \mathbb{G}_1$, $H_2 : \{0,1\}^* \rightarrow \mathbb{Z}_q^*$, $H_3 : \mathbb{G}_2 \rightarrow \{0,1\}^*$. Here $\mathbb{G}_1$ and $\mathbb{G}_2$ are two cyclic groups with the same prime order q, and P is a generator of $\mathbb{G}_1$. KGC chooses three random numbers: $\alpha \in \mathbb{Z}_q^*$; $P_1, P_2 \in \mathbb{G}_1$. System master secret key $msk = \alpha$, master public key $P_k = \alpha P$, and a token $Z = \alpha P_1$. KGC keeps msk secret and publishes system parameters $prms = (\mathbb{G}_1, \mathbb{G}_2, q, \hat{e}, P_k, P, P_1, P_2, Z, H_1, H_2, H_3)$.

3.2 *PartialKeyGen*($prms, msk, ID_i$)

$User_i$ sends ID_i to KGC. KGC computes $User_i$'s partial private key $U_i = \alpha H_1(ID_i)$, then sends U_i to $User_i$.

3.3 $FullKeyGen(prms, U_i)$

$User_i$ selects $x_i \in \mathbb{Z}_q^*$ randomly. $User_i$ sets it's full private key $sk_i = (x_i, U_i)$, and public key $pk_i = (H_1(ID_i), x_iP)$.

3.4 $Encrypt(prms, sk_1, m)$

$User_1$ selects a random number $r \in \mathbb{Z}_q^*$ and extracts the keyword w of m. $User_1$ sends w to KGC to get symmetric encryption parameter $K_j = \alpha H_1(w||ID_1)$ and computes $C_0 = m \oplus H_3(\hat{e}(K_j, rP_1))$, $C_1 = \hat{e}(P_k, rH_2(w)P_1)$, $C_2 = rH_1(ID_1)$, $C_3 = rP$, $C_4 = rP_2, C_5 = rP_1)\}$. Then sets $CT = \{C_0, C_1, C_2, C_3, C_4, C_5\}$.

3.5 $Sign(prms, sk_1, M)$

$User_1$ calculates $\sigma = x_iH_1(m) + rU_1$. Finally, $User_1$ sends the full ciphertext $CT = (C_0, C_1, C_2, C_3, C_4, C_5)$ and the signature σ to $CloudServer$.

3.6 $TrapDoorGen(prms, U_2, w')$

If $User_2$ want to get m, $User_2$ will chose keyword w', and randomly selects $v \in \mathbb{Z}_q^*$, then calculates $T_1 = H_2(w')Z + U_2 - vx_2(H_1(ID_1) + H_1(ID_2))$, $T_2 = P_k - vx_2P, T_3 = vx_2H_1(ID_2) - U_2$. Then $User_2$ sends the trapdoor $TD = (T_1, T_2)$ to $CloudServer$ and T_3 to $User_1$.

3.7 $ReKeyGen(prms, U_1, ID_1, ID_2)$

$User_1$ randomly selects ρ, then computes $rk_1 = \rho P_2 + T_3 + U_1$, $rk_2 = \rho P$, $rk_3 = \hat{e}(r(H_1(w||ID_1) + H_1(ID_2)), Z)$. $User_1$ sends $RK = (rk_1, rk_2, rk_3)$ to $CloudServer$.

3.8 $ReEncrypt(prms, CT, TD, RK)$

$CloudServer$ computes $C_1' = C_1 \cdot \hat{e}(C_4, rk_2)/\hat{e}(C_3, rk_1)$ and calculates whether the Eq. (1) holds.

$$\hat{e}(T_1, C_3) = \hat{e}(T_2, C_2) \cdot C_1' \tag{1}$$

If Eq. (1) holds, $CloudServer$ will return $CT' = (C_0, C_1', C_2, C_3, C_4, C_5, rk_3)$ and σ, otherwise return $\perp$.

3.9 $Decrypt(U_2, CT')$

$User_2$ computes

$$m' = C_0 \oplus H_3(rk_3 \cdot \hat{e}(U_2, C_5)^{-1}). \tag{2}$$

3.10 $Verify(prms, \sigma, m')$

$User_2$ verifies the signature as the Eq. (3):

$$\hat{e}(\sigma, Pk) = \hat{e}(H_1(m), Pk_1) \cdot \hat{e}(C_2, Pk) \quad (3)$$

4 Security Analysis

4.1 IND-CKA

Theorem 1. *Let the hash functions H_1, H_2 be random oracle machines, and if the BDH problem is hard on the group generated by P, then our scheme is IND-CKA secure.*

Correctness: Suppose that $w = w'_j$, the correctness is shown as follows.

$$\begin{aligned}
&\hat{e}(T_1, C_3) \\
&= \hat{e}(H_2(w'_j)Z + U_2 - vx_2(H_1(ID_1) + H_1(ID_2)), rP) \\
&= \hat{e}(\alpha H_2(w'_j)P_1 + \alpha H_1(ID_2) - vx_2 H_1(ID_2) - vx_2 H_1(ID_1), rP) \\
&= \hat{e}((\alpha - vx_2)H_1(ID_2), rP) \cdot \hat{e}(vH_1(ID_1), rP)^{-1} \\
&\quad \cdot \hat{e}(\alpha H_2(w'_j)P_1, rP) \\
&= \hat{e}((\alpha - vx_2)H_1(ID_1), rP) \cdot \hat{e}((\alpha - v)(H_1(ID_2) - H_1(ID_1)), rP) \\
&\quad \cdot \hat{e}(vH_1(ID_1), rP)^{-1} \cdot \hat{e}(\alpha H_2(w')P_1, rP) \\
&= \hat{e}((\alpha - vx_2)H_1(ID_1), rP) \cdot \hat{e}((\alpha - v)(H_1(ID_2) - H_1(ID_1)), rP) \\
&\quad \cdot \hat{e}(vH_1(ID_1), rP)^{-1} \cdot \hat{e}(\alpha H_2(w)P_1, rP) \\
&= \hat{e}((\alpha - vx_2)P, rH_1(ID_1)) \cdot \hat{e}((\alpha - v)H_1(ID_2) - \alpha H_1(ID_1), rP) \cdot \hat{e}(\alpha P, rH_2(w)P_1) \\
&= \hat{e}((\alpha - vx_2)P, rH_1(ID_1)) \cdot \hat{e}(\alpha P, rH_2(w)P_1) \\
&\quad \cdot \frac{\hat{e}(\rho P_2, rP)}{\hat{e}(\rho P_2, rP) \cdot \hat{e}(\alpha H_1(ID_1) + (v - \alpha)H_1(ID_2), rP)} \\
&= \hat{e}(\alpha P - vx_2 P, rH_1(ID_1)) \cdot \hat{e}(\alpha P, rH_2(w)P_1) \cdot \frac{\hat{e}(rP_2, \rho P)}{\hat{e}(\rho P_2 + vH_1(ID_2) - U_2 + U_1, rP)} \\
&= \hat{e}(P_k - vx_2 P, rH_1(ID_1)) \cdot \hat{e}(P_k, rH_2(w)P_1) \cdot \frac{\hat{e}(rP_2, \rho P)}{\hat{e}(\rho P_2 + T_3 + U_1, rP)} \\
&= \hat{e}(T_2, C_2) \cdot C_1 \cdot \frac{\hat{e}(C_4, rk_2)}{\hat{e}(C_3, rk_1)} \\
&= \hat{e}(T_2, C_2) \cdot C'_1
\end{aligned}$$

Lemma 1. *Let the hash functions H_1, H_2 be random oracle machines, and if the BDH problem is hard on the group generated by P, then our scheme is IND-CKA-I secure.*

Proof. Assuming that it exists as a PPT adversary $\mathcal{A}_\mathcal{I}$ can break the IND-CKA GAME with a non-negligible advantage $\epsilon(\lambda)$, and $\mathcal{A}_\mathcal{I}$ conducts at most $q_s > 0$ times Partial Key Queries, $q_f > 0$ times Full Key Queries, q_r times Re-Key Queries, $q_t > 0$ times Trapdoor Queries, and $q_{test} > 0$ times Test Queries, then there must exist an challenger $\mathcal{C}$ who can solve the BDH problem in the group generated by P with advantage

$$ADV_{P,\mathcal{C}}^{BDH} \geq \frac{2\epsilon(\lambda)}{e(1+q_s+q_f+q_r+q_t)q_{test}},$$

where e is the base of the natural logarithm.

1. Setup: $\mathcal{C}$ inputs the security parameter λ and performs *Setup* algorithm to generate system parameters $prms$ and master secret key msk. $\mathcal{C}$ randomly chooses $ID_{\mathcal{A}_\mathcal{I}} \in \{0,1\}^*$ as the identity of $\mathcal{A}_\mathcal{I}$.
   In response to $\mathcal{A}_\mathcal{I}$'s discontinuous queries, $\mathcal{C}$ needs to maintain four initially empty lists:
   - $List^{H_1}$ of H_1 with elements of type 4-tuple $(ID_i, Q_i, b_i, coin)$ in response to $\mathcal{A}_\mathcal{I}$'s H_1 Queries.
   - $List^{H_2}$ of H_2 with elements of type 2-tuple (w_i, t_i) in response to $\mathcal{A}_\mathcal{I}$'s H_2 Queries.
   - $List^{key}$ with elements of type 3-tuple (ID_i, Pk_i, x_i) in response to $\mathcal{A}_\mathcal{I}$'s Public Key Queries and Full Key Queries.
   - $List^{test}$ with elements of type 2-tuple (T, X) in response to $\mathcal{A}_\mathcal{I}$'s Test Queries.

   Then $\mathcal{C}$ sends $prms = (\mathbb{G}_1, \mathbb{G}_2, q, \hat{e}, P_k, P, P_1, P_2, Z, H_1, H_2)$ to the adversary $\mathcal{A}_\mathcal{I}$ and keeps msk secret.
2. H_1 Queries: $\mathcal{A}_\mathcal{I}$ can make queries at any time. When $\mathcal{A}_\mathcal{I}$ queries H_1 oracle with identity ID_i, $\mathcal{C}$ will respond in the following way:
   - $\mathcal{C}$ first checks whether the ID_i exists in the list $List^{H_1}$. If it exists, the corresponding $H_1(ID_i) = Q_i \in \mathbb{G}_1$ is returned to $\mathcal{A}_\mathcal{I}$.
   - Otherwise if $ID_i = ID_{\mathcal{A}_\mathcal{I}}$, $\mathcal{C}$ randomly chooses $Q_{\mathcal{A}_\mathcal{I}} \in \mathbb{G}_1$ and sets $Q_i = Q_{\mathcal{A}_\mathcal{I}}$
   - Otherwise $\mathcal{C}$ will produce $coin \in \{0,1\}$ with probability distribution satisfying $Pr[coin = 0] = \delta$. $\mathcal{C}$ chooses a random number $b_i \in \mathbb{Z}_q^*$. If $coin = 0$, $\mathcal{C}$ will compute $Q_i = b_i Q_{\mathcal{A}_\mathcal{I}} \in \mathbb{G}_1$, and if $coin = 1$, $\mathcal{C}$ will compute $Q_i = b_i P \in \mathbb{G}_1$. Finally, $\mathcal{C}$ saves $(ID_i, Q_i, b_i, coin)$ to the $List^{H_1}$ and returns $H_1(ID_i) = Q_i$ to $\mathcal{A}_\mathcal{I}$.

   Here, the coin serves as a guess for $\mathcal{C}$: $coin = 0$ denotes $\mathcal{A}_\mathcal{I}$ will launch an attack on the ID_i of this query.
3. H_2 Queries: $\mathcal{A}_\mathcal{I}$ can make queries at any time. $\mathcal{A}_\mathcal{I}$ queries H_2 oracle with keyword w_i. $\mathcal{C}$ will respond in the following way: $\mathcal{C}$ first checks whether the w_i exists in the list $List^{H_2}$. If it exists, then corresponding t_i is returned to $\mathcal{A}_\mathcal{I}$ as $H_2(w_i)$. If w_i does not exist, $\mathcal{C}$ will select $t_i \in \mathbb{Z}_q^*$ and saves (w_i, t_i) to the $List^{H_2}$ and returns t_i to $\mathcal{A}_\mathcal{I}$ as $H_2(w_i)$.

4. Query phase I:
   - Partial Key Queries: $\mathcal{A_I}$ chooses identity $ID_i \neq ID_{\mathcal{A_I}}$ to query. $\mathcal{C}$ will get the corresponding $(ID_i, Q_i, b_i, coin)$ in $List^{H_1}$ first.
   If $coin = 0$, $\mathcal{C}$ aborts the query and quits. At this point, $\mathcal{C}$ was initially intended to use $\mathcal{A_I}$'s attack on IND-CKA-I to attack the BDH problem. In the query phase, $\mathcal{C}$ cannot use $\mathcal{A_I}$ to launch an attack.
   If coin=1, $\mathcal{C}$ computes ID_i's partial private key $U_i = b_i P_k$ and sends U_i to $\mathcal{A_I}$. Here, the private key of $User_i$ has a good representation: $U_i = \alpha Q_i = \alpha b_i P = b_i P_k$.
   - Public Key Queries: $\mathcal{A_I}$ chooses identity ID_i to query. $\mathcal{C}$ will respond in the following way: $\mathcal{C}$ first checks whether the ID_i exists in the list $List^{key}$. If it exists, then the corresponding Pk_i is returned to $\mathcal{A_I}$. If ID_i does not exist, $\mathcal{C}$ will select $x_i \in \mathbb{Z}_q^*$ and computes $Pk_i = x_i P$, and saves (ID_i, Pk_i, x_i) to the $List^{key}$. Then $\mathcal{C}$ returns Pk_i to $\mathcal{A_I}$.
   - Full Key Queries: $\mathcal{A_I}$ chooses identity $ID_i \neq ID_{\mathcal{A_I}}$ to query. $\mathcal{C}$ makes a Partial Key Query (corresponding coin=1) and a Public Key Query with ID_i to get the corresponding $sk_i = (U_i, x_i)$.
5. Public Key Replacement: $\mathcal{A_I}$ chooses $x_i' \in \mathbb{Z}_q^*$ and sends (ID_i, x_i'), then $\mathcal{C}$ saves (ID_i, Pk_i', x_i') into $List^{key}$.
6. Challenge: $\mathcal{A_I}$ launches a challenge.
   $\mathcal{A_I}$ chooses a random number $v \in \mathbb{Z}_q^*$ and (ID_o, ID^*, w_0, w_1) it wants to challenge. $\mathcal{C}$ looks up item $(ID_o, Q_o, b_o, coin)$ in $List^{H_1}$ where $coin = 0$. And looks up item $(ID_i, Q_i, b_i, coin)$ in $List^{H_1}$ and (ID_i, Pk_i, x_i) in $List^{key}$, such that $ID_i = ID^*$. If coin $= 1$, $\mathcal{C}$ aborts $\perp$ and quits. If coin $= 0$, $\mathcal{C}$ selects $\beta \in \{0,1\}$ and $r, \rho \in \mathbb{Z}_q^*$ randomly, and gets (w_β, t_β) from $List^{H_2}$, then computes $C^* = (C_{1,\beta} = \hat{e}(P_k, rt_\beta Z) \cdot \hat{e}(rP, (\alpha - vx_i)Q_i - \alpha Q_o), C_2 = b_i^{-1} r Q_o, C_3 = b_i^{-1} rP, C_4 = b_i^{-1} r P_2, rk_1 = \rho P_1 + (vx_i - \alpha) Q_i + \alpha Q_o, rk_2 = \rho P)$. That's because the $Q_{\mathcal{A_I}}$ needs to be restored when $coin = 0$, e.g. $\hat{e}(U_i, C_3) = \hat{e}(\alpha b_i Q_{\mathcal{A_I}}, b_i^{-1} rP) = \hat{e}(\alpha Q_{\mathcal{A_I}}, rP)$.
   $\mathcal{C}$ defines $C^* = (C_{1,\beta}, C_2, C_3)$. Let $a = \alpha, bP = Q_{\mathcal{A_I}}, c = r$. $\mathcal{C}$ has embedded the BDH problem $\hat{e}(P,P)^{abc}$ into $C_{1,\beta}$:

$$
\begin{aligned}
& C_{1,\beta} \\
&= \hat{e}(b_i^{-1} rP, (\alpha - vx_i)Q_i - \alpha Q_o) \cdot \hat{e}(Pk, rt_\beta P_1) \\
&= \hat{e}(b_i^{-1} rP, \alpha Q_i - vx_i Q_i - \alpha Q_o) \cdot \hat{e}(Pk, rt_\beta P_1) \\
&= \hat{e}(b_i^{-1} rP, \alpha Q_i) \cdot \hat{e}(b_i^{-1} rP, vx_i Q_i + \alpha Q_o)^{-1} \cdot \hat{e}(Pk, rt_\beta P_1) \\
&= \hat{e}(rP, \alpha Q_{\mathcal{A_I}}) \cdot \hat{e}(b_i^{-1} rP, vx_i Q_i + \alpha Q_o)^{-1} \cdot \hat{e}(Pk, rt_\beta P_1) \\
&= \hat{e}(P,P)^{abc} \cdot \hat{e}(b_i^{-1} rP, vx_i Q_i + \alpha Q_o)^{-1} \cdot \hat{e}(Pk, rt_\beta P_1).
\end{aligned}
$$

7. Query phase II.
   - Queries are the same as in query phase 1.
   - Re-Key Queries: $\mathcal{A_I}$ chooses identity ID_i to query. $\mathcal{C}$ will get the corresponding $(ID_i, Q_i, b_i, coin)$ in $List^{H_1}$ first. If $coin = 0$, $\mathcal{C}$ aborts the query and quits. If $coin = 1$, $\mathcal{C}$ computes $rk_1 = \rho P_1 + vx_i Q_i - \alpha Q_i, rk_2 = \rho P$ and sends (rk_1, rk_2) to $\mathcal{A_I}$.

- Trapdoor Queries: $\mathcal{A}_{\mathcal{I}}$ chooses identity ID_i and $w \neq (w_0|w_1)$ to query. $\mathcal{C}$ will get the corresponding $(ID_i, Q_i, b_i, coin)$ in $List^{H_1}$ first. If $coin = 0$, $\mathcal{C}$ aborts the query and quits. If $coin = 1$, $\mathcal{C}$ computes TD to $\mathcal{A}_{\mathcal{I}}$.
- Test queries: When $\mathcal{A}_{\mathcal{I}}$ queries with TD to know whether the trapdoor is validated, $\mathcal{C}$ will respond in the following way: $\mathcal{C}$ first checks whether the T_1 exists in the list $List^{test}$ such that $T_1 = T$. If it exists, then it gets the corresponding X. And if T_1 does not exist, $\mathcal{C}$ computes

$$\begin{aligned} T &= T_1, \\ X &= \hat{e}(T_1, C_3) \\ &= \hat{e}(b_i^{-1} rP, (\alpha - vx_i)Q_i - vx_i Q_o) \cdot \hat{e}(rP, \alpha H_2(w_j)P_1) \\ &= \hat{e}(rP, (\alpha - vx_i)Q_{\mathcal{A}_{\mathcal{I}}}) \cdot \hat{e}(b_i^{-1} rP, Q_o)^{-vx_i} \cdot \hat{e}(rP, \alpha H_2(w_j)P_1) \\ &= \hat{e}(P, P)^(\alpha - vx_i)br \cdot \hat{e}(b_i^{-1} rP, Q_o)^{-vx_i} \cdot \hat{e}(rP, \alpha H_2(w_j)P_1) \\ &= \hat{e}(P, P)^{abc - bcvx_i} \cdot \hat{e}(b_i^{-1} rP, Q_o)^{-vx_i} \cdot \hat{e}(cP, aH_2(w_j)P_1) \\ &= D \cdot \hat{e}(P, P)^{-bcvx_i} \cdot \hat{e}(b_i^{-1} rP, Q_o)^{-vx_i} \cdot \hat{e}(cP, aH_2(w_j)P_1), \end{aligned}$$

and saves (T, X) to the $List^{test}$. If $X = \hat{e}(C_2, T_2) \cdot C^*$, then the challenger C returns 1; otherwise, 0 is returned to the $\mathcal{A}_{\mathcal{I}}$.

8. Guess: $\mathcal{A}_{\mathcal{I}}$ outputs a guess $\beta' \in \{0, 1\}$. $\mathcal{A}_{\mathcal{I}}$ wins the above game if $\beta' = \beta$. $\mathcal{C}$ randomly takes (T, X) from the $List^{test}$ and takes $D = X \cdot \hat{e}(P, P)^{bcvx_i} \cdot \hat{e}(b_i^{-1} rP, Q_o)^{vx_i} \cdot \hat{e}(cP, aH_2(w_j)P_1)^{-1}$ as the solution of BDH.

The advantage of $\mathcal{A}_{\mathcal{I}}$ in the simulation attack is

$$\mathrm{Adv}_{\mathrm{Sim}, \mathcal{A}_{\mathcal{I}}}^{\mathrm{IND-CKA-I}}(\lambda) \geq \epsilon(\lambda) = \left| \Pr\left[\mathrm{Exp}_{\Pi, \mathcal{A}_{\mathcal{I}}}^{\mathrm{IND-CKA-I}}(\lambda) = 1\right] - \frac{1}{2} \right|.$$

If $\mathcal{A}_{\mathcal{I}}$'s guess is correct, $\mathcal{A}_{\mathcal{I}}$ succeeds in attacking the IND-CKA-I at step 8. Since the probability that $\mathcal{C}$ does not interrupt at steps 4 and 7 is $(1-\delta)^{q_s+q_f+q_r+q_t}$, and the probability that $\mathcal{C}$ does not interrupt at steps 6 is δ, the probability that $\mathcal{C}$ does not interrupt is $(1-\delta)^{q_s+q_f+q_r+q_t}\delta$, and the advantage of $\mathcal{C} \geq \epsilon(\lambda)$ is as follows:

$$\begin{aligned} \mathrm{Adv}_{\mathrm{Sim}, \mathcal{C}}^{\mathrm{IND-CKA-}I}(\lambda) &\geq \epsilon'(\lambda) \\ &= (1-\delta)^{q_s+q_f+q_r+q_t} \cdot \delta \cdot \mathrm{Adv}_{\mathrm{Sim}, \mathcal{A}_{\mathcal{I}}}^{\mathrm{IND-CKA-}I}(\lambda) \\ &\geq (1-\delta)^{q_s+q_f+q_r+q_t} \cdot \delta \cdot \epsilon(\lambda). \end{aligned}$$

When $\delta = \frac{1}{q_s+q_f+q_r+q_t+1}$, $(1-\delta)^{q_s+q_f+q_r+q_t} \cdot \delta \cdot \epsilon(\lambda)$ reaches its maximum, and the maximum value is $\frac{\epsilon(\lambda)}{e(q_s+q_f+q_r+q_t+1)}$.

Let $\mathcal{H}$ denote the event: $\mathcal{A}_{\mathcal{I}}$ issues Test query that makes X appears in the $List^{test}$.

Since we have:

$$\Pr\left[\mathrm{Exp}_{\Pi, \mathcal{C}}^{\mathrm{IND-CKA-}I}(\lambda) = 1 | \neg\mathcal{H}\right] = \frac{1}{2},$$

we have:

$$\begin{aligned}
&\Pr\left[\mathrm{Exp}_{\Pi,\mathcal{C}}^{\mathrm{IND-CKA}}(\lambda)=1\right]\\
=&\Pr\left[\mathrm{Exp}_{\Pi,\mathcal{C}}^{\mathrm{IND-CKA}}(\lambda)=1|\neg\mathcal{H}\right]\Pr\left[\neg\mathcal{H}\right]\\
&+\Pr\left[\mathrm{Exp}_{\Pi,\mathcal{C}}^{\mathrm{IND-CKA}}(\lambda)=1|\mathcal{H}\right]\Pr\left[\mathcal{H}\right]\\
\leq&\Pr\left[\mathrm{Exp}_{\Pi,\mathcal{C}}^{\mathrm{IND-CKA}}(\lambda)=1|\neg\mathcal{H}\right]\Pr\left[\neg\mathcal{H}\right]+\Pr\left[\mathcal{H}\right]\\
=&\frac{1}{2}\Pr\left[\neg\mathcal{H}\right]+\Pr\left[\mathcal{H}\right]\\
=&\frac{1}{2}(1-\Pr\left[\mathcal{H}\right])+\Pr\left[\mathcal{H}\right]=\frac{1}{2}+\frac{1}{2}\Pr\left[\mathcal{H}\right],
\end{aligned}$$

$$\begin{aligned}
&\Pr\left[\mathrm{Exp}_{\Pi,\mathcal{C}}^{\mathrm{IND-CKA-I}}(\lambda)=1\right]\\
\geq&\Pr\left[\mathrm{Exp}_{\Pi,\mathcal{C}}^{\mathrm{IND-CKA-I}}(\lambda)=1|\neg\mathcal{H}\right]\Pr\left[\neg\mathcal{H}\right]\\
=&\frac{1}{2}(1-\Pr\left[\mathcal{H}\right])=\frac{1}{2}-\frac{1}{2}\Pr\left[\mathcal{H}\right].
\end{aligned}$$

Since we have:

$$\left|\Pr\left[\mathrm{Exp}_{\Pi,\mathcal{C}}^{\mathrm{IND-CKA-}I}(\lambda)=1\right]-\frac{1}{2}\right|\geq\epsilon'(\lambda),$$

we have:

$$\epsilon'(\lambda)\leq\left|\Pr\left[\mathrm{Exp}_{\Pi,\rfloor}^{\mathrm{IND-CKA}}(\lambda)=1\right]-\frac{1}{2}\right|\leq\frac{1}{2}\Pr\left[\mathcal{H}\right].$$

So:

$$Pr[\mathcal{H}]\geq 2\epsilon'(\lambda).$$

The X containing D appears in $List^{test}$ with probability at least $2\epsilon'(\lambda)$ after the simulation ends. $\mathcal{A}_{\mathcal{I}}$ queries Test at least q_{test} times, and $List^{test}$ established by $\mathcal{C}$ has at least q_{test} entries, so $\mathcal{C}$ randomly picks an entry in $List^{test}$ as D with probability

$$\epsilon''(\lambda)=\frac{2\epsilon'(\lambda)}{q_{test}}.$$

We have $\epsilon'(\lambda)\geq\frac{\epsilon(\lambda)}{e(q_s+q_f+q_r+q_t+1)}$, so there exists $\mathcal{C}$ that solves the BDH problem with advantage

$$ADV_{P,\mathcal{C}}^{BDH}\geq\epsilon''(\lambda)\geq\frac{2\epsilon(\lambda)}{e(q_s+q_f+q_r+q_t+1)q_{test}}.$$

Lemma 2. *Let the hash functions H_1, H_2 be random oracle machines, and if the BDH problem is hard on the group generated by P, then our scheme is IND-CKA-II secure.*

Proof. Assuming that it exists as a PPT adversary $\mathcal{A}_{\mathcal{II}}$ can break the IND-CKA-II GAME with a non-negligible advantage $\epsilon(\lambda)$, and $\mathcal{A}_{\mathcal{II}}$ conducts at most $q_s > 0$ times Partial Key Queries, $q_f > 0$ times Full Key Queries, q_r times Re-Key Queries, $q_t > 0$ times Trapdoor Queries, and $q_{test} > 0$ times Test Queries, then there must exist an challenger $\mathcal{C}$ who can solve the BDH problem in the group generated by P with advantage

$$ADV_{P,\mathcal{C}}^{BDH} \geq \frac{2\epsilon(\lambda)}{e(1 + q_s + q_f + q_r + q_t)q_{test}},$$

where e is the base of the natural logarithm.

1. Setup: $\mathcal{A}_{\mathcal{II}}$ randomly selects $msk = \alpha \in \mathbb{G}_1$ and sends it to $\mathcal{C}$. $\mathcal{C}$ inputs the security parameter λ and performs *Setup* algorithm to generate system parameters $prms$. $\mathcal{C}$ randomly chooses $ID_{\mathcal{A}_{\mathcal{II}}} \in \{0,1\}^*$ as the identity of $\mathcal{A}_{\mathcal{II}}$.
   In response to $\mathcal{A}_{\mathcal{II}}$'s discontinuous queries, $\mathcal{C}$ needs to maintain four initially empty lists:
   - $List^{H_1}$ of H_1 with elements of type 4-tuple $(ID_i, Q_i, b_i, coin)$ in response to $\mathcal{A}_{\mathcal{II}}$'s H_1 Queries.
   - $List^{H_2}$ of H_2 with elements of type 2-tuple (w_i, t_i) in response to $\mathcal{A}_{\mathcal{II}}$'s H_2 Queries.
   - $List^{key}$ with elements of type 3-tuple (ID_i, Pk_i, x_i) in response to $\mathcal{A}_{\mathcal{II}}$'s Public Key Queries and Full Key Queries.
   - $List^{test}$ with elements of type 2-tuple (T, X) in response to $\mathcal{A}_{\mathcal{II}}$'s Test Queries.

   Then $\mathcal{C}$ sends $prms = (\mathbb{G}_1, \mathbb{G}_2, q, \hat{e}, P_k, P, P_1, P_2, Z, H_1, H_2)$ to the adversary $\mathcal{A}_{\mathcal{II}}$ and keeps msk secret.
2. H_1 Queries: $\mathcal{A}_{\mathcal{II}}$ can make queries at any time. When $\mathcal{A}_{\mathcal{II}}$ queries H_1 oracle with identity ID_i, $\mathcal{C}$ will respond in the following way:
   - $\mathcal{C}$ first checks whether the ID_i exists in the list $List^{H_1}$. If it exists, the corresponding $H_1(ID_i) = Q_i \in \mathbb{G}_1$ is returned to $\mathcal{A}_{\mathcal{II}}$.
   - Otherwise if $ID_i = ID_{\mathcal{A}_{\mathcal{II}}}$, $\mathcal{C}$ randomly chooses $Q_{\mathcal{A}_{\mathcal{II}}} \in \mathbb{G}_1$ and sets $Q_i = Q_{\mathcal{A}_{\mathcal{II}}}$
   - Otherwise $\mathcal{C}$ will produce $coin \in \{0,1\}$ with probability distribution satisfying $Pr[coin = 0] = \delta$. $\mathcal{C}$ chooses a random number $b_i \in \mathbb{Z}_q^*$. If $coin = 0$, $\mathcal{C}$ will compute $Q_i = b_i Q_{\mathcal{A}_{\mathcal{II}}} \in \mathbb{G}_1$, and if $coin = 1$, $\mathcal{C}$ will compute $Q_i = b_i P \in \mathbb{G}_1$. Finally, $\mathcal{C}$ saves $(ID_i, Q_i, b_i, coin)$ to the $List^{H_1}$ and returns $H_1(ID_i) = Q_i$ to $\mathcal{A}_{\mathcal{II}}$.

   Here, the coin serves as a guess for $\mathcal{C}$: $coin = 0$ denotes $\mathcal{A}_{\mathcal{II}}$ will launch an attack on the ID_i of this query.
3. H_2 Queries: $\mathcal{A}_{\mathcal{II}}$ can make queries at any time. $\mathcal{A}_{\mathcal{II}}$ queries H_2 oracle with keyword w_i. $\mathcal{C}$ will respond in the following way: $\mathcal{C}$ first checks whether the w_i exists in the list $List^{H_2}$. If it exists, then corresponding t_i is returned to $\mathcal{A}_{\mathcal{II}}$ as $H_2(w_i)$. If w_i does not exist, $\mathcal{C}$ will select $t_i \in \mathbb{Z}_q^*$ and saves (w_i, t_i) to the $List^{H_2}$ and returns t_i to $\mathcal{A}_{\mathcal{II}}$ as $H_2(w_i)$.
4. Query phase I:

- Partial Key Queries: $\mathcal{A}_{II}$ chooses identity $ID_i \neq ID_{\mathcal{A}_{II}}$ to query. $\mathcal{C}$ will get the corresponding $(ID_i, Q_i, b_i, coin)$ in $List^{H_1}$ first.
  If $coin = 0$, $\mathcal{C}$ aborts the query and quits. At this point, $\mathcal{C}$ was initially intended to use $\mathcal{A}_{II}$'s attack on IND-CKA to attack the BDH problem. In the query phase, $\mathcal{C}$ cannot use $\mathcal{A}_{II}$ to launch an attack.
  If coin=1, $\mathcal{C}$ computes ID_i's partial private key $U_i = b_i P_k$ and sends U_i to $\mathcal{A}_{II}$. Here, the private key of $User_i$ has a good representation: $U_i = \alpha Q_i = \alpha b_i P = b_i P_k$.
- Public Key Queries: $\mathcal{A}_{II}$ chooses identity ID_i to query. $\mathcal{C}$ will respond in the following way: $\mathcal{C}$ first checks whether the ID_i exists in the list $List^{key}$. If it exists, then the corresponding Pk_i is returned to $\mathcal{A}_{II}$. If ID_i does not exist, $\mathcal{C}$ will select $x_i \in \mathbb{Z}_q^*$ and computes $Pk_i = x_i P$, and saves (ID_i, Pk_i, x_i) to the $List^{key}$. Then $\mathcal{C}$ returns Pk_i to $\mathcal{A}_{II}$.
- Full Key Queries: $\mathcal{A}_{II}$ chooses identity $ID_i \neq ID_{\mathcal{A}_{II}}$ to query. $\mathcal{C}$ makes a Partial Key Query (corresponding coin=1) and a Public Key Query with ID_i to get the corresponding $sk_i = (U_i, x_i)$.

5. Master key Replacement: $\mathcal{A}_{II}$ has already selected the msk by itself in the Setup phase, and $\mathcal{C}$ uses this msk for Setup.
6. Challenge: $\mathcal{A}_{II}$ launches a challenge.
   $\mathcal{A}_{II}$ chooses a random number $v \in \mathbb{Z}_q^*$ and (ID_o, ID^*, w_0, w_1) it wants to challenge. $\mathcal{C}$ looks up item $(ID_o, Q_o, b_o, coin)$ in $List^{H_1}$ where $coin = 0$. And looks up item $(ID_i, Q_i, b_i, coin)$ in $List^{H_1}$ and (ID_i, Pk_i, x_i) in $List^{key}$, such that $ID_i = ID^*$. If coin $= 1$, $\mathcal{C}$ aborts $\perp$ and quits. If coin $= 0$, $\mathcal{C}$ selects $\beta \in \{0,1\}$ and $r, \rho \in \mathbb{Z}_q^*$ randomly, and gets (w_β, t_β) from $List^{H_2}$, then computes $C^* = (C_{1,\beta} = \hat{e}(P_k, rt_\beta Z) \cdot \hat{e}(rP, (\alpha - vx_i)Q_i - \alpha Q_o), C_2 = b_i^{-1} r Q_o, C_3 = b_i^{-1} rP, C_4 = b_i^{-1} r P_2, rk_1 = \rho P_1 + (vx_i - \alpha) Q_i + \alpha Q_o, rk_2 = \rho P)$. That's because the $Q_{\mathcal{A}_{II}}$ needs to be restored when $coin = 0$, e.g. $\hat{e}(U_i, C_3) = \hat{e}(\alpha b_i Q_{\mathcal{A}_{II}}, b_i^{-1} rP) = \hat{e}(\alpha Q_{\mathcal{A}_{II}}, rP)$.
   $\mathcal{C}$ defines $C^* = (C_{1,\beta}, C_2, C_3)$. Let $a = x_i, bP = Q_{\mathcal{A}_{II}}, c = r$. $\mathcal{C}$ has embedded the BDH problem $\hat{e}(P,P)^{abc}$ into $C_{1,\beta}$:

$$
\begin{aligned}
&C_{1,\beta} \\
&= \hat{e}(b_i^{-1} rP, (\alpha - vx_i)Q_i - \alpha Q_o) \cdot \hat{e}(Pk, rt_\beta P_1) \\
&= \hat{e}(b^{-1} rP, (\alpha Q_i - vx_i Q_i - \alpha Q_o) \cdot \hat{e}(Pk, rt_\beta P_1) \\
&= \hat{e}(b^{-1} rP, vx_i Q_i^{-1}) \cdot \hat{e}(b^{-1} rP, \alpha(Q_i - Q_o)) \cdot \hat{e}(Pk, rt_\beta P_1) \\
&= \hat{e}(rP, vx_i Q_{\mathcal{A}_{II}}^{-1}) \cdot \hat{e}(b^{-1} rP, \alpha(Q_i - Q_o)) \cdot \hat{e}(Pk, rt_\beta P_1) \\
&= \hat{e}(P,P)^{abc \cdot (-v)} \cdot \hat{e}(b^{-1} rP, \alpha(Q_i - Q_o)) \cdot \hat{e}(Pk, rt_\beta P_1).
\end{aligned}
$$

7. Query phase II.
   - Queries are the same as in query phase 1.
   - Re-Key Queries: $\mathcal{A}_{II}$ chooses identity ID_i to query. $\mathcal{C}$ will get the corresponding $(ID_i, Q_i, b_i, coin)$ in $List^{H_1}$ first. If $coin = 0$, $\mathcal{C}$ aborts the query and quits. If $coin = 1$, $\mathcal{C}$ computes $rk_1 = \rho P_1 + vx_i Q_i - \alpha Q_i, rk_2 = \rho P$ and sends (rk_1, rk_2) to $\mathcal{A}_{II}$.

- Trapdoor Queries: $\mathcal{A}_{II}$ chooses identity ID_i and $w \neq (w_0|w_1)$ to query. $\mathcal{C}$ will get the corresponding $(ID_i, Q_i, b_i, coin)$ in $List^{H_1}$ first. If $coin = 0$, $\mathcal{C}$ aborts the query and quits. If $coin = 1$, $\mathcal{C}$ computes TD to $\mathcal{A}_{II}$.
- Test queries: When $\mathcal{A}_{II}$ queries with TD to know whether the trapdoor is validated, $\mathcal{C}$ will respond in the following way: $\mathcal{C}$ first checks whether the T_1 exists in the list $List^{test}$ such that $T_1 = T$. If it exists, then it gets the corresponding X. And if T_1 does not exist, $\mathcal{C}$ computes

$$\begin{aligned}
&T = T_1, \\
&X = \hat{e}(T_1, C_3) \\
&= \hat{e}(b_i^{-1} rP, (\alpha - vx_i)Q_i - vx_i Q_o) \cdot \hat{e}(rP, \alpha H_2(w_j)P_1) \\
&= \hat{e}(rP, (\alpha - vx_i)Q_{\mathcal{A}_{II}}) \cdot \hat{e}(b_i^{-1} rP, Q_o)^{-vx_i} \cdot \hat{e}(rP, \alpha H_2(w_j)P_1) \\
&= \hat{e}(P, P)^(\alpha - vx_i)br \cdot \hat{e}(b_i^{-1} rP, Q_o)^{-vx_i} \cdot \hat{e}(rP, \alpha H_2(w_j)P_1) \\
&= \hat{e}(P, P)^{\alpha br - abcv} \cdot \hat{e}(b_i^{-1} rP, Q_o)^{-vx_i} \cdot \hat{e}(cP, aH_2(w_j)P_1) \\
&= D^{-v} \cdot \hat{e}(P, P)^{\alpha br} \cdot \hat{e}(b_i^{-1} rP, Q_o)^{-vx_i} \cdot \hat{e}(cP, aH_2(w_j)P_1),
\end{aligned}$$

and saves (T, X) to the $List^{test}$. If $X = \hat{e}(C_2, T_2) \cdot C^*$, then the challenger C returns 1; otherwise, 0 is returned to the $\mathcal{A}_{II}$.

8. Guess: $\mathcal{A}_{II}$ outputs a guess $\beta' \in \{0, 1\}$. $\mathcal{A}_{II}$ wins the above game if $\beta' = \beta$. $\mathcal{C}$ randomly takes (T, X) from the $List^{test}$ and computes D as the solution of BDH.

The advantage of $\mathcal{A}_{II}$ in the simulation attack is

$$\mathrm{Adv}_{\mathsf{Sim},\mathcal{A}_{II}}^{\mathrm{IND-CKA-II}}(\lambda) \geq \epsilon(\lambda) = \left| \Pr\left[\mathrm{Exp}_{\Pi,\mathcal{A}_{II}}^{\mathrm{IND-CKA-II}}(\lambda) = 1\right] - \frac{1}{2} \right|.$$

If $\mathcal{A}_{II}$'s guess is correct, $\mathcal{A}_{II}$ succeeds in attacking the IND-CKA-II at step 8. Since the probability that $\mathcal{C}$ does not interrupt at steps 4 and 7 is $(1-\delta)^{q_s+q_f+q_r+q_t}$, and the probability that $\mathcal{C}$ does not interrupt at steps 6 is δ, the probability that $\mathcal{C}$ does not interrupt is $(1-\delta)^{q_s+q_f+q_r+q_t}\delta$, and the advantage of $\mathcal{C} \geq \epsilon(\lambda)$ is as follows:

$$\begin{aligned}
\mathrm{Adv}_{\mathsf{Sim},\mathcal{C}}^{\mathrm{IND-CKA-}II}(\lambda) &\geq \epsilon'(\lambda) \\
&= (1-\delta)^{q_s+q_f+q_r+q_t} \cdot \delta \cdot \mathrm{Adv}_{\mathsf{Sim},\mathcal{A}_{II}}^{\mathrm{IND-CKA-}II}(\lambda) \\
&\geq (1-\delta)^{q_s+q_f+q_r+q_t} \cdot \delta \cdot \epsilon(\lambda).
\end{aligned}$$

When $\delta = \frac{1}{q_s+q_f+q_r+q_t+1}$, $(1-\delta)^{q_s+q_f+q_r+q_t} \cdot \delta \cdot \epsilon(\lambda)$ reaches its maximum, and the maximum value is $\frac{\epsilon(\lambda)}{e(q_s+q_f+q_r+q_t+1)}$.

Let $\mathcal{H}$ denote the event: $\mathcal{A}_{II}$ issues Test query that makes X appears in the $List^{test}$.

Since we have:

$$\Pr\left[\mathrm{Exp}_{\Pi,\mathcal{C}}^{\mathrm{IND-CKA-}II}(\lambda) = 1 | \neg\mathcal{H}\right] = \frac{1}{2},$$

we have:

$$
\begin{aligned}
&\Pr\left[\mathrm{Exp}_{\Pi,\mathcal{C}}^{\mathrm{IND-CKA-}II}(\lambda)=1\right]\\
=&\Pr\left[\mathrm{Exp}_{\Pi,\mathcal{C}}^{\mathrm{IND-CKA-}II}(\lambda)=1|\neg\mathcal{H}\right]\Pr\left[\neg\mathcal{H}\right]\\
&+\Pr\left[\mathrm{Exp}_{\Pi,\mathcal{C}}^{\mathrm{IND-CKA-}II}(\lambda)=1|\mathcal{H}\right]\Pr\left[\mathcal{H}\right]\\
\leq&\Pr\left[\mathrm{Exp}_{\Pi,\mathcal{C}}^{\mathrm{IND-CKA-}II}(\lambda)=1|\neg\mathcal{H}\right]\Pr\left[\neg\mathcal{H}\right]+\Pr\left[\mathcal{H}\right]\\
=&\frac{1}{2}\Pr\left[\neg\mathcal{H}\right]+\Pr\left[\mathcal{H}\right]\\
=&\frac{1}{2}(1-\Pr\left[\mathcal{H}\right])+\Pr\left[\mathcal{H}\right]=\frac{1}{2}+\frac{1}{2}\Pr\left[\mathcal{H}\right],
\end{aligned}
$$

$$
\begin{aligned}
&\Pr\left[\mathrm{Exp}_{\Pi,\mathcal{C}}^{\mathrm{IND-CKA-II}}(\lambda)=1\right]\\
\geq&\Pr\left[\mathrm{Exp}_{\Pi,\mathcal{C}}^{\mathrm{IND-CKA-II}}(\lambda)=1|\neg\mathcal{H}\right]\Pr\left[\neg\mathcal{H}\right]\\
=&\frac{1}{2}(1-\Pr\left[\mathcal{H}\right])=\frac{1}{2}-\frac{1}{2}\Pr\left[\mathcal{H}\right].
\end{aligned}
$$

Since we have:

$$\left|\Pr\left[\mathrm{Exp}_{\Pi,\mathcal{C}}^{\mathrm{IND-CKA-}II}(\lambda)=1\right]-\frac{1}{2}\right|\geq\epsilon'(\lambda),$$

we have:

$$\epsilon'(\lambda)\leq\left|\Pr\left[\mathrm{Exp}_{\Pi,\mathrm{J}}^{\mathrm{IND-CKA-II}}(\lambda)=1\right]-\frac{1}{2}\right|\leq\frac{1}{2}\Pr\left[\mathcal{H}\right].$$

So:

$$Pr[\mathcal{H}]\geq 2\epsilon'(\lambda).$$

The X containing D appears in $List^{test}$ with probability at least $2\epsilon'(\lambda)$ after the simulation ends. $\mathcal{A}_{\mathcal{II}}$ queries Test at least q_{test} times, and $List^{test}$ established by $\mathcal{C}$ has at least q_{test} entries, so $\mathcal{C}$ randomly picks an entry in $List^{test}$ as D with probability

$$\epsilon''(\lambda)=\frac{2\epsilon'(\lambda)}{q_{test}}.$$

We have $\epsilon'(\lambda)\geq\frac{\epsilon(\lambda)}{e(q_s+q_f+q_r+q_t+1)}$, so there exists $\mathcal{C}$ that solves the BDH problem with advantage

$$ADV_{P,\mathcal{C}}^{BDH}\geq\epsilon''(\lambda)\geq\frac{2\epsilon(\lambda)}{e(q_s+q_f+q_r+q_t+1)q_{test}}.$$

4.2 IND-CCA

Theorem 2. *Let the hash functions H_1, H_2 be random oracle machines, and if the BDH problem is hard on the group generated by P, then our scheme is IND-CCA secure.*

Correctness:

$$
\begin{aligned}
& C_0 \oplus H_3(rk_3 \cdot \hat{e}(U_2, C_5)^{-1}) \\
= & m \oplus H_3(\hat{e}(K_j, rP_1)) \oplus H_3(rk_3 \cdot \hat{e}(U_2, C_5)^{-1})) \\
= & m \oplus H_3(\hat{e}(K_j, rP_1)) \oplus H_3(\hat{e}(r(H_1(w||ID_1) + H_1(ID_2)), Z) \cdot \hat{e}(U_2, C_5)^{-1})) \\
= & m \oplus H_3(\hat{e}(\alpha H_1(w||ID_1), rP_1)) \oplus H_3(\hat{e}(r(H_1(w||ID_1) + H_1(ID_2)), \alpha P_1) \\
& \cdot \hat{e}(\alpha H_1(ID_2), rP_1)^{-1})) \\
= & m \oplus H_3(\hat{e}(\alpha H_1(w||ID_1), rP_1)) \oplus H_3(\hat{e}(\alpha(H_1(w||ID_1) + H_1(ID_2)), rP_1) \\
& \cdot \hat{e}(\alpha H_1(ID_2), rP_1)^{-1})) \\
= & m \oplus H_3(\hat{e}(\alpha H_1(w||ID_1), rP_1)) \oplus H_3(\hat{e}(\alpha H_1(w||ID_1), rP_1) \\
= & m
\end{aligned}
$$

The proof proceeds similarly to Subsect. 4.1. To make $\mathcal{A}$ a true guess, adversary $\mathcal{A}$ needs to figure out the value of $\hat{e}(\alpha H_1(w||ID_1), rP_1)$. Let $aP = H_1(w||ID_1), b = r, cP = P_1$, the solution to the BDH problem will appear in $\hat{e}(\alpha H_1(w||ID_1), rP_1) = \hat{e}(P, P)^{abc \cdot \alpha}$.

The difficulty of winning this game for either $\mathcal{A}_{\mathcal{I}}$ or $\mathcal{A}_{\mathcal{II}}$ depends on the BDH problem.

4.3 SUF-CMA

Theorem 3. *Let the hash functions H_1, H_2 be random oracle machines, and if the CDH problem is hard on the group generated by P, then our scheme is SUF-CMA secure.*

Correctness:

$$
\begin{aligned}
& \hat{e}(H_1(m), Pk_1) \cdot \hat{e}(C_2, P_k) \\
= & \hat{e}(H_1(m), x_1 P) \cdot \hat{e}(rH_1(ID_1), \alpha P) \\
= & \hat{e}(x_1 H_1(m), P) \cdot \hat{e}(r\alpha H_1(ID_1), P) \\
= & \hat{e}(x_1 H_1(m) + rU_1, P) \\
= & \hat{e}(\sigma, P)
\end{aligned}
$$

The proof proceeds similarly to Subsect. 4.1. If $\mathcal{A}$ can make a forgery, $\mathcal{A}$ needs to figure out the value of $x_1 H_1(m) + rU_1$. When $\mathcal{A} = \mathcal{A}_{\mathcal{I}}$, Let $aP = U_1, bP = rP$, the solution abP to the CDH problem will appear in $x_1 H_1(m) + rU_1 = x_1 H_1(m) + abP$. When $\mathcal{A} = \mathcal{A}_{\mathcal{II}}$ Let $aP = H_1(m), bP = x_1 P$, the solution abP to the CDH problem will appear in $x_1 H_1(m) + rU_1 = abP + rU_1$.

The difficulty of winning this game for either $\mathcal{A}_{\mathcal{I}}$ or $\mathcal{A}_{\mathcal{II}}$ depends on the CDH problem.

5 Conclusion

In cloud environments, ensuring data security, privacy-preserving sharing, and efficient authentication is paramount for practical application. To address these challenges, this paper proposes a certificateless searchable proxy re-encryption scheme, which enables secure and searchable data sharing while eliminating certificate management and key escrow issues. By integrating searchable encryption with proxy re-encryption, our scheme allows data owners to delegate search permissions to authorized users without compromising confidentiality. The embedded authentication mechanism ensures data integrity and prevents unauthorized access during sharing.

However, a limitation of our scheme is that data requesters must obtain re-encryption keys from the data owner online, which may introduce latency in dynamic sharing scenarios. In future work, we plan to further enhance flexibility in private data sharing.

References

1. Song, D.X., Wagner, D., Perrig, A.: Practical techniques for searches on encrypted data. In: Proceeding 2000 IEEE Symposium on Security and Privacy, S&P 2000, pp. 44–55. IEEE (2000)
2. Boneh, D., Di Crescenzo, G., Ostrovsky, R., Persiano, G.: Public key encryption with keyword search. In: Cachin, C., Camenisch, J.L. (eds.) EUROCRYPT 2004. LNCS, vol. 3027, pp. 506–522. Springer, Heidelberg (2004). https://doi.org/10.1007/978-3-540-24676-3_30
3. Byun, J.W., Rhee, H.S., Park, H.-A., Lee, D.H.: Off-line keyword guessing attacks on recent keyword search schemes over encrypted data. In: Jonker, W., Petković, M. (eds.) SDM 2006. LNCS, vol. 4165, pp. 75–83. Springer, Heidelberg (2006). https://doi.org/10.1007/11844662_6
4. Chen, R., Yi, M., Yang, G., Guo, F., Wang, X.: Dual-server public-key encryption with keyword search for secure cloud storage. IEEE Trans. Inf. Forensics Secur. **11**(4), 789–798 (2015)
5. Emura, K.: Generic construction of public-key authenticated encryption with keyword search revisited: stronger security and efficient construction. In: Proceedings of the 9th ACM on ASIA Public-Key Cryptography Workshop, pp. 39–49 (2022)
6. Liu, Z.-Y., Tseng, Y.-F., Tso, R., Mambo, M., Chen, Y.-C.: Public-key authenticated encryption with keyword search: cryptanalysis, enhanced security, and quantum-resistant instantiation. In: Proceedings of the 2022 ACM on Asia Conference on Computer and Communications Security, pp. 423–436 (2022)
7. Xu, X.: Method for protecting privacy data security in smart healthcare under searchable encryption. Inf. Syst. Eng. **07**, 24–27 (2025)
8. Blaze, M., Bleumer, G., Strauss, M.: Divertible protocols and atomic proxy cryptography. In: Nyberg, K. (eds.) EUROCRYPT 1998. LNCS, vol. 1403, pp. 127–144. Springer, Heidelberg (1998). https://doi.org/10.1007/bfb0054122
9. Abdalla, M., et al.: Searchable encryption revisited: consistency properties, relation to anonymous IBE, and extensions. In: Shoup, V. (ed.) CRYPTO 2005. LNCS, vol. 3621, pp. 205–222. Springer, Heidelberg (2005). https://doi.org/10.1007/11535218_13

10. Isshiki, T., Nguyen, M.H., Tanaka, K.: Proxy re-encryption in a stronger security model extended from CT-RSA2012. In: Dawson, E. (ed.) CT-RSA 2013. LNCS, vol. 7779, pp. 277–292. Springer, Heidelberg (2013). https://doi.org/10.1007/978-3-642-36095-4_18
11. Tang, F., Li, H., Chang, J.: Multi-hop unidirectional proxy re-encryption from multilinear maps. IEICE Trans. Fundam. Electron. Commun. Comput. Sci. **98**(2), 762–766 (2015)
12. Deng, H., Qin, Z., Qianhong, W., Guan, Z., Zhou, Y.: Flexible attribute-based proxy re-encryption for efficient data sharing. Inf. Sci. **511**, 94–113 (2020)
13. Feng, J., Li, Y., Wang, T., Liu, S.: A certificateless threshold proxy re-encrypted data sharing scheme with cloud-chain collaboration in industrial internet environments. IEEE Internet Things J. (2024)
14. Shao, J., Cao, Z., Liang, X., Lin, H.: Proxy re-encryption with keyword search. Inf. Sci. **180**(13), 2576–2587 (2010)
15. Yau, W.-C., Phan, R.C.-W., Heng, S.-H., Goi, B.-M.: Proxy re-encryption with keyword search: new definitions and algorithms. In: Kim, T., Fang, W., Khan, M.K., Arnett, K.P., Kang, H., Ślęzak, D. (eds.) SecTech 2010. CCIS, vol. 122, pp. 149–160. Springer, Heidelberg (2010). https://doi.org/10.1007/978-3-642-17610-4_17
16. Chen, X., Li, Y.: Efficient proxy re-encryption with private keyword searching in untrusted storage. Int. J. Comput. Netw. Inf. Secur. **3**(2), 50 (2011)
17. Wang, X.A., Huang, X., Yang, X., Liu, L., Wu, X.: Further observation on proxy re-encryption with keyword search. J. Syst. Softw. **85**(3), 643–654 (2012)
18. Fang, L., Susilo, W., Ge, C., Wang, J.: Chosen-ciphertext secure anonymous conditional proxy re-encryption with keyword search. Theoret. Comput. Sci. **462**, 39–58 (2012)
19. Yang, Y., Ma, M., Lin, B.: Proxy re-encryption conjunctive keyword search against keyword guessing attack. In: 2013 Computing, Communications and IT Applications Conference (ComComAp), pp. 125–130. IEEE (2013)
20. Chen, Z., Li, S., Huang, Q., Wang, Y., Zhou, S.: A restricted proxy re-encryption with keyword search for fine-grained data access control in cloud storage. Concurr. Comput.: Pract. Exp. **28**(10), 2858–2876 (2016)
21. Yang, Y., Ma, M.: Conjunctive keyword search with designated tester and timing enabled proxy re-encryption function for e-health clouds. IEEE Trans. Inf. Forensics Secur. **11**(4), 746–759 (2015)
22. Yang, Y., Zheng, X., Chang, V., Tang, C.: Semantic keyword searchable proxy re-encryption for postquantum secure cloud storage. Concurr. Comput.: Pract. Exp. **29**(19), e4211 (2017)
23. Manzoor, A., Liyanage, M., Braeke, A., Kanhere, S.S., Ylianttila, M.: Blockchain based proxy re-encryption scheme for secure IoT data sharing. In: 2019 IEEE International Conference on Blockchain and Cryptocurrency (ICBC), pp. 99–103. IEEE (2019)
24. Xiaoling, Yu., Chungen, X., Dou, B., Wang, Y.: Multi-user search on the encrypted multimedia database: lattice-based searchable encryption scheme with time-controlled proxy re-encryption. Multimedia Tools Appl. **80**(2), 3193–3211 (2021)
25. Fugkeaw, S., Deevijit, J.: Se-collab: achieving fine-grained and efficiently verifiable searchable encryption with Boolean multi-keyword search for collaborative IIoT data sharing. IEEE Access (2025)

Federated Learning Backdoor Attack Using Dynamic Parameter Adjustment and Collaborative Trigger Optimization

Shijie Chen and Bo Dai(✉)

University of Electronic Science and Technology of China (UESTC), Chengdu 611731, Sichuan, China
daibo@uestc.edu.cn
https://www.uestc.edu.cn

Abstract. Federated learning is a distributed machine learning approach that enables clients to train models on their local datasets. However, its distributed architecture and heterogeneous data distribution create vulnerabilities for backdoor attacks, where malicious clients manipulate local updates to implant hidden backdoors into the global model. Existing backdoor attack methods in federated learning typically lack adaptability to dynamic training environments—often failing to maintain stealth and effectiveness when confronted with defense mechanisms, leading to compromised attack performance or detection. To address these limitations of current attack strategies, this paper introduces a backdoor attack method based on dynamic parameter adjustment and collaborative trigger optimization. By employing carefully designed evaluation metrics, the attack selects parameters with minimal impact on the primary task for dynamic adjustment, thereby minimizing interference with the global model's performance. Simultaneously, the attack trigger is optimized to decentralize the attack, enhancing both its stealth and success rate. Extensive experiments on the CIFAR-10 and Tiny-ImageNet datasets show that this method successfully evades advanced defense mechanisms, achieving an attack success rate exceeding 98%.

Keywords: Federated Learning · Backdoor Attack · Dynamic Parameter Adjustment · Collaborative Trigger Optimization

1 Introduction

To clarify the research context, we first define the "dynamic training environment" in federated learning (FL) backdoor scenarios, which aligns with the characteristics of real-world FL deployments. This dynamicity manifests in three key dimensions: (1) Adaptive defense updates: Servers may adjust defense strategies during training (e.g., switching from simple model pruning to advanced anomaly detection based on historical updates), or enhance defense intensity as suspicious activities are detected; (2) Fluctuating client participation: The

J. Yu and Y. Chen (Eds.): FCS 2025, CCIS 2805, pp. 425–440, 2027.
https://doi.org/10.1007/978-981-95-9514-3_28

set of participating clients (including malicious ones) changes across training rounds, altering the distribution of local updates; (3) Data distribution shifts: Local datasets evolve over time (e.g., new classes added or data characteristics changed), which affects both legitimate model training and backdoor implantation. Existing attacks often rely on fixed parameters and triggers, failing to adapt to such dynamic changes—this is the core problem addressed in our work (Fig. 1).

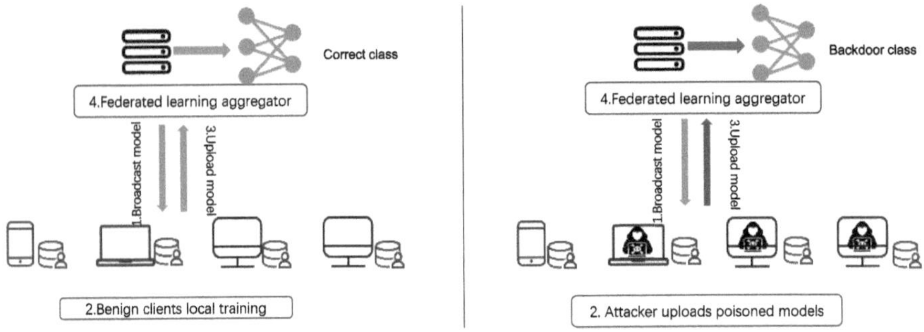

Fig. 1. Federated Learning Backdoor Attack

With the rapid advancement of artificial intelligence and machine learning, federated learning has emerged as a widely applied distributed learning paradigm. In the traditional framework, the server distributes the global model to each client, which then trains on local data and uploads the updated models back to the server. The server aggregates the model parameters according to specified rules and updates the global model until convergence. However, since the server cannot directly monitor the clients' training processes, it cannot detect potential attacks within the model, thereby providing an opportunity for malicious clients. Backdoor attacks, a type of model poisoning attack, aim to manipulate the global model such that it activates backdoor behavior in response to specific input triggers. Attackers design trigger conditions to implant backdoor logic while preserving the global model's prediction accuracy, such as through specific image labels or particular audio frequencies. Once the attack succeeds, the model will output the result predefined by the attacker whenever the trigger condition is met.

Bagdasaryan et al. [2] were the first to highlight the issue of backdoor attacks in federated learning. Since then, numerous studies have investigated the feasibility of implanting backdoors in federated learning environments. However, existing research on backdoor attacks in federated learning reveals a significant limitation: the lack of effective adjustment strategies for dynamic training environments. This limitation manifests in two key aspects. First, when adjusting model parameters, the dynamic changes in the training process are often overlooked, and a uniform adjustment method is applied. This can lead to excessive interference with normal parameters, resulting in significant performance degradation and premature exposure of the attack [21]. For instance, in image recognition tasks, a sharp decline in performance and accuracy could trigger defense mechanisms, causing the attack to fail. Second, most current attack methods rely on fixed trigger patterns [22], which are vulnerable in the complex and dynamic environment of federated learning and can be easily detected. When a large number of benign models are aggregated with a few malicious models, fixed trigger features may be overridden by benign features, rendering the attack ineffective.

This paper presents a backdoor attack strategy that combines dynamic parameter adjustment with collaborative trigger optimization, aiming to address the challenges posed by existing methods in dynamic environments while balancing the dynamic nature and diversity of trigger patterns. We introduce a novel evaluation metric that comprehensively accounts for both the proportion of local model parameters and the dynamic changes occurring during the training process. This approach enables the modification of parameters with minimal impact on training, thereby injecting a backdoor while reducing interference with the global model's normal performance and mitigating the issue of unreasonable parameter adjustments. Additionally, we propose a collaborative trigger optimization method that distributes centralized attacks across multiple malicious clients and implements the attack cooperatively during the aggregation process, thereby increasing the diversity of trigger patterns. Dynamic parameter adjustment enhances the model's sensitivity and adaptability to subsequent collaborative trigger attacks, while the collaborative trigger optimization further amplifies the impact of dynamic parameter adjustments.

The main contributions of this paper are as follows:

- We propose a dynamic parameter adjustment attack method that operates during the federated learning process by evaluating and dynamically adjusting model parameters. This approach precisely selects and modifies parameters with minimal impact on the primary task, effectively mitigating the performance degradation typically caused by excessive parameter adjustments.
- We introduce a novel collaborative trigger optimization strategy that decomposes the centralized attack across multiple malicious clients, allowing the local triggers to collaborate. This enables a more efficient attack during the global model aggregation while achieving diversification of trigger patterns.
- The proposed attack method has been rigorously tested and compared with the latest backdoor attack techniques on the CIFAR-10 and TinyImageNet datasets. Experimental results demonstrate that, in comparison to other backdoor attack methods, our approach successfully evades advanced backdoor defense mechanisms, achieving an attack success rate exceeding 98%.

2 Related Work

In this section, we discuss common backdoor attacks and defense methods in federated learning.

Backdoor Attacks in Federated Learning: Backdoor attacks have become a prominent focus in the field of federated learning, with many researchers investigating the issue from various perspectives. Bagdasaryan et al. [2] introduced backdoors by predicting global model updates and replacing them with models embedded with backdoors. Wang et al. [5] demonstrated that the robustness of a backdoor refers to the model's resilience to adversarial examples and the edge-case backdoors they proposed. Xie et al. [3] decomposed the trigger pattern into sub-patterns and distributed them across multiple malicious clients for implantation. Building on DBA, Gong et al. [15] proposed a coordinated backdoor attack in which the local trigger is tied to the model. Fang et al. [1] introduced a focus-flip backdoor attack in federated learning. Semantic backdoors, a variant of FL backdoor attacks, use inherent features in the target image as triggers, such as a car with green stripes. Wang et al. [5] proposed an edge-case backdoor similar to a semantic backdoor but with backdoor data located at the tail end of the global dataset distribution.

Backdoor Defense in Federated Learning: Several defense mechanisms have been developed to provide empirical robustness against poisoning attacks. Deepsight [9] conducts deep model checks on each model by inspecting its Normalized Update Energy and Division Difference. FL-Detector [11] predicts the global model based on model update consistency and detects outliers by measuring their distance from the predicted model. RFLBAT [12] utilizes principal component updates to reduce the dimensionality of gradient updates. FLAME [13] adapts weak differential privacy (DP) methods through noise boundary proofs and dynamic pruning boundaries, which help mitigate backdoor attacks while preserving high accuracy for the main task. Xie et al. [14] proposed a certified defense that uses clipping and smoothing techniques to enhance model smoothness. Rieger et al. [9] identify suspicious model updates through cluster-based similarity estimation.

Threat Model: This paper assumes that N clients participate in the federated learning training process, with attackers having full control over their local training processes. This includes injecting backdoor data and adjusting local training hyperparameters. However, attackers cannot influence the central server's authority, such as altering aggregation rules, nor can they tamper with other parties' training processes or model updates. Furthermore, since federated learning involves dynamic updates, attackers can only carry out attacks during a limited number of update rounds. The goal of the attackers in this study is to mislead the model so that it simultaneously fits both the main task and the backdoor task. This ensures that the global model achieves a high success rate for backdoor data samples while maintaining normal performance for unaltered data samples.

3 Proposed Approach

3.1 Overall Framework

This paper adopts the classic federated learning (FL) paradigm, where the central server broadcasts the global model and receives gradient updates from n clients, each of which trains on its local dataset.

Federated Learning. In the t-th round, the central server sends the shared model G_t to n clients. Each client i in this round first creates a local model $L_{i,t}$, then uses its dataset D_i and learning rate lr with an optimization algorithm to compute the local loss function and obtain the updated local model $L_{i,t+1}$.

$$L_{i,t} = G_t - \eta \cdot \nabla \mathcal{L}_{\text{task}}(G_t, \mathcal{D}_i) \tag{1}$$

After each user uploads $L_{i,t+1}$, the central server aggregates all the local models using an aggregation rule to generate the new global model G_{t+1}.

$$G_{t+1} = G_t + \frac{\eta}{n} \sum_{i=1}^{n} (L_{i,t+1} - G_t) \tag{2}$$

3.2 Dynamic Parameter Adjustment Federated Backdoor Attack

In this section, we propose a backdoor attack strategy in which malicious clients utilize a specific weight adjustment method to manipulate only those model parameters that have minimal impact on the main task. The objective is to induce significant activation changes at each layer of the model when a specific trigger pattern is encountered. Throughout the implementation, we aim to minimize the disruption to the original structure and performance of the model, making the modifications harder to detect while minimizing the negative impact on model accuracy. The diagram illustrating the proposed attack strategy is shown in Fig. 2.

Step 1: Parameter Evaluation. In the complex architecture of federated learning, implementing backdoor attacks requires careful design to avoid detection while maintaining the model's predictive performance on the main task. To achieve this, we propose a parameter manipulation strategy aimed at exerting a subtle yet effective influence on the model.

Our attack manipulates only a small subset of parameters that are relatively insignificant to the main task, thereby minimizing interference with the model's overall accuracy. Traditional evaluations of model parameters often overlook the dynamic changes that occur during the training process. In contrast, we employ a motion-based scoring mechanism that comprehensively considers both the parameter's weight and its gradient, constructing a more dynamic and adaptive evaluation metric. Specifically, the evaluation metric M_t for each parameter

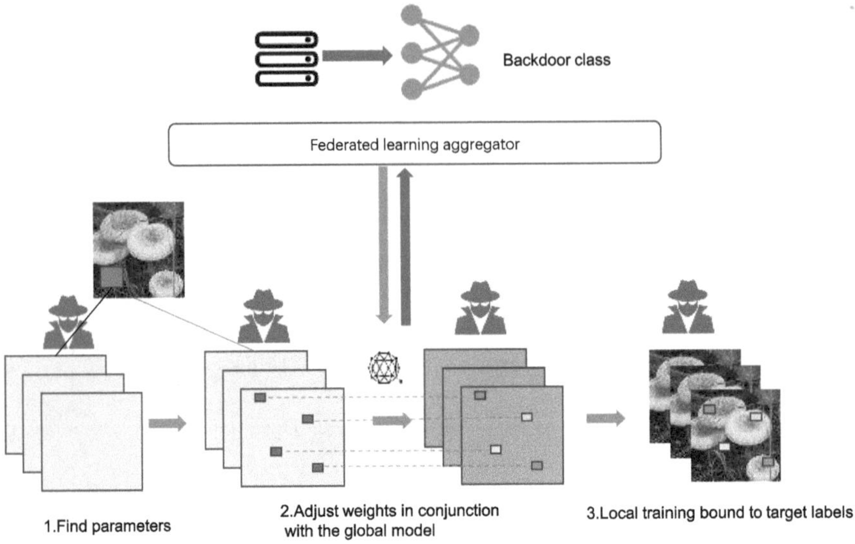

Fig. 2. Diagram of the Dynamic Parameter Adjustment Federated Backdoor Attack

is defined as the negative product of its weight and the partial derivative of the gradient with respect to that weight.

$$M_t = -\frac{\partial \mathcal{L}_g}{\partial w_t} \odot w_t \tag{3}$$

where w_t represents the weight of the parameter, and L_g denotes the global training loss. In the context of the federated learning backdoor attack, we have optimized and adapted the scoring mechanism to suit the specific characteristics of this environment, as outlined below:

- In federated learning, due to the distributed nature of information across clients, attackers are unable to directly access the global loss. To address this, we approximate the partial derivative using the model difference between consecutive training rounds, as follows:

$$-\frac{\partial \mathcal{L}_g}{\partial w_t} \approx w_t - w_{t-1} \tag{4}$$

  At the initial stage ($t = 0$), we randomly generate the importance score M_0.
- To accommodate different defense mechanisms in federated learning, such as directional and non-directional standards, we define the directional standard as focusing on the alignment between parameter changes and the model's training direction. The corresponding formula is:

$$M_t = (w_t - w_{t-1}) \odot w_t \tag{5}$$

This calculates the parameters that align with the overall trend and have low importance during training. The non-directional standard, on the other hand, emphasizes the magnitude of parameter changes. In this case, the formula is:

$$M_t = |(w_t - w_{t-1}) \odot w_t| \tag{6}$$

This approach identifies parameters with ambiguous directional changes but small magnitudes of variation.

After obtaining the evaluation metric M_t, we select the parameters with the lowest $\%k$ value in each layer as the target parameters for further manipulation. During the attack, we dynamically choose the most appropriate criterion based on the defense mechanism's characteristics, thereby optimizing both the success rate and stealth of the attack.

Step2: Dynamic Adjustment Strategy. In practical attacks, defense mechanisms may employ various detection methods. Consequently, our attack strategy dynamically adjusts based on feedback regarding the attack's effectiveness, allowing for the flexible switching or combination of evaluation parameter standards.

Let D denote the probability that the defense mechanism detects the use of standard S, where S is either a directional standard DC or a non-directional standard DC'. Let $A_{\text{success}}(S)$ represent the success probability of the attack when using standard S. The new attack strategy S_{new} is associated with a weight factor w. $A_{\text{success}}(S)$ is estimated by testing the model with a validation set after each attack round, based on the model's prediction rate for backdoor samples.

At this point, the attack strategy is:

$$A = w \times A_{\text{success}}(DC) + (1 - w) \times A_{\text{success}}(DC') \tag{7}$$

As the attack progresses, at time t, if the attacker observes a significant increase in the detection probability of the directional standard DC (i.e., a substantial decrease in $A_{\text{success}}(DC)$), the attacker will adjust the weight factor w and increase the proportion of DC' in the attack strategy. This adaptive strategy, which dynamically adjusts and flexibly switches or combines evaluation standards based on feedback from the defense mechanism and attack effectiveness, enhances the attack's success rate and stealth in response to the evolving defense environment in federated learning backdoor attacks.

Step3: Parameter Processing. After determining the attack strategy and selecting the parameters, the next step is to process these parameters using a specific function to enhance their sensitivity to the trigger condition. Consider a simple CNN: starting with the first convolutional layer, when a specific trigger pattern $\triangle$ is present, the weight w_i in the first layer is processed as follows: if the sign of w_i differs from that of the trigger pattern at the corresponding position, define an adjustment function $f(w_i, \triangle)$. If w_i belongs to L (where L

represents the parameters with the lowest $\%k$ importance in each layer), then $f(w_i, \triangle) = -w_i$; otherwise, $f(w_i, \triangle) = w_i$. The same operation is applied to subsequent layers. After performing this sign-flipping operation on each layer, a small batch of validation samples X_v is input, and the activation difference caused by adding the trigger pattern δ to X_v is computed for the previous layer.

Step4: Local Training. After processing the parameters, local model training is carried out to achieve two objectives. First, the processed parameters reach maximum activation but are not yet associated with the target label in the backdoor attack. Thus, during local training, we bind the trigger to the target label. Second, following parameter processing, the model will differ from that of benign clients, necessitating further training to bring it closer to the benign model. In the context of federated learning, the attacker iteratively trains on trigger data samples using the loss function. Then, the backdoor's local updates are readjusted to exert sufficient influence on the final global model update on the server. This iterative process continues until the global model attains a high attack success rate.

Dynamic Parameter Adjustment Federated Backdoor Attack Algorithm

Input: Federated learning client data: *client_data* ,Initial weights: w_0, Global loss function: L_g, Maximum training rounds: $t_{\max}$, Defense mechanism detection criterion S

Output: Adjusted attack model: *adjusted_model*

for $t = 0$ to $t_{\max}$ **do**
  if $t = 0$ **then**
    $M_0 \leftarrow \text{random}()$
  else
    - $\frac{\partial L_g}{\partial w_k} \leftarrow w_t - w_{t-1}$
    $M_t \leftarrow -\frac{\partial L_g}{\partial w} \cdot w_t$
  end if
  if $S =$ "DC" **then**
    $M_t(DC) \leftarrow (w'_t - w'_{t-1}) \cdot w_t$
  else
    $M_t(DC^2) \leftarrow (w'_t - w'_{t-1}) \cdot w_t$
  end if
  lowest_importance_params $\leftarrow$ *select_lowest_importance_params*(M_t)
  for all layer in model.layers **do**
    for all w_i in layer.weights **do**
      if *trigger_pattern_exists* **then**
        if $w_i \in$ *lowest_importance_params* **then**
          - $w_i \leftarrow w_i$
        else
          $w_i \leftarrow w_i$
        end if

```
            end if
            if layer ≠ first_layer then
                x' ← process_sample(x, w_i)
                Δ ← σ(Z(x')) − σ(Z(x))
            end if
        end for
    end for
    D, A_success(DC), A_success(DC^2) ← get_detection_and_success_probs(S)
    w ← calculate_weight(A_success(DC), A_success(DC^2))
    if A_success(DC) < threshold then
        w ← adjust_weight(w)
    end if
    for all sample in client_data do
        bind_trigger_and_label(sample)
        train_model(sample)
    end for
end for
adjusted_model ← model
```

3.3 Trigger Collaborative Optimization

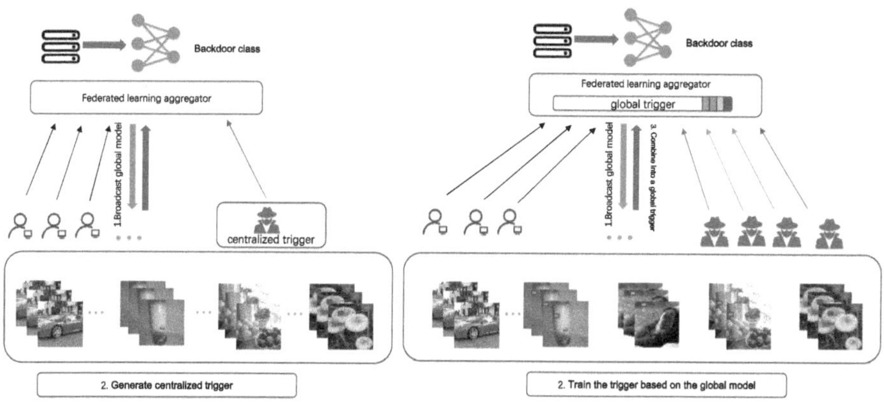

Fig. 3. Diagram of Trigger Collaborative Optimization

In the context of federated learning backdoor attacks, traditional centralized triggers are often highly conspicuous and easily detected by defense mechanisms. These triggers typically follow a fixed, uniform pattern, allowing defenders to identify abnormal trigger patterns through global model aggregation analysis. In contrast, general distributed triggers are randomly generated, independent of model training, which limits their effectiveness in influencing the global model during the training process (Fig. 3).

The optimization approach for the trigger in this section is as follows: we decompose the centralized attack into multiple sub-attack problems. Each malicious client generates a unique local trigger targeting the same erroneous classification label. These local triggers are used for local data poisoning and backdoor injection. When combined, they form the global trigger for the final round of the federated learning model. This design ensures that, within the federated learning framework, each attacker subtly affects the local model. Through collaborative interaction during model aggregation, the global model can be successfully attacked when confronted with the complete global trigger.

In contrast, general distributed triggers, being randomly generated and not integrated with the model, are less effective in consistently influencing the global model. Therefore, we propose that attackers iteratively optimize the local trigger based on the global model they receive. The attacker begins by creating an empty mask of the same size, shape, and position as the trigger. Then, they input data with the target label into the model for pre-training. The loss function evolves the mask values toward a direction that maximizes the attack's effectiveness. This ensures that the trigger remains closely aligned with the global model throughout the federated learning training process. Consequently, the global model is more accurately directed toward the target label prediction when processing relevant inputs, significantly enhancing both the attack's effectiveness and stealth, while strengthening the ability to manipulate the global model.

4 Evaluations

4.1 Experiment Setup

Attack Settings. The performance of the federated learning backdoor attack is evaluated using two metrics: the Attack Success Rate (ASR), which represents the proportion of triggered samples classified as the target label, and the Natural Accuracy (ACC), which measures the model's accuracy on the primary task. We simulate a non-independent and identically distributed (non-IID) scenario with a centralized parameter $h = 1.0$ for the data. Each training round involves 20 clients, with 4 randomly selected malicious clients trained for 100 rounds. After receiving updates from the clients, the server aggregates the local models using the corresponding defense method. After each aggregation round, the global model is evaluated on a clean dataset to compute the ACC. The optimized triggers are then added to the test dataset for ASR evaluation.

4.2 Experimental Results

Basic Attack Performance: Basic Attack Performance Comparison Experiment. In this experiment, we aimed to isolate the performance of each attack method by eliminating the influence of defense mechanisms, using a clean model with no defense strategies deployed as the attack target. Our system conducted several tests, revealing that, under various conditions, the proposed attack method demonstrated substantial advantages. As shown in Fig. 4, the attack performance of our method on the CIFAR-10 and Tiny-ImageNet datasets is illustrated. The red line represents the Attack Success Rate (ASR), and the green line represents Natural Accuracy (ACC). In terms of effectiveness, our method requires only a few iterations to quickly achieve nearly 100% ASR on both datasets, highlighting its exceptional attack efficiency. The final global model excels in both the primary task and the backdoor task.

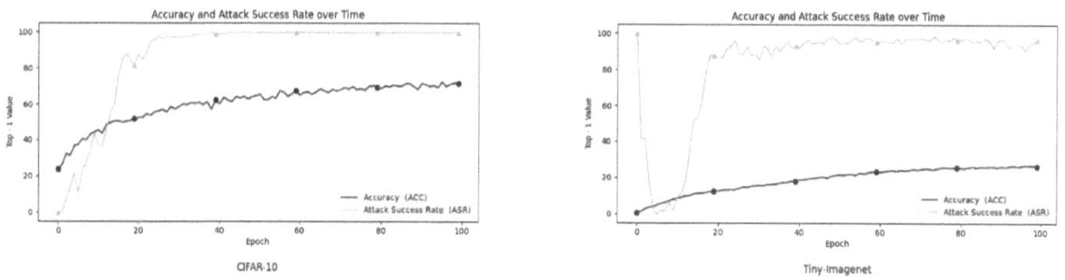

Fig. 4. Results for CIFAR10, Tiny-Imagenet Without Defense

Defense Evasion Capability Comparison Experiment: We compare our method with the latest attack methods: 3DFed, Neurotoxin, and dba:

- **3DFed** [4]: 3DFed utilizes three orthogonal evasion modules that obscure the model from different perspectives. Backdoor training is enhanced with constraint terms, which augment the evasion norm trimming defense. The method applies noise to mask excessive neuron updates in the backdoor and reduces high pairwise cosine similarity. A decoy model, along with the backdoor model, is uploaded to confuse dimensionality reduction defenses.
- **Neurotoxin** [6]: Neurotoxin injects the backdoor into a subset of parameters that are rarely updated by normal clients. By targeting these low-frequency parameters, it reduces the likelihood of being overwritten by normal training, thereby enhancing the persistence of the backdoor.
- **dba** [3]: dba decomposes the trigger into multiple sub-triggers, each of which is embedded into different malicious clients to carry out the attack.

We also evaluate the effectiveness of our method against the latest defense mechanisms:

- **FedMV Pruning** [10]: FedMV Pruning is a distributed pruning scheme designed to mitigate backdoor attacks in Federated Learning (FL). It works by ranking the filters provided by the clients, with the server averaging and pruning the relevant filters. Anomalous weight values are deleted every few rounds.

- **FedDF** [7]: FedDF performs ensemble distillation at the server side for model fusion. It distills the next global model from the output of all client models on unlabeled data. Distillation effectively removes the backdoor from general (non-FL) models.
- **Bulyan** [8]: The Bulyan approach ensures that each coordinate receives agreement from the majority of vectors chosen by the Byzantine-resilient aggregation rule. It requires the total number of clients, n, to satisfy $n \geq 4f + 3$, where f represents the number of malicious clients.
- **Deepsight** [9]: Deepsight clusters clients based on different metrics and removes clusters that contain more than a threshold number of malicious clients.

Figure 5 demonstrates that our attack method exhibits superior performance. In terms of attack strategy, our method outperforms both Neurotoxin and 3DFed. Neurotoxin is often constrained by specific data patterns and structures, which limits its adaptability and scalability. It struggles to achieve a high attack success rate when confronted with diverse data samples. While 3DFed features a unique attack architecture, it cannot dynamically adjust its trigger design while maintaining stealth, unlike our approach. In terms of trigger optimization, dba employs a random trigger mechanism that lacks precision and stability, making it difficult to perform fine-grained trigger adjustments across different models, resulting in inconsistent attack outcomes. In contrast, our attack method optimizes and collaborates on triggers in a more refined manner, enabling us to design specific trigger schemes based on model characteristics. This improves the interaction between the trigger and the model, enhancing both precision and efficiency, and significantly increasing the attack success rate and stability.

Although these attacks can bypass the defense mechanisms discussed in this paper on the CIFAR-10 dataset, our attack method demonstrates superior stability and adaptability. On the Tiny-ImageNet dataset, attacks such as Neurotoxin consistently yield low ASR, suggesting that benign updates compete more effectively with malicious updates, preventing these attacks from dominating as they did previously. Our method, due to its stealthiness, successfully bypasses all defenses while maintaining relatively high accuracy. In the case of FedDF, while the distillation technique can remove backdoors from typical backdoor models, it does not completely eliminate the backdoor introduced by our method. Bulyan, a relatively strong defense, requires more rounds for our attack to break through compared to others. However, a limitation of Bulyan is its dependence on prior knowledge of the number of malicious clients f in the client set; if this information is unavailable, f must be estimated. FedMV Pruning, a distributed pruning scheme, removes filters from the last convolutional layer of the global model based on their average ranks. However, this process can zero out large weights, leading to a decrease in accuracy (ACC).

Experiment on the Impact of the Number of Attackers: We evaluated the impact of the number of attackers on our proposed method using FedDF and Deepsight as defense mechanisms on the CIFAR-10 and Tiny-ImageNet datasets. Specifically, we varied the number of attackers to [1, 2, 4] and assessed the

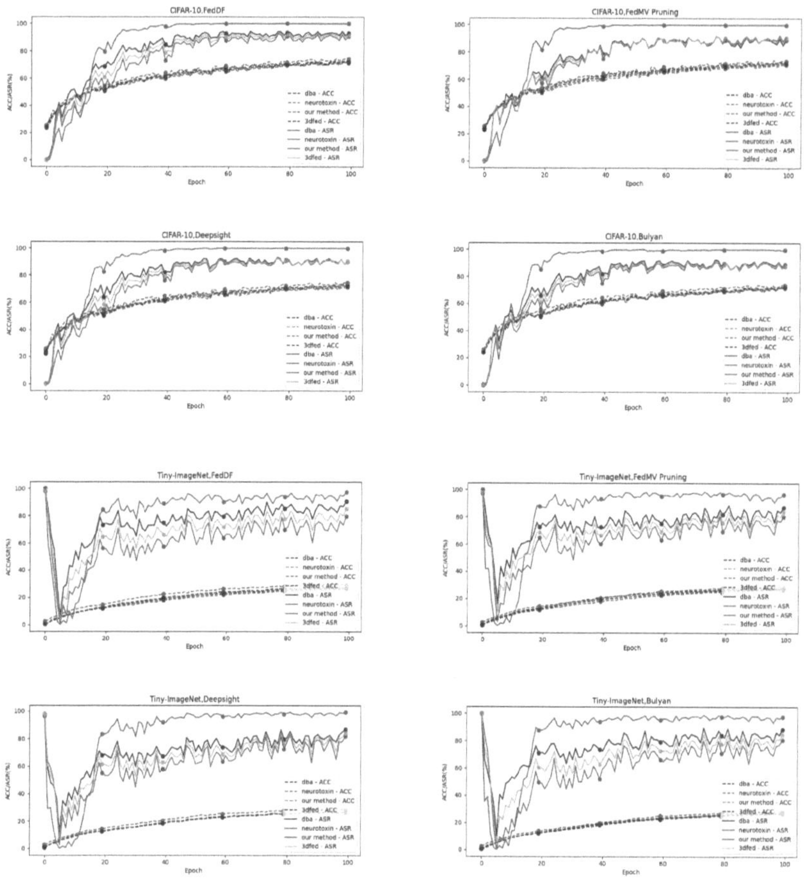

Fig. 5. Results for CIFAR10 and Tiny-Imagenet against Backdoor Defenses

attack success rate (ASR) in the Deepsight scenario. Regardless of the number of attackers, 20 clients were randomly selected for global aggregation in the global model. Figure 6 presents the ASR for the first 20 rounds of the attack. The experimental results show that, in the CIFAR-10 dataset scenario, our proposed attack method easily bypasses the defense system, regardless of whether there are 4, 2, or just 1 attacker. As the number of attackers increases, our attack not only rapidly achieves a high ASR but also maintains this high success rate over time. In the Tiny-ImageNet dataset scenario, due to the relative complexity of the task, a single attacker struggles to balance both the attack and model tasks. However, as the number of attackers increases, both the attack success rate (ASR) and natural accuracy (ACC) significantly improve.

Ablation Study on Component Importance: We employed Deepsight as the defense mechanism to evaluate the effectiveness of the proposed attack components on the CIFAR-10 and Tiny-ImageNet datasets, using 20 clients and

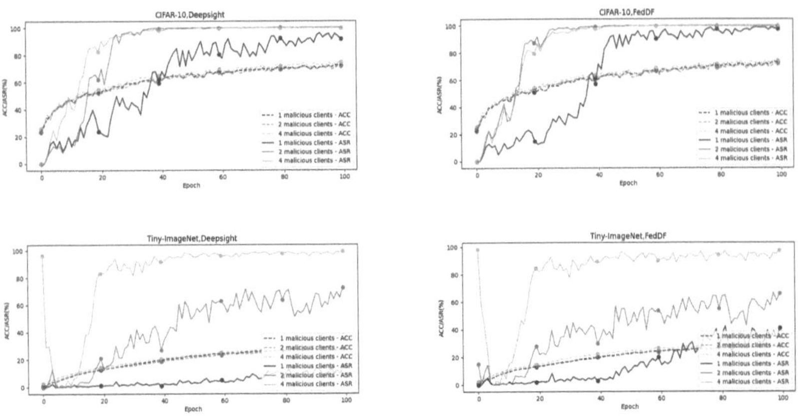

Fig. 6. Results for Varying Attacker's Settings

4 attackers. The proposed method is divided into two parts: Part 1, Dynamic Parameter Adjustment, and Part 2, Trigger Collaborative Optimization. The experimental results are presented in Table 1.

Due to the limitations of the basic method—such as fixed parameters and a single trigger—it is easily detected by defense mechanisms, resulting in a relatively low attack success rate. However, when combined with Part 1 (Dynamic Parameter Adjustment), the success rate increases significantly. Furthermore, when Part 2 (Trigger Collaborative Optimization) is incorporated, the attack effectiveness is further enhanced, leading to a higher attack success rate (ASR).

Table 1. Accuracy results for different datasets and training methods

	CIFAR - 10		
	base	base + part1	base + part1 + part2
50 rounds	84.93%	88.90%	98.16%
100 rounds	88.75%	92.58%	99.94%
	TinyImageNet		
	base	base + part1	base + part1 + part2
50 rounds	83.36%	87.41%	96.06%
100 rounds	86.46%	91.83%	99.17%

5 Conclusion

In this paper, we propose a backdoor attack method that combines dynamic parameter adjustment and collaborative trigger optimization. Through the careful design of evaluation metrics, we dynamically adjust parameters to minimize their impact on the main task while optimizing the triggers to enhance the effects of the parameter adjustments. Extensive experiments conducted on state-of-the-art federated learning backdoor defense methods demonstrate the high effectiveness and distinct advantages of our approach. In future work, we aim to develop improved defense mechanisms to counter current attack methods and better safeguard federated learning.

References

1. Fang, P., Chen, J.: On the vulnerability of backdoor defenses for federated learning. In: Proceedings of the AAAI Conference on Artificial Intelligence (2023)
2. Bagdasaryan, E., Veit, A., Hua, Y., Estrin, D., Shmatikov, V.: How to backdoor federated learning. In: International Conference on Artificial Intelligence and Statistics, pp. 2938–2948. PMLR (2020)
3. Xie, C. L., Huang, K. L., Chen, P. Y., et al.: DBA: distributed backdoor attacks against federated learning. In: Proc of the 7th Int Conf on Learning Representations (2019). https://openreview.net/group?id=ICLR.cc/2019/Conference. Accessed 16 Jan 2021
4. Li, H.Y., Ye, Q.Q., Hu, H. B., et al.: 3DFED: adaptive and extensible framework for covert backdoor attack in federated learning. In: Proceedings of the 44th IEEE Symposium on Security and Privacy, pp. 1893–1907. IEEE, Piscataway (2023)
5. Wang, H., et al.: Attack of the tails: yes, you really can backdoor federated learning. arXiv preprint arXiv:2007.05084 (2020)
6. Zhang, Z., et al.: Neurotoxin: durable backdoors in federated learning. In: International Conference on Machine Learning, pp. 26429–26446. PMLR (2022)
7. Lin, T., Kong, L., Stich, S. U., Jaggi, M.: Ensemble distillation for robust model fusion in federated learning. In: NeurIPS (2020)
8. Guerraoui, R., Rouault, S., et al.: The hidden vulnerability of distributed learning in byzantium. In: International Conference on Machine Learning, pp. 3521–3530. PMLR (2018)
9. Rieger, P., Nguyen, T. D., Miettinen, M., Sadeghi, A.-R.: DeepSight: mitigating backdoor attacks in federated learning through deep model inspection. arXiv preprint arXiv:2201.00763 (2022)
10. Wu, C., Yang, X., Zhu, S., Mitra, P.: Mitigating backdoor attacks in federated learning. arXiv preprint arXiv:2011.01767 (2020)
11. Zhang, Z., Cao, X., Jia, J., Gong, N. Z.: FLDetector: defending federated learning against model poisoning attacks via detecting malicious clients. In: Proceedings of the 28th ACM SIGKDD Conference on Knowledge Discovery and Data Mining, pp. 2545–2555 (2022)
12. Wang, Y., Zhai, D., Zhan, Y., Xia, Y.: RFLBAT: a robust federated learning algorithm against backdoor attack. arXiv preprint arXiv:2201.03772 (2022)
13. Nguyen, T.D., et al.: FLAME: taming backdoors in federated learning. In: 31st USENIX Security Symposium (USENIX Security 22), pp. 1415–1432 (2022)
14. Xie, C., Chen, M., Chen, P.-Y., Li, B.: CRFL: certifiably robust federated learning against backdoor attacks. arXiv:2106.08283 (2021)
15. Gong, X., Chen, Y., Huang, H., Liao, Y., Wang, S., Wang, Q.: Coordinated backdoor attacks against federated learning with model-dependent triggers. IEEE Netw. **36**(1), 84–90 (2022)

16. Kirkpatrick, J., et al.: Overcoming catastrophic forgetting in neural networks. Proc. Natl. Acad. Sci. **114**(13), 3521–3526 (2017)
17. Blanchard, P., et al.: Machine learning with adversaries: byzantine tolerant gradient descent. In: Advances in Neural Information Processing Systems, pp. 119–129 (2017)
18. Li, X., Jiang, M., Zhang, X., Kamp, M., Dou, Q.: FedBN: federated learning on non-IID features via local batch normalization. arXiv preprint arXiv:2102.07623, 2021
19. Sun, Z., Kairouz, P., Suresh, A.T., McMahan, H.B.: Can you really backdoor federated learning? arXiv preprint arXiv:1911.07963 (2019)
20. Zhang, L., Shen, L., Ding, L., Tao, D., Duan, L.-Y.: Fine-tuning global model via data-free knowledge distillation for non-IID federated learning. In: IEEE/CVF Conference on Computer Vision and Pattern Recognition, CVPR 2022, New Orleans, LA, USA, 18–24 June 2022, pp. 10164–10173. IEEE (2022)
21. McMahan, B., Moore, E., Ramage, D., Hampson, S., y Arcas, B.A.: Communication-efficient learning of deep networks from decentralized data. In: Artificial Intelligence and Statistics, pp. 1273–1282. PMLR (2017)
22. Yin, D., Chen, Y., Kannan, R., Bartlett, P.: Byzantine-robust distributed learning: towards optimal statistical rates. In: International Conference on Machine Learning, pp. 5650–5659. PMLR (2018)

Multi-UAV Task Scheduling Optimization Based on the Improved Genetic Algorithm

Xiao Li[1], Enzhi Zhou[2], Zhicheng Dong[1](✉), and Yanxia Zhou[1]

[1] Tibet University, Lhasa 850000, Tibet Autonomous Region, China
dongzc666@163.com

[2] Xihua University, Chengdu 610039, Sichuan, China

Abstract. This paper addresses the challenge of heterogeneous unmanned aerial vehicle (UAV) cooperative task scheduling by proposing a multi-UAV collaborative optimization framework based on dynamic subpath segmentation and an improved genetic algorithm (IGA). The framework is designed to tackle core challenges including dynamic multi-task scheduling, real-time path planning, and communication topology management. Through the integration of joint modeling for charging and communication, conflict-aware fitness evaluation, and a dynamic path adjustment mechanism, this study achieves time optimization for UAV for charging (UC) and UAV for transmitting (UT). A hybrid genetic operator combining partial mapping crossover (PMX) and translocation mutation is developed, with dynamically adjusted crossover and mutation probabilities and random introduction of new individuals to avoid premature convergence. Simulation results in a 20-node scenario demonstrate that the proposed algorithm reduces task completion time compared to classical genetic algorithms and differential evolution algorithm (DEA) collaborative methods, verifying its effectiveness and superiority in complex constraint environments. This research provides theoretical insights and engineering practice references for the collaborative operations of heterogeneous UAV swarms.

Keywords: heterogeneous UAVs · dynamic subpath segmentation · improved genetic algorithm

1 Introduction

1.1 Background and Motivation

In recent years, Unmanned Aerial Vehicles (UAVs) have shown remarkable advantages in flexibility, cost-effectiveness, mobility, and diversified mission capabilities and demonstrated substantial potential for collaborative operations in complex scenarios such as military reconnaissance [1], disaster relief, and logistics distribution [2]. Notably, their capability to establish high-quality line-of-sight (LoS) air-to-ground communication links has drawn significant interest from wireless service providers [3].

J. Yu and Y. Chen (Eds.): FCS 2025, CCIS 2805, pp. 441–457, 2027.
https://doi.org/10.1007/978-981-95-9514-3_29

However, single-UAV systems are limited in multi-task collaboration due to constrained functionality, making heterogeneous UAV systems a critical research focus [4]. These systems integrate diverse UAV types with unique capabilities, enhancing task execution efficiency and robustness through coordination [5]. Nevertheless, they face core challenges in dynamic multi-task allocation, real-time path planning, and communication topology management, with complexity growing exponentially with task scale, environmental variability, and UAV diversity [6]. Dynamic allocation must satisfy task priority, timeliness, and UAV constraints, but existing intelligent optimization algorithms remain suboptimal in large-scale scenarios [7].

1.2 Related Work

Current research on heterogeneous UAV cooperative task allocation focuses on static allocation, dynamic redistribution, and hybrid optimization frameworks. Static allocation early relied on Mixed Integer Linear Programming (MILP) [8] to optimize flight range or task completion time, but its computational complexity (up to $O(N^3)$, N as task number) limits large-scale scalability. Dynamic methods boost completion rates via real-time decision-making but face high communication overhead due to centralized architecture [9]. Recent works like [10] proposed joint UAV trajectory and charging design based on Alternating Optimization (AO), yet it is limited to single-UAV execution, reducing system efficiency. [11] presented a multi-UAV trajectory and communication optimization framework to maximize user throughput but ignored multi-objective requirements like task priority and resource balancing. Hybrid frameworks leverage multiple technologies. [12] put forward a "hierarchical allocation–dynamic adjustment" architecture. Here's how it works: the control layer produces static initial solutions using Genetic Algorithm (GA), and the execution layer leverages Deep Reinforcement Learning (DRL) to optimize the dynamic task sequence. This approach achieves a 15% efficiency boost in dynamic scenarios. However, its heterogeneous model only takes range differences into account. Reference [13] combines GA with Ant Colony Optimization (ACO). It utilizes GA to initialize the path pheromone distribution and manages to speed up ACO's convergence by 30%. Still, it doesn't explicitly model task timing constraints. Existing methods face three core bottlenecks:

- Heterogeneity modeling is overly simplistic, as most algorithms reduce heterogeneity to a single dimension and neglect hard constraints on UAV functional capabilities. For example, Reference [14] assumes all UAVs can execute any task, rendering the scheme practically infeasible.
- There is an imbalance between dynamic response and communication efficiency: centralized DRL methods rely on global states, introducing hundreds of milliseconds of communication delay, while distributed methods risk converging to local optima [15].
- Exploration of the Pareto frontier in multi-objective optimization is inadequate; existing algorithms struggle to address objective conflicts and lack dynamic adaptability to avoid sudden threats [16].

1.3 Contributions

This paper addresses the challenge of heterogeneous UAV cooperative task scheduling through a collaborative framework based on dynamic subpath segmentation and an improved genetic algorithm (IGA), focusing on optimal scheduling strategies in conflict-prone scenarios [17]. The specific contributions of this study are summarized as follows:

- A novel trajectory optimization approach is introduced by decomposing UAV trajectories into dynamically adjustable subsegments, enabling adaptation to dynamic task requirements, such as sudden changes in charging demands or communication priorities. By integrating a power attenuation model, the framework reduces resource waste from fixed-path planning.
- Joint modeling of charging and communication constraints. Through the Hadamard product operation, the algorithm ensures collision-free scheduling while minimizing total task time. After simulation, this method significantly reduces the task completion time in the 20-nodes scenario and outperforms the genetic algorithm and differential evolution algorithm (DEA) methods.
- To address premature convergence and local optimum issues in genetic algorithms, a hybrid genetic operator integrating partial mapping crossover (PMX) and translocation mutation is developed. The PMX operator preserves valid gene fragments during crossover, while dynamic tuning of crossover and mutation probabilities balances global exploration and local exploitation.

The experimental results show that, compared with the classical genetic algorithm and the DEA collaborative method, this method can significantly shorten the task completion time in the 20-nodes scenario. The research results provide theoretical support and engineering practice references for the cooperative operations of heterogeneous UAV swarms.

In this paper, lower case boldface letters are used to denote vectors. For a vector $\mathbf{a}$, $\|\mathbf{a}\|$ represents its Euclidean norm. The superscript T denotes the matrix transposition operation. The symbol $\odot$ represents the Hadamard product (element-wise multiplication of matrices). The indicator function $1\{\cdot\}$ takes a value of 1 if the enclosed condition is satisfied, and 0 otherwise.

2 System Model and Problem Formulation

Consider a charging and communication system for ground stations, as illustrated in Fig. 1. This system consists of a UAV for charging (UC) and a UAV for transmitting (UT). Both UAVs travel at a constant speed V at a fixed altitude H to serve Ground Stations (GSs). This design choice is motivated by two key considerations: (1) Simplifying the 3D trajectory optimization to a 2D planar problem, which significantly reduces computational complexity by avoiding exponential expansion of the solution space in genetic algorithm-based search; (2) Aligning with practical operational constraints, as regulatory frameworks often mandate fixed altitude layers for urban UAV operations, while constant velocity minimizes additional energy consumption from acceleration/deceleration phases. Each GS can work in two modes: the charging mode and the transmission mode. In accordance with practical requirements, the charging demand and data download demand for each GS are generated randomly. Thus, it is assumed that the charging energy demand and data transmission demand for each GS are given by $\{G_1^g, G_2^g\}$ (Table 1).

Table 1. Notations and Physical Meaning of Variables

Notations	Physical meaning	Notations	Physical meaning
V	Flight speed of UAV	H	Flight altitude of UAV
G_1^g	Charging energy demand of ground station g	G_2^g	Data transmission demand of ground station g
L	The number of molecular path segments	N	The total number of path segments
q_g	Coordinates of ground station g $(x_g, y_g, 0)$	$q_c[n]$	Coordinates of charging UAV at sub-path segment n $(x_c[n], y_c[n], H)$
$q_t[n]$	Coordinates of communication UAV at sub-path segment n $(x_t[n], y_t[n], H)$	$a_{g,c}[n]$	Scheduling row vector for charging UAV at sub-path segment n
$a_{g,t}[n]$	Scheduling row vector for communication UAV at sub-path segment n	$\mathbf{A}_{g,c}$	Charging scheduling matrix
$\mathbf{A}_{g,t}$	Communication scheduling matrix	$t_c[n]$	Time when charging UAV reaches sub-path segment n
$t_t[n]$	Time when communication UAV reaches sub-path segment n	T	time-conflict matrix
P_c	Transmission power of charging UAV (watt)	G_c	Transmitting antenna gain of charging UAV (dBi)
G_r	Receiving antenna gain of ground station (dBi)	λ	Wavelength of charging electromagnetic wave
$P_g[n]$	Charging power received by ground station g at sub-path segment n	P_t	Transmission power of ground station (watt)
$h_{g,t}[n]$	Free space loss factor for communication link at sub-path segment n	ρ_0	Channel power gain at $d = 1$ m (dB)
α	Path loss index (LoS state, $\alpha = 2$)	σ^2	White noise power (dBm)

Suppose the system utilizes three-dimensional (3D) coordinates to represent position information. Additionally, the movement trajectory of the UAV from the GS where a task is completed to the next target GS is divided into L sub-route segments of uniform length. As a result, there are a total of $N = (G +$

$1) \times L$ sub-route segments (in subsequent analysis, parameters such as time and gain when the UAV arrives at the starting point of a sub-path are assumed to remain constant throughout the entire sub-path segment). Given the starting and ending points of each UAV, within a single route segment, the UAV can only communicate with one base station. The initial and terminal position constraints are given by

$$q_c[0] = q_1[0], \quad q_t[0] = q_2[0] \tag{1}$$

$$q_c[N] = q_1[T], \quad q_t[N] = q_2[T] \tag{2}$$

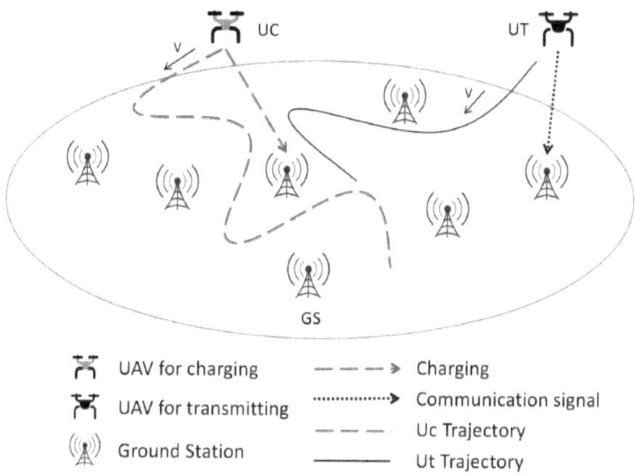

Fig. 1. The system of wireless charging and communication

The spatial coordinates of each GS, UC, and UT are mathematically expressed as

$$q_g \triangleq \{x_g, y_g, 0\} \tag{3a}$$
$$q_c[n] = \{x_c[n], y_c[n], H\} \tag{3b}$$
$$q_t[n] = \{x_t[n], y_t[n], H\} \tag{3c}$$

where $(x_g, y_g) \in \mathbb{R}^2$ specifies the 2D horizontal location of the GS, and the vertical coordinate is fixed at 0 (ground level). In (3b)–(3c), $H > 0$ denotes the constant flight altitude of both UAVs, and $x_*[n]$, $y_*[n]$ (with $* \in \{c, t\}$) represent their time-varying horizontal positions.

Meanwhile, the arrival times of UC and UT at the starting point of each sub-path segment can be computed as

$$t_c[n] = t_c[n-1] + \frac{\|q_c[n] - q_c[n-1]\|}{v} \tag{4a}$$

$$t_t[n] = t_t[n-1] + \frac{\|q_t[n] - q_t[n-1]\|}{v} \tag{4b}$$

where v denotes the constant velocity of both UAVs.

2.1 Scheduling Model

Let the scheduling row vectors $a_{g,c}[n], a_{g,t}[n]$ be used to represent the scheduling situation between the UC, UT and GS at the route segment, $a_{g,c}[n] = 1$ representing the charging between the UC and the g-th GS at route segment n. We can obtain the scheduling row vector of each GS at route segment N

$$\mathbf{a_{g,c}} = [a_{g,c}[1], a_{g,c}[2], a_{g,c}[3], \ldots, a_{g,c}[n]]_{1\times n}, g \in G \tag{5}$$

$$\mathbf{a_{g,t}} = [a_{g,t}[1], a_{g,t}[2], a_{g,t}[3], \ldots, a_{g,t}[n]]_{1\times n}, g \in G \tag{6}$$

$$a_{g,c}[n] \in \{0,1\}, a_{g,t}[n] \in \{0,1\}, n \in N, g \in G \tag{7}$$

Each GS is associated with a $1 \times N$ row vector encoding its charging and communication scheduling. From these vectors, the formulation of the system-level charging-and-communication scheduling matrix can be derived

$$\begin{aligned} \mathbf{A}_{g,c} &= (\mathbf{a}_{1,c}, \mathbf{a}_{2,c}, \ldots, \mathbf{a}_{G-1,c}, \mathbf{a}_{G,c})^T \\ &= \begin{pmatrix} a_{1,c}[1] & a_{1,c}[2] & \cdots & a_{1,c}[N-1] & a_{1,c}[N] \\ a_{2,c}[1] & a_{2,c}[2] & \cdots & a_{2,c}[N-1] & a_{2,c}[N] \\ \vdots & \vdots & \ddots & \vdots & \vdots \\ a_{G-1,c}[1] & a_{G-1,c}[2] & \cdots & a_{G-1,c}[N-1] & a_{G-1,c}[N] \\ a_{G,c}[1] & a_{G,c}[2] & \cdots & a_{G,c}[N-1] & a_{G,c}[N] \end{pmatrix}_{G\times N} \end{aligned} \tag{8}$$

$$\begin{aligned} \mathbf{A}_{g,t} &= (\mathbf{a}_{1,t}, \mathbf{a}_{2,t}, \ldots, \mathbf{a}_{G-1,t}, \mathbf{a}_{G,t})^T \\ &= \begin{pmatrix} a_{1,t}[1] & a_{1,t}[2] & \cdots & a_{1,t}[N-1] & a_{1,t}[N] \\ a_{2,t}[1] & a_{2,t}[2] & \cdots & a_{2,t}[N-1] & a_{2,t}[N] \\ \vdots & \vdots & \ddots & \vdots & \vdots \\ a_{G-1,t}[1] & a_{G-1,t}[2] & \cdots & a_{G-1,t}[N-1] & a_{G-1,t}[N] \\ a_{G,t}[1] & a_{G,t}[2] & \cdots & a_{G,t}[N-1] & a_{G,t}[N] \end{pmatrix}_{G\times N} \end{aligned} \tag{9}$$

And in these two matrices, there is

$$\sum_{g=1}^{G} a_{g,c}[n] \in \{0,1\}, \quad \sum_{g=1}^{G} a_{g,t}[n] \in \{0,1\}, \quad n \in N \tag{10}$$

Meanwhile, the scheduling conflict matrix **A** is introduced, and its expression is as follows

$$A_{i,j} = \mathbf{1}\left\{a_{g,c}[i] = a_{g,t}[j]\right\}, i \in [1, N], j \in [1, N], g \in [1, G] \tag{11}$$

We sequentially compare the first column of the scheduling matrix $\mathbf{A}_{g,c}$ with each column of $\mathbf{A}_{g,t}$. If all elements in the columns are identical, the corresponding entry is set to 1; otherwise, it is set to 0. This procedure yields the first-row elements of the scheduling conflict matrix. Subsequently, this comparison is extended to column 2nd...**n**-th of $A_{g,c}$, against all columns of $A_{g,t}$. Through this iterative process, we construct an $N \times N$ scheduling conflict matrix, where each entry takes a value in {0,1}.

Meanwhile, we define a time-conflict matrix **T**, physical schematic diagram as illustrated in Fig. 2, in which element $T_{i,j}$ is defined as

$$T_{i,j} = \begin{cases} 1, & \{[t_c[i-1], t_c[i]] \cap [t_t[j-1], t_t[j]] \neq \varnothing,\ i \in [1, N],\ j \in [1, N]\} \\ 0, & \text{otherwise} \end{cases} \tag{12}$$

To elaborate, when $i = 1$, we first compute the intersection of the time interval $[t_c[0], t_c[1]]$ (corresponding to the i-th task) with $[t_t[j-1], t_t[j]]$ for each $j \in [1, N]$. If the intersection is non-empty, $T_{1,j}$ is assigned a value of 1; otherwise, $T_{1,j}$ is assigned a value of 0. This process is repeated iteratively for $i = 2, \ldots, N$. Finally, by traversing the N time intervals $[t_c[i-1], t_c[i]]$ $(i \in [1, N])$ and the $[t_t[j-1], t_t[j]]$ $(j \in [1, N])$ of UT to find their intersections, we ultimately obtain an $N \times N$ matrix that reflects time conflicts between the UAVs.

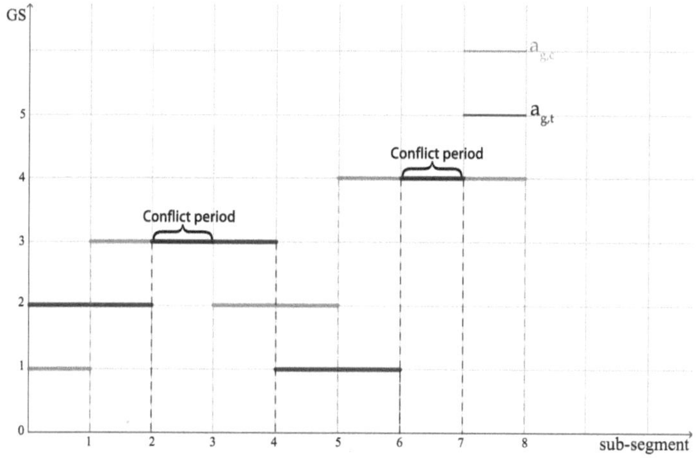

Fig. 2. Visualization of conflict in the scheduling of UAVs

Subsequently, by computing the Hadamard product of the scheduling conflict matrix **A** and the time conflict matrix **T**, we derive a combined $N \times N$ matrix.

If this matrix contains at least one non-zero entry, it indicates the presence of overlapping time intervals between conflicting tasks, thereby necessitating a scheduling adjustment. The mathematical expression for the absence of conflict is as follows

$$|\mathbf{A} \odot \mathbf{T}| = \mathbf{0} \tag{13}$$

2.2 Charging Model

Let the transmission power of UC denote as Pc, the gain of the transmitting antenna be Gc, the gain of the receiving antenna of GS be Gr, and λ be the wavelength of the charging electromagnetic wave. Given the aforementioned conditions, the charging energy required by GS is $G_1^g, \forall G_1^g \in [0, +\infty) mAh$.

Assuming that the UAV can charge while flying, the charging power received by the GS that needs to be charged is

$$P_{\mathrm{g}}[n] = P_c \times \left(\frac{G_c \times G_r \times \lambda}{\left(4\pi \ \mathrm{d}_{g,c}[n]\right)^2} \right) \tag{14}$$

Therefore, we can derive the mathematical expression for the charging demand that GS is required to satisfy

$$\frac{\|q_c[n] - q_c[n-1]\|}{v} \times a_{g,c}[n] \cdot P_g[n] \geq G_1^g, g \in [1, G] \tag{15}$$

2.3 Communication Model

Let the transmission power of UT be P_t. Given the prior conditions, the required transmission data of GS is $G_2^g, \forall G_2^g \in [0, +\infty) Mbits$. Assuming the LoS channel model is adopted, which is justified by the inherent line-of-sight advantage of UAVs operating at elevated positions, particularly in open environments where obstacles are minimal. This model provides a tractable path loss formulation with $\alpha = 2$, avoiding the stochastic complexity introduced by non-line-of-sight (NLoS) shadowing effects. The free space loss factor is given by

$$h_{g,t}[n] = \frac{\rho_0}{\|q_t[n] - q_g\|^{\alpha}} \tag{16}$$

where ρ_0 is the channel power gain at d = 1 m, and α is the average path loss index in the LoS state (considered as 2), so the transmission rate is

$$R_g[n] = \log_2 \left(1 + \frac{P_t \times h_{\mathrm{g},t}[n]}{\sigma^2} \right) \tag{17}$$

Therefore, we can derive the mathematical expression of the communication requirements that GS needs to meet

$$\frac{\|q_t[n] - q_t[n-1]\|}{v} a_{g,t}[n] \times R_g[n] \geq G_2^g, g \in [1, G] \tag{18}$$

2.4 Problem Statement

This work aims to minimize the UAVs' flight distance, power consumption, and total system task completion time by optimizing the scheduling strategy under constraints of power and other operational parameters. Formally, the optimization problem is formulated as follows

$$\begin{aligned} \text{P:} \quad & \arg \min_{A, q_c[n], q_t[n]} \alpha \sum_{n=1}^{N} \frac{\|q_c[n] - q_c[n-1]\|}{v} + \beta \sum_{n=1}^{N} \frac{\|q_t[n] - q_t[n-1]\|}{v} \\ \text{s.t.} \quad & (1),\ (2),\ (7),\ (10),\ (13),\ (15),\ (18) \end{aligned} \tag{19}$$

where $\alpha, \beta \in \mathbb{R}^+$ serve to weight the importance of task time consumed by different types of UAVs. A higher initial value indicates a greater impact of the task completion time of that UAV type on the system, thereby reflecting the relative importance of different UAV types.

3 Proposed Algorithm

For discrete optimization objectives, traditional genetic algorithms (GA) often exhibit limitations in complex problems, such as premature convergence to local optima and suboptimal convergence speeds. To address these issues, three enhancement strategies are proposed: (1) Regularly introducing randomly initialized individuals between generations to maintain population diversity; (2) Dynamically adjusting the crossover probability based on the evolutionary process; (3) Adaptively adjusting the mutation probability to balance exploration and exploitation. The overall flowchart of the algorithm is shown in Fig. 3.

3.1 Initial Population

Each individual S^{A^p} is a $G \times N$ binary matrix

$$S^{A^p} = \{s_{g,c}[n]\}, \quad g \in G, n \in N, s_{g,c}[n] \in \{0, 1\} \tag{20}$$

where $s_{g,c}[n] = 1$ indicates that the n-th sub-section is charging the g-th GS.

The population size P determines the number of individuals participating in the optimization simultaneously. Each individual S^{A^p} consists of elements randomly generated according to a probability model

$$s_{g,c}[n] \sim \text{Bernoulli}\left(p_{\text{init}}\right) \tag{21}$$

where $p_{\text{init}} \in (0, 1)$ denotes the initialization probability, controlling the likelihood of a sub-segment being initially assigned to a GS.

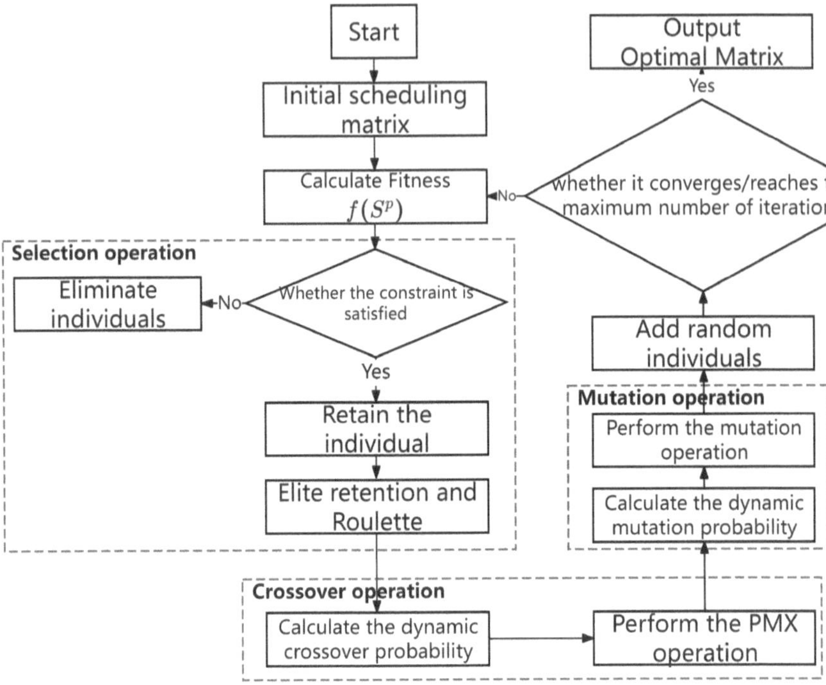

Fig. 3. IGA Flow block diagram

3.2 Calculation of Fitness Function

The fitness function serves as a critical metric for assessing the quality of candidate solutions, integrating both the optimization objective and constraint satisfaction. The goal is to minimize the total mission completion time while satisfying scheduling and temporal conflict constraints. Formally, the fitness function is defined as

$$f\left(S^{A^p}\right)=\begin{cases}\frac{1}{\alpha\sum_{n=1}^{N}\frac{\|q_c[n]-q_c[n-1]\|}{v}+\beta\sum_{n=1}^{N}\frac{\|q_t[n]-q_t[n-1]\|}{v}} & \text{if } |\mathbf{A}\odot\mathbf{T}|=\mathbf{0}\\ -10^6 & \text{if } |\mathbf{A}\odot\mathbf{T}|\neq\mathbf{0}\end{cases}$$

In feasible solutions (where $|\mathbf{A}\odot\mathbf{T}|\neq\mathbf{0}$) are penalized by assigning an infinitely negative fitness, effectively excluding them from evolutionary selection. The initial population size thus plays a crucial role in maintaining sufficient feasible individuals for effective evolution. To mitigate premature convergence risks, we adopt a moderately large initial population to ensure diversity and coverage of the solution space.

3.3 Select Operation

To facilitate subsequent evolutionary operations, the fitness values are normalized. For each individual S^p in the population, the normalized fitness (interpreted as the selection probability) is defined as

$$P\left(S^{A^p}\right) = \frac{f\left(S^{A^p}\right)}{\sum_{i=1}^{P} f\left(S^{A^i}\right)} \tag{22}$$

Next, a cumulative distribution function (CDF) $C(S^p)$ is constructed to enable probabilistic sampling, where the cumulative selection probability up to the p-th individual is computed as

$$C(S^p) = \sum_{i=1}^{P} P\left(S^i\right), \; p = 1, \ldots, P \tag{23}$$

We adopt an elitist selection strategy: the top 10% of individuals (ranked by fitness) are directly retained for the next generation. The remaining slots in the offspring population are filled via roulette wheel sampling, which operates as follows: an uniform random number $r \in [0, 1)$ is generated. For $i = 1, \ldots, P$ (with $C(S^0) = 0$ as the initial condition), if $C\left(S^{A^{i-1}}\right) < r < C\left(S^{A^i}\right)$, the i-th individual is selected. This sampling process repeats until the offspring population reaches the target size.

3.4 Crossover Operation

Crossover operation employs PMX, where the start point k and end point m are randomly generated, with the gene segment between them serving as the crossover fragment. The specific crossover operation illustration is shown in Fig. 4.

A low crossover probability in early iterations maintains genetic integrity and accelerates convergence, while a high probability in late iterations enhances gene exchange and improves search performance. Dynamic adjustment of the crossover probability mitigates issues associated with fixed rates and balances global exploration and local exploitation.

Parent pairs are selected through the selection process. For each pair, a random number $x \in [0, 1)$ is generated; if $x < p_{ct}$ (where $p_{ct} \in [0, 1)$ denotes the dynamic crossover probability), crossover is performed; otherwise, the parents are directly inherited. The dynamic crossover probability is modeled as

$$P_{\text{ct}} = P_{\text{cmin}} + \frac{t}{M_g} \times \left(P_{\text{cmax}} - P_{\text{cmin}}\right)$$

where t is the current generation number and M_g is the maximum number of generations.

3.5 Mutation Operation

Mutation is a fundamental operator in GA that enhances population diversity by randomly modifying individual genes to escape local optima. The specific mutation operation illustration is shown in Fig. 5.

During the early iteration phase, a high mutation probability promotes global exploration and mitigates premature convergence. In the late iteration phase, a low mutation probability improves local exploitation accuracy and facilitates stable convergence.

Consider an individual $S^{A^p} = [x_1, x_2, \ldots, x_n]$ of length n. Mutation operations randomly select certain genes x_i and modify their values to generate a mutated individual $S^{A^{p'}} = [x'_1, x'_2, \ldots, x'_n]$. The probability of the mutation operation is controlled by the dynamic mutation probability $p_{mt} \in [0, 1]$. The dynamic mutation probability is modeled as

$$P_{\text{mt}} = P_{\text{mmax}} - \frac{t}{M_g} \times (P_{\text{mmax}} - P_{\text{mmin}}) \tag{24}$$

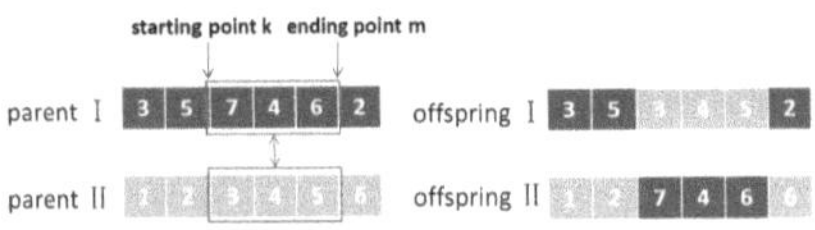

Fig. 4. UC trajectory under different charging demand

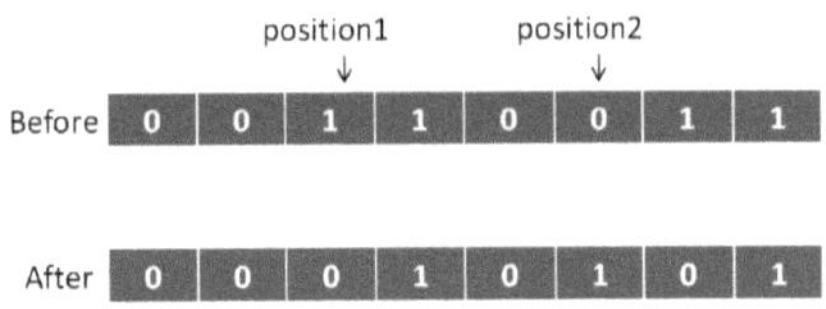

Fig. 5. UC trajectory under different path segments

After completing fitness evaluation, selection, crossover, and mutation operations, a proportional number of randomly generated new individuals are introduced during each generation's population update to mitigate the decline in population diversity. Specifically, the top 80% of elite individuals are retained, while 20% of randomly initialized individuals are injected into the population. This strategy continuously replenishes diversity to reduce the risk of premature convergence to local optima. The algorithm terminates upon reaching the maximum number of generations.

The individual with the highest fitness value selected from the final generation's population is designated as the optimal scheduling solution A_c^* for the charging model.

4 Simulation Results

This chapter presents simulations investigating the impact of the proposed IGA on UAV trajectory scheduling. Numerical experiment results demonstrate that the genetic algorithm exhibits favorable performance in UAV scheduling, which can substantially enhance the overall system efficiency.

The system comprises 20 GSs, which are randomly and uniformly distributed within a 2D area of 10×10 km^2. All UAVs are assumed to fly at a fixed altitude of $H = 50$ m and a speed of $V = 10$ m/s. The charging power is set as $P_c = 1$ W, communication power as $P_t = 0.1$ W, transmitting antenna gain as $G_t = 2$ dBi, receiving antenna gain as $G_r = 2$ dBi, path loss index as $\alpha = 2$, channel power gain as $\rho_0 = -60$ dB, and white noise power as $\sigma^2 = -119$ dBm.

For the IGA, the population size is set as $P = 2000$, with the minimum and maximum crossover probability given by $[p_{\text{cmin}}, p_{\text{cmax}}] \in [0.5, 0.9]$, and the minimum and maximum mutation probability as $[p_{\text{mmin}}, p_{\text{mmax}}] \in [0.01, 0.1]$.

4.1 The Influence of Demand and the Number of Segments on the Trajectory

The above two figures illustrate the effects of varying demand levels and segment counts on UAV trajectories under a single GS scenario. As shown in Fig. 6, low GS demand enables the UAV to complete tasks ahead of schedule, after which it either proceeds to the next GS or returns to the destination, with trajectory turning points aligning closely with operational waypoints. Conversely, higher demand drives the UAV to loiter above the GS for extended periods, as additional flight time toward the GS is required to satisfy communication requirements.

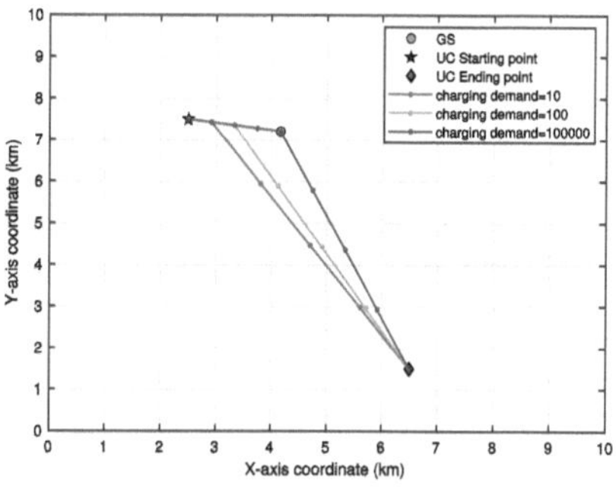

Fig. 6. UC trajectory under different charging demand

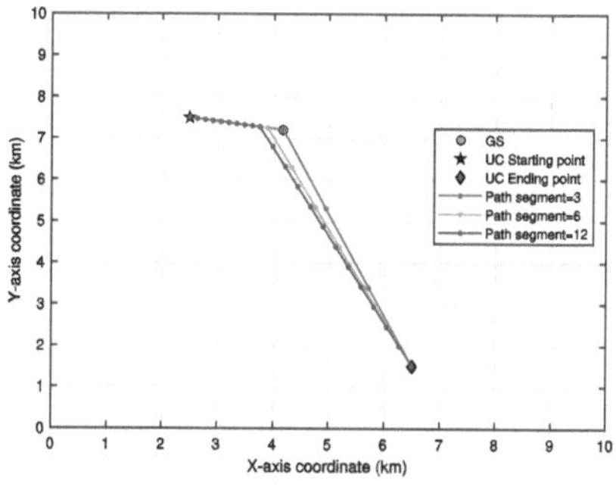

Fig. 7. UC trajectory under different path segments

Figure 7 systematically analyzes the nuanced impact of flight segment count on UAV operational behavior under constant single-GS demand, revealing a direct correlation between finer spatial discretization and trajectory accuracy. Empirical results demonstrate that increasing the segment count enhances waypoint alignment precision and maneuver timing. Notably, fewer segments impose rigid constraints on UAV movement, necessitating task completion exclusively at segment midpoints even after primary objectives are met. By contrast, higher segment counts mitigate such inefficiencies by enabling earlier maneuver initiation, streamlining task execution and improving overall operational efficiency through the reduction of redundant flight paths.

4.2 The Influence of Genetic Algorithm on Scheduling

Figure 8 illustrates the distribution of GSs within a two-dimensional area of $10 \times 10\ \text{km}^2$ in a multi-base station scenario. The initial and terminal positions of UC and UT are marked in the figure. This figure enables intuitive observation of the UAVs' initial path planning in multi-mission scenarios, serving as a benchmark for performance comparison in complex multi-base station scenarios.

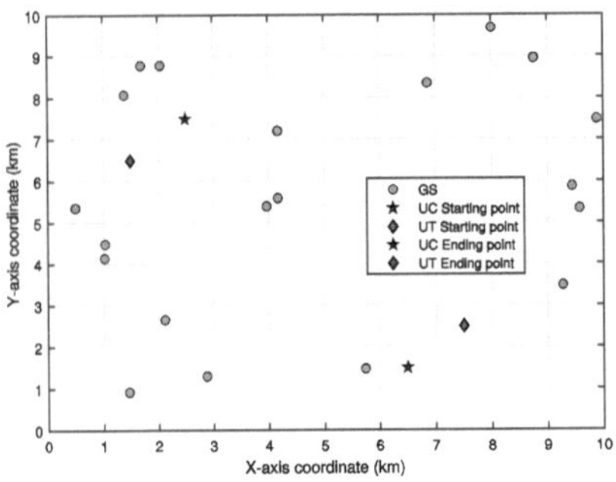

Fig. 8. GS distribution map

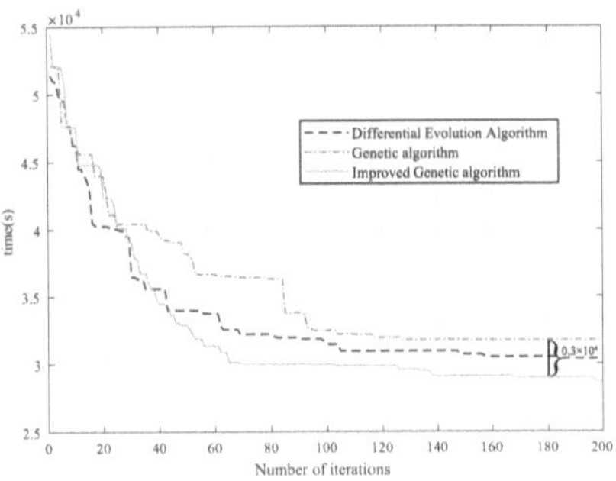

Fig. 9. System time iteration diagram

Figure 9 compares the convergence behaviors of the objective function values for IGA, DEA, and GA. IGA exhibits a rapid initial decline, mid-stage fine-tuning, and stabilizes within 200 iterations, balancing global exploration and local exploitation via adaptive parameter adjustments.

GA demonstrates mid-term oscillations and slower convergence owing to fixed probabilities, resulting in higher final objective function values and trapping in local optima due to insufficient population diversity. DEA, which relies on local search mechanisms, shows weak global exploration capability and fails to achieve complete convergence. IGA outperforms the benchmark algorithms (GA and DEA) in both efficiency and accuracy, attributed to dynamic sub-path segmentation, conflict matrix constraints, and diversity preservation strategies. It exhibits robust performance in multi-UAV scheduling scenarios, and its stability confirms effective mitigation of the premature convergence issue inherent in traditional GA.

This translates to a 30% improvement in convergence speed. From the perspective of convergence outcomes, at the 180th iteration, the time consumption of IGA is 2.9×10^4, representing an approximate 10% reduction compared to that of the other two algorithms. Additionally, IGA exhibits superior path smoothness and enhanced coverage performance.

Figure 10 illustrates UC's flight trajectory: departing from the starting position, it navigates around multiple GSs in a zigzag path, with dense turning points aligning closely with GS locations. The GA framework refines flight segments through increased spatial discretization, maintaining trajectory continuity to dynamically approach GSs and enhance charging efficiency. The trajectory flattens near the endpoint before returning to the destination, which validates the objective function's penalty mechanism for "invalid flight distance".

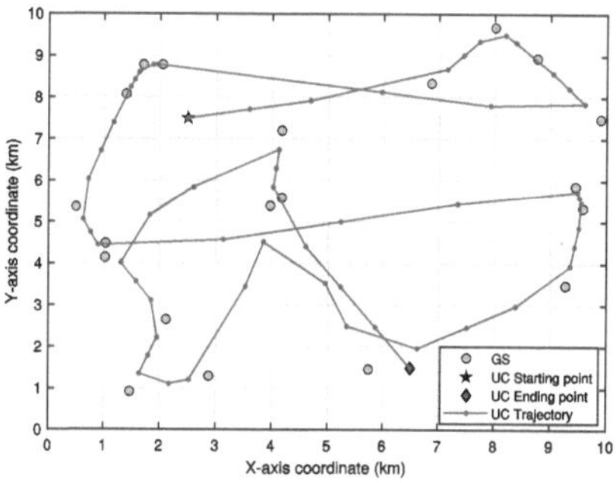

Fig. 10. The trajectory of UC

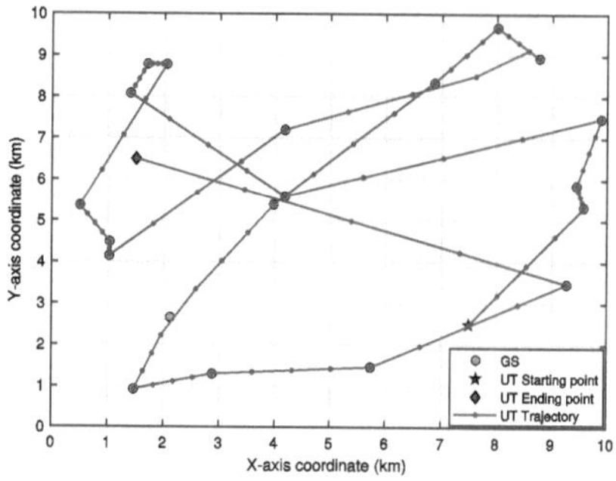

Fig. 11. The trajectory of UT

Figure 11 depicts UT's smooth trajectory, which avoids densely distributed charging areas via straight-line segments and large-radius arcs to minimize spatial conflicts. A mid-section fluctuation reflects trajectory adjustments driven by the "maximizing transmission rate" objective—specifically, a brief approach toward the base station to achieve improved communication quality.

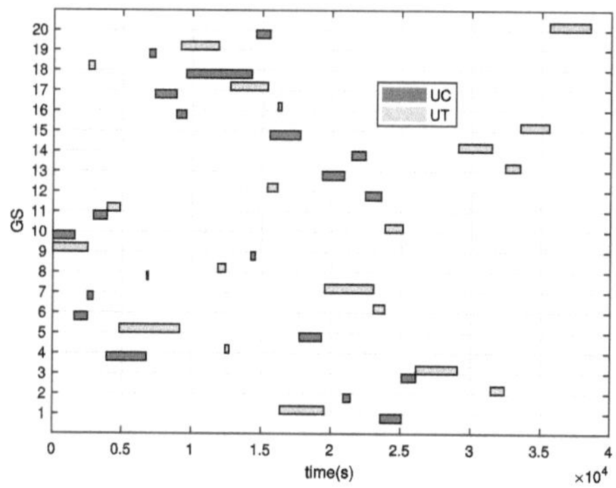

Fig. 12. UAVs schedule Gantt chart

Figure 12 illustrates the task scheduling sequence of UC and UT on the time axis. After optimization via IGA, their time intervals exhibit no overlap, indicating that the Hadamard product of the scheduling conflict matrix and time conflict matrix effectively resolves resource contention. UC's task segments consist of multiple short time slots, reflecting fine-grained scheduling enabled by path segmentation to minimize individual charging cycles. This visualization validates the temporal coordination capability of the joint optimization algorithm, establishing a foundation for ensuring the real-time performance of the system.

5 Conclusion

This paper focuses on the collaborative scheduling of heterogeneous UAVs, particularly the synergy between UC and UT in complex scenarios. The core challenges lie in balancing spatiotemporal conflicts, optimizing task allocation efficiency, and enhancing algorithm convergence for dual-UAV operations.

An IGA framework tailored to this scenario is proposed: the task space is modeled to define constraints, conflicts, and objectives, and the IGA integrates specialized encoding, a time-minimization fitness function, and enhanced genetic operations—including the PMX operator for effective gene inheritance.

The experimental results demonstrate that the IGA algorithm outperforms the other two benchmark algorithms significantly. As observed from the simulation plots, IGA tends to converge and level off around the 65th iteration, whereas GA and DEA do not stabilize until the 105th and 85th iterations, respectively. It provides a strictly feasible solution for dual-heterogeneous UAV scheduling and lays a solid foundation for the deployment of multi-UAV systems in complex application scenarios.

Future work will enhance the framework by integrating anti-interference protocols to improve environmental adaptability, extending it to multi-type UAV clusters for cluster collaboration and energy management, and adopting edge computing and federated learning to address the limitations of centralized optimization in large-scale scenarios.

Acknowledgments. This study was funded by the Science and Technology Major Project of Tibetan Autonomous Region of China (XZ202201ZD0006G); 2024 Xizang Autonomous Region Central Guided Local Science and Technology Development Fund Project (XZ202401YD0015).

References

1. Zeng, Y., Zhang, R., Lim, T.J.: Wireless communications with unmanned aerial vehicles: opportunities and challenges. IEEE Commun. Mag. **54**(5), 36–42 (2016)
2. Li, J., Zhao, H., Wang, H., et al.: Joint optimization on trajectory, altitude, velocity, and link scheduling for minimum mission time in UAV-aided data collection. IEEE Internet Things J. **7**(2), 1464–1475 (2019)
3. Lin, X., Yajnanarayana, V., Muruganathan, S.D., et al.: The sky is not the limit: LTE for unmanned aerial vehicles. IEEE Commun. Mag. **56**(4), 204–210 (2018)
4. Mozaffari, M., Saad, W., Bennis, M., et al.: Unmanned aerial vehicle with underlaid device-to-device communications: performance and tradeoffs. IEEE Trans. Wireless Commun. **15**(6), 3949–3963 (2016)
5. Zeng, Y., Zhang, R.: Energy-efficient UAV communication with trajectory optimization. IEEE Trans. Wireless Commun. **16**(6), 3747–3760 (2017)

6. Petryshyn, L., Petryshyn, M.: Service mechanism and queue discipline in multitasking UAV systems. In: 2024 IEEE 7th International Conference on Actual Problems of Unmanned Aerial Vehicles Development (APUAVD), Kyiv, Ukraine, pp. 246–249 (2024). https://doi.org/10.1109/APUAVD64488.2024.10765886
7. Yi, M., Wang, X., Liu, J., Zhang, Y., Hou, R.: Multitask transfer deep reinforcement learning for timely data collection in rechargeable-UAV-aided IoT networks. IEEE Internet Things J. **10**(23), 20545–20559 (2023). https://doi.org/10.1109/JIOT.2023.3300927
8. Lee, J., Friderikos, V.: Interference-aware path planning optimization for multiple UAVs in beyond 5G networks. J. Commun. Netw. **24**(2), 125–138 (2022)
9. Jia, X., Wu, S., Sun, J.: UAV formation collaborative search and dynamic task allocation method under uncertain environment. In: Proceedings of the 33rd Chinese Control Conference, pp. 947–953. IEEE (2014)
10. Gupta, N., Agarwal, S., Mishra, D.: Multi-UAV replacement and trajectory design for coverage continuity. In: ICC 2022-IEEE International Conference on Communications, pp. 1–6. IEEE (2022)
11. Wu, Q., Zeng, Y., Zhang, R.: Joint trajectory and communication design for multi-UAV enabled wireless networks. IEEE Trans. Wireless Commun. **17**(3), 2109–2121 (2018)
12. Jin, R., Lin, R.: Optimizing dual-mode UAV-assisted remote IoT data collection with decentralized DRL in 6G-enabled space-air-ground networks. In: ICASSP 2025–2025 IEEE International Conference on Acoustics, Speech and Signal Processing (ICASSP), Hyderabad, India, pp. 1–5 (2025). https://doi.org/10.1109/ICASSP49660.2025.10888179
13. Zhu, Z., Qian, Y., Zhang, W.: Research on UAV searching path planning based on improved ant colony optimization algorithm. In: 2021 IEEE 3rd International Conference on Civil Aviation Safety and Information Technology (ICCASIT), Changsha, China, pp. 1319–1323 (2021). https://doi.org/10.1109/ICCASIT53235.2021.9633591
14. Du, Y., Liu, C., Chen, D.: Research on UAV task allocation algorithm based on mixed task grouping. In: 2025 IEEE 8th Information Technology and Mechatronics Engineering Conference (ITOEC), Chongqing, China, pp. 446–451 (2025). https://doi.org/10.1109/ITOEC63606.2025.10967693
15. Zhang, J., Wen, P., Xiong, A.: Application of improved quantum particle swarm optimization algorithm to multi-task assignment for heterogeneous UAVs. In: 2022 6th Asian Conference on Artificial Intelligence Technology (ACAIT), pp. 1–5. IEEE (2022)
16. Zhen, Z., Xing, D., Gao, C.: Cooperative search-attack mission planning for multi-UAV based on intelligent self-organized algorithm. Aerosp. Sci. Technol. **76**, 402–411 (2018)
17. Scorrano, L., Dinoi, L.: A platform-aware approach to the design of asymmetric arrays for UAVs based on GA optimization. In: 2016 IEEE International Symposium on Antennas and Propagation (APSURSI), Fajardo, PR, USA, pp. 1905–1906 (2016). https://doi.org/10.1109/APS.2016.7696659

Data Integrity Auditing Supporting Multi-ownership Transfer in Cloud Storage

Yating Gu[1], Lunzhi Deng[1,2(✉)], and Huan Li[1]

[1] School of Mathematical Sciences, Guizhou Normal University, Guiyang 550025, China
denglunzhi@163.com

[2] Guizhou Provincial Specialized Key Laboratory of Information Security Technology in Higher Education Institutions, Guiyang 550025, China

Abstract. In cloud storage environments, data owners leverage integrity auditing mechanisms to authorize third-party auditors (TPAs) for real-time verification of cloud data integrity. A significant breakthrough in this domain is the development of multi-ownership transfer auditing schemes, which enable simultaneous transfer of both cloud data and their corresponding integrity verification authenticators. This capability proves particularly vital in multi-ownership scenarios such as corporate mergers and acquisitions, or collaborative healthcare systems with shared electronic medical records, where multiple stakeholders require verifiable control over shared data assets. While several auditing schemes supporting multi-ownership data transfer have been proposed to address these needs, existing solutions predominantly depend on secure communication channels during ownership transfer - a major limitation given the practical impossibility of guaranteeing absolute channel security in real-world implementations. To overcome this limitation, we present an innovative data integrity auditing mechanism based on the discrete logarithm problem, which achieves secure multi-ownership transfer without relying on additional secure communication channels. Our solution offers three groundbreaking advantages: it ensures end-to-end security in multi-ownership transfers while completely eliminating secure channel requirements; security analysis confirms the scheme achieves unforgeable authenticators, undeniable proof information, and robust protection against data theft; and comprehensive performance evaluations demonstrate significant efficiency improvements over existing approaches, particularly during the critical multi-ownership transfer phase.

Keywords: Cloud storage · Privacy preservation · Multi-ownership transfer · Data integrity auditing

1 Introduction

Data integrity auditing is a critical cloud security technology that enables third-party auditors to verify the integrity and authenticity of cloud-stored data without accessing the original content through cryptographic methods. The core

J. Yu and Y. Chen (Eds.): FCS 2025, CCIS 2805, pp. 458–476, 2027.
https://doi.org/10.1007/978-981-95-9514-3_30

value of this technology lies in addressing trusted verification challenges in data outsourcing scenarios. Supporting data ownership transfer, integrity auditing represents a significant evolution of the former. Its innovation is reflected in securely converting verification tags generated by the original owner into those of the new owner without downloading the entire dataset during ownership changes (e.g., corporate mergers, business transfers). This conversion process must satisfy two key security properties: ensuring the confidentiality of non-transferred data and maintaining the verifiability of transferred data. Furthermore, integrity auditing schemes supporting multi-ownership transfer advance dynamic data governance in complex business environments. This technology allows multiple entities to take ownership of the same dataset at different stages, such as multi-party supervision in supply chain finance and cross-institutional sharing of electronic medical records.

1.1 Related Work

To verify data integrity, Ateniese et al. first introduced the concept of Provable Data Possession (PDP) [1]. Subsequently, to support dynamic data outsourcing, Ateniese's team further proposed a dynamic PDP scheme supporting data deletion and modification operations [2]. In certain application scenarios, privacy protection has become an indispensable element of PDP. Since the Third Party Auditor's (TPA) responsibility is limited to verifying cloud file integrity, it should not access file contents or user identity information. Addressing the risk of TPAs potentially inferring file contents, Wang et al. first proposed a privacy-preserving PDP scheme combining public key homomorphic authenticators with random masking [15]. For fine-grained privacy in multi-user scenarios, Erway et al. presented location privacy protection for dynamic data updates [8], while Dhakad et al. developed a low-cost verifiably secure cloud data possession proof scheme [7], aiming to balance privacy protection, integrity verification, and computational efficiency.

In research on system performance enhancement and scalability improvement, Chen et al. proposed a dynamic auditing scheme for Hadoop big data platforms. They designed a Data Block Index Table to support dynamic data operations on HDFS (Hadoop Distributed File System) and developed a parallel auditing algorithm integrated with the MapReduce framework [3]. Addressing the certificate management complexity in traditional Public Key Infrastructure (PKI), scholars have proposed innovative solutions using identity-based cryptography [14,17]. To achieve security proofs in the Standard Model (SM), Deng et al. designed an identity-based auditing protocol [5] with SM security proofs, though key escrow issues remained. To simplify certificate management and eliminate key escrow, He et al. proposed a Certificateless Provable Data Possession (CL-PDP) scheme in 2017 [10], which Deng et al. later found vulnerable. Deng's team systematically improved this, proposing a novel data integrity verification scheme for smart grid cloud management systems [4]. Huang et al. developed a privacy-preserving certificateless public auditing scheme supporting variable audit frequencies [11], using permutation techniques to obfuscate file indices for user privacy. Zhao et al.'s certificateless public auditing scheme effectively protects stored data privacy against designated verifier misbehavior [20], while Deng et al. also proposed an SM-secure certificateless public auditing scheme [6].

In cloud storage scenarios, data ownership may transfer during events like corporate mergers. When an acquiring company takes over specific business units, relevant data ownership should simultaneously transfer. To maintain confidentiality of non-transferred data, the acquired company must prevent encryption key leakage to the acquirer. The most straightforward approach requires the acquirer to download all original data, self-sign with its keys, and perform cloud verification, imposing substantial communication and computation overhead [18]. Wang et al. proposed a PKI-based PDP scheme supporting ownership transfer [16], though still requiring distributors to download and convert authenticators for transferred data blocks. Shen et al. proposed another PKI-based transferable PDP scheme [13] delegating authenticator conversion to the cloud, eliminating re-download/upload needs and reducing communication costs. Both schemes suffered complex certificate management. Shen et al. introduced a blockchain-based accountable auditing system supporting multi-ownership transfer [12], combining homomorphic verification with multisignatures to ensure verifiability, accountability, and low overhead during corporate mergers and medical record migrations. However, transfer overhead grew linearly with data blocks. In 2023, Gai et al. proposed a privacy-preserving identity-based public auditing scheme for cloud-based IoT data [9], delegating transfer tasks to TPA, making overhead data-block-independent, but raising practicality concerns. To eliminate certificate management and key escrow, Huang et al. proposed in 2024 a certificateless cloud storage auditing scheme supporting ownership transfer [11], though with high transfer overhead. In 2025, Zhang et al. introduced a blockchain-enabled dynamic certificateless outsourced data auditing mechanism supporting multi-ownership transfer [19], yet without achieving authenticator unforgeability and proof information non-deceptiveness.

1.2 Motivations

While the aforementioned schemes have achieved notable progress in specific application domains, they still suffer from three critical limitations:

1. Dependence on secure channels: Beyond the private key generation phase, these schemes require additional secure channels during data ownership transfer—a requirement that is notoriously difficult to fulfill in practice, severely compromising their practicality (e.g., Schemes [9,11,14,19]).
2. Lack of data privacy preservation: Adversaries can reconstruct data owners' original datasets by solving systems of linear equations using proof information, exposing a fundamental vulnerability (e.g., Schemes [11–13,19]).
3. Linear computational overhead: The computational cost during ownership transfer scales linearly with the number of data blocks, resulting in prohibitively high operational overhead (e.g., Schemes [11–13]).

1.3 Contributions

To address these limitations, we propose a data integrity auditing scheme supporting multi-ownership transfer (DIA-MOT), featuring three key contributions:

1. Dual-functional auditing framework: Our scheme not only verifies cloud-stored data integrity but also enables secure multi-ownership transfer. Crucially, it eliminates the need for additional secure channels beyond the initial private key generation phase.
2. Provable security guarantees: Under the hardness assumption of the Diffie-Hellman problem, we formally prove our scheme achieves three fundamental security properties: authenticator unforgeability, proof information non-repudiation, and data theft resistance.
3. Constant-time efficiency: The computational cost during ownership transfer remains constant regardless of data block quantity, demonstrating significantly lower overhead compared to existing schemes [11–13].

2 Preliminaries

2.1 Notations

Table 1 presents the definitions of the symbols utilized in DIA-MOT.

2.2 Bilinear Pairing

G_1, G_2 are two multiplicative cyclic group, both having the same prime order q. The mapping $e : G_1 \times G_1 \rightarrow G_2$ is defined as a bilinear map when the following conditions are satisfied.

1. Bilinearity: The mapping $e(u^a, u^b) = e(u, u)^{ab}$ holds true for all $u \in G$ and $a, b \in Z_q^*$.
2. Nondegenearacy: $\exists\, u, v \in G_1$ such that $e(u, v) \neq 1_{G_2}$.
3. Computability: For all $u, v \in G_1$, there exists an efficient algorithm to compute $e(u, v)$.

Definition 1. *Computation Diffie-Hellman (CDH) problem. Consider a multiplicative cyclic group G_1 of prime order q with generator g. For randomly chosen $a, b \in Z_q^*$, given the elements g^a and g^b, the challenge is to compute g^{ab}.*

3 Scheme Framework

3.1 System Model

Figure 1 illustrates the system model of DIA-MOT, which consists of five entities, described as follows.

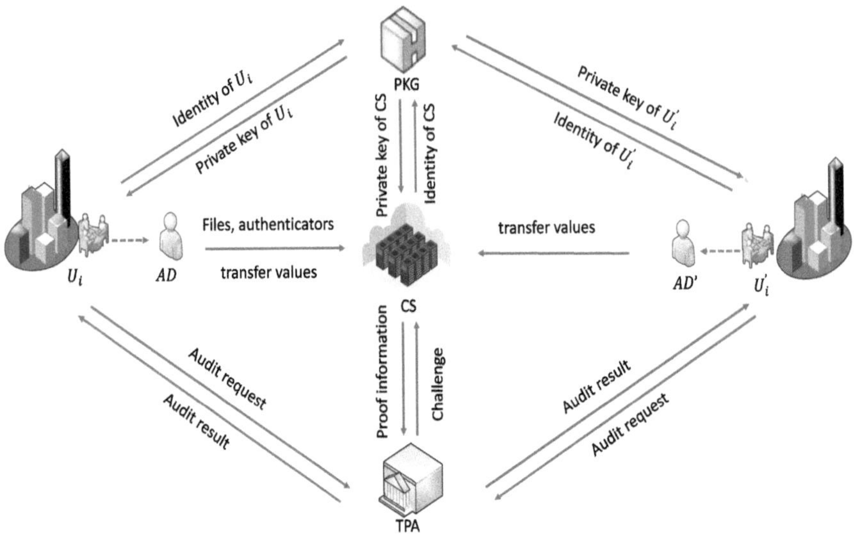

Fig. 1. System Model

Table 1. NOTATIONS

Notation	Description
G_1, G_2	Two different multiplicative cyclic group of the large prime order q
q	A large prime number
e	Bilinear mapping $e : G_1 \times G_1 \rightarrow G_2$
g, u	Two generators of group G_1
$H, H_{1\sim6}$	Seven secure hash functions
ID_i	The U_i's identity
ID_i'	The U_i''s identity
ID_{CS}	The CS's identity
sk	The aggregated private key of Group A
sk'	The aggregated private key of Group B
sk_{CS}	The private key of CS
PK	The aggregated public key of Group A
PK'	The aggregated public key of Group B
PK_{CS}	The public key of CS
$chal$	Challenge information
prf	Proof information

- Private Key Generator (PKG): Generates the system's public parameters and private keys for Group A, Group B, and CS.
- User U_i: Denotes the w users in Group A who jointly retain ownership of the target file.
- User U_i': Represents the w' users in Group B designated as recipients via Group A's multi-ownership transfer protocol.
- Group A Administrator (AD): Communicates with AD' and the Cloud Server (CS) to orchestrate multi-ownership transfer to U_i'.
- Group B Administrator (AD'): Interacts with AD and CS to execute transfer of ownership rights (originally held by U_i) to U_i'.
- Cloud Service (CS): It has vast storage capacity and interacts with AD and AD' to facilitate the transfer of data ownership.
- Third-party Auditor (TPA): With powerful computing capabilities, it assists U_i or U_i' in performing data audits.

DIA-MOT comprises the following eight algorithms:

- *Setup*: PKG generates the system parameters params and the master secret key msk x.
- *Registration*: The system assigns unique identifiers to two user categories: U_i who originally co-own the target file, and U_i' acquiring ownership through a transfer protocol.
- *Key Generation*: The PKG generates distinct private keys for U_i, U_i', and the Cloud Server (CS), then securely distributes each key to its respective recipient.
- *Authenticator Generation*: Let a file$F$ be composed of n data blocks, denoted as $F = \{m_1, m_2, \ldots, m_n\}$. AD generates the authenticator (θ_j, ζ) for the data block m_j and uploads to the cloud for storage.
- *Multi-Ownership Transformation*: AD, AD' and CS perform information exchange to facilitate the transfer of the authenticator.
- *Challenge Generation*: TPA generates the challenge information chal and sends it to CS.
- *Proof Generation*: CS generates the proof information prf and forwards it to TPA
- *Proof verification*: The TPA verifies prf and returns the result to U_i before ownership transfer or to U_i' after transfer.

3.2 Security Requirements

DIA-MOT must consider three distinct types of adversaries.

- Adversary $\mathcal{A}_1$: Played by an external malicious entity. He/she attempts to forge non-owner signatures, with the objectives of depleting storage resources and provoking CS-user disputes.
- Adversary $\mathcal{A}_2$: Played by a compromised or dishonest CS. He/she attempts to extract private data from audit transcripts, with the objectives of compromising data confidentiality and inferring sensitive information.

- Adversary $\mathcal{A}_3$: Played by a semi-trusted TPA. He/she attempts to access the user's data in order to compromise their privacy.

The security properties are defined through the following three games, unforgeability of authenticators, unforgeability of proof information, and non-stealability of data respectively.

To simplify the description, the terms identity, public key, and private key in the security analysis refer specifically to the group's identity, aggregate public key, and aggregate private key, respectively.

Theorem 1. *An DIA-MOT scheme achieves security against adversary* $\mathcal{A}_1$ *if* $\mathcal{A}_1$*'s advantage in Game I is a negligible.*

Game I: The security experiment involves a challenger $\mathcal{C}$ and an adversary $\mathcal{A}_1$ interacting through the following phases:

Initialization: $\mathcal{C}$ executes the setup algorithm to generate public system parameters, and sends it to $\mathcal{A}_1$.

Query: $\mathcal{A}_1$ perform the following query.

1. Query-Public key: $\mathcal{A}_1$ can query the public key of any ID_i.
2. Query-Hash: $\mathcal{A}_1$ can query any hash value for any input.
3. Query-Private Key: $\mathcal{A}_1$ submits an identity ID_i and receives the corresponding private key from $\mathcal{C}$.
4. Query-Authenticator: $\mathcal{A}_1$ provides a data block m_j and receives its authenticator.

Forge: If $\mathcal{A}_1$ wins, a tuple $(m^*, \phi^* = (\theta^*, \zeta^*), ID^*)$ output by $\mathcal{A}_1$ satisfies the following conditions.

1. ϕ^* is a valid authenticator associated with the tuple (m^*, ID^*).
2. $\mathcal{A}_1$ has not queried the private key corresponding to the identity ID^*.
3. θ^* was not obtained by Query-Authenticator.

The advantage of $\mathcal{A}_1$ is defined as: $Adv_{\mathcal{A}_1} = \Pr[\mathcal{A}_1 wins]$.

Theorem 2. *An DIA-MOT scheme achieves security against adversary* $\mathcal{A}_2$ *if* $\mathcal{A}_2$*'s advantage in Game II is a negligible.*

Game II: This game is played between the challenger $\mathcal{C}$ and adversary $\mathcal{A}_2$.

Initialization: Similar to Game I.

Phase I: $\mathcal{A}_2$ makes various queries as those in Game I.

Challenge: $\mathcal{C}$ outputs a challenge message $Chal^*$ that satisfies the following conditions:

1. $\mathcal{C}$ has not made a query for the private key associated with identity ID^*.

2. There is at least one data block m_h such that $\mathcal{A}_2$ has not queried the authenticator for the tuple (m_h, ID^*).

Phase II: $\mathcal{A}_2$ executes the same queries as in Phase I,and satisfies the following conditions:

1. $\mathcal{A}_2$ has not made a query for the private key associated with identity ID^*.
2. $\mathcal{A}_2$ did not query the authenticator of the tuple (m_h, ID^*)

Response: Return the proof Prf^*. If the verification passes, the adversary $\mathcal{A}_2$ wins the game.

The advantage of $\mathcal{A}_2$ is expressed as: $Adv_{\mathcal{A}_2} = \Pr[\mathcal{A}_2 wins]$.

Theorem 3. *An DIA-MOT scheme achieves security against adversary $\mathcal{A}_3$ if $\mathcal{A}_3$'s advantage in Game III is a negligible.*

Game III: This game is played between the challenger $\mathcal{C}$ and adversary $\mathcal{A}_3$.

Initialization: Similar to Game I.

Phase I: $\mathcal{A}_3$ performs the same queries as those in Game I.

1. Public key queries, private key queries, and authenticator queries are handled identically to Game 1.
2. Submit a challenge message $Chal$ to obtain corresponding proof information Prf.

Challenge: $\mathcal{A}_3$ submit a tuple (ID^*, Δ^*, ρ^*). A Perform the following operations.

1. Randomly select a new data file $F^* = \{m_1^*, m_2^*, \cdots, m_n^*\}$.
2. Execute the authenticator generation algorithm on the tuple (ID^*, Δ^*, ρ^*) to generate an authenticator θ^*.
3. $\mathcal{C}$ send filename *name* to $\mathcal{A}_3$, where only the filename of the file is known.

Phase II: $\mathcal{A}_3$ performs the same queries as those in Game II.

Extract: $\mathcal{A}_3$ obtain a data block $m' \in Z_q^*$.

The advantage of $\mathcal{A}_3$ is expressed as: $Adv_{\mathcal{A}_3} = \Pr[m' \in F^*]$.

4 New Scheme

Next, we present the concrete algorithmic construction.

Setup: Given a security parameter ν, the PKG execute the following procedures:

1. Choose two multiplicative cyclic group G_1 and G_2, both of which have order equal to a prime number q. g and u are two generators of G_1. Select a bilinear map $e : G_1 \times G_1 \rightarrow G_2$.

2. Randomly choose $x \in Z_q^*$ as the master secret key $msk = \{x\}$, and compute $U = g^x$.
3. Choose six hash functions: $H : \{0,1\}^* \rightarrow Z_q^*$, $H_1 : \{0,1\}^* \rightarrow Z_q^*$, $H_2 : \{0,1\}^* \rightarrow G_1$, $H_{3\sim 6} : \{0,1\}^* \rightarrow Z_q^*$.
4. Outputs the public parameters $paras = \{G_1, G_2, q, e, g, u, H, H_{1\sim 6}, U\}$.

Registration: In Group A, the system assigns sequential identifiers to w users based on chronological registration order, where the earliest registrant receives identifier 1 with each subsequent user's identifier incremented by 1. Group B follows an identical numbering protocol for its w' users.

Key Generation: Upon receiving the identity credentials ID_i from user U_i, ID_i' from user U_i', and ID_{CS} from the CS, the PKG executes the following steps to generate respective private keys for U_i, U_i', and CS:

1. PKG generates private keys for each user U_i in Group A:
   (a) Each U_i submits identity ID_i to the PKG, which randomly selects $\mu_i \in Z_q^*$, computes $r_i = \mu_i + xH(ID_i, PK_i)$ and $PK_i = g^{\mu_i}$, then sends r_i to U_i while publishing PK_i.
   (b) Upon receiving their private key, all U_i except AD computes $\lambda_i = r_i + H_1(PK_{i+1}^{r_i}) - H_1(PK_{i-1}^{r_i})$, and publishes λ_i. Specially, we define $1-1 = w$ and $w+1 = 1$.
   (c) The AD computes $sk = \sum_{i=1}^{w} \lambda_i$, and $PK = g^{sk}$.
2. PKG generates private keys for each user U_i' in Group B:
   (a) Each U_i' submits identity ID_i' to the PKG, which randomly selects $\mu_i' \in Z_q^*$, computes $r_i' = \mu_i' + xH(ID_i', PK_i')$ and $PK_i' = g^{\mu_i'}$, then sends r_i' to U_i' while publishing PK_i'.
   (b) Upon receiving their private key, all U_i' except AD' computes $\lambda_i' = r_i' + H_1(PK_{i+1}'^{r_i'}) - H_1(PK_{i-1}'^{r_i'})$, and publishes λ_i'. Specially, we define $1-1 = w'$ and $w'+1 = 1$.
   (c) The AD' computes $sk' = \sum_{i=1}^{w'} \lambda_i'$, and $PK' = g^{sk'}$.
3. The CS submits identity ID_{CS} to the PKG, which randomly selects μ_{CS}, computes $sk_{CS} = \mu_{CS} + xH(ID_{CS}, PK_{CS})$ and $PK_{CS} = g^{\mu_{CS}}$, then sends sk_{CS} to CS while publishing PK_{CS}.

Authenticators Generation: The AD generates authenticators for data blocks m_j through the following procedure:

1. Randomly picks $\delta \in Z_q^*$ and computes $\Delta = g^{\delta}$.
2. Computes the authenticators $\theta_j = (H_2(name||j)u^{m_j})^{\delta}$, $j \in [1, n]$.
3. Computes $h_{AD} = H_3(name, ID, PK, \Delta)$ and $\zeta = \delta + sk \cdot h_{AD}$, where $ID = \{ID_1, ID_2, \cdots, ID_w\}$.
4. Outputs $\sigma = \{\theta_j\}_{1 \leq j \leq n}$.
5. Sends $\{F, \zeta, \sigma\}$ to CS.

Authenticators Verify: CS verifies whether the following two equations hold. If both equations are valid, the CS accepts the file and its authenticators; otherwise, it rejects them.

$$g^{\zeta} = \Delta \cdot PK^{H_3(name,ID,PK,\Delta)} \tag{1}$$

$$e(\prod_{j=1}^{n} \theta_j, g) = e(\prod_{j=1}^{n} H_2(name||j) \cdot u^{\sum_{j=1}^{n} m_j}, \Delta) \tag{2}$$

Muti-ownership Transformation: AD, AD' and CS perform information exchange to facilitate the transfer of the authenticator.

1. The AD computes $\alpha = \delta^{-1}H_4(PK'^{sk}) - H_5(PK_{CS}{}^{sk})$, sends α to CS.
2. The AD' randomly picks $\delta' \in Z_q^*$ and computes $\Delta' = g^{\delta'}$, $\alpha' = \delta'(H_4(PK^{sk'}))^{-1} - H_5(PK_{CS}{}^{sk'})$, sends α' to CS.
3. The CS computes $\beta = \alpha + H_5(PK^{sk_{CS}}) = \delta^{-1}H_4(PK'^{sk})$, $\beta' = \alpha' + H_5(PK'^{sk_{CS}}) = \delta'(H_4(PK^{sk'}))^{-1}$, $k = \beta\beta' = \delta^{-1}H_4(PK'^{sk}) \cdot \delta'(H_4(PK^{sk'}))^{-1} = \delta'\delta^{-1}$.
4. AD' computes $h_{AD'} = H_3(name, ID', PK', \Delta')$ and $\zeta' = \delta' + sk' \cdot h_{AD'}$, where $ID = \{ID'_1, ID'_2, \cdots, ID'_w\}$. Send ζ' to CS.
5. CS saves $\{F, \zeta', \sigma' = \{\theta_j{}^k\}_{1 \le j \le n}\}$ and deletes $\{F, \zeta, \sigma\}$.

Chal-Gen: TPA generates a challenge information *chal* as follows.

1. Randomly select a subset $J \subseteq \{1, ..., s\}$ that contains a elements.
2. Randomly selects $c_j \in Z_q^*$ for every $j \in J$.
3. Sends $chal = \{j, c_j\}_{j \in J}$ to the CS.

Prf-Gen: After receiving the challenge message *chal*, CS performs the following procedure to generate the proof message.

1. When data is not transferred, the cloud calculates
   (a) Computes $\Lambda = \prod_{j \in J} \theta_j^{c_j}$, and $M' = \sum_{j \in J} m_j c_j$.
   (b) Randomly picks a value $\tau \in Z_q^*$, computes $\rho = g^{\tau}$ and $M = M' - \tau H_6(name, \rho)$.
   (c) Sends $Prf = \{\Lambda, M, \rho\}$ to the TPA.
2. When data is not transferred, the cloud calculates
   (a) Computes $\Lambda' = \prod_{j \in J} (\theta_j^k)^{c_j}$, and $M' = \sum_{j \in J} m_j c_j$.
   (b) Randomly picks a value $\tau \in Z_q^*$,compute $\rho = g^{\tau}$ and $M = M' - \tau H_6(name, \rho)$.
   (c) Sends $Prf = \{\Lambda', M, \rho\}$ to the TPA.

Prf-Verify: TPA verifies the proof information using the following equation, outputs 1 if the equation holds, 0 otherwise.

1. When data is not transferred, the TPA verifies prf via the following equation:

$$e(\Lambda, g) = e(\prod_{j \in J} H_2(name||j)^{c_j} u^M \rho^{H_6(name,\rho)}, \Delta) \tag{3}$$

2. When data is transferred, the TPA verifies prf via the following equation:

$$e(\Lambda', g) = e(\prod_{j \in J} H_2(name||j)^{c_j} u^M \rho^{H_6(name,\rho)}, \Delta') \tag{4}$$

5 Scheme Analysis

5.1 Correctness

Next, we provide a detailed analysis of the correctness of the *Authenticators-Verify* algorithm and *Prf-Verify* algorithm.

1. *Correctness of Authenticator-Verify*:

$$g^{\zeta} = g^{\delta + sk \cdot h_{AD}} = \Delta \cdot PK^{H_3(name,ID,R,\Delta)} \tag{5}$$

$$e(\prod_{j=1}^{n} \theta_j, g) = e(\prod_{j=1}^{n} (H_2(name||j) \cdot u^{m_j})^{\delta}, g) = e(\prod_{j=1}^{n} H_2(name||j) \cdot u^{\sum_{j=1}^{n} m_j}, \Delta) \tag{6}$$

2. *Correctness of Prf-Verify*:

$$\begin{aligned} e(\Lambda, g) &= e(\prod_{j \in J} \theta_j^{c_j}, g) \\ &= e(\prod_{j \in J} ((H_2(name||j) u^{m_j})^{\delta})^{c_j}, g) \\ &= e(\prod_{j \in J} H_2(name||j)^{c_j} \cdot u^{\sum_{j \in J} m_j c_j}, g^{\delta}) \\ &= e(\prod_{j \in J} H_2(name||j)^{c_j} u^M \rho^{H_6(name,\rho)}, \Delta) \end{aligned}$$

5.2 Security Analysis

This section proves the security of our scheme against three types of adversary attacks in the random oracle model.

Theorem 4. *If the Computational Diffie-Hellman (CDH) problem is hard, then the DIA-MOT scheme achieves authenticator unforgeability.*

Proof. Given the tuple (g^a, g^b), the challenger $\mathcal{C}$ in Game 1 will simulate the protocol and compute g^{ab}.

Initialisation: $\mathcal{C}$ executes the setup algorithm to generate public system parameters $paras = \{G_1, G_2, q, e, , g, u, H, H_{1\sim 6}\}$, and sends it to $\mathcal{A}_1$.

Queries: For each U_i, $\mathcal{A}_1$ first executes a Query-PK on it. $\mathcal{C}$ establishes several lists to store queries and answers.

1. Query-Public key: $\mathcal{C}$ maintains a list l_{PK}, consisting of the tuple $(ID_j, \delta_j, \mu_j, \Delta_j, PK_j)$. When $\mathcal{A}_1$ submits an identity ID_j, the operation of $\mathcal{C}$ is detailed as follows.
   (a) On the Υ-th query, set $DO_\Upsilon = DO^\nabla$ and $\Delta_\Upsilon = g^a, PK_\Upsilon = g^{\mu_\Upsilon}$, adds $(\Delta_\Upsilon, PK_\Upsilon)$ to the list l_{PK}.
   (b) Otherwise, randomly select $\delta_j, \mu_j \in Z_q^*$ and calculate $\Delta_j = g^{\delta_j}, PK_j = g^{\mu_j}$, adds (Δ_j, PK_j) to the list l_{PK}.
2. Query-H: $\mathcal{C}$ maintains a list l consisting of tuples (π_j, h). When $\mathcal{A}_1$ query H for π, $\mathcal{C}$ randomly selects h_j, set $H(\pi_j) = h_j$, and adds the tuple (π_j, h_j) to the list l.
3. Query-H_1: $\mathcal{C}$ maintains a list l_1 consisting of tuples (ε_j, h_1). When $\mathcal{A}_1$ query H_1 for ε, $\mathcal{C}$ randomly selects h_{1_j}, set $H_1(\varepsilon_j) = h_{1_j}$, and adds the tuple (ς_j, h_{1_j}) to the list l_1.
4. Query-H_2: $\mathcal{C}$ maintains a list l_2 consisting of tuples (ϖ_j, h_{2_j}). When $\mathcal{A}_1$ query H_2 for ϖ_j, $\mathcal{C}$ randomly selects h_{2_j}, set $H_2(\varpi_j) = h_{2_j}$, and adds the tuple (ϖ_j, h_{2_j}) to the list l_2.
5. Query-H_3: $\mathcal{C}$ maintains a list l_3 consisting of tuples (χ_j, h_{3_j}). When $\mathcal{A}_1$ query H_3 for χ_j, $\mathcal{C}$ randomly selects h_{3_j}, set $H_3(\chi_j) = h_{3_j}$, and adds the tuple (χ_j, h_{3_j}) to the list l_3.
6. Query-H_4: $\mathcal{C}$ maintains a list l_4 consisting of tuples (E_j, h_{4_j}). When $\mathcal{A}_1$ query H_4 for E_j , $\mathcal{C}$ randomly selects h_{4_j}, set $H_4(E_j) = h_{4_j}$, and adds the tuple (E_j, h_{4_j}) to the list l_4.
7. Query-H_5: $\mathcal{C}$ maintains a list l_5 consisting of tuples (κ_j, h_{5_j}). When $\mathcal{A}_1$ query H_5 for κ_j, $\mathcal{C}$ randomly selects h_{5_j}, set $H_5(\kappa_j) = h_{5_j}$, and adds the tuple (κ_j, h_{5_j}) to the list l_5.
8. Query-H_6: $\mathcal{C}$ maintains a list l_6 consisting of tuples (o_j, h_{6_j}). When $\mathcal{A}_1$ query H_5 for o_i, $\mathcal{C}$ randomly selects h_{6_j}, set $H_6(o_j) = h_{6_j}$, and adds the tuple (o_j, h_{6_j}) to the list l_6.
9. Query-Key: When $\mathcal{A}_1$ submits an identity ID_j, $\mathcal{C}$'s operation is as follows.
   (a) $\mathcal{C}$ fails if $ID_j = ID^\nabla$.
   (b) Otherwise,$\mathcal{C}$ find the tuple (ID_j, δ_j, μ_j) from n_{PK} and performs the Key-Gen algorithm to generate (δ_j, μ_j), then adds the tuple (ID_j, δ_j, μ_j) to list l_{PK}.
10. Query-Authenticator: When $\mathcal{A}_1$ ask for an authenticator for (ID_j, m_j), $\mathcal{C}$ return a authenticator according to the next steps.
    (a) If $ID_j \neq ID^\nabla$, then $\mathcal{C}$ run the Authenticator-Gen algorithm to produce an authenticator $\phi_j = (\theta_j, \zeta_j)$.

(b) If $ID_j = ID^{\nabla}$, $\mathcal{C}$'s operation is as follows.
   i. Randomly picks $\varrho \in Z_q^*$.
   ii. Append $g^{\varrho} \cdot u^{-m_j} = H_2(name||j)$ to list l_2 and repeat i-ii if collision occur.
   iii. output $\theta_j = \Delta^{\varrho}$.

Forge: $\mathcal{A}_1$ output a tuple $(m^*, \phi^* = (\theta^*, \zeta^*), ID^*)$ and that satisfies the requirement as defined in the Game I

Solving CDH problems: The forged authenticator $\phi^* = (\theta^*, \zeta^*)$ constitutes a valid authenticator on data block m^* under identity ID^*. By rewinding adversary $\mathcal{A}_1$ with identical randomness except for substituting the output $h_{2_j}*$ of the H_2 query for the forged data block, $\mathcal{C}$ obtains two valid authenticators: $\phi^* = (\theta^*, \zeta^*)$ and $\phi^{*'} = (\theta^{*'}, \zeta^*)$. If $ID^* = ID^{\nabla}$, then $\theta^* = ((g^b)^{\varrho - \varrho'} \cdot u^{m^*})^a$ and $\theta^{*'} = ((g^b)^{\varrho'} \cdot u^{m^*})^a$. C can solve the CDH problem with the following calculation:

1. $\mathcal{C}$ computes $\theta^*/\theta^{*'} = ((g^b)^{\varrho - \varrho'})^a = (g^{ab})^{\varrho - \varrho'}$.
2. $g^{ab} = (\theta^*/\theta^{*'})^{(\varrho - \varrho')^{-1}}$.

Probability: Let Y_{PK} and Y_K denote the number of Query-Public key and Query-Key operations, respectively. These two events are defined as follows:

Ξ_1: The private key of ID^{∇} cannot be queried.

Ξ_2: $ID^* = ID^{\nabla}$.

The following results can be easily derived.

$\Pr[\Xi_1] = \frac{Y_{PK} - Y_K}{Y_{PK}}$, $\Pr[\Xi_2] = \frac{1}{Y_{PK} - Y_K}$.

$\Pr[\mathcal{C}win] = \Pr[E_1 \wedge E_2] = \Pr[E_1] \cdot \Pr[E_2|E_1] = \frac{Y_{PK} - Y_K}{Y_{PK}} \cdot \frac{1}{Y_{PK} - Y_K} = \frac{1}{Y_{PK}}$

In summary, assuming the probability of $\mathcal{A}_1$ forging a valid authenticator is B, then the probability of $\mathcal{C}$ solving the DL problem is $\frac{B}{Y_{PK}}$, this contradicts the intractability of the CDH problem.

Theorem 5. *If H_2 is a collision-resistant hash functions, then DIA-MOT is proof-information non-deceptive.*

Proof. In Game II, CS acts as the adversary.

Initialisation: Same as that in Game I .

Phase I: $\mathcal{A}_2$ performs the same initialization as in *Proof of Theorem 1* and satisfies the conditions specified in Game II.

Challenge: $\mathcal{A}_2$ receives a challenge information $chal^* = \{\{j, c_j\}_{j \in J}\}$, and adheres to the criteria outlined in Game II.

Phase II: $\mathcal{A}_2$ executes the same queries as in Phase I

Response:$\mathcal{A}_2$ does not have the authenticator θ_h, where $l \in I_\eta, h \in J$. According to *Theorem 4*, $\mathcal{A}_2$ cannot forge the authenticator θ_h, so he attempts to replace authenticator $\theta_{h'}$ with θ_h, where $h \neq h'$. $\mathcal{A}_2$ provides proof information as follow:

1. Computes $\Lambda = \prod_{j \in J, j \neq h} \theta_j^{c_j} \cdot \theta_{h'}^{c_h}$
2. Computes $M^{*'} = \sum_{j \in J, j \neq h} m_j c_j + m_{h'} c_h$, and $M^* = M^{*'} - \tau H_6(name, \rho)$.
3. Send $Prf^* = \{\Lambda^*, M^*, \rho\}\}$ to C.

$\mathcal{C}$ can easily get the following equation:

$$
\begin{aligned}
e(\Lambda^*, g) &= e(\prod_{j \in J, j \neq h} \theta_j^{c_j} \cdot \theta_{h'}^{c_h}, g) \\
&= e(\prod_{j \in J, j \neq h} ((H_2(name||j)u^{m_j})^{\delta_*})^{c_j} \cdot (H_2(name||h')u^{m_{h'}})^{\delta_*})^{c_h}, g) \\
&= e(\prod_{j \in J, j \neq l} H_2(name||j)^{c_j} \cdot (H_2(name||l'))^{c_l} \cdot u^{\sum_{j \in J, j \neq l} m_j c_j + m_{h'} c_h}, g^{\delta_*}) \\
&= e(\prod_{j \in J} H_2(name||j)^{c_j} (H_2(name||h))^{-c_h} (H_2(name||h'))^{c_h} u^{M^*} V^{H_6(name,\rho)}, \Delta_*)
\end{aligned}
$$

To passing verification, the relation $(H_2(name||h))^{-c_h}(H_2(name||h'))^{c_h}$ must hold, which contradicts with H_2 is a secure cryptographic hash function. The advantage of adversary $\mathcal{A}_2$ is negligible in Game II.

Theorem 6. *DIA-MOT possesses the property of data non-stealability.*

Proof. In Game III, the TPA acts as the adversary.

Initialisation: $\mathcal{C}$ generate public parameters and forward them to $\mathcal{A}_3$.

Phase I: $\mathcal{A}_3$ executes multiple queries.

Query-Public Key: $\mathcal{C}$ maintain a list l_{PK} containing $(ID_j, \delta_j, \mu_j, \Delta_j, PK_j)$. When ID_j is submitted by $\mathcal{A}_3$, $\mathcal{C}$ randomly select $\delta_j, \mu_j \in Z_q^*$, compute $\Delta_j = g^{\delta_j}, PK_j = g^{\mu_j}$, and add (ID_j, Δ_j, PK_j) to l_{PK}.

Query-Private Key: $\mathcal{C}$ maintain a list l_s storing (ID_j, δ_j, μ_j). When ID_j is submitted by $\mathcal{A}_3$, retrieve (ID_j, δ_j, μ_j) from l_{PK} and generate the private key (δ_j, μ_j), then add the tuple (ID_j, δ_j, μ_j) to l_s.

Query-Authenticator: Maintain a list L_{auth} containing $(ID_j, \Delta_j, PK_j, m_j, \phi_j = (\theta_j, \zeta_j))$. When authenticator for m_j is requested, execute the authenticator generation algorithm to produce $\phi_j = (\theta_j, \zeta_j)$, and add $(ID_j, \Delta_j, PK_j, m_j, \phi_j = (\theta_j, \zeta_j))$ to l_{auth}.

Query-Proof Generation: $\mathcal{A}_3$ submit a challenge *chal*, and return a proof prf according to the proof generation algorithm.

Challenge Generation: Same as in Game III.

Phase II: $\mathcal{A}_3$ execute various queries identical to Phase I.

Extract: Obtain the data block $m' \in Z_q^*$.

According to the proof generation algorithm, $\mathcal{A}_3$ obtain $prf^* = \{\Lambda^*, M^*, \rho\}$, where $\Lambda^* = \prod_{j\in J}(\theta_j^*)^{c_j}$ and $M^* = \sum_{j\in J} m_j^* c_j - \tau H_6(name, \rho)$. m^* can be obtained in two ways:

Method I: $\mathcal{A}_3$ attempt to compute m^* from the value of $M^* = \sum_{j\in J} m_j^* c_j - \tau H_6(name, \rho)$. Due to the Discrete Logarithm (DL) Problem , it is infeasible to derive m^* from M^*, hence m^* cannot be obtained from M^*.

Method II: $\mathcal{A}_3$ attempt to compute m^* from $\Lambda^* = \prod_{j\in J}((H_2(name||j)u^{m_j})^{\delta})^{c_j}$. By executing public key and private key queries, $\mathcal{A}_3$ can obtain (Δ^*, PK^*) and (δ^*, μ^*). $\mathcal{A}_3$ can obtain $\prod_{j\in J} u^{m_j}$ by compute Λ^*. However, due to DL Problem, it is infeasible to derive m^* from Λ^*. In a word, the advantage of $\mathcal{A}_3$ in the Game III is negligible.

6 Performance Analysis

6.1 Features Comparison

Table 2. Comparison of functionality between different schemes.

Scheme	Privacy Protection	Secure-Channel-Free	Multi-ownership
[9]	Yes	No	Yes
[11]	No	No	Yes
[12]	No	No	Yes
[13]	No	No	No
DIA-MOT	Yes	Yes	Yes

Table 3. Explanation of operation symbols.

symbol	meaning	operation time (ms)
T_p	A pairing operation	9.92672
T_H	A hash-to-point operation	11.0386
T_E	A scale multiplication operation in G_1	4.37832
T_h	A hash-to-point operation	0.004

Table 4. Computational costs between different schemes.

Scheme	Authen-Gen	Proof-Gen and Proof-Verifiy	Mti-Ownership Transfer
[9]	$n(T_H + 2T_E)$	$2T_p + (2a+5)T_E + aT_H + 3T_h$	–
[11]	$n(T_H + 2T_E)$	$3T_p + (2a+1)T_E + (a+1)T_H$	$n(2T_H + 3T + E + 2T_h)$
[12]	$n(2T_H + 4T_E + T_h)$	$3T_p + 2(a+1)T_E + (a+1)T_H$	$(n+1)T_H + (5n+6)T_h$
[13]	$n(T_H + 3T_E)$	$3T_p + (2a+3)T_E + aT_H + 2T_h$	$n(T_H + 8T_E + T_h)$
DIA-MOT	$n(T_H + 2T_E) + T_h$	$2T_p + 2(a+1)T_E + aT_H + 2T_h$	$12(T_E + T_h)$

The DIA-MOT scheme achieves data integrity auditing with multi-ownership transfer capability in cloud storage. Table 2 compares various ownership-transferable data integrity auditing schemes. Unlike schemes [9,11–13], DIA-MOT uniquely operates without requiring secure channels. While only DIA-MOT and [9] implement proper data privacy protection among the compared schemes, it should be noted that [13] additionally fails to support multi-ownership transfer. Crucially, DIA-MOT successfully incorporates all functional features that remain unattained by these referenced schemes.

6.2 Computational Costs

We benchmarked the computational costs of cryptographic operations using the MIRACL library on a Windows 10 system (Intel i5-6500 @ 3.19 GHz, 8GB RAM). The experiments were conducted under the $MR_PAIRING_SSP$ parameter set, where G_1 is a multiplicative cyclic subgroup defined over the elliptic curve $E/F_p : y^2 = x^3 + (-3) mod p$, with the bit lengths of q and p set to 160 bits and 512 bits, respectively. Table 3 presents the measured execution times for all critical operations.

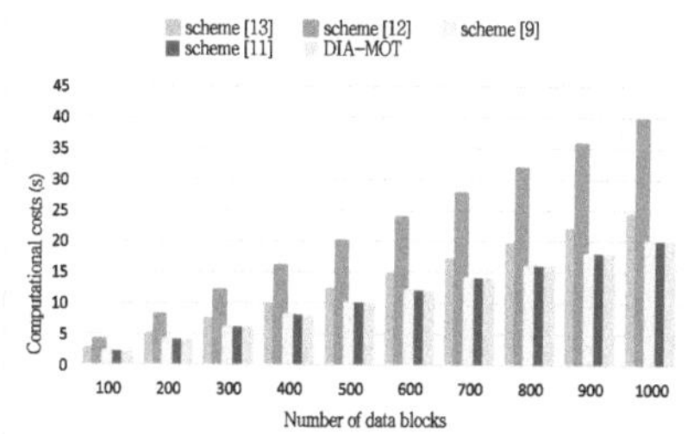

Fig. 2. Comparison for Authenticator Generation

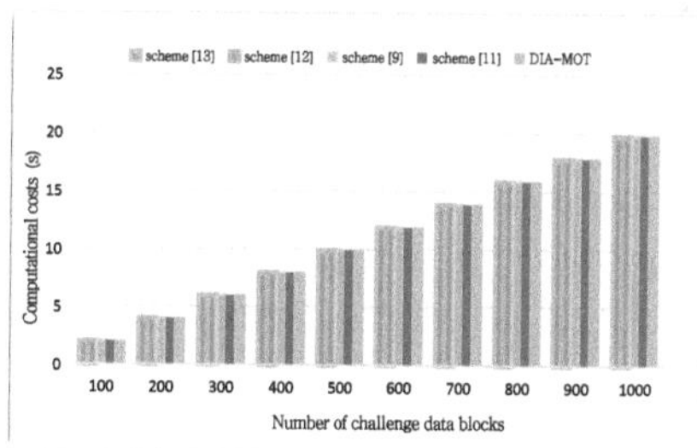

Fig. 3. Comparison for Proof Generation and Proof Verification

Table 4 compares the computational costs of five schemes across three critical phases: authenticator generation, proof generation and verification, and multi-ownership transfer, where n denotes the total number of data blocks and a

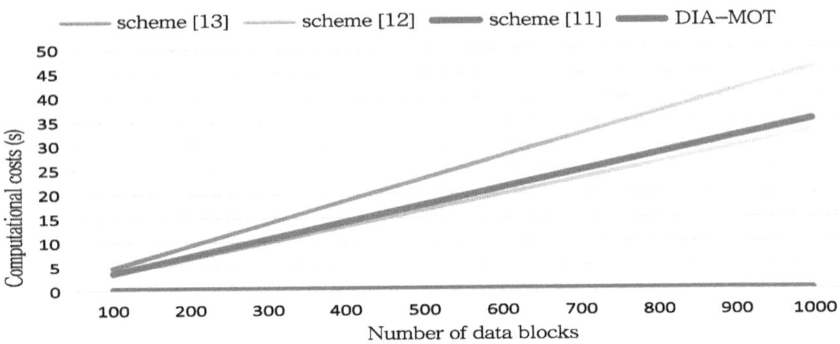

Fig. 4. Comparison for Multi-Ownership

represents the TPA's challenged block count. DIA-MOT achieves constant-time complexity during ownership transfer, independent of the data block volume. Notably, Scheme [9] is excluded from this comparison due to its inherent lack of multi-ownership transfer capability.

Figure 2 presents the computational costs during the authenticator generation phase. Notably, scheme [14] exhibits the highest overhead, while DIA-MOT achieves comparable efficiency to scheme [11], with both outperforming Schemes [13,14]. Figure 3 demonstrates the costs for proof generation and verification. Here, DIA-MOT shows similar performance to schemes [9,11], all of which are significantly more efficient than schemes [13,14]. Figure 4 evaluates the multi-ownership transfer phase, where DIA-MOT reduces computational costs by a substantial margin compared to all counterparts, while scheme [13] incurs the highest overhead.

7 Conclusion

Most existing data integrity auditing schemes supporting multi-ownership transfer rely on secure channels during the ownership transfer phase—an idealized assumption often unattainable in practical deployments. To address this limitation, we propose a novel multi-ownership-transferable auditing scheme that eliminates secure channel requirements while provably guaranteeing security under the hardness of the Computational Diffie-Hellman (CDH) assumptions. Efficiency analysis confirms our scheme's significant computational advantages during ownership transfer. Future work will focus on developing certificateless ownership-transferable auditing mechanisms to mitigate the need for key escrow and complex certificate management.

Acknowledgments. This research is supported by the National Natural Science Foundation of China with grants 62462014, the Guizhou Province Hundred-level Innovative Talent Project with grant number GCC[2022]018-1, the Natural Science

Research Project of the Guizhou Provincial Department of Education with grant number [2023]010, and the Guizhou Provincial Science and Technology Plan Project with key grant number ZK[2024]058.

References

1. Ateniese, G., et al..: Provable data possession at untrusted stores. In: Proceedings of the 14th ACM Conference on Computer and Communications Security, pp. 598–609 (2007)
2. Ateniese, G., Kamara, S., Katz, J.: Proofs of storage from homomorphic identification protocols. In: International Conference on the Theory and Application of Cryptology and Information Security, pp. 319–333. Springer (2009)
3. Chen, X., Shang, T., Zhang, F., Liu, J., Guan, Z.: Dynamic data auditing scheme for big data storage. Front. Comp. Sci. **14**(1), 219–229 (2020)
4. Deng, L., Chen, Z., Ruan, Y., Zhou, H., Li, S.: Certificateless provable data possession scheme suitable for smart grid management systems. IEEE Syst. J. **17**(3), 4245–4256 (2023)
5. Deng, L., Feng, S., Wang, T., Hu, Z., Li, S.: Identity-based data auditing scheme with provable security in the standard model suitable for cloud storage. IEEE Trans. Dependable Secure Comput. **21**(4), 3644–3655 (2023)
6. Deng, L., Wang, B., Wang, T., Feng, S., Li, S.: Certificateless provable data possession scheme with provable security in the standard model suitable for cloud storage. IEEE Trans. Serv. Comput. **16**(6), 3986–3998 (2023)
7. Dhakad, N., Kar, J.: EPPDP: an efficient privacy-preserving data possession with provable security in cloud storage. IEEE Syst. J. **16**(4), 6658–6668 (2022)
8. Erway, C.C., Küpçü, A., Papamanthou, C., Tamassia, R.: Dynamic provable data possession. ACM Trans. Inf. Syst. Secur. (TISSEC) **17**(4), 1–29 (2015)
9. Gai, C., Shen, W., Yang, M., Yu, J.: PPADT: privacy-preserving identity-based public auditing with efficient data transfer for cloud-based IoT data. IEEE Internet Things J. **10**(22), 20065–20079 (2023)
10. He, D., Kumar, N., Zeadally, S., Wang, H.: Certificateless provable data possession scheme for cloud-based smart grid data management systems. IEEE Trans. Industr. Inf. **14**(3), 1232–1241 (2017)
11. Huang, Y., Shen, W., Qin, J.: Certificateless cloud storage auditing supporting data ownership transfer. Comput. Secur. **139**, 103738 (2024)
12. Shen, J., Chen, X., Wei, J., Guo, F., Susilo, W.: Blockchain-based accountable auditing with multi-ownership transfer. IEEE Trans. Cloud Comput. **11**(3), 2711–2724 (2022)
13. Shen, J., Guo, F., Chen, X., Susilo, W.: Secure cloud auditing with efficient ownership transfer. In: Chen, L., Li, N., Liang, K., Schneider, S. (eds.) ESORICS 2020. LNCS, vol. 12308, pp. 611–631. Springer, Cham (2020). https://doi.org/10.1007/978-3-030-58951-6_30
14. Shen, W., Yu, J., Yang, M., Hu, J.: Efficient identity-based data integrity auditing with key-exposure resistance for cloud storage. IEEE Trans. Dependable Secure Comput. **20**(6), 4593–4606 (2022)
15. Wang, C., Wang, Q., Ren, K., Lou, W.: Privacy-preserving public auditing for data storage security in cloud computing. In: 2010 Proceedings IEEE Infocom, pp. 1–9. IEEE (2010)

16. Wang, H., He, D., Fu, A., Li, Q., Wang, Q.: Provable data possession with outsourced data transfer. IEEE Trans. Serv. Comput. **14**(6), 1929–1939 (2019)
17. Yawalkar, P.M., Paithankar, D.N., Pabale, A.R., Kolhe, R.V., William, P.: Integrated identity and auditing management using blockchain mechanism. Measur.: Sens. **27**, 100732 (2023)
18. Yu, J., Shen, W.: Secure cloud storage auditing with deduplication and efficient data transfer. Clust. Comput. **27**(2), 2203–2215 (2024)
19. Zhang, X., Liu, Q., Liu, B., Zhang, Y., Xue, J.: Dynamic certificateless outsourced data auditing mechanism supporting multi-ownership transfer via blockchain systems. IEEE Trans. Netw. Serv. Manage. (2025)
20. Zhao, Y., Chang, J.: Certificateless public auditing scheme with designated verifier and privacy-preserving property in cloud storage. Comput. Netw. **216**, 109270 (2022)

Incentive-Driven Zero-Knowledge Proof-Based Gradient Aggregation for Federated Learning

Chelsea Kowouvi[1,2], Emmanuel Ahene[1,2], Joojo Walker[1,2](✉), Rosemary Akyere[1,2], Dominic Kingsford Opoku Agyili[1,2], Desmond Nani[1,2], and Jerry John Kponyo[1,2]

[1] Department of Computer Science, Kwame Nkrumah University of Science and Technology (KNUST), Kumasi, Ghana
aheneemmanuel@knust.edu.gh

[2] Chengdu University of Technology, Oxford Brookes College, Chengdu, China
joojo.walker@zy.cdut.edu.cn

Abstract. Federated learning (FL) enables collaborative model training across distributed clients without direct data sharing. However, verifying the integrity of aggregated updates and ensuring fair incentive distribution among clients remain open challenges, particularly under untrusted aggregation. This paper presents Incentive-Driven Zero-Knowledge Proof-based Gradient Aggregation (IZKGA) for federated learning. IZKGA is a blockchain-integrated framework that ensures verifiable and privacy-preserving FL aggregation with client-level contribution incentives. Each client commits to its model update using Pedersen vector commitments, and a zero-knowledge succinct non-interactive argument of knowledge (zk-SNARK) proves the correctness of the weighted aggregation (FedAvg) without revealing the underlying gradients. A Commitment Integrity (CI) mechanism, inspired by Shapley value estimation, quantifies each client's contribution efficiently in one round, enabling fair and transparent reward distribution through smart contracts. The proposed system eliminates the need to expose model updates while enabling public verification of aggregation correctness on-chain. Experimental evaluations on CIFAR-10 dataset show that IZKGA preserves model accuracy while adding only modest computational overhead for proof generation and verification. Compared with baseline FL and incentive mechanisms, IZKGA achieves verifiable, privacy-preserving aggregation and stable incentive alignment, demonstrating the feasibility of auditable and incentive-compatible federated learning in decentralized environments.

Keywords: Federated learning · zero-knowledge proof · gradient aggregation · incentive mechanism

1 Introduction

Federated Learning (FL) has emerged as a promising paradigm for collaboratively training machine learning models across multiple distributed clients with-

J. Yu and Y. Chen (Eds.): FCS 2025, CCIS 2805, pp. 477–493, 2027.
https://doi.org/10.1007/978-981-95-9514-3_31

out sharing their raw data [1–4]. By keeping data local, FL mitigates privacy risks and regulatory challenges associated with centralized data collection [5]. However, despite these advantages, current FL frameworks still rely on a central aggregator to combine model updates. This dependency introduces two persistent challenges: *(i)* the inability to verify whether the aggregator performs the aggregation honestly, and *(ii)* the lack of transparent and fair incentive mechanisms that motivate clients to contribute high-quality updates. Addressing these challenges is essential for building trustworthy and sustainable federated ecosystems [2,6–9].

In conventional FL, the aggregator computes a weighted average of client updates (e.g., FedAvg) to produce the global model [1]. Yet, the aggregator may behave dishonestly—dropping, altering, or fabricating client updates—without any means for clients or third parties to verify correctness. This risk becomes acute in cross-organizational or decentralized settings where mutual trust cannot be assumed. Ensuring verifiable aggregation is therefore crucial. Unfortunately, existing blockchain-based FL frameworks, while offering auditability, often require clients to reveal their updates publicly or to execute expensive on-chain computations. Similarly, zero-knowledge-based solutions protect model privacy but typically verify only partial computations (e.g., gradient norms or signatures) rather than the full weighted aggregation process [10,11]. Consequently, they fail to guarantee that the final model truly reflects all legitimate client contributions.

Another unresolved issue lies in incentivizing participation. Clients in FL settings invest local computation, storage, and communication resources but rarely receive measurable rewards proportional to their contribution. Existing incentive mechanisms either depend on repetitive retraining to compute Shapley values—which is computationally infeasible—or rely on heuristic scoring functions that do not capture true data utility. Moreover, these mechanisms are not integrated with verifiable aggregation, leaving room for manipulation and unfair reward allocation [12–14].

To address these limitations, we propose Incentive-Driven Zero-Knowledge Gradient Aggregation (IZKGA), a blockchain-integrated FL framework that unifies verifiable aggregation, privacy preservation, and incentive compatibility. In IZKGA, each client first computes a local update and commits to it using *Pedersen vector commitments*, ensuring that the update remains hidden while allowing homomorphic combination by the aggregator. The aggregator then generates a *zero-knowledge succinct non-interactive argument of knowledge (zk-SNARK)* that proves the correctness of the weighted aggregation—specifically, that the committed local models were averaged according to FedAvg rules—without revealing the underlying parameters. Verification of this proof can be performed by any participant or smart contract on the blockchain, ensuring transparency and immutability of results.

To promote fairness and participation, IZKGA introduces a lightweight *Commitment Integrity (CI)* mechanism inspired by Shapley value estimation. CI quantifies each client's contribution in a single training round based on commitment differentials, avoiding retraining or complex marginal evaluations. Smart contracts leverage CI scores to distribute rewards automatically, achieving transparent, tamper-resistant incentive management across decentralized clients. Comprehensive experiments on CIFAR-10 dataset demonstrate that IZKGA maintains accuracy çomparable to standard FL while providing verifiable aggregation. The zk-SNARK proof generation and verification introduce only modest computational overhead. CI correlates strongly with actual data utility, outperforming baseline reputation and gradient-norm methods. The framework thus achieves a practical balance between verifiability, privacy, and incentive efficiency. The key contributions of this work are summarized as follows:

- We design a novel aggregation protocol that employs Pedersen vector commitments and zk-SNARK proofs to verify weighted averaging without revealing model updates.
- We propose the Commitment Integrity (CI) metric, a one-round approximation of Shapley value that quantifies individual client contributions with minimal cost.
- We integrate CI into smart contracts to automate transparent, tamper-resistant reward allocation in decentralized FL environments.
- We implement IZKGA and conduct extensive experiments demonstrating verifiability, fairness, and low computation overhead under heterogeneous conditions.

Paper Organization: The remainder of this paper is organized as follows. Section 2 discusses the related work. Section 3 details the system overview and architecture, the threat model and system goals. Section 4 describes the concrete cryptographic foundations and protocol design, including the proposed incentive mechanism. Section 5 outlines the experimental setup and results. Finally, Sect. 6 concludes the paper and highlights future work.

2 Related Work

The literature most relevant to this study spans three intersecting areas: blockchain-assisted federated learning, zero-knowledge and cryptographic verification in FL, and incentive mechanisms for client contribution. While each domain has achieved partial progress, none offers a unified framework that simultaneously ensures privacy-preserving verifiable aggregation and transparent incentive distribution.

2.1 Blockchain-Assisted Federated Learning

Blockchain has been widely explored as an immutable and decentralized ledger for recording model updates in FL [15–17]. Early frameworks such as *BlockFLow* [18], *FedChain* [19], and FLEDGE [17] integrated blockchain to audit local updates and ensure tamper resistance. However, these designs often required uploading model gradients or weights directly to the chain, exposing sensitive information and causing significant latency and storage overhead. Other research efforts reduced on-chain [20] load by storing only model hashes or update digests, but they still relied on a trusted aggregator or failed to verify whether aggregation was performed correctly. Recent Hyperledger-based designs improved efficiency and governance through permissioned consensus but maintained the same limitation: they only guarantee traceability, not computational correctness. Hence, blockchain alone cannot guarantee verifiable model aggregation without compromising privacy.

2.2 Zero-Knowledge Proofs and Verifiable Computation in FL

Zero-knowledge proofs (ZKPs) and related cryptographic primitives have recently been adopted to enable verifiable yet privacy-preserving machine learning [10]. Earlier zero-knowledge based federated learning protocols apply zk-SNARKs or homomorphic encryption to validate local training integrity or verify simple operations like gradient summation [10,21–23]. Nevertheless, these systems typically either (1) verify only scalar operations or model-signature checks rather than the complete weighted aggregation rule (*FedAvg*), (2) assume a semi-trusted server that has partial access to updates, or (3) incur prohibitively high computation and proof generation costs due to the large dimensionality of model parameters. Recent cryptographic accelerations mitigate cost but still lack scalable vector-commitment integration [24]. Thus, no existing ZK-based FL system provides an efficient proof of correct weighted aggregation that remains consistent with FedAvg under realistic client numbers. Our proposed IZKGA framework bridges this gap by combining *Pedersen vector commitments* [25] with a succinct zk-SNARK circuit that proves aggregation correctness homomorphically. This ensures that no raw model updates are revealed and that no trusted intermediary is required.

2.3 Incentive Mechanisms in Federated Learning

Incentive mechanisms aim to encourage honest and sustained client participation [14]. Traditional approaches adopt reputation-based or auction-based schemes where rewards depend on task completion or reputation history [13,14,26,27]. While computationally lightweight, these methods are often vulnerable to manipulation and fail to reflect the actual contribution quality of each client's data. Another line of work uses data valuation frameworks such as the *Shapley value* or influence-function-based methods to quantify each client's marginal utility [12]. Although theoretically sound, exact Shapley estimation requires exponential retraining, and even approximation algorithms incur substantial computational overhead, making them unsuitable for large-scale FL. Other heuristic metrics—such as gradient magnitude or model divergence—capture only partial information about contribution and do not account for aggregation verifiability or model correctness. IZKGA overcomes these limitations through its *Commitment Integrity (CI)* mechanism. CI approximates client contribution directly from commitment differentials in a single round, avoiding retraining and integrating naturally into the verifiable aggregation protocol. This tight coupling of cryptographic verifiability and economic incentives ensures that only legitimate and provably contributing clients are rewarded, mitigating free-riding and data-poisoning incentives.

3 System Overview and Architecture

The proposed **IZKGA** framework extends the standard federated learning (FL) workflow with cryptographic verifiability and blockchain-based incentive enforcement. It involves four principal entities and proceeds through five interrelated phases.

3.1 System Entities

1. **Clients** ($\mathcal{C} = \{C_1, C_2, \ldots, C_n\}$): Distributed data owners that train local models on private datasets. Each client produces a model update $\mathbf{w}_i$ after local training and computes a *Pedersen vector commitment* $\mathrm{Com}(\mathbf{w}_i; r_i)$ to conceal its update while allowing homomorphic aggregation.
2. **Aggregator** ($\mathcal{A}$): Coordinates each FL round by collecting client commitments (with sample counts n_i and digital signatures) and computing the weighted aggregate. The aggregator then generates a *zero-knowledge succinct non-interactive argument of knowledge (zk-SNARK)* proving that the aggregation was performed correctly without revealing $\mathbf{w}_i$.
3. **Blockchain Network / Verifiers** ($\mathcal{B}$): A permissioned blockchain (e.g., Hyperledger Fabric) that stores commitments, zk-proofs, and smart-contract state. Peers verify proofs and maintain an immutable audit trail of model updates and rewards.
4. **Smart-Contract Module** ($\mathcal{I}$): Implements the incentive logic and the *Commitment Integrity (CI)* computation for each round. It executes reward distribution and maintains client reputation and stake records.

3.2 Protocol Phases

(1) *Client Registration and Initialization.* Clients register public keys with the blockchain and obtain system parameters, including generators (g, h), the zk-SNARK common reference string (CRS), and training configuration. The set of eligible clients (roster) is anchored on-chain. Each client initializes the shared global model $\mathbf{w}^{(0)}$.

(2) *Local Training and Commitment Generation.* Client C_i trains on local data D_i to produce an update $\mathbf{w}_i$ and computes

$$\mathrm{Com}(\mathbf{w}_i; r_i) = h^{r_i} \prod_{j=1}^{d} g_j^{w_{ij}}, \tag{1}$$

where $r_i \leftarrow \mathbb{F}_q$ is random blinding and d is the model dimension. By additively homomorphic composition, $\mathrm{Com}(a+b) = \mathrm{Com}(a)\,\mathrm{Com}(b)$, the aggregator can combine commitments without accessing plaintext parameters.

(3) *Aggregation and Proof Generation.* The aggregator computes the aggregate commitment using integer-weighted (or fixed-point encoded) coefficients derived from each client's local sample count:

$$\mathrm{Com}(N\mathbf{w}^*) = \prod_{i=1}^{n} \mathrm{Com}(\mathbf{w}_i)^{n_i}, \qquad N = \sum_{i=1}^{n} n_i, \tag{2}$$

where n_i denotes the number of training samples held by client C_i. This ensures exponentiation uses integers in the commitment group; the FedAvg update is then $\mathbf{w}^* = \frac{1}{N}\sum_{i=1}^{n} n_i \mathbf{w}_i$ inside the zero-knowledge circuit via fixed-point encoding. The zk-SNARK relation R_{agg} proves that: (i) the integer-weighted aggregation over committed updates is correct, and (ii) each $(\mathrm{Com}(\mathbf{w}_i), n_i)$ is roster-authorized and bound by the client's signature. The aggregator outputs a succinct proof π. Roster membership and client signatures over $(\mathsf{round\_id}, \mathrm{Com}(w_i), n_i)$ are verified before accepting contributions.

(4) *On-Chain Verification and Model Publication.* Blockchain peers execute $\mathrm{Verify}(vk, x, \pi)$ using the public verification key vk. On success, the transaction containing $\mathrm{Com}(N\mathbf{w}^*)$ and π is committed to the ledger along with the round identifier and roster root. The verified global model (or its hash) is then distributed to all clients for the next FL round.

(5) *Commitment Integrity and Reward Distribution.* The smart contract computes for each client C_i a CI score ϕ_i from the quotient of aggregate commitments and a hash-to-field:

$$\Delta_i = \mathrm{Com}(N\mathbf{w}^*) \cdot \mathrm{Com}(N\mathbf{w}^*_{(-i)})^{-1}, \qquad \phi_i = \mathsf{H}_{\mathbb{F}}(\Delta_i \parallel \mathsf{round\_id} \parallel i)\,,$$

where $\mathrm{Com}(N\mathbf{w}^*_{(-i)})$ is the aggregate commitment excluding C_i, and $\mathsf{H}_{\mathbb{F}}$ is a collision-resistant hash-to-field. Rewards are distributed proportionally $(R_i = \beta\phi_i / \sum_j \phi_j)$ and logged on-chain.

3.3 Security and Privacy Guarantees

- **Privacy Preservation:** Pedersen commitments hide all local model parameters under the discrete-log assumption; no gradient is exposed to the aggregator or blockchain.
- **Verifiable Correctness:** zk-SNARK proofs guarantee that the aggregator performs the exact weighted averaging specified by FedAvg.
- **Transparency and Auditability:** All commitments, proofs, and rewards are recorded immutably on the blockchain, enabling independent verification.
- **Incentive Compatibility:** CI-based rewards correlate with true contribution quality, discouraging malicious or low-effort participation.

Formal assumptions and analyses appear in Sect. 4.4 (Fig. 1).

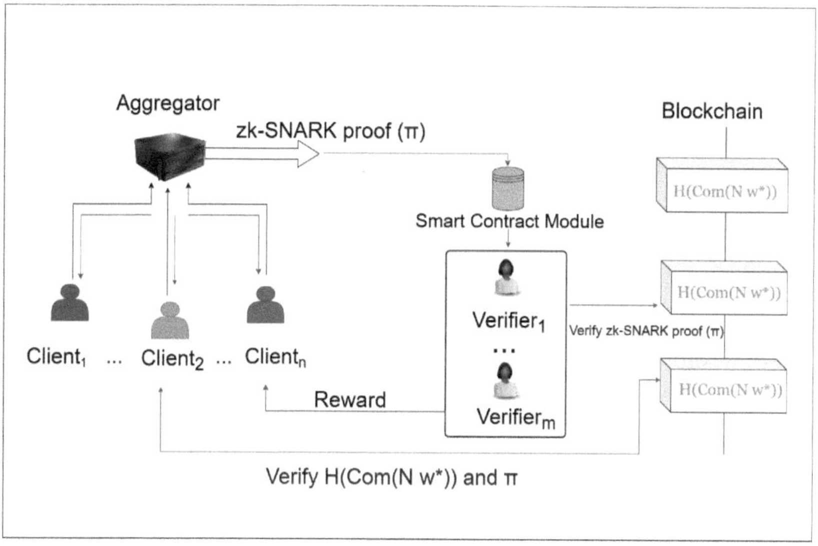

Fig. 1. System architecture of IZKGA. Clients send $(\text{Com}(\mathbf{w}_i), n_i)$ to the aggregator, which performs integer-weighted FedAvg, generates a zk-SNARK proof π, and submits $(\text{Com}(N\mathbf{w}^*), \pi)$ to the blockchain verifier network. Smart contracts verify π, compute CI scores ϕ_i, distribute rewards R_i, and log all commitments immutably for the next round.

3.4 Threat Model

The IZKGA framework considers an FL environment with multiple clients ($\mathcal{C}$), an aggregator ($\mathcal{A}$), and a permissioned blockchain network ($\mathcal{B}$) acting as a public verifier. Adversaries may include a semi-honest aggregator and a subset of malicious clients.

- **Aggregator:** $\mathcal{A}$ is *honest-but-curious* (semi-honest): it follows the protocol but may attempt to infer private client information or bias aggregation. It may collude with a limited subset of clients.
- **Clients:** Malicious clients may submit poisoned updates, fabricate participation, or collude to inflate rewards.
- **Blockchain:** The verifier network executes smart contracts correctly and maintains an immutable ledger; it does not trust the aggregator or clients.
- **Network:** Channels are authenticated and encrypted; denial-of-service attacks are out of scope.

Adversaries can observe protocol messages but cannot break standard assumptions (discrete logarithm hardness, zk-SNARK soundness). Adversarial objectives include leaking private model information, falsifying aggregation results, or manipulating reward outcomes.

3.5 System Goals

IZKGA targets the following objectives:

1. **Privacy Preservation:** Local model parameters remain confidential; neither the aggregator nor verifiers can recover $\mathbf{w}_i$ from commitments.
2. **Verifiable Aggregation:** The aggregator must produce a valid zk-SNARK proof that the global model corresponds to the correct integer-weighted average of all committed updates.
3. **Integrity and Accountability:** All commitments, proofs, and reward transactions are immutably recorded on-chain, enabling public auditability and dispute resolution.
4. **Fair Incentive Distribution:** The CI mechanism ensures that clients receive rewards proportional to their verifiable contribution, discouraging dishonest or idle behavior.
5. **Efficiency and Scalability:** The protocol maintains practical overhead, with proof generation and verification adding moderate latency per FL round in our experiments.

Under these assumptions and goals, IZKGA delivers privacy-preserving, verifiable, and incentive-compatible federated learning suitable for deployment in decentralized and cross-organizational environments.

4 Cryptographic Foundations and Protocol Design

This section formalizes the cryptographic primitives underpinning the **IZKGA** framework and details how they realize privacy-preserving and verifiable federated aggregation under the threat model in Sect. 3.

4.1 Cryptographic Preliminaries

Let $\mathbb{G}$ be a cyclic group of prime order q generated by g, where the discrete logarithm assumption holds.

Pedersen Vector Commitments. Each client commits to its local model update $\mathbf{w}_i = (w_{i1}, w_{i2}, \ldots, w_{id}) \in \mathbb{F}_q^d$ using a Pedersen vector commitment:

$$\mathrm{Com}(\mathbf{w}_i; r_i) = h^{r_i} \prod_{j=1}^{d} g_j^{w_{ij}}, \tag{3}$$

where $r_i \leftarrow \mathbb{F}_q$ is a random blinding factor and $\{g_1, \ldots, g_d, h\}$ are fixed public generators.

Properties. Pedersen commitments are *hiding* (concealing $\mathbf{w}_i$), *binding* (preventing multiple openings), and *additively homomorphic*:

$$\text{Com}(a + b; r_1 + r_2) = \text{Com}(a; r_1)\,\text{Com}(b; r_2). \tag{4}$$

This property enables aggregation over commitments without exposing underlying model parameters.

Zero-Knowledge Proofs. IZKGA employs a zero-knowledge succinct non-interactive argument of knowledge (zk-SNARK) defined by algorithms $(\mathsf{Setup}, \mathsf{Prove}, \mathsf{Verify})$:

- $\mathsf{Setup}(\lambda, R) \to (pk, vk)$: generates proving and verification keys for relation R under security parameter λ;
- $\mathsf{Prove}(pk, x, w) \to \pi$: produces proof π that witness w satisfies $R(x, w) = 1$;
- $\mathsf{Verify}(vk, x, \pi) \to \{0, 1\}$: publicly verifies the proof.

Completeness, soundness, and zero-knowledge ensure that only valid aggregations can be proven and that no private information is revealed.

4.2 Federated Aggregation Relation

During each FL round t, clients compute local updates $\mathbf{w}_i$ and report sample counts n_i. The standard FedAvg update is

$$\mathbf{w}^{t+1} = \frac{1}{N} \sum_{i=1}^{n} n_i \mathbf{w}_i, \qquad N = \sum_{i=1}^{n} n_i. \tag{5}$$

To verify that the aggregator $\mathcal{A}$ performs this operation correctly on hidden updates, aggregation is expressed over commitments using integer weights:

$$\text{Com}(N\mathbf{w}^{t+1}) = \prod_{i=1}^{n} \text{Com}(\mathbf{w}_i)^{n_i}. \tag{6}$$

This avoids fractional exponents and preserves the group's homomorphic property. Division by N is implemented inside the zk-SNARK circuit through fixed-point encoding.

The zero-knowledge relation for verifiable aggregation is

$$\begin{aligned} R_{\text{agg}} = \Big\{ (\{\text{Com}(\mathbf{w}_i)\}, \{n_i\}, \text{Com}(N\mathbf{w}^{t+1})) \;\big|\; \exists \{\mathbf{w}_i, r_i\} : \\ \text{Com}(N\mathbf{w}^{t+1}) = \prod_{i=1}^{n} \text{Com}(\mathbf{w}_i)^{n_i} \Big\}. \end{aligned} \tag{7}$$

The public input x contains $\{\text{Com}(\mathbf{w}_i)\}$, $\{n_i\}$, and $\text{Com}(N\mathbf{w}^{t+1})$; the private witness w includes each opening $(\mathbf{w}_i, r_i)$. The aggregator generates proof π showing that it executed integer-weighted FedAvg correctly and that all participating commitments are authorized under the registered client roster (see Sect. 3). For each participant C_i, the tuple $(\mathsf{round\_id}, \mathrm{Com}(\mathbf{w}_i), n_i)$ is signed under pk_i; the circuit (or chaincode) verifies membership in the roster and signature validity before accepting $(\mathrm{Com}(\mathbf{w}_i), n_i)$.

4.3 Commitment Integrity and Incentive Mechanism

To measure each client's contribution, IZKGA introduces the *Commitment Integrity (CI)* score. After successful aggregation verification, the smart contract computes for every client C_i:

$$\Delta_i = \mathrm{Com}(N\mathbf{w}^{t+1}) \cdot \mathrm{Com}(N\mathbf{w}^{t+1}_{(-i)})^{-1}, \qquad \phi_i = \mathsf{H}_{\mathbb{F}}(\Delta_i \parallel \mathsf{round\_id} \parallel i), \tag{8}$$

where $\mathrm{Com}(N\mathbf{w}^{t+1}_{(-i)})$ is the aggregate commitment excluding C_i, and $\mathsf{H}_{\mathbb{F}}$ is a collision-resistant hash-to-field function modeled as a random oracle. Note that, ϕ_i approximates the marginal impact of C_i on the global model and is publicly verifiable on-chain.

Rewards are distributed proportionally:

$$R_i = \beta \frac{\phi_i}{\sum_j \phi_j}, \tag{9}$$

where β denotes the total reward pool per round. Smart contracts also enforce stake and reputation rules to deter Sybil and free-riding behaviour.

4.4 Security and Rationality Analysis

- **Privacy:** Pedersen commitments ensure that no participant, including the aggregator, can learn any $\mathbf{w}_i$ without solving the discrete-log problem.
- **Verifiability:** zk-SNARK soundness guarantees that a forged aggregation proof is computationally infeasible; completeness ensures all valid proofs are accepted.
- **Integrity and Auditability:** Every commitment, proof, and reward transaction is immutably stored on the blockchain, enabling third-party auditability.
- **Rational Participation:** CI-based rewards align client utility with truthful participation; a rational client maximizes its payoff by submitting valid, high-quality updates.
- **Scope:** We do not attest client-side training correctness; mechanisms such as gradient clipping, differential privacy, and robust aggregation are orthogonal, and can be integrated independently.

4.5 Design Advantages

The IZKGA design provides:

- End-to-end verifiable FedAvg aggregation without exposing gradients.
- Constant-size zk-proofs independent of model dimension, supporting scalability.
- Tight integration of cryptographic verifiability and economic incentives for trustworthy and sustainable federated learning.

In summary, IZKGA unifies Pedersen commitments, zk-SNARK verification, and commitment-based incentives to deliver a privacy-preserving, verifiable, and incentive-compatible federated learning ecosystem.

5 Experimental Setup and Results

5.1 Testbed

Experiments were conducted on an Apple M1 MacBook Pro equipped with 32 GB RAM. This setup provides sufficient computational capacity to run both the Flower federated learning server and multiple simulated client instances concurrently, alongside the demands of Zero-Knowledge Proof (ZKP) generation. The Flower framework (v1.14) orchestrates client selection, aggregation, and round synchronization, while PyTorch is used to define and train local neural network models. Zero-knowledge components are integrated using the `ezkl` library, which exports PyTorch models to ONNX format, compiles them into ZKP circuits, and generates/validates proofs natively on Apple Silicon. Although production deployment will leverage a live Hyperledger Fabric network for immutable storage and incentive execution, a mock blockchain API replicating Fabric interactions (e.g., submission of model hashes, ZKPs, and CI records) was implemented to enable faster prototyping and evaluation.

5.2 Dataset and Model

Unless otherwise noted, data are partitioned non-IID via a Dirichlet sampler ($\alpha \in \{0.1, 0.5\}$) with partial participation $p \in \{0.2, 0.5, 1.0\}$ per round. We evaluate the framework on the CIFAR-10 dataset, consisting of 60,000 color images (32×32 pixels) grouped into 10 classes (50,000 for training and 10,000 for testing). A lightweight Convolutional Neural Network (CNN) is employed, composed of two convolutional layers for feature extraction, pooling layers for dimensionality reduction, and fully connected layers for classification. Each client trains locally for a fixed number of epochs per round, after which the local model updates $\mathbf{w}_i$ are committed and submitted for secure aggregation.

5.3 Zero-Knowledge Proof Integration

The `ezkl` library is used to compile the FedAvg aggregation function into a zero-knowledge proof (ZKP) circuit. During an initial setup phase, the aggregator $\mathcal{A}$ executes the Setup procedure to generate the proving and verification keys (pk, vk) corresponding to the aggregation relation defined in Sect. 4.2. After each training round, $\mathcal{A}$ collects the client commitments $\{\mathrm{Com}(\mathbf{w}_i)\}$ and their local sample counts $\{n_i\}$, and computes the aggregate commitment

$$\mathrm{Com}(N\mathbf{w}^*) = \prod_{i=1}^{n} \mathrm{Com}(\mathbf{w}_i)^{n_i}, \qquad N = \sum_{i=1}^{n} n_i. \tag{10}$$

The FedAvg update $\mathbf{w}^* = \frac{1}{N}\sum_{i=1}^{n} n_i \mathbf{w}_i$ is then enforced within the circuit using fixed-point arithmetic to handle the division by N. Using `ezkl`, the aggregator compiles this computation into a zk-SNARK circuit and runs the Prove algorithm to produce a succinct proof π attesting that valid openings $(\mathbf{w}_i, r_i)$ exist satisfying the aggregate commitment relation. The proof confirms that the aggregation was performed exactly as specified by the FedAvg algorithm, without revealing any local model parameters. The aggregator submits to the blockchain (or mock API) the tuple $(\mathrm{Com}(N\mathbf{w}^*), \pi, \mathsf{round\_id})$. Blockchain peers invoke $\mathsf{Verify}(vk, x, \pi)$ to confirm proof validity before appending the transaction to the ledger. Once verified, the hash of $\mathbf{w}^*$ is broadcast to all clients for the next round, ensuring verifiable and auditable global model updates.

5.4 Contribution Index Recording

To ensure transparent and auditable participation, the system logs lightweight metadata for each client C_i after every training round:

- Hash of the client's local model commitment $\mathrm{Com}(\mathbf{w}_i; r_i)$.
- Local sample count n_i used for training.
- Hash of the verified global commitment $\mathrm{Com}(N\mathbf{w}^*)$ received at the beginning of the round.

These on-chain records create an immutable participation trace linking each client's contribution to the verified aggregation proof. The smart contract later references these entries to compute the *Commitment Integrity (CI)* score ϕ_i using the aggregate-commitment differential ($\Delta_i = \mathrm{Com}(N\mathbf{w}^*)\cdot\mathrm{Com}(N\mathbf{w}^*_{(-i)})^{-1}$) and the hash-to-field mapping defined in Sect. 4.3. This ensures that all rewards are derived from verifiable, tamper-resistant ledger data. This linkage enables transparent attribution and reproducible computation of rewards across all rounds.

5.5 Experimental Results

The results demonstrate that integrating CI-based incentives and ZKPs maintains model accuracy, introduces minimal overhead, and enables transparent reward allocation.

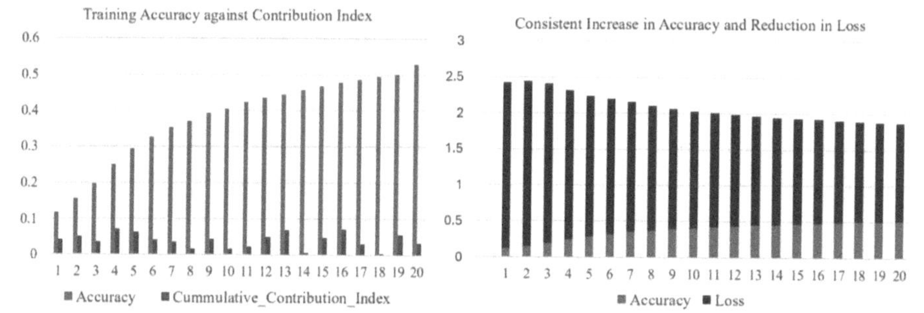

Fig. 2. Global accuracy (%) trend showing that CI computation does not degrade model performance.

(1) *CI Does Not Affect Model Accuracy.* Figure 2 shows steady improvement in global model accuracy across training rounds. The inclusion of CI computation does not alter the convergence behaviour, confirming that the incentive mechanism operates orthogonally to the learning process.

(2) *CI Computation Overhead Is Negligible.* As shown in Fig. 3, ZKP proof generation dominates the computational cost, while CI calculation incurs only marginal overhead since it reuses intermediate gradients from training. Overall, the additional cost of CI remains significantly below that of ZKP operations, validating the scalability of the incentive mechanism.

(3) *CI Enables Fair and Transparent Rewards.* Figure 4 illustrates that CI effectively differentiates client contributions: participants improving model accuracy receive higher CI scores, whereas those degrading performance are penalized with lower or negative values. This allows equitable, performance-based incentive distribution and sustains client engagement across rounds.

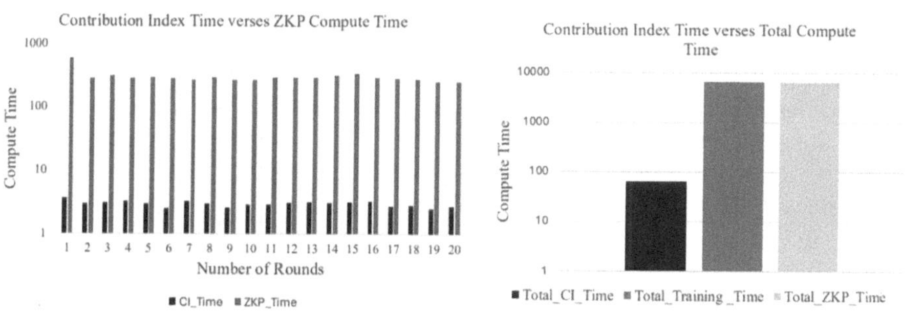

Fig. 3. Comparison of compute times (ms) for ZKP and CI operations.

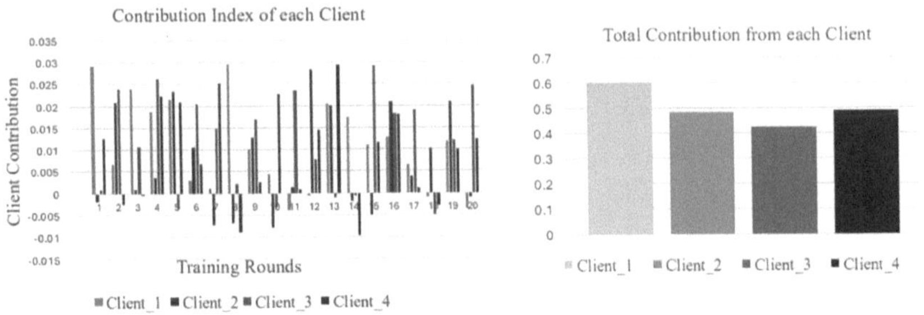

Fig. 4. Client CI scores demonstrating fair contribution-based reward allocation.

Adversarial Sanity Checks. We simulated (i) aggregator omission of k clients, (ii) client sign-flip updates, and (iii) two-client collusion. The zk roster check rejected (i); CI penalized (ii) with limited accuracy impact; and (iii) showed separable CI scores, indicating robustness to simple collusion.

5.6 Discussion

Overall, results confirm that IZKGA achieves verifiable aggregation and incentive alignment without sacrificing model performance or computational efficiency. CI computation remains lightweight, accuracy improves steadily, and client participation remains consistent, supporting the framework's goal of secure, fair, and scalable federated learning. It is important to note that this work focuses on verifiable aggregation and does not attest to client-side training correctness. Mechanisms such as gradient clipping, differential privacy, or robust aggregation are orthogonal and can be incorporated independently to mitigate poisoning and fairness concerns.

6 Conclusion and Future Work

This work presented the IZKGA framework, a unified approach that integrates zero-knowledge proofs, blockchain technology, and contribution-based incentives into federated learning. By embedding zk-SNARK–based aggregation, IZKGA guarantees the correctness of the global model update while preserving the confidentiality of client parameters. The proposed Contribution Integrity (CI) metric, inspired by Shapley values, provides an efficient and interpretable measure of each client's influence on the global model with negligible computational overhead compared to proof generation. Smart contracts automate CI-based reward allocation, establishing transparency and fairness in otherwise trust-deficient FL environments. Experimental evaluations demonstrate that IZKGA achieves verifiable aggregation, preserves model accuracy, and maintains low computational overhead, confirming its practicality for privacy-sensitive and decentralized applications. The strength of the framework lies in its integration of verifiability, privacy preservation, and incentive compatibility into a single, auditable system. Future work will extend IZKGA with collusion-resistant verification among clients, aggregators, and verifiers, enhancing robustness against poisoning and inference attacks toward a fully trustworthy and scalable federated ecosystem.

References

1. McMahan, B., Moore, E., Ramage, D., Hampson, S., y Arcas, B.A.: Communication-efficient learning of deep networks from decentralized data. In: Proceedings of the 20th International Conference on Artificial Intelligence and Statistics, AISTATS 2017, 20–22 April 2017, Fort Lauderdale, FL, USA, ser. Proceedings of Machine Learning Research, vol. 54, pp. 1273–1282. PMLR (2017)
2. Kairouz, P., McMahan, H.B., Avent, B., et al.: Advances and open problems in federated learning. Found. Trends Mach. Learn. **14**(1–2), 1–210 (2021). https://doi.org/10.1561/2200000083
3. Mondal, A., More, Y., Rooparaghunath, R.H., Gupta, D.: Poster: FLATEE: federated learning across trusted execution environments. In: IEEE European Symposium on Security and Privacy, EuroS&P 2021, Vienna, Austria, 6–10 September 2021, pp. 707–709. IEEE (2021). https://doi.org/10.1109/EUROSP51992.2021.00054
4. Mo, F., Haddadi, H., Katevas, K., Marin, E., Perino, D., Kourtellis, N.: PPFL: privacy-preserving federated learning with trusted execution environments. In: Banerjee, S., Mottola, L., Zhou, X. (eds.) MobiSys '21: The 19th Annual International Conference on Mobile Systems, Applications, and Services, Virtual Event, Wisconsin, USA, 24 June – 2 July 2021, pp. 94–108. ACM (2021). https://doi.org/10.1145/3458864.3466628
5. Pejó, B., Biczók, G.: Quality inference in federated learning with secure aggregation. IEEE Trans. Big Data **9**(5), 1430–1437 (2023). https://doi.org/10.1109/TBDATA.2023.3280406
6. Miao, L., Yang, W., Hu, R., Li, L., Huang, L.: Against backdoor attacks in federated learning with differential privacy. In: IEEE International Conference on Acoustics, Speech and Signal Processing, ICASSP 2022, Virtual and Singapore, 23–27 May 2022, pp. 2999–3003. IEEE (2022). https://doi.org/10.1109/ICASSP43922.2022.9747653
7. Threats, attacks and defenses to federated learning: issues, taxonomy and perspectives. Cybersecur. **5**(1), 4 (2022). https://doi.org/10.1186/S42400-021-00105-6
8. Xia, G., Chen, J., Yu, C., Ma, J.: Poisoning attacks in federated learning: a survey. IEEE Access **11**, 10708–10722 (2023). https://doi.org/10.1109/ACCESS.2023.3238823
9. Rieger, P., Krauß, T., Miettinen, M., Dmitrienko, A., Sadeghi, A.: CrowdGuard: federated backdoor detection in federated learning (2024)
10. Wang, Z., Dong, N., Sun, J., Knottenbelt, W., Guo, Y.: ZkFL: zeroKnowledge proof-based gradient aggregation for federated learning. IEEE Trans. Big Data **11**(2), 447–460 (2025). https://doi.org/10.1109/TBDATA.2024.3403370
11. Xing, Z., Zhang, Z., Li, M., et al.: Zero-knowledge proof-based practical federated learning on blockchain. CoRR, vol. abs/2304.05590 (2023). https://doi.org/10.48550/ARXIV.2304.05590

12. Song, T., Tong, Y., Wei, S.: Profit allocation for federated learning. In: Proceedings of the 2019 IEEE International Conference on Big Data (Big-Data 2019), pp. 2577–2586. IEEE (2019). https://doi.org/10.1109/BigData47090.2019.9006327
13. Lim, W.Y.B., Xiong, Z., Miao, C., et al.: Hierarchical incentive mechanism design for federated machine learning in mobile networks. IEEE Internet Things J. **7**(10), 9575–9588 (2020). https://doi.org/10.1109/JIOT.2020.2985694
14. Tu, X., Zhu, K., Luong, N.C., Niyato, D., Zhang, Y., Li, J.: Incentive mechanisms for federated learning: From economic and game theoretic perspective. IEEE Trans. Cogn. Commun. Netw. **8**(3), 1566–1593 (2022). https://doi.org/10.1109/TCCN.2022.3177522
15. Ali, I., Gervais, M., Ahene, E., Li, F.: A blockchain-based certificateless public key signature scheme for vehicle-to-infrastructure communication in VANETs. J. Syst. Archit. **99** (2019). https://doi.org/10.1016/J.SYSARC.2019.101636
16. Hua, G., Zhu, L., Wu, J., Shen, C., Zhou, L., Lin, Q.: Blockchain-based federated learning for intelligent control in heavy haul railway. IEEE Access **8**, 176830–176839 (2020). https://doi.org/10.1109/ACCESS.2020.3021253
17. Castillo, J., Rieger, P., Fereidooni, H., Chen, Q., Sadeghi, A.: FLEDGE: ledger-based federated learning resilient to inference and backdoor attacks. In: Annual Computer Security Applications Conference, ACSAC 2023, Austin, TX, USA, 4–8 December 2023, pp. 647–661. ACM (2023). https://doi.org/10.1145/3627106.3627194
18. Mugunthan, V., Rahman, R., Kagal, L.: BlockFlow: decentralized, privacy-preserving, and accountable federated machine learning. In: 3rd International Congress, BLOCKCHAIN 2021, Salamanca, Spain, vol. 320, pp. 233–242. Springer (2021). https://doi.org/10.1007/978-3-030-86162-9_23
19. Nguyen, C.T., Hoang, D.T., Nguyen, D.N., et al.: FedChain: secure proof-of-stake-based framework for federated-blockchain systems. IEEE Trans. Serv. Comput. **16**(4), 2642–2656 (2023). https://doi.org/10.1109/TSC.2023.3240235
20. Fereidooni, H., Dmitrienko, A., Rieger, P., Miettinen, M., Sadeghi, A., Madlener, F.: FedCRI: federated mobile cyber-risk intelligence. In: 29th Annual Network and Distributed System Security Symposium, NDSS 2022, San Diego, California, USA, 24–28 April 2022, The Internet Society (2022)
21. Mugunthan, V., Peraire-Bueno, A., Kagal, L.: PrivacyFL: a simulator for privacy-preserving and secure federated learning. In: CIKM '20: The 29th ACM International Conference on Information and Knowledge Management, Virtual Event, Ireland, 19–23 October 2020, pp. 3085–3092. ACM (2020). https://doi.org/10.1145/3340531.3412771
22. Ma, R., Hwang, K., Li, M., Miao, Y.: Trusted model aggregation with zero-knowledge proofs in federated learning. IEEE Trans. Parallel Distributed Syst. **35**(11), 2284–2296 (2024). https://doi.org/10.1109/TPDS.2024.3455762
23. Ma, J., Liu, H., Zhang, M., Liu, Z.: VPFL: enabling verifiability and privacy in federated learning with zero-knowledge proofs. Knowl. Based Syst. **299**, 112115 (2024). https://doi.org/10.1016/J.KNOSYS.2024.112115
24. Jin, Y., Wang, T., Yang, Q., Shi, L., Zhang, S.: Zero-knowledge federated learning: a new trustworthy and privacy-preserving distributed learning paradigm (2025). arXiv: 2503.15550 [cs.CR]
25. Demirel, D., Lancrenon, J.: How to securely prolong the computational bindingness of Pedersen commitments. IACR Cryptol. ePrint Arch., 584 (2015). http://eprint.iacr.org/2015/584

26. Zhan, Y., Li, P., Qu, Z., Zeng, D., Guo, S.: A learning-based incentive mechanism for federated learning. IEEE Internet Things J. **7**(7), 6360–6368 (2020). https://doi.org/10.1109/JIOT.2020.2967772
27. Kaheni, M., Lippi, M., Gasparri, A., Franceschelli, M.: Selective trimmed average: a resilient federated learning algorithm with deterministic guarantees on the optimality approximation. IEEE Trans. Cybern. **54**(8), 4402–4415 (2024). https://doi.org/10.1109/TCYB.2024.3350387

Author Index

J. Yu and Y. Chen (Eds.): FCS 2025, CCIS 2805, pp. 495–496, 2027.
https://doi.org/10.1007/978-981-95-9514-3

Kolibri 360 GmbH
Ferdinand-Jühlke-Straße 7
99095 Erfurt, Deutschland
produktsicherheit@kolibri360.de